ORIGINS OF UNIVERSAL SYSTEMS:

A Brief History of the Right Answers…Simple and Beautiful

Alexander Alan Scarborough

Featuring

The Revolutionary LB-FLINE-BEC Model Based on
The New Fourth and Fifth Laws of Planetary Motion:
How the planets attained their orbital spacing around our Sun.

Order this book online at www.trafford.com
or email orders@trafford.com

Most Trafford titles are also available at major online book retailers.

© Copyright 2008 Alexander A. Scarborough.
All rights reserved. No part of this publication may be reproduced, stored in a retrieval system, or transmitted, in any form or by any means, electronic, mechanical, photocopying, recording, or otherwise, without the written prior permission of the author.

Edited by: Karol Kay Scarborough
Photography by: JPL / Nasa Max-Planck-Institut
Butler, Marcy / SFSV

Printed in the United States of America.

ISBN: 978-1-4251-6585-7 (sc)
ISBN: 978-1-4669-6447-1 (e)

Trafford rev. 10/12/2012

www.trafford.com

North America & international
toll-free: 1 888 232 4444 (USA & Canada)
phone: 250 383 6864 • fax: 812 355 4082

Dedicated to the quest for truth

This book of new concepts belongs in every library and in every science classroom. It establishes a revolutionary trend in scientific thought on creation of matter, our solar system and systems universal. Its concepts seem destined to obsolete current theories about such creations.

Richard Arena

"The right answers will be simple and beautiful." Einstein
"And they are!" Author

We must forsake such beliefs that Earth is flat, that its core is iron or rock, that fuels were made from fossils, that energy is finite, and that planets are created from dust and gases. Such concepts have too long misled scientists down blind alleys. © 1980

A Modification (2006):

We must forsake such beliefs that Earth and the Universe are flat, that active planetary cores are iron, rock, or metal, that hydrocarbon fuels were made from fossils, that energy is finite, that Pluto is not a planet, that planets are created from gaseous dust-clouds, and that comets are dirty snowballs. Such concepts too long have misled scientists down blind alleys.

ACKNOWLEDGEMENTS

So many references have been used throughout the years of research, learning, and piecing together the concepts in this book, that it would be impossible to list them all. My thanks go to everyone involved in any facet of these theories, from the ancients to Ptolemy to the personnel involved in the space probes to Venus, Mars, Jupiter, Saturn, etc.

However, a special acknowledgement of one man's contribution to this book should be noted. The faith and foresight of Ralph Howard, Jr., of Kleen-Tex International, Inc., in LaGrange, Georgia helped make this book possible when its future looked bleak.

Special recognition and thanks go to Kay Scarborough for help in putting together the manuscript and book.

A large volume of research on the book's contents was done at Coleman Library (Callaway Educational Association), at LaGrange Memorial Library, at LaGrange College Library and in the author's home library during the years 1973-2008.

Much credit goes to the National Philosophical Alliance (NPA) of international scientists who allowed the preservation of the book's new theories.

The reader will find this tenth cumulative edition of the Energy Series, Origins of Universal Systems, distinctly different in style of presentation of contents. The reason can be traced to the nature of its intertwined, well-substantiated evidence in support of a scientific revolution that appears destined to displace a major portion of the Big Bang/Accretion hypotheses.

Of necessity, many crucial discoveries of the past 2500 years are intertwined with historical letters and memos that record in real time the spirit and frustrations encountered along the way.

ENERGY SERIES

First Edition: Fuels: A New Theory (1973)
Second Edition: Fuels: A New Theory (1975)
Third Edition: Undermining the Energy Crisis (1977)
Fourth Edition: Undermining the Energy Crisis (1979)
Fifth Edition: From Void to Energy to Universe (1980)
(Manuscript unpublished)
Sixth Edition: New Concepts of Origins: With White Fire Laden (1986)
Seventh Edition: The I-T-E-M Connection: How Planet Earth and Its Systems Were Made by Means of Natural Laws (1991)
Eighth Edition: The Spacing of Planets: The Solution to a 400-Year Mystery (1996)
Ninth Edition: New Principles of Origins and Evolution: Revolutionary Paradigms of Beauty, Power and Precision (2002)
Tenth Edition: Origins of Universal Systems: A Brief History of "The Right Answers… Simple and Beautiful" (2007-2008)

Order from Ander Publications, e-mail: dubuissonk@bellsouth.net

(First-Ninth Editions – Limited Remainders, Out of Print)

AUTHOR'S PHILOSOPHY OF SCIENCE AND RELIGION

Einstein's famous 1905 formula, $E=mc^2$, reveals that the Universe is composed of only two basic, and totally interchangeable, substances: energy and matter. An ever fuller appreciation of this most fundamental principle of the Universe will enable scientists to understand the intimate relationships of the precisely interwoven laws of chemistry, physics and mathematics that eventually will reveal how the Universe functions.

God established these inviolate laws, and then instilled in mankind the brainpower to discover them one by one—a process we call science. Mankind's souls were instilled to discover and enact the purposes of our existence in the Universe—a process we call religion.

About the author:

Alexander Alan Scarborough, a native of Macon, Georgia (1924-1944), now resides in LaGrange, Georgia. He holds a Chemistry degree from the University of Georgia (1944), and a Chemical Engineering degree from Georgia Tech (1948-1950). In between, he served two years of Army duty (1944-1946), including eight months in Italy (1944-1945). Po Vally Campaign.

A career in industrial research and development proved highly satisfactory. For example, his breakthrough idea for a double-density, washable industrial walk-off mat, along with solutions to subsequent problems, grew into a multi-billion-dollar industry on an international scale. Problem-solving proved to be his forté.

This strong point was carried over into his great lifetime interest in astronomy and geology. By way of these pursuits, he was able to develop the Little Bangs/FLINE model of universal origins, now soundly based on his discovery of the fourth and fifth laws of planetary motion and Kepler's first three laws, plus $E=mc^2$.

This tenth edition of his Energy Series (1975-2007/08), at age 84, may be his final, all-inclusive volume of how everything came into being: a well-substantiated concept that appears to point the way to the long-sought theory of everything.

Alexander is married to the former Mary Frances Akins of Statesboro, Georgia. They have three grown children, and currently reside in LaGrange, Georgia.

CONTENTS

FOREWORD

During the past 33 years (1973-2006), a revolutionary concept of dynamic origins and evolution has been painstakingly put together. Like pieces of a jigsaw puzzle, a multitude of facts interlock precisely to reveal the sheer beauty, power and precision of definitive and testable, but unorthodox, concepts of universal origins and evolution: the Little Bangs (LB) and the FLINE model of origins of solar systems and evolution of planets, moons, stars, etc.

An important lesson to be learned from the new LB/FLINE model is that any discovery in the physical sciences should be interpreted in more than one perspective before judging which interpretation has truer scientific validity. Examples of this lesson spring from the discoveries of Copernicus, Galileo, Descartes, Dutton, Darwin *et al.*—men whose discoveries have outlived the popular beliefs of their time.

Upton Sinclair once wrote, "It is difficult to get a man to understand something when his salary depends on his not understanding it." Unfortunately, in the strict peer review system of science, new ideas in opposition to beliefs of reviewers schooled in current dogma are rejected out-of-hand in spite of sufficient substantiated evidence and the absence of speculation. Thus, prevailing beliefs are shielded against the intrusion of new ideas that indicate the need for a change in the direction of scientific thought; scientists are permitted little, if any, opportunity to challenge the scientific validity of new ideas or to benefit from them. The damages wrought by such policy, as history teaches, can be huge.

To quote *World* magazine: "As Thomas Kuhn has shown in his book, *The Structure of Scientific Revolutions,* when scientists discover more data, they must discard old models, as new models take shape. But unfortunately, between the models comes virulent controversy. Whenever the old model becomes exhausted, Kuhn showed, the scientific establishment acts in a predictable way. First, it stretches the old model beyond plausibility in an attempt to account for the new data. Then the scientific community lashes out at those who dare challenge the existing model"—a procedure aptly named the Galilean treatment.

Today, new data accumulated during the past quarter-century poses a serious challenge to the Big Bang/ Accretion model, and to important discoveries and critical data that always fit more logically and precisely into the LB/FLINE model. But scientists usually cannot see the evidence differently until they change their interpretive framework. The new concepts in this revolutionary publication provide a convincing basis for that change.

Throughout the Universe, nothing escapes Nature's strict laws that make it possible for mankind eventually to understand how everything came into being. Essential to that understanding is sufficient knowledge to eliminate the need for speculation that too often leads to misinterpretations of important findings. As history teaches, new ideas, generally known as breakthroughs, precede new knowledge, and knowledge begets knowledge. Adhering to the precise laws of Nature and soundly backed by substantiated evidence, the revolutionary LB/ FLINE model is the only known concept of planetary origins, orbital spacing and evolution that seems capable of surviving the test of time.

The struggles involved in researching and preserving these new findings in the face of strongly entrenched beliefs and under adverse conditions, therefore, seem worthy of recording for posterity. To give the reader a broader historical and insightful picture of the slowly developing situation in which discouraging rejections and expensive struggles were common, the contents of some typical correspondence, usually concerning new discoveries, is presented in its original letters form in each chapter. But most of the newer discoveries are blended precisely into the cumulative text of the past quarter century (1975-2006). Advocates of the definitive LB/FLINE model foresee the time when it will have an impact on scientific beliefs equal to that of the Copernican idea of our heliocentric Solar System described in his masterpiece, *Concerning the Revolutions of the Celestial Spheres* (1543). If so, building onto these new findings during the 21st century should bring the exciting knowledge of our origins (i.e., the Copernican Revolution) full circle.

PREFACE

In his 1946 autobiography Einstein states two criteria for a valid physical theory—two criteria for scientific truth: external justification and internal perfection. The first requires agreement between observations and experiments; the second, the possibility of inferring a given theory by scientific means from general principles without the need of additional propositions.

Judged strictly on the basis of these criteria, and in view of recent knowledge gained in the fields of astronomy, geography, physics, geology, and chemistry, current theories of creation are on very weak foundations. For example, the Fossil Fuels Theory, since its inception in the 1830s, has served its purpose beyond its time in furnishing an unsound explanation of the way Earth's fuels were formed. This theory (actually a hypothesis) and others are challenged in this book to determine each one's validity.

The Big Bang Theory served its purpose well in the interim of the search for solutions to origins. Now in the light of today's knowledge, too many facts do not fit into it.

Many scientists are always in the process of abandoning unsound theories in favor of more plausible ones that can better explain recent discoveries. Their new facts are forcing theories that are ever closer to the concepts of origins described in this text.

Scientists recognize that things we can't see because they are beyond horizons, are in principle non-falsifiable, and therefore, metaphysical; i.e., a valid scientific concept must be falsifiable. The revolutionary Little Bangs (LB)/FLINE model of universal origins meets the essential criteria that remain elusive in the BBT.

Einstein put things in proper perspective with his statement: "The right answers will be simple and beautiful." One ambitious purpose of this 10th edition is to express the sheer beauty of its answers in as simple a manner as possible.

As Douglas Hofstadter wisely stated, "...creativity is the ability to discover things that are in some sense there for anyone to find: things that the rest of humanity will appreciate because they are beautiful or because they are insightful or because they are truths about nature." The new model once again meets the criteria.

The history of scientific progress teaches us that each and every epoch-making discovery always emanates from the singular mind of one person, and that any consensus rule by majority is always, always, tragically wrong; e.g., the Big Bang.

This treatise is the most comprehensive book of epoch-making scientific discoveries of the past 2500 years.

Here are the keys to a definitive understanding of origins and functions of universal systems: mysteries that remain enigmatic only in the prevailing Big Bang concept.

The author's new Fourth and Fifth Laws of Planetary Motion open the floodgates to understanding the past, present, and future of our Solar System, its Planet Earth, and all weird planets near and far—and the spectacular galaxies that spawn them.

A high priority of this book is to the preservation for posterity a brief history in real time of the ongoing revolution in scientific thought on universal origins: the discoveries of brilliant minds during the past 2500 years; the processes involved; the time frames, some forgotten, some ignored, yet vital to the reactions of scientific organizations and news media; the trials and tribulations of the author; the saving grace of the NPA; and the ongoing fulfillment of the predictions of Albert Einstein and Thomas Kuhn.

The book presents a brief history of epoch-making discoveries that are new, bringing mankind ever closer to understanding origins of universal systems and the Unified Field Theory. Surprisingly, a small number of critical, intermeshing discoveries are spread over a time frame of 2500 years; each is crucial to structuring the revolutionary Little Bangs/Five Laws/Internal Nucleosynthesis/Evolution (LB/FLINE) model of universal origins.

The new model provides the right answers…simple and beautiful—precisely as Einstein predicted.

CHAPTER 1

ORIGINS OF UNIVERSAL SYSTEMS
A BRIEF HISTORY OF "THE RIGHT ANSWERS...SIMPLE AND BEAUTIFUL"

An Overview

The most fundamental mystery in science and in human imagination is the nature of our origins. Throughout civilization, humankind has been obsessed with trying to understand how and why everything came into being. The concepts in this book build on the solid foundation of understanding established by many great names from the past. Each of their discoveries dispelled established myths pertaining to our origins.

This revolutionary work takes us from the brilliant knowledge of the Pythagoreans (2500 B.C.) to the Copernican idea of our Sun-centered Solar System (SS) and to the recent discoveries of giant planets in other distant solar systems as it seeks to do no less than rebuild the foundation of knowledge of the origins and evolution of planets. It offers new answers to questions that have puzzled philosophers for centuries: How did Planet Earth come into being? How and why did the planetary orbits of our Solar System form in a mathematical pattern? Why did this enigmatic solution to the spacing of our planets remain a mystery for 400 years after it first eluded Johannes Kepler in 1595? Will this solution cause scientists to rethink their beliefs about the origins and evolution of planets?

These profound questions are addressed with powerful substantiated evidence that voids the need for speculative uncertainties now common in current theories of planet formation. The book aims to advance scholarship and to enlighten readers with no background in science and philosophy. The first chapter opens with an insightful and startling account of the enigmatic solution to the new Fourth Law of Planetary Motion explaining how the nebulous planetary masses attained their orbital spacing around the Sun. In conjunction with Kepler's Three Laws of Planetary Motion, the Four Laws (FL) reveal the explosive, dynamic origin of our Solar System some five billion years ago, and thus challenge the modified Laplace accretion concept of planet formation. The new Fifth Law explains small changes in orbital positions.

The book gives definitive insights into how and why each planet evolves through five common stages of planetary evolution in full accord with Einstein's famous formula and all natural laws. This new FLINE model makes it easy for readers to understand the origin and evolution of planetary matter; e.g., hydrocarbon fuels (gas, petroleum, coal) are formed via IN and E—not from fossils—and thereby illustrate how Planet Earth and all other planets evolve via IN and E: Nature's two inseparable principles.

Major myths are challenged and decisive discoveries are brought full circle into the 21st century. The nature of comets, planetary rings and asteroids are explored in depth. The FLINE Paradigm is a profound work that provides what philosophers have sought through the ages: a definitive foundation for understanding the origins of solar systems and the evolution of planets via the new FLINE model. To understand anything and everything about planets, we must first understand their three governing and inseparable principles: The Five Laws of Planetary Motion (FL), Internal Nucleosynthesis (IN) and Evolution (E).

For the first time, the geometric, fiery origin of the SS, the spacing of planets and the subsequent evolution of all planets and planetary systems via Internal Nucleosynthesis (IN) can be explained within the scope (and by the means) of natural laws in a beautiful continuity of substantiated evidence. The FLINE model is the only concept capable of making such claims and backing them with facts.

The ideas in this book are the culmination of 33 years of painstaking research into past and current beliefs about universal and planetary origins. Surprisingly, some little-known, but accurate interpretations and a few brilliant ideas, emerge from among the relevant hypotheses suggested throughout the pages of scientific history. Sadly, some of the most astute observations and accurate interpretations have lain buried beneath the onslaught of more popular, but erroneous beliefs.

During the past 26 years of its ongoing development and presentations of evolving facets to scientific organizations, the definitive FLINE model has not encountered any valid argument against its basic principles—nor does that seem likely to happen.

A Brief History of "The Right Answers ... Simple and Beautiful"

A high priority of this book is to preserve for posterity a brief 2500-year history in real time of the ongoing revolution in scientific thought on origins of universal systems.

Surprisingly, a small number of epoch-making discoveries crucial to understanding our Universe have been ignored, overlooked or perhaps, at times, forgotten during the 25-century time frame. Discoveries by the Pythagoreans, Aristotle, Kepler, Descartes, Olbers, Chladni, and more recently, Einstein, Heaston and Scarborough are crucial to understanding origins and ongoing functions of universal systems.

The definitive works of these discoverers, when intermeshed with those of Copernicus, Newton, Hubble and a number of astronomers, geologists, physicists, *et al.*, present clear insight into a definitive concept of the origins, evolution, and ongoing functions of universal systems: the revolutionary Little Bangs/FLINE model. Solutions to universal anomalies readily proven to be impossible via the prevailing Big Bang hypothesis are inherent factors in the new model. Just as Einstein predicted, "The right answers will be simple and beautiful"—and are exciting keys to the long-sought ultimate theory of everything.

The book presents a brief history of epoch-making discoveries that are bringing mankind ever closer to understanding origins of universal systems and the Unified Field Theory. Surprisingly, a small number of critical, intermeshing discoveries are spread over a time frame of 2500 years; each is crucial to structuring the revolutionary Little Bangs/Five Laws/Internal Nucleosynthesis/Evolution (LB/FLINE) model of universal origins.

The new model provides the right answers...simple and beautiful—precisely as Einstein predicted—to questions such as these 12 examples:

1. What is the source of the inexhaustible supply of energy needed to sustain the speed-of-light expansion and the functions of our Universe, perhaps forever?
2. Why the illusion of a static Universe?
3. How were the original 10 or more planets of our Solar System placed in geometrically–spaced orbits?
4. What are the five observable stages of planetary evolution?

5. Why do all galaxies contain large numbers and varieties of solar systems?
6. How did our planet become livable Earth?
7. Why will the true size of our Universe always remain impossible to determine?
8. Why is the size of our Universe increasing at accelerating rates of expansion in all directions?
9. Why does our Milky Way galaxy appear to be at the center of the Universe?
10. Why and how are galaxies formed in pinwheel fashion?
11. What holds each galaxy together in its spinning pinwheel formation during the accelerating rate of universal expansion?
12. Why is our geometrically-spaced Solar System uniquely different from all known (and always weird) extra-solar systems?

Why is the Big Bang incapable of definitive answers to these and all other universal anomalies?

The second priority of the book is to preserve the history of the trials and tribulations, rejections and successes, of the author during the time frame (1973-2007) of his efforts to break through barriers firmly embedded in prevailing beliefs, primarily the Big Bang. Sample letters that illustrate the frustrating efforts and some exciting breakthroughs are legitimate testimony to the accurate predictions made by Thomas Kuhn in his *Structure of Scientific Revolutions* (1956). The letters are distributed throughout the book, along with sample comparisons of two different interpretations of the new discovery in question: the wrong and the right answers.

This tenth edition and all previous editions were accomplished without use of a single dollar of taxpayer or grant monies, while the Big Bang was, and is, promoted to the tune of billions.

The author sincerely hopes that the gist and excitement of the ongoing revolution in scientific thought will be conveyed to each and every reader, and that the theory of everything can be advanced by the revolutionary LB/FLINE model of universal systems.

Background (Dec. 14, 2000)

The revolutionary concepts in this book are the culmination of a lifetime of interest in how everything came into being. Several things happened in the mid-30s to spark the author's curiosity. A seventh grade lesson taught that coal was made from plants. Since very soon after plants on the farm died, and turned into dust and dirt, this seemed impossible to the student. The lumps of coal used in fireplaces often contained the brightly colored pattern of traces of oil, and they looked like lumps of solidified petroleum. When burning, they occasionally spewed out small, beautiful blue jets of flame. Why did these lumps of coal contain both oil and gas? Surely, coal had to be the solidified petroleum it appeared to be.

News headlines of the time boldly announced that our world's petroleum reserves would be exhausted by 1960. In the 1960s-1970s, headlines announced the same dire fate for 1990. Yet today our fuel reserves are greater than ever. One must wonder why. The complete answers are detailed in the chapter on fuels.

During the 1930s, it was not unusual to play baseball in the summertime until dark, then lie on the hillside to watch the stars appear. It seemed natural to wonder what these brilliant dots were made of and how they were made and placed in their positions. Later, studies of Johannes Kepler's discovery of the First Three Laws of Planetary Motion proved fascinating, but even more intriguing was his failed attempt to discover the Fourth Law explaining the geometric spacing of the six then-known planets. The author was hooked for life, and his studies were selected with the idea of learning enough to be able to understand the mysteries of science as they pertain to universal origins.

In 1973, the pieces began falling into place; each new piece of the puzzle became intermeshed with all previous ones. Efforts became concentrated toward separating the myths and facts of science, while eliminating any and all speculation, and avoiding any and all force-fitting. The first breakthrough was published in 1973 as *Fuels: A New Theory,* describing the non-biological origins of the hydrocarbon fuels (gas, petroleum, coal). As research progressed, substantiated evidence mounted dramatically, soon to the point where the new abiogenic origins concept, underpinned solely with incontrovertible evidence, seemed irrefutable. *Undermining the Energy Crisis,* published in 1977 and updated in 1979, told of the vast fuels reserves and the reasons they were predictable via the new Energy Fuels concept. A copy of the later book was given to Ronald Reagan during his presidential campaign through Louisiana. After his election, President Reagan issued an edict to the oil companies: Produce, produce, produce. They did, and our most serious energy crisis soon vanished.

Since it remained too far ahead of its time, the original revolutionary Energy Fuels Theory was forced into self-publications in 1975, 1977, 1979. It initially appeared in the official science literature under the title of *Evolution of Gas, Oil and Coal* (1975-1983) in *Alternative Energy Sources VI, Volume 3, Wind/Ocean/Nuclear/ Hydrogen,* p.337, published as *Proceedings of the Sixth Miami International Conference on Alternative Energy Sources,* December 12-14, 1983, Miami Beach, Florida, sponsored by the Clean Energy Research Institute, University of Miami, Coral Gables, Florida *et al.*

Nature's processes for creating energy fuels led to the realization that all planetary matter was created in like manner: via Internal Nucleosynthesis (the transformation of energy to atomic matter) and thence to all molecular matter comprising any planet's composition throughout its five-stage evolution from energy to inactive spheres. Nature's two inseparable principles, Internal Nucleosynthesis (IN) and Evolution (E) of planets, still lacked a critical piece of the puzzle: How did the nebulous energy masses attain their orbital spacing around the Sun before beginning their five-stage evolution into the planets we see today?

This crucial question was answered by the solution to the enigmatic Fourth Law of Planetary Motion (1980-1995) that first eluded Kepler in 1595. Together, the Four Laws conclusively reveal the dynamic manner in which our Solar System (and all solar systems) came into being. In early 2006, the Fifth Law was extracted from the revised Fourth Law; together they reveal the origin, history and functioning of our Solar System since its dynamic layout some five billion years ago.

Origin and Evolution of Hydrocarbon Fuels (Gas, Oil, Coal)

The Atlanta Journal-Constitution / Sunday, Nov. 14, 2004

Crude abounds; the only problem is how to get to it. By: Ed Porter

"In recent months, numerous observers have claimed that the recent spike in global oil prices is an early warning sign of impending resource exhaustion. They join those who have been predicting such exhaustion for over a century. The data suggests all of these observers are wrong."

"Proven reserves of crude oil worldwide stand at more than 1.1 trillion barrels. In addition, the U.S. Geological Survey estimates that more than another trillion barrels of conventional oil remain to be discovered worldwide."

Response

Predictions of impending oil resource exhaustion have indeed been made periodically for over a century.

Each time the next prediction was made, oil reserves were much larger than when the previous one was made. The author remembers the false predictions in 1935, 1960s, 1970s, 1980s, and 1990s — and, of course, the current ones.

The reasons all predictions have been wrong are elucidated in Chapter I of *Origins of Universal Systems.* This chapter provides additional substantiated evidence.

Dispelling the Myth of Fossil Fuels (1973-1979)

The next phase of the new FLINE model describes how planetary systems evolve from energy into the matter that makes Earth such an ideal planet for all of us. The example of the origin of hydrocarbon fuels (gas, petroleum, coal) best illustrates the processes whereby Nature makes all things by means of natural laws. A little background is essential.

During the 1830s, William E. Logan, a graduate of Edinburgh University, managed his family's coal and smelting interests in Wales. Logan's great interest in the origin of coal led him to study some 100 coal beds and seams. In every case he made note of three observations:

1. A bed of bleached clay lay under each coal bed.
2. Within the clay beds were tangled masses of long, slender, fibrous root systems with a thin coating of carbonaceous matter.
3. Well preserved imprints of ferns and other plants were scattered throughout the coal.

Logan concluded that all plants, specifically the Stigmaria ficoides, had turned into coal. Stigmaria structures, some microscopic, some full size, plus larger fossils (branches and tree trunks) were found in the coal. He recorded that "in Stigmaria ficoides we have the plant to which the earth is mainly indebted for those vast stores of fossils fuels."

In the 1920s, J.B.S. Haldane's hypothesis of petroleum created from tiny marine organisms added more credibility to the concept of fuels made from fossils. So it seemed logical that all natural gas is a product of decomposition of plants and animals. Thus, the Fossils Fuels Theory (FFT) became firmly entrenched in scientific literature.

Since Logan's time, advocates have attempted, in difficult and costly efforts, to explain their relevant discoveries in its perspective. In reality, as we shall see, much time, frustration and money could be saved if scientists have opportunities to interpret their findings in the new perspective of the FLINE concept.

Contrary-wise, through the years, there have been a number of opponents who kept faith in the belief of non-biological (abiogenic) origins of fuels. The list includes some historical names: Berthelot (1866), Mendele'ev (1877, 1920), Humbolt, Gay-Lussac and others. In more recent times, a significant volume of evidence that argues against the validity of the Fossil Fuels Theory (FFT) has accumulated in scientific literature. Through intensive library searches, the author has gleaned and condensed much evidence into six critical points:

1. The tremendous volumes and extreme depths of fuels (especially gas) contradict, rather than enhance, the FFT.
2. The patterns of distribution, size and thickness of coal beds and seams simply do not fit into the FFT.
3. No multi-layers of root systems are found in either thick or thin coal beds; carbonized root systems are found only in under-bed clay.
4. Plants yield vegetable oils; coal and petroleum are mineral-oil based.
5. Peat, credited as a transitional stage of some plant-to-coal processes, actually turns into black dirt that retains no imprints of Stigmaria ficoides.
6. The structural integrity of plant imprints can be preserved only through rapid encapsulation of live plants

before decay begins.

Each point adds weight to the powerful evidence accumulating against the FFT. For example, today we know that a fossil imprint found in rock was still in its original life form when encapsulated by softer material that later solidified around it. The imprint remained in the rock after dissipation of the encased specimen, which obviously cannot be credited with creation of its encapsulating rock. In this perspective, we can reasonably conclude that the plants whose live imprints were discovered by Logan could not have created the coal; they simply were victims of encapsulation.

Logan erroneously concluded that coal was made from plants. His three astute observations were misinterpreted. In reality, they argue strongly against his hypothesis. Such integrity of plant structure could not have been preserved via the decaying leaf interpretation made by Logan *et al.* Further, his findings can be explained more reasonably by the FLINE concept; only in this revolutionary perspective can his three observations be interpreted accurately.

Only hot, chemically-contained petroleum could have bleached the clay beds and seeped down the root systems to preserve them via carbonization. Only the sudden encasement of live swamp plants by an encapsulating medium (petroleum) could have preserved their live structural integrity, discovered as imprints of live plants. How did it all happen?

Leaves imprinted in rocks. Material encapsulated the live leaves and solidified around it as rock. Similar imprints found in coal were preserved in the same manner - encapsulated by petroleum that solidified as coal. Plants did not decay and create either the rock or the coal.

How Hydrocarbon Fuels Formed in Earth's Crust

As confirmed in a research project at Neely Nuclear Research Center at Georgia Tech in 1985-86, the three hydrocarbon fuels (gas, petroleum and coal) that abound throughout Earth's crust generally exist in overlapping three-layer systems (Fig. 4). Coal is found in the surface or within the top mile of the crust. Petroleum dominates farther down at medium levels, while methane gas is found vastly more abundantly at deeper depths.

At the center of Earth lies its nuclear core, the driving source of the energy that is transformed into the atomic elements that comprise the hot material of the mantle from which the crust is being formed continuously. Carbon and hydrogen are among the most abundant elements created in the nucleosynthesis processes. They are the building blocks of methane gas; the two elements combine to form the vast quantities of methane gas found throughout Earth's crust, much of it in the form of methane hydrates.

During the rocky stage of evolution, virgin elements combine in similar manner to form a large variety of compounds that become the crust of planetary matter comprising Earth's ever-thickening shell. In our fuels example, much of the methane links together, through the process known as polymerization to create other gases: ethane, propane and butane.

When five of the methane molecules linked together, they formed the next product in the carbon-chain series: pentane, the first liquid and the lightest component of light-weight petroleum. Over time, polymerization continued to forge larger and larger molecules of thin and medium-weight oils, and finally, thicker crude petroleum. While unknown to Logan *et al.*, scientists now know that all crude petroleum contains tiny particles of coal—the beginning of Nature's final transitional phase from gas to petroleum to coal.

When huge quantities of crude petroleum were forced onto Earth's surface, filling caves and tunnels, large areas of swamplands were inundated with hot encapsulating oils that eventually polymerized, cross-linked and solidified as beds and veins of coal.

From this evidence, we can reasonably conclude that the hydrocarbon fuels, gas, petroleum, and coal, should be called 'energy' fuels rather than 'fossil' fuels.

Strong evidence supporting this conclusion abounds. From the scientific literature, the five recognized facts that dramatically reveal the very close relationship of the three energy fuels could be summarized as follows:

1. All wet gases contain lightweight oils (condensate; e.g., pentane), illustrating the first transitional phases via polymerization from gas to petroleum.
2. All petroleum contains gases in varying amounts inversely proportional to the degree of polymerization.
3. All crude petroleum contains tiny particles of coal, illustrating the second transitional phase via polymerization from petroleum to coal.
4. Gummy coals (bogshead and cannel) can be classified as either petroleum or coal, illustrating an advanced stage of the second transitional phase (additional polymerization and some cross-linking).
5. Every lump of coal contains oils and gases, illustrating all three phases and the degrees of polymerization and cross-linking of the three fuels, while confirming their close relationship.

A close study of the evidence clearly reveals that gas, petroleum and coal are the first, second and third phases in the formation of hydrocarbon (energy) fuels, and that polymerization is the key to evolutionary changes from gas to petroleum to coal.

The Three-Layer Systems in Earth's Crust

Why would coal, if formed from great swamps of plants and trees, generally be found among and under the rocky layers of mountainous regions like West Virginia? Why are coal beds located at or very near the surface? Why is coal found in veins, many of which are smaller in diameter and relatively short in length? Would not the predominately mountainous distribution pattern of coal and the shapes and sizes of coal veins indicate that great forces pressured gases and liquids into such locations and formations, where polymerization to petroleum and solidification to coal occurred? If so, how could one explain the process, in terms of natural laws, that created these situations?

Why does petroleum exist generally at intermediate depths ranging from near the surface down to 30,000 or more feet? Why is it that the deeper the drilling of wells, the greater the proportion of gas to petroleum? Why should gas be the deepest of the three hydrocarbon fuels? Why do no fossils exist in the deeper oil and gas sites? And how did gas form at such great depths in unimaginable quantities under tremendous pressures at such extremely high temperatures?

Through literature searches in many libraries over the ensuing years, confirmations of the imagined answers to the persistent questions rattling in my brain were sought and found. The excitement of discovering in the literature the facts that substantiate the wild imagination serve to spur one to the next phase of the problem. There can be no turning back or termination of the quest for truth: one can become obsessed with it. This, in spite of the discountenance certain to be encountered along any and all ways that is in opposition to established beliefs. While that can be traumatic, it cannot be a deterrent to progress. History is replete with such examples, and time will never alter the cycle.

To summarize briefly, the energy fuels are arranged generally in three overlapping layers in Earth's crust. Coal is found on or near the surface, while petroleum is found at lower (medium) depths. Deeper drillings result in higher ratios of gas to petroleum. Finally, the deepest drillings may yield only gas. These overlapping layers of gas-to-liquid-to-solid fuels are arranged roughly in ascending order from Earth's interior to its surface by natural laws of physics and chemistry.

The first publication of the three-layered system of fuels in 1973 (revised 1975) *(Fuels: A New Theory)* by the author understandably met apathy and some opposition from disciples of the FFT. Additional publications in 1977, 1979, and 1986 managed to gain a few converts, but the numbers remain small at this writing.

Meanwhile, in February 1981, an article on *World Energy Resources* was published in a special report in *National Geographic.* Included were three maps of the North American continent showing the distribution pattern for each of the three fuels. When superimposed, the three maps revealed very similar distribution patterns for gas, petroleum and coal over the whole continent, thereby adding much credibility to the three-layer concept.

With this background, let's take another look at Logan's three astute observations of the 1830s, and, working in reverse of Nature's processes, continue developing the new theory of formation of these hydrocarbon fuels.

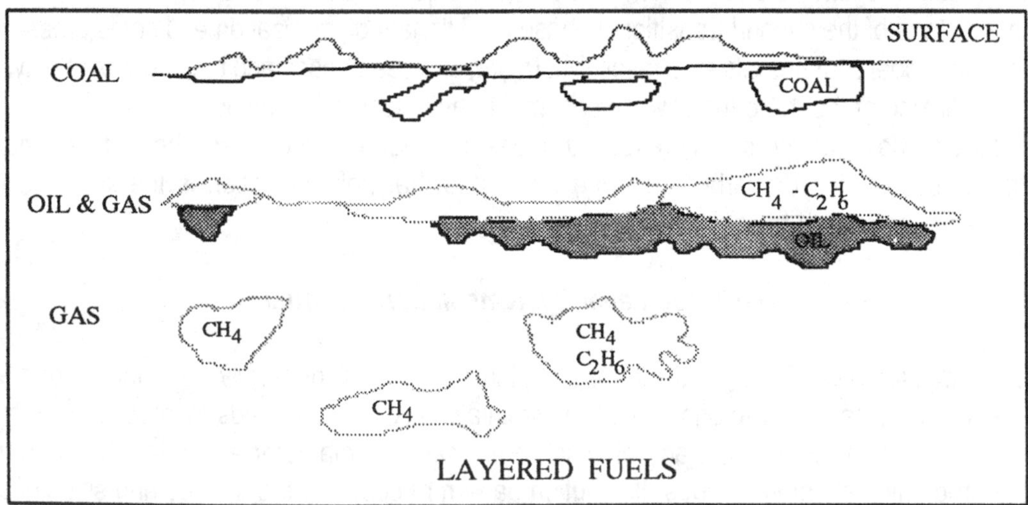

Coal, oil and gas are generally layered in the descending order shown. Deeper drillings uncover higher ratios of gas:petroleum.

How Coal Formed from Petroleum

When Logan saw the bleached clay under every bed of coal, he should have questioned why it was bleached. Bleaching is usually accomplished by a combination of certain chemicals at high temperatures. Imagine what might occur if large volumes of petroleum were forced from the ground under thousands of pounds of pressure and at very high temperatures. Such a familiar gusher happened in 1979 in the Gulf of Mexico when an oil well flowed threateningly for nine months before being forcibly closed off.

Hundreds of millions of years ago, such hot oil erupting onto land filled all the low areas (generally swamplands) within reach. It inundated low-lying areas where Stigmaria was usually the dominant plant life. The high heat and chemicals in the petroleum bleached the clay underneath, thereby leaving the first clue for Logan to discover.

By seeping downward and penetrating the root system of plants, the oil preserved the "tangled mass of long, slender fibrous root systems with a thin coating of carbonaceous matter" in their original form. The carbonaceous matter found by Logan was the remains of the oils that had preserved the root systems. If not for this preservative, the root systems would have disappeared rapidly from the scene, as all roots normally do.

Ferns, leaves, branches and even tree trunks were scattered, suspended throughout the lake of hot viscous oil. Thus they were preserved in situ, destined to leave their distinctive imprints, porphyrins and carbonized skeletons in the mass as the oil cooled, thickened, solidified, cross-linked and polymerized into solid coal. Time, temperature and pressure inevitably changed the oil channels and deep ponds into coal veins and coal beds, some as much as hundreds of feet thick. Much of the gas and oil was forced by the tremendous pressures into buried crevices, deep sand formations, porous rocks and strata where it transformed into vast quantities of shale and tar sands.

Thus, the world's coals were created when extremely hot petroleum containing gases from Earth's interior poured out over the swamplands, encasing the plants, seeping down the root systems, and bleaching the under-beds before solidifying into solid coal.

But what could be the source of such vast quantities of petroleum? Could these oils have formed from deeper gases that polymerized?

How Petroleum Formed from Methane Gas

If coal formed from petroleum, then the next step is to find the source of the oils. Nature creates petroleum by the process of molecular chain-building; i.e., methane gas molecules are joined together, much like a chain, by natural processes (polymerization) into longer, larger, molecules known as light oils, or condensate. The lightest of these is identified as pentane, because each molecule contains five atoms of carbon. Each time another molecule of methane is polymerized into the chain, the larger molecule is renamed in ascending order: hexane, heptane, octane, etc. Various proportions of these condensates always are found in all wet gases and in all lightweight petroleum.

As polymerization continues, the oils become heavier and thicker, more viscous. At a higher viscosity, the petroleum becomes known as crude oil. During these transformations, a process called cross-linking is initiated to make ever larger and more rigid molecules. While the term is self-explanatory to many readers, its meaning should be clarified: the molecular chains link together side-to-side, thus becoming less mobile, more solid. Finally, when sufficient polymerization and cross-linking occur, the petroleum begins to solidify into the tiny particles of coal always found in heavy crudes.

Where and why does polymerization of gas into petroleum occur? Starting deep inside Earth, methane gas, under tremendous pressure, seeks the paths of least resistance on its upward journey. Most of it becomes trapped either in porous rock or strata or beneath impervious barriers. Here the high heat and pressure, in conjunction with the contaminants (trace metals) in the gas, initiate and sustain the well-known chemical process of polymerization.

During the past two decades, duplication of the process has been accomplished in a number of chemical labs, and patents have been issued covering the catalytic conversion of methane into gasoline (a mixture of lighter weight oils). The process is not commercially feasible at this time.

A fantastic example of such lightweight oils created by natural processes was discovered on June 5, 1989, near a town south of Riyadh called Hawtah in Saudi Arabia. The gusher tested at 8,000 barrels a day. Later, four more prolific oil wells and one gas well were drilled there. The vast deposits contain rare, super light oil that could be used in a car engine without being refined. The new oil has the color of gasoline and the consistency of water. Its gravity varies from 42 to 49 degrees, as measured by the standard API method. The Hawtah oil contains almost no sulfur or other impurities. The oil field may prove to be the world's largest.

Only a week before Iraq invaded Kuwait, Aramco announced the size of the prospective drilling area: 1,440 square miles. The discovery raises questions that may never be answered: Was this field the next target envisioned by Saddam Hussein? Was it the primary reason he gave up the objectives of the 8-year war with Iran and turned his full attention southward? Was his primary objective the control of oil prices worldwide?

One would expect the Hawtah oils to be free of the contaminants found in crude oils, and they are. Petroleum crudes usually contain numerous substances, both organic and inorganic, including many trace metals, salts, lignite, and coal. Microscopic studies reveal fragments of petrified wood, spores, algae, insect scales, tiny shells and fragments. In contrast to Haldane's theory, it is safe to conclude that these substances are contaminants trapped in the oil; they did not create the oil! These contaminants remain in the mass during and after its transition into coal, thereby often participating in the reactions that create the many chemicals found in coal.

It is important to remember that not all gas polymerizes into petroleum and coal. Molecules of gas are found in every lump of coal and every drop of petroleum. Pockets of these gases in coal mines, when released from entrapment in the black gold, present great dangers of explosion and suffocation.

Source of the Methane Gas

If coal was made from gushing petroleum that had been created from deeper gas, what is the source of the methane? How was it made in such unimaginable quantities under such extreme conditions of temperature and pressure? Is it still being created? The answers to these questions are profound, revealing a logical concept in which the creation of all matter comprising our planet can be explained with very little, if any, speculation.

Before delving into these questions, we need to take a closer look at the vast resources of gas in Earth's crust. Geo-pressured methane, an example of one type of gas, lies in the untapped gas-laden briny waters buried deep beneath the Gulf Coast. A Baton Rouge hydrogeologist, a leading authority on the subject, calculated the supply at 50,000 trillion cubic feet (tcf) of methane in Louisiana and Texas. The annual use rate of 20 tcf in the USA equates to 2,500 years of reserves in this one location. One tcf is the energy equivalent of approximately 180 million barrels of oil.

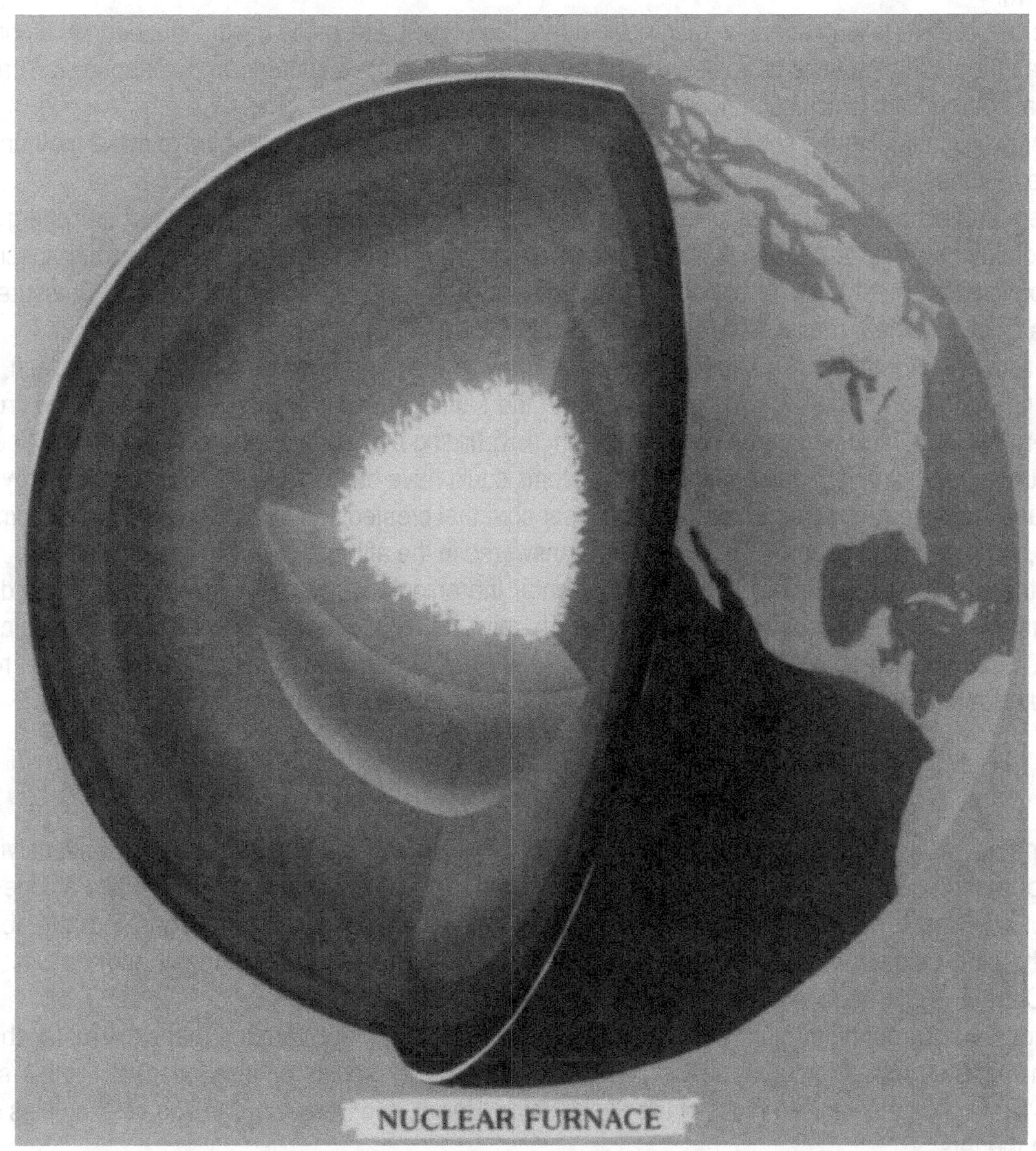

Earth's nuclear core: A hot, searing, violent, mysterious force that both builds and destroys. Why does mankind seem incapable of grasping its significance in Nature's scheme of creation, of realizing its existence and scope?

In addition to the Gulf Coast reserves, methane is known to exist in a large basin area some 500 miles in length in California, and other states, including Oklahoma, Washington, Oregon and Alaska (250 tcf). These resources can be multiplied many times in the worldwide perspective, since there is no reason to believe that the listed areas have a monopoly on gas supplies. As in past decades, even these estimates will prove far too conservative.

Therefore, are we talking about unimaginable quantities of methane inside Earth - stupendous amounts far too deep and too voluminous to have been made from fossils, or to have come from outer space? What might be a logical source from which such huge volumes could evolve?

One clue lies in finding the source of the carbon and hydrogen atoms that combine to make methane. How and where can these atoms be created?

Einstein and his contemporaries proved that atoms could be forced to release their nuclear energy under specific conditions. This tells us that Nature made atoms from nuclear energy under various specific and extreme conditions, most likely at millions of degrees and fewer than millions of pounds of pressure. Such conditions exist in nuclear masses like our Sun and stars.

Scientists have identified a number of types of atoms on the surface of the Sun, from hydrogen to iron. The glowing hot gases of the Sun are made of about the same chemical elements composing the crust and atmosphere of Earth, including carbon and hydrogen, the building blocks for methane. Since these clues from the Sun suggest the strong possibility that Earth's atoms could have been created from nuclear energy in situ, the key question becomes: Does Earth have a nuclear core that created and still creates the atoms comprising its mantle, crust and atmosphere? This question is answered in the affirmative in Chapter II.

With the concept of a nuclear core in each planet, the enigmatic mystery of planetary origins dropped its impregnable veil, and the dawn of reality came sharply into focus. Too good to be true; something must be wrong. But in perfection, where every piece has fitted precisely into the concept, its first flaw yet remains undiscovered after 27 years of piecing it all together and testing it superficially in scientific meetings.

Evidence for the Making of Abiogenic Methane Gas

Fortunately, it is possible to differentiate between biogenic and abiogenic methane by identifying two isotopes of carbon comprising the gas. Biogenic (biological) methane is enriched in ^{12}C, while abiogenic gas consists of ^{13}C (Hoefs, 1980). The high ratios of ^{13}C to ^{12}C in methane, especially deep gas, indicate abiogenic origins. The high ratios in methane "hot spots" of the Red Sea, Lake Kivu (East Africa), and the East Pacific Rise (McDonald, 1983) suggest abiogenic origins.

Additional background literature by Peyve (1956) and Subbottin (1966) argues that sub-crustal abiogenic petroleum migrates up major faults to be trapped in sedimentary basins or dissipated at Earth's surface. Profir'ev (1974) cites the flanking faults of the Suez, Rhine, Baikal, and Barguzin grabens as examples of such petroleum feeders.

This school of thought gained support in later publications; e.g., *The Deep-Earth-Gas Hypothesis* by Gold and Soter *(Scientific American,* June, 1980). The article presented much evidence that earthquakes and volcanoes release gases from Earth's mantle, and such gases may include methane of a non-biological origin. In a later publication, Gold argues that earthquake out-gassing along faults allows methane to escape from the mantle, a process that gives rise to deep gas reservoirs and via polymerization, to petroleum at shallower levels. Further, the *Reader's Digest* (April 1981) published the article *Bonanza, America Strikes Gas,* which tells of geologists hitting field after field of natural gas deep within the nation's bedrock.

These articles and a number of other discoveries and arguments strongly support both the Energy Fuels

Theory and its all-embracing concept of Internal Nucleosynthesis (first introduced in 1975 as the TIFFE concept: The Internal Formation of Fuels and Elements, later given the I-T-E-M concept label). They add powerful support to the one small, lone voice that argued against the hysteria of the energy crisis in the 1970s in attempts to convince the establishment of the true nature of the vast reserves of hydrocarbon fuels in Earth's crust worldwide—a message not heeded until the 1980s.

Today's glutted oil market and reasonable prices attest to the warranted faith in the concept. But even at this writing (1996), the origin of these energy fuels remains erroneously attributed to fossils.

The Rise and Decline of the Abiogenic Fuels Theory

The taproot of the energy crisis of the 1970s can be traced back to the original theories of the 1830s and 1920s: gas, petroleum and coal were made from fossils. This misconception proved to be one of the most tragic scientific myths of our time. Its implications of very limited fuel supplies had traumatic effects during and after the 1970s debacle.

During 1973-1976 the price of petroleum tripled, and fuel shortages resulted in long lines at gasoline pumps, high inflation and soaring interest rates. Some experts predicted that gasoline prices would climb as high as $5.00 per gallon. OPEC seemed in complete control.

It was a time of genuine, irrational fear that fuel reserves would be depleted within a few years. In April 1977, President Jimmy Carter expressed his alarm to a national television audience, declaring the moral equivalent of war on the energy situation. By 1979, oil prices had nearly tripled again, with devastating effects on the economy.

Many causes and effects were debated. Two of the primary reasons were attributed to: (1) the steadily declining exploration and production of petroleum in the USA between1955-1973, a decline of nearly 70%; and (2) the pervasive fear that the world's fossil fuel reserves were very limited and were being rapidly exhausted. The heightened impact of low supply and high demand, intensified by depletion woes, was traumatizing too many people. Fear and dire predictions ran rampant, at times bordering on hysteria that was duly captured by the news media: experts spreading doom and gloom—all based on the false gospel of very limited reserves of fuels made from fossils.

Much of the sanity that survived remained in relative obscurity. For example, a copy of my 1979 book, *Undermining the Energy Crisis,* was given to Ronald Reagan during his presidential campaign in Louisiana. Its energy fuels concept argued for maximum production of fuels from the vastly underestimated reserves in the USA, while alleviating the unwarranted fear of rapid depletion of the world's supplies. Further, it predicted that the price of gasoline in the 1980s would stabilize at about $1.00 per gallon whenever supplies met demand.

It is not known for certain that he read the book, but immediately after Reagan's election, the new President issued a three-word order to oil companies: Produce, produce, produce. Consequently, the book's predictions proved deadly accurate. What had made it possible to make such precise short and long-term forecasts in opposition to so many prevailing expert opinions? The answer came from understanding how these vast stores of energy fuels were created in Earth's crust—not from fossils, but via the ongoing processes of Internal Nucleosynthesis and polymerization of these hydrocarbon fuels.

As predicted by the new "energy fuels" theory, and quoting from *Financial World* magazine of November 13, 1990: "...the worldwide glut of oil in the ground, estimated to be over one trillion barrels and rising yearly, will find its way to the market, driving prices inexorably lower." Other estimates run as high as six trillion barrels; even that figure will prove to be much too conservative.

The latest verification of the prediction made by my first publications in 1973 and 1975 on the origin of unimaginably vast amounts of a biogenic natural gas in Earth's crust appeared in *Science* magazine, 28 June 1991. In this article, *Fire and Ice under the Deep-Sea Floor,* Tim Appenzeller describes ubiquitous gas hydrates occurring naturally in deep sea sediments and under the Arctic permafrost in staggering amounts. The total amount of gas hydrates worldwide has been estimated as equivalent to some 10,000 billion metric tons of carbon—twice the carbon of all known reserves of gas, oil and coal.

And who can imagine the vast reserves and the source of the fuel that exists beneath these gas hydrates? Once again, these new estimates of the world's gas reserves will prove woefully inadequate.

The most significant and exciting aspect of these discoveries is the powerful evidence they offer of the manner in which these fuels were made (and are still being made, and will continue to be made throughout future eons, possibly longer than mankind will survive on Earth). These and all future findings will fully vindicate the warranted faith in the 1973 concept of Internal Nucleosynthesis and polymerization of hydrocarbon fuels.

However, the fossil fuels and condensation/planetesimals/accretion beliefs are firmly entrenched in the scientific literature. They will be very difficult to displace. But time and evidence are definitely on the side of the FLINE model; it does seem only a question of time.

Interest in the origin of natural gas peaked in the mid 1980s, primarily as a result of the annual spring meeting in 1985 of the American Association for the Advancement of Science (AAAS). A session on origins of fuels featured Professor Thomas Gold's excellent presentation of strong evidence for the abiogenic origin of deep methane gas. Attempts made through AAAS channels to permit presentation of my findings since 1973 to reinforce Gold's findings proved futile, in spite of the fact that several of the program's scheduled speakers failed to show.

Although given its due publicity, the concept was destined to fall out of favor because of two reasons. First, the information I could have contributed in the session would have completed the big picture of the abiogenic origins of all hydrocarbon fuels (gas, petroleum, coal). However, Gold's singular attempts to fit his version of the origin in with prevailing beliefs—one being that methane came from outer space and was trapped deep inside during Earth's early formation—were doomed to failure, simply because they are unrealistic, and they simply fail to mesh with facts. Second, through Gold's persuasiveness, a deep well was drilled later in Sweden to test the theory. Since no fuel was found, interest in the abiogenic origin waned considerably.

The concept of the abiogenic origin of methane and subsequent polymerization to petroleum erroneously became known as "Gold's Theory".

Until the facts in the situation are sorted out and faith in the concept is restored by some future breakthrough, this vital key to the origins of atmospheric, crustal and internal compounds of all planetary spheres via Internal Nucleosynthesis now faces an even tougher road to the full credibility it warrants within the community. But this can be accomplished only if scientists are permitted to hear the full, factual story. It is hoped that this book is a giant step along that winding trail to the restored credibility warranted by these findings.

Even without having a chance for a full understanding of the now-completed FLINE model (formerly I-T-E-M) concept, Gold and many others have added significant corroborative evidence to the new Energy Fuels Theory (EFT) since its conception in 1973. Now structured with an ever-increasing multitude of interlocking, incontrovertible facts, the revolutionary concept does indeed seem destined eventually to displace the prevailing erroneous beliefs about the origins of hydrocarbon fuels and other matter comprising our planetary systems.

Revival of the Abiogenic Fuels Theory

Perhaps the breakthrough hoped for above has begun. In the same project study at Georgia Tech (referenced previously), nickel and a number of other metals were found in close association with the hydrocarbon fuels. In the common knowledge that metal catalysts generally play a prominent role in polymerization processes, one or more of these metals seemed to be involved here. The presence of these metals, helium, other gases and oils usually associated with natural gas deposits is explained by the nucleosynthesis processes within Earth. For obvious reasons, such magmatic mixtures are more the rule than the exception.

Additional strong evidence on the abiogenic origin of hydrocarbon fuels was presented by Frank Mango at the August, 1995 meeting of the American Chemical Society. Mango, a research scientist in the geology/geophysics and chemical engineering departments at Houston's Rice University, disclosed that metals such as nickel played a major role in the generation of natural gas.

It all began when the particular light hydrocarbons he was examining "had structures that were fundamentally a contradiction to existing views on the origin of petroleum." Mango's conclusion that "the origin of light [and crude] hydrocarbons could not be the thermal breakdown of biological molecules" is further confirmation of the abiogenic Energy Fuels Theory of 1973.

However, due again to the prevailing belief in the FFT, Mango was thrown off track when he concluded that the nickel promotes "the conversion of decomposing organic debris into natural gas."

Evidence indicates that natural gas forms via the natural affinities of its component elements, carbon and hydrogen, which later polymerize (usually with the aid of metallic catalysts) into even larger and varied molecules of gas and petroleum (light to crude). Then, in reaching the third and final stage, the crude petroleum, always in the presence of numerous other elements, polymerizes, cross-links and solidifies into coal.

The key point in Mango's finding is the further confirmation that the origin of hydrocarbon fuels could not be of a biological nature, but is indeed of an abiological nature. Perhaps this could be the spark that re-ignites the interest of scientists in the true origin of abiogenic fuels: the seed theory of 1973 that eventually grew into the FLINE model during the next two decades.

From Energy to Matter to Life: A Reasonable Continuity

Scientists have much evidence to show that Nature starts with the simpler elements as building blocks to construct ever-larger molecules. Atoms and molecules with mutual affinities arrange themselves properly and assemble together, under proper conditions, into more complex molecules. Nature's self-assembly process has been recognized by biologists as one commonly found throughout biology. Much like a jigsaw puzzle coming together to make the correct picture, the pieces stick together by hydrophobic and electrostatic interactions, according to David S. Lawrence of State University of New York in Buffalo.

Other laboratories have simulated Earth's primitive atmosphere (methane, nitrogen, water vapor) in experiments yielding all five bases that make up the sophisticated building blocks of the genetic code. Four of these bases, (cytosine, quinine, thiamine, adenine) form DNA, the double-helix molecule that spells out the instructions for all living things. The fifth critical base, uracil, substitutes for thiamine to make RNA, which acts as a master slave to carry out DNA's orders.

The key point here is that chemicals are formed easily by duplicating Nature's primitive codes. Thus, the process of chemical evolution must have been relatively simple. This lends strong credibility to the EFT theme that all matter is built with basic building blocks (atomic elements) under proper conditions of temperature, pressure and time specific to the matter in question.

Located with an underwater sonar system in 1993, a natural gas deposit was discovered 170 miles east of Charleston, South Carolina. Based on preliminary mappings, USGS scientists estimated that the area could contain more than 1,300 trillion cubic feet of methane gas locked up in hydrate deposits. If extracted, that volume would serve the country's needs for more than 70 years, based on 1989 consumption levels. However, extracting gas from the deposit located more than a mile beneath the ocean surface would not be a simple matter.

The deep-sea gas vent is ringed by an unusual formation of mussels, clams and other mollusks. Apparently, these life forms have adapted to their dark gaseous environments by deriving energy from the vented methane and other chemicals. This situation illustrates three important points:

1. The continuity in the chain of dependency in the linked cycles of evolution: the energy-to-matter-to-life relationship in which everything is connected to everything else.
2. The environmental survival principle: adapt or die.
3. These findings are indicative of the huge volumes of abiogenic gas already discovered and the vast deposits yet to be discovered in the crust worldwide.

All three factors are vital aspects of the FLINE concept of an inter-meshing continuity of events in which our SS was created dynamically in the energy form from which our planets and their planetary systems are evolving through five stages of evolution in full accord with the laws of physics and chemistry.

More on Hydrocarbon Fuels (1973-1986)
The Order of Energy Fuels

The energy fuels are generally arranged in three overlapping layers in Earth's crust. Coal is found on or near the surface of the earth, while petroleum is found at lower (medium) depths. Drilling deeper results in higher ratios of natural gas to petroleum. These layers of gas-to-liquid-to-solid fuels are arranged in ascending order from the earth's interior to its surface by natural laws of physics and chemistry.

To quote from *Elements of Petroleum Geology* (R.C. Selley): "Oil and gas tend to occur in sedimentary basins in a regular pattern. Considered vertically, oil gravity decreases with depth. Heavy oils tend to be shallow and, with increasing depth, pass down into light oils, condensate, and finally gas, until the point at which hydrocarbons and porosity are absent." Hunt (1979) produced graphs from data in the 1975 *International Petroleum Encyclopedia* that show this statement to be true, further verifying my earlier findings.

Oils also follow the graph pattern in their lateral movements: heavy oils occur around basin margins and are progressively lighter toward the basin center. The reason for these vertical and lateral progressive variations of oil and gas are readily explained in the perspective of basic chemical action accelerated by heat: <u>polymerization of molecules.</u> From gas, to oils (light to crude), to coals (lignite to anthracite), the beautiful, simple basic chemistry runs its natural course. This is the basic claim of the Energy Fuels Theory as first postulated in 1973.

Nature avoids the problems inherent in Gussow's principle of step-entrapment over large distances with discontinuous reservoirs that supposedly explain this progression of variations in fuels. Actually, the patterned zonation of gas, light oil, and heavy oil from basin center to margin in any reservoir, especially in large, discontinuous reservoirs, can be accomplished <u>only by polymerization.</u> The differentiation of oil and gas in basins reflects the degree of polymerization of the methane building blocks into light oils, then into heavier oils, and finally into coals. Coal is solidified, polymerized oil; oil is polymerized gas, which comes from deep within Earth's interior where its elements and molecules are created from nuclear energy.

The next significant question concerns the locale of nature's manufacturing plant for all the elements. Current theory teaches that all the elements of Earth were created in distant stars; the EFT teaches that each planet creates elements from its interior source of energy particles, then uses these elements as building blocks for all its matter.

Earth's Nuclear Core

Earth's nuclear core reactions will someday be duplicated by scientists. Laboratory officials at Sandia National Laboratories fired the Particle Beam Fusion Accelerator II (completed in the autumn of 1985) at 70 percent capacity, delivering a brief burst of ions into a diode in a pulse that lasted about 50 billionth of a second. The procedure consists of irradiating pea-sized pellets of hydrogen isotopes with intense beams of ions, which ultimately compress the hydrogen fuel and heat it to 100 million degrees Celsius, causing the atoms to fuse together to form larger atomic elements. The heat emanating from the fusion cycle is identical to the heat of our sun (but on a smaller scale, of course). This same process occurs within all suns, planets and large moons, and when duplicated successfully by mankind, will some day allow scientists to create valuable metals, such as gold, from lesser elements, thus fulfilling the ancient dreams of the alchemists who sought in vain the secret used by nature to create its materials.

It is logical to believe that practically all Earth's crustal and atmospheric elements were created in like manner and in accordance with natural laws of physics and chemistry. Logic becomes hazy in attempts to visualize creation of Earth and its solar system mates by any other concept.

Evidence indicates that Earth's core (approximately 2700 miles in diameter) is still extremely hot, perhaps composed of dense energy particles where no elements can exist. Where the core meets the surrounding magma, the elements emerge as atoms. Countless hydrogen and carbon atoms combined to form methane gas. (All earth's crust was similarly created: countless other atoms of hydrogen and oxygen combined to create water, for example.)

From Gas to Coal

The methane gases are forced up through the paths of least resistance until trapped beneath impenetrable barriers. Here, the slow process of molecular chain-building forms heavier gases and oils. These oil molecules gradually cross-link to form heavier oils and branch-chain solids—tiny particles of such coals are always found in crude oils. Under proper conditions, the ratio of coal-to-oil increases until very soft, gummy coals (cannel and bogshead) are formed. In time, such coals may be compacted into drier types, progressing through the lignite and bituminous stages into anthracite coal.

Of course, these fuels are generally found in various combinations of their transitional stages. For example, coal, as mentioned earlier, always contains oil and gas, which can be easily converted back into their original free states. Oil and gas are constant companions in oil wells; the original gas has been only partially converted (by molecular chain-building) into petroleum oils. The degrees of conversion are the results of pressures, temperatures, contaminants, and time, to which the gas and oil molecules are subjected during their transformation within the earth by natural molecular chain-building processes. Thus, coal becomes the product of a natural chain that begins with energy particles and moves through gaseous, liquid and solid states. No natural laws can be violated in these processes of energy and matter transformations from hot-to-cold states.

Examples of all stages of these transformation processes are known to humankind. Scientists know that many types of gases and oils exist in a variety of thicknesses and viscosities; coal may be found in any equal number of stages. Coals exist in stages ranging from very soft to very hard, from soggy to almost dry. There are certain bituminous coals (bogshead and cannel coals) which are difficult to classify as coal rather than petroleum, and vice versa. Such a close chemical relationship can only be explained by a closely related origin.

More on Evolution of Oil from Gas

Petroleum crude oil usually contains numerous miscellaneous substances, both organic and inorganic. Microscopic studies reveal fragments of petrified wood, spores, algae, insect scales, tiny shells and fragments. These substances are contaminants that were trapped in the oil; they did not create the oil.

Nature creates petroleum by the process of molecular chain-building, i.e., the methane (gas) molecules are used as chain links that are joined together by natural processes into larger molecules man calls oils. These changes occur within Earth's crust where the methane is trapped for a long time under impervious rocks. The high heat and tremendous pressures initiate and sustain the molecular chain-building processes that change methane into petroleum oils, beginning with the lightweight condensate found with gases and proceeding through ever-thicker stages of crude oils.

Crude oils usually contain tiny particles of lignite and coal, many metals, and salts. The presence of lignite and coal points to the logical conclusion that the oil has begun its transition into coal. The many metals and salts remain in the coal during and after its transition from crude oil. Science reverses the process when it changes the coal back into oil, gas and chemicals, molecules and atoms, and, finally, atomic energy.

The Energy Fuels Theory

Under the prevailing beliefs, primordial materials buried deep within the forming Earth have been giving off hydrocarbons as a result of the planet's heated interior. Oil molecules (hydrocarbons) under these hot conditions would break down into methane gas, the most stable hydrocarbons.

The Energy Fuels Theory differs from this concept in several ways. First, petroleum products (gas, oil, and coal) are not primordial matter; each was created successively by natural laws. Methane gas is formed from virgin atoms of carbon and hydrogen created in Earth's hot core of energy particles. Under tremendous pressures, the methane seeks paths of least resistance on its upward journey.

When no more paths are available, the gas accumulates in reservoirs of various sizes. Here, two probable alternatives occur: (1) methane may begin its polymerization into heavy gases and light oils (condensate); or (2) it may break through its overlying cap in an explosive violence that can fracture granite into gravel. We call these explosions "earthquakes" or "moonquakes" or other planet-name quakes. Such quakes still occur on all spherical planets and moons that have any nuclear energy core not yet exhausted into matter.

A study of the Sun and its huge flares can teach us much about planetary cores and the creation of atomic elements in nuclear furnaces, as we shall see later.

In summary, the EFT states that time, temperature, and pressure work their transformations on planetary cores of energy particles, making them into atoms, then molecules, then gases, then oils and coals. Each fuel product is a result of these molecular chain-building processes, beginning with methane molecules (gas) forming into oils. Huge quantities of the oils were forcibly spilled out over the lowlands, bleaching the landscape they buried, encasing the plants they covered (be they land or swamp plants), and finally solidifying into coals—true energy fuels that were mistakenly called "fossil fuels".

Today, scientific evidence, logic and time weigh heavily in favor of the new Energy Fuels Theory (EFT). It is an idea whose time has come, a concept that is destined to shake the foundation of scientific thought concerning creation of matter. When proven more conclusively, it will cast new perspectives on the gas, oil, and coal industries, greatly altering the world's economic outlook for many generations to come.

Recent literature offers further proof of vast new fuel supplies in Earth's crust. *Scientific American* (June, 1980) published an article entitled *The-Deep-Earth-Gas Hypothesis* by Thomas Gold and Steve Soter, which presented much evidence indicating that earthquakes and volcanoes release gases from deep in the Earth's mantle, and such gases may include methane of a non-biological origin. The *Reader's Digest* (April, 1981) published the article, *Bonanza! America Strikes Gas*, which tells of geologists hitting field after field of natural gas deep within the nation's bedrock.

Biological vs. Non-Biological

Basic hypotheses for the abiogenic origin of gas and/or petroleum were postulated by Aristotle, Berthelot (1866), Mendele'ev (1877, 1902), Studier, *et al.* (1965), Humbolt, Gay-Lussac, etc. Jupiter and Saturn are known to contain methane (theoretically, all planets and large moons should contain methane). The carbonaceous chondrites, a class of meteorites, contain traces of various hydrocarbons, including complex amino acids and the isoprenoids, phytane and pristane (Mueller, 1963). The presence of extraterrestrial hydrocarbons is interpreted by most authorities as evidence of inorganic formation (Studier, Hayatsu, and Anders, 1965). The conclusion that hydrocarbons of abiological origin exist in and on all planetary bodies was predictable by the Energy Fuels Theory (1973).

It is possible to differentiate biogenic from abiogenic methane. Carbon has two isotopes: ^{12}C and ^{13}C. The ^{12}C—^{13}C ratio varies for different compounds, and this ratio is especially large for the different types of methane. Biogenic (biological) methane is enriched in ^{12}C; abiogenic (non-biological) methane contains a higher proportion of ^{13}C (Hoefs, 1980). ^{12}C—^{13}C ratios of methane from the "hot spots" of the Red Sea, Lake Kivu (East Africa), and the East Pacific Rise suggest an abiogenic origin (McDonald, 1983).

Additional background literature shows that Peyve (1956) and Subbottin (1966) have argued that sub-crustal abiogenic petroleum migrates up major faults to be trapped in sedimentary basins or dissipated at Earth's surface. Porfir'ev (1974) cites the flanking faults of the Suez, Rhine, Baikal, and Barguzin grabens as examples of such petroleum feeders. Mueller (1963) cites the presence of hydrocarbons in some meteorites.

This school of thought has been supported by more recent papers (Gold, 1979; Gold and Soter, 1982). Gold argues that earthquake out-gassing along faults allows methane to escape from the mantle. This process gives rise to deep gas reservoirs and, by polymerization, to oil at shallower levels.

All of these arguments strongly support the Energy Fuels Theory of 1973 by Scarborough (1975, 1977, and 1979) as expounded in his book, *Undermining the Energy Crisis*. For many years (1973-1979) his was one small voice against the hysteria of the "energy crisis", trying to convince the establishment that limitless quantities of oil and gas exist worldwide—simply drill deeply enough and often enough. Today's oil glut verifies the faith in the concept that is now attracting a bandwagon of believers.

Molecular Chain Building (MCB); Polymerization

Molecules of gas are found in every lump of coal and every drop of petroleum. Heat and chemical processes release these molecules from the coal and oil to produce "manufactured gas". In reality, such gas is not manufactured; it is merely separated from the other two forms of fuel. Coal always contains both oil and gas, both of which can be separated by relatively simple processes.

Methane is the simplest form of gas, consisting of one carbon atom with four hydrogen atoms attached. As one carbon at a time becomes attached so that chains consisting of 2, 3, 4, 5, 6, 7, 8, etc., carbon atoms are formed, other types of gases and oils are created. The following series illustrates such hydrocarbon gases and condensate liquids that represent the initial phases of MCB from gas into oil.

NAME	NUMBER OF CARBON ATOMS
Gases	
Methane	1
Ethane	2
Propane	3
Butane	4
Liquids	
Pentane	5
Hexane	6
Heptane	7
Octane	8

The principle hydrocarbon comprising natural gas is methane. Butane and propane are found in smaller amounts, along with such chemical impurities as carbon dioxide, nitrogen, hydrogen, sulfide, and sometimes helium. Natural gas is classified according to the chemical impurities it contains.

If methane is considered as a chain-link building block, any configuration and size chain can be constructed by continually adding one or more methane blocks. The three variables necessary to create each product from the basic ingredients are temperature, pressure, and time. Given the proper set of variables, any gas, oil, or coal can be manufactured by man or by nature. However, man usually prefers to reverse the process, starting with coal and producing oil and gas as end products.

It is probable that the presence of salt water is necessary for nature to work such transformation. Briny water is almost always found associated with gases, oils, and coals at various stages of their existence. It is logical to suspect that the chlorine atoms from the salt may act as linkage intermediates for the methane blocks. (The use of chlorine is a common method of block-building used by chemists in their laboratories).

Vast Reserves

The real key to our nation's energy future might well be gas. Unimaginable vast resources of this fuel exist in the Earth's crust. One example of one type of gas is "geo-pressured methane", known to lie in the untapped gas-laden briny waters buried deep beneath the Gulf Coast. At one time, a Baton Rouge hydrogeologist, a leading authority on the subject, calculated the supply at 50,000 trillion cubic feet (tcf) of methane in Louisiana and Texas. The U.S. currently uses about 20 tcf annually, which equates to 2,500 years of known reserves at that rate. One tcf is the energy equivalent of approximately 180 million barrels of oil.

In addition to the Gulf Coast reserves, methane is known to exist in a large basin area some 500 miles in length in California, and in other states, including Oklahoma, Washington, Oregon, and Alaska (250 tcf).

There is no reason to believe that such areas have a monopoly on the world's supplies, either current or future. Similar reserves will be found in many locations throughout the crust of our planet. If properly utilized, such gas resources will not become exhausted before the <u>need</u> for them vanishes. Proper utilization means getting maximum efficient use of all three types of energy from the sources—heat, methane, and the high pressure of the water as it exits—as well as minimizing waste during end use.

The depth (10,000 to 20,000 feet) at which current sources seem most abundant is fairly insignificant, except that this probably represents near minimum depths for such discoveries. Deeper drillings should result in many more producing wells in a wide variety of locations.

Month by month, we seem to become more inundated by vast reserves of known and newly discovered energy fuels. The figures are becoming astronomical: more than 50,000 trillion cubic feet gas in one area alone. And many such areas will eventually be discovered around the world. Almost any reasonable, current guess will prove, eventually, to have been too conservative; for example, currently known reserves are many times larger than was imagined even fifteen years ago. (*The World Book Encyclopedia* of 1965 quoted the maximum estimate at 1700 tcf of natural gas in the nation's reserves.)

Such huge quantities of gas, oil, and coal found around the world, under oceans and mountains, almost anywhere and at any depth, would lead one to suspect that such fuels could not have been created by fossils and plants. The only logical method of creating such quantities (trapped in this manner in the Earth's crust) is through a hot nuclear source from below. Since all matter (oceans, land, air, elements, etc.) was created from nuclear energy, according to natural laws, one could then only conclude that conditions ideal for converting nuclear energy into matter must, of necessity, be present at the site of such conversions.

More Evidence of Earth's Nuclear Energy Core

Evidence that such a conclusion is accurate is found everywhere. Volcanoes, earthquakes, the rise and fall of great land masses, widening cracks in the earth's crust, the separation of continents, extrusion of material from the underwater seams of mountains in the Pacific and Atlantic Oceans, the disappearance of beaches and the build-up of others, the warmth of the Gulf Stream, geysers, springs, and mountain streams are some of the more noticeable items of evidence. The raised domes that geologists look for as initial evidence of oil reserves below ground, and the tremendous heat and pressure that usually accompany such oil when a drilling blow-out occurs, add strong evidence to the conclusion.

Strong arguments for creation of virgin materials that form Earth's crust include the "black smokers" discovered in 1979 by a team of American, French, and Mexican investigators. They observed turbulent black clouds of fluid billowing up from chimney-like vents, much like factory smokestacks. The venting fluid, metal-rich hydrothermal solution was measured at 350°C. Mixing with the ambient seawater causes copper-iron-zinc sulfides to precipitate as fine black particles suspended in the plumes.

The chimney-like vents rise more than 30 feet above mounds of similar height and composition. The grade of the metals is comparable to that of many ancient massive sulfides on land. A typical analysis shows 31 percent zinc, 14 percent iron, 1 percent copper, and small amounts of silver and gold. Nearly all the metals discharged in this manner are dispersed by oceanic currents.

In a 2,000-square-kilometer area on the island of Cyprus some 90 large deposits of copper-iron-zinc sulfides occur as saucer-shaped bodies up to hundreds of meters in diameter. They fill depressions in volcanic lavas that erupted on the sea floor some 85 million years ago. Still farther back in geological time, similar hydrothermal convection systems were active some 2.7 billion years ago in rocks of the Archean period, now exposed in the eastern Canadian shield.

Minerals and lavas, water and land, air, gases, oils constantly pouring from an internal source over the eons! Such observations, and logic, led to the Theory of Internal Formation of Fuels and Elements (TIFFE) in 1973, and served as the basic foundation for other theories expounded in this book. The next chapter explores the TIFFE in more detail.

Fuel for All

In the future, the question may not concern where to drill for oil, so much as how deeply to drill. Many companies, including General Motors, American Standard, and Coors, have successfully drilled their private supplies of these fuels, further substantiating this energy fuels concept as first examined in a pamphlet entitled *Fuels: A New Theory* (1975). Colleges and municipalities are also warming to the idea. In the future almost every large city will have its own gas well. This trend is destined to continue until anyone who can afford to drill deeply enough for fuels (gas primarily) will do so with an excellent chance of success.

Many exciting reports are coming from the pages of leading publications. *Time* magazine (December 22, 1980), in its article, *Backyard Fuel*, opened many eyes and gave even more validity to the truth about abundant fuels in Earth's crust. It tells of nearly 200 companies that have entered the do-it-yourself scramble for fuels beneath their own yards, and it describes how even the nuns of Mount Saint Benedict's Priory in Erie, Pennsylvania, successfully drilled for gas on their one hundred acres. The drilling took four days and cost $105,000.

An article in the *Wall Street Journal* gave another interesting account of a successful drilling, this time on the campus of Wells College in Aurora, New York, a 500-student liberal arts school for women. Natural gas was found at a depth of 2,600 feet and at a pressure of about 600 pounds in the well.

In September, 1981, government scientists began analyzing metal-bearing chunks spewed out of an undersea volcano 270 miles west of Oregon. H.E. Clifton, Chief of the U.S. Geological Survey's Pacific-Arctic Branch of Marine Geology, reported that "new earth crust is actually being made" by the volcano.

The evidence of huge quantities of natural gases, fuels, metals, water, and other matter begins to mount overwhelmingly in favor of the new theory that these fuels (and other matter) were created within Planet Earth. As the facts accumulate in the literature, the sound logic of this new theory of evolution of gas, oil, and coal emerges more and more clearly. The Fossil Fuels Theory as formulated by Dr. Logan in the 1830s is being revealed as a fallacy. It may well prove to be among the greatest fairy-tale conclusions of all time—one that altered the course of both economics and science, and contributed to an international trauma over energy supply shortages.

But as the new Energy Fuels Theory becomes better recognized by scientists and the general public, trauma will subside. Energy prices will stabilize, and perhaps even retreat to lower levels. It would not be unreasonable to expect future gasoline prices of less than $1.00 per gallon—if reasonable political controls are exercised and taxing does not become excessive.

Most scientists recognize that matter was, and is indeed, transformed from energy particles. But it may be some time before they agree that Earth created its own atmosphere and crust with atoms made within its own interior nuclear furnace (much in the manner that Jupiter is revealing to scientists today). And in forming that crust, Earth created (and probably still is creating) its own energy fuels—beginning with hydrogen and methane, then higher gases that formed oils, some of which inundated lowlands, then cross-linked and solidified into coals. Since these events occurred worldwide, it is reasonable to expect that such fuels can be found almost anywhere in the world if one drills deeply enough.

The possibilities of finding future energy fuels may be almost unlimited. Indeed, man may become extinct before energy fuels are exhausted.

New Vistas Opened—Infinite Resources (?) and the Ramifications

In view of the new concept of creation of coal and its predecessors, oil and gas, whole new energy avenues have been opened to mankind. The probability that such energy fuels might not have the limitations of "finite fossil fuels" has worldwide ramifications. When proven more conclusively, the EFT will cast new perspectives on the fuel industries while greatly altering the world's economic and political outlooks for many centuries to come.

Our supply of energy fuels may be nearly limitless. This notion is a certain deathblow to OPEC, and to attendant high fuel prices. In the near future, many more private gas wells will be drilled by corporations, universities and municipalities across the country. In the long term, the most popular sources of energy should be natural gas and nuclear fusion. Coal and oil, both less plentiful and more slowly renewable than gas, should not be burned, but utilized for more sophisticated ends: the manufacture of chemicals, or the insurance of transportation. Certainly the fear of running out of fuel would be forgotten. The notion of "fossil" fuels is itself a fossil of antiquated thinking, the relic of an outdated concept.

Unga Island: A Petrified Forest

On Unga Island, near the tip of the Alaska Peninsula, a forest of petrified tree stumps has been uncovered by wind and water wearing away their hiding place: a 50-foot bluff. The petrified forest, much of it below tide line, covers about five miles of beach. The giant trees are believed to be either California-type sequoia or metasequoai, now found mostly in China. Diameters of the trees range between two and nine feet. Geologists believe they appeared about 25 million years ago, when Earth was much warmer.

This discovery is powerful evidence of the ongoing evolution of Planet Earth: a basic principle of the LB/FLINE model of origins and evolution of planetary systems. The trees did indeed grow in warmer and wetter climate during an earlier period of Earth's fourth (rocky) stage of evolution, eventually to be covered by water, as the planet's nuclear energy core produced (and is producing) water about twice as fast as it created (and creates) its land surface.

None of these events would be possible without ongoing evolution driven by an energy source: Earth's nuclear core. One cannot exist without the other.

In the short article, *Spending A Legacy (U.S. News & World Report,* Nov 10, 2003, p69), Tim Appenzeller writes: "Today's oil fields formed as microscopic plants living in ancient seas died and sank, then were buried deeply enough for heat and pressure to cook the organic matter into oil." Ecologist Jeffrey Dukes had "used published figures to calculate that every day, humanity burns up oil equivalent to nearly a year's worth of plant growth across the entire planet." Dukes' calculation was based on a conversion rate of almost 100 tons of ancient plants to make one gallon of gasoline.

This is another example of the sophistry forced upon writers who are fed ludicrous information about the origin and evolution of hydrocarbon fuels, (gas, oil, coal), erroneously known as fossil fuels. This myth was initiated in the 1830s by William Logan's false conclusion that coal was made from plants, and was enhanced in the 1920s by J. B. S. Haldane's hypothesis that petroleum was made from marine organisms.

From 1973 to 1980, the fallacious nature of these claims clearly was exposed by the author's in-depth research and subsequent publications, beginning with *Fuels: A New Theory* (1973 & 1975), later followed by *Undermining the Energy Crisis* (1977, 1979). Continual research finally led to discovery of the Fourth Law of Planetary Motion (1980-1995), then to the definitive LB/FLINE model of planetary origins and evolution (1996).

At this writing (July 2004), the fully substantiated results of 31 years of research and forced self-publications are still being suppressed, while the Big Bang/Accretion myth and its antiquated fossil fuels hypothesis continue to prevail in the scientific literature and in the minds of the public, and are still being taught in educational institutions. Obviously and unfortunately, this does not speak well of science.

Asphalt deposits cover parts of Gulf of Mexico (Science News, May 15, 2004)

Like the LaBrea Tar Pits in California, such extrusions of asphalt are powerful examples of how Earth's internal conditions dictate the type and quantity of hydrocarbon (and all) material formed (and continuously form) in any given time-frame by nucleosynthesis and subsequent polymerization and cross-linking; e.g., from methane to petroleum to coal. Coal always contains some asphalt and a variety of other chemicals, and always forms on, or within one mile of Earth's surface.

Other related products created in like manner within Earth's nuclear core, magma, and crust include the gases, methane, ethane, propane, and butane. Pentane, a lightweight oil is always found in lightweight petroleum deposits. These hydrocarbon fuels are intimately related, and are generally arranged in ascending order from Earth's core to its surface as they evolve from energy to atoms to gases to petroleum to coal. A description of these processes is detailed in the author's previous writings, beginning with the 1975 edition, *Fuels: A New Theory.*

The Universal Law of Creation of Matter

Nature's procedure for making its huge stores of energy fuels can be expressed as follows:

ENERGY to ATOMS to MOLECULES to GAS to OIL to COAL

These processes of Nature are completely reversible by mankind. Beginning with coal, scientists can extract its original petroleum, reduce it to gas molecules, which can be separated into atoms. And as Einstein and the atomic bomb illustrated, atoms can be forced to release the nuclear energy from which they were created.

By expanding this formula to include all matter made from energy, we can derive a simple Universal Law of Creation of Matter (ULCM):

ENERGY to ATOMS to MOLECULES to GAS to LIQUID to SOLID

Planetary cores supply the nuclear energy under various conditions of extreme temperatures and pressures for forging atomic elements, the building blocks of matter. These atoms eventually combine and evolve into countless molecular configurations of Nature's handiworks. To cite a basic example other than methane, the waters of Earth were made by Internal Nucleosynthesis of the hydrogen and oxygen atoms that subsequently combined to form the countless molecules of water comprising Earth's vast oceans and other water systems.

Internal conditions vary from planet to planet, primarily as a function of core size. The large gaseous spheres composed of lighter elements are products of huge, medium-density, open-to-space cores, while rocky planets containing both lighter and heavier elements are products of smaller, higher-density, encapsulated cores. Additionally, distances of planets from the Sun play a role in shaping their surface characteristics; e.g., frozen gases versus liquid surfaces. Thus, at any given time, planets will differ in composition and outward appearances as functions of size and distance from the Sun.

Connections: Gaia, Natural Selection and the LB/FLINE Model

A review of James Lovelock's autobiography, *Homage to Gaia (Science* March 2001) gives some valuable insight into his Gaia concept, along with typical reactions of his peers to the revolutionary theory. To quote the reviewer: "Emerging from Lovelock's interaction with microbiologist, Lynn Margulis, the Gaia hypothesis embraces the notion that Earth's living and nonliving components constitute a set of interactive feedback processes that reflect whole-system scale emergent properties—phenomena not likely to be revealed by disciplinary study of Earth's subsystems alone. Lovelock and Margulis suggested that through these interactions the biota made the physical environment more fit for life, a clear departure from earlier scientific ideas... Gaia is a 'large theory', one further inflated by the claims of some of its advocates that it supplants Darwin's theory of natural selection."

"At the first American Geophysical Union Chapman Conference on Gaia (1988), a controversial meeting because it addressed the topic as a serious science [which it is!!], geologist and philosopher, James Kirchner, quoted from various Lovelock and Margulis works over 15 years to demonstrate inconsistent implicit definitions of Gaia." Although Lovelock handled the situation with grace, he has since let his feelings be known, and in *Homage* he correctly calls Kirchner's arguments "sophistry".

The completion in 1996 of the FLINE model initiated in 1973 brought the works of Lovelock and Margulis full circle. This revolutionary concept brings definitive understanding to the connections between Gaia's teachings and Darwin's theory of natural selection. Both have their taproots deeply embedded in the principles of the FLINE paradigm in which the origins and evolution of planetary systems, including its ever-changing environment, are driven by Earth's energy core. Species are spawned continually, and when they eventually cannot adapt to the ongoing environmental changes they become extinct.

While the FLINE model's basic principles are supported by incontrovertible evidence that makes them easy to comprehend, they have yet to win the battle to view Earth as a self-sustaining entity—one that, since its initial orbit around our Sun, has never depended on objects from outer space for the ongoing evolution of itself and its multitudes of systems. Earth's living and non-living components do indeed constitute a set of interactive feedback processes that reflect whole-system scale emergent properties—phenomena clearly revealed in the FLINE model by disciplinary studies of planetary origins and Earth's sub-systems.

The connections between Lovelock's Gaia, Darwin's natural selection, and the encompassing LB/FLINE model bring a clearer understanding of John Muir's famous quote: "When we try to pick out anything by itself, we find it is hitched to everything else in the Universe." Muir's great foresight is fully vindicated by the irrefutable connections between these three incontrovertible concepts that bring clear understanding to how and why Earth's systems (and all other planetary systems) function interactively during evolution from energy mass to inactive sphere.

From: George Brawner, Rte.2, Box 234-K, West Point, GA 31833 March 20, 1977

Thank you so much for coming to see me last Friday, the 18th. I enjoyed talking to you and regret that we didn't have more time. My wife has expressed regret too that she was not up to meeting you, but her condition turned out to be a virus which became worse in the night, but is gone today.

I want to thank you for giving me one of your books. I read "Undermining the Energy Crisis" that night, and can only say that it is a profound disclosure of logical explanations which have been right under mankind's feet all this time!

You are way ahead of my own theoretical evolution in thought.

On your next edition, I would suggest a hard cover, with your present cover as the "wrap" in glossy paper, and expounding on the more complicated subjects and portions, for the layman, to help fill out the contents and justify the hard cover, and increased price. It should sell for at least $4.85 or $6.75 or $7.00 due to the nature and value of its contents. Even if it only costs $2.00, you are selling something as startling as the Atomic Bomb for $1.20 and $2.00. Not right nor doing justice to your own achievement.

Howsoever, you did excellent, for your first book, and beyond measure for your theories. I am glad for your achievement, proud of being associated with you in the past, and a little envious, but inspired to try and get my own on the market as soon as I can. My own is like a science-fiction novel, but true, and will take many pages to tell.

I think you should be nominated for the Nobel Prize and I will do my best to achieve that for you.

Best regards to your family.

A Secure Conclusion

This condensed book of the FLINE model presents a fundamentally sound, factual argument for a revolutionary version of the origin of our Solar System and the evolution of its planets and their planetary systems by means of natural laws. As with a giant jigsaw puzzle in which all pieces interlock precisely in place, it presents a beautiful and complete picture that offers a new perspective in which the enigmatic anomalies of the SS can be understood, and all relevant discoveries of the space probes interpreted more logically.

In the new perspective, scientists will be better able to understand the true cause of planetary quakes (e.g., earthquakes): explosions that cause land movements on those planets and moons during their active stages.

Subsequently, the source and cause of lightning will be understood. Recently, scientists have recognized that ground-to-cloud lightning is a reality—actually the norm, rather than a figment of the author's imagination. We can look to Earth's core for the supplies of electrons that make lightning possible. Inevitably, the understanding of all phenomena of Earth and the SS will follow.

But, the path will not be easy. Obviously, the strong entrenchment of the prevailing Accretion Disk theory (the dust aggregation/planetesimals/accretion hypothesis) in the scientific literature presents one of history's most formidable barriers to change in direction of scientific thought. However, history teaches that more satisfactory concepts eventually do displace unsatisfactory concepts.

Inevitably, time will correctly judge the two viewpoints.

Historical Letters, Memos and New Releases

To: Gordon Rehberg, Dalton, Ga. **October 23, 2000**

The article, *The World Has More Oil, Not Less* by Alan Caruba, promotes some truly exciting news. He is doing the world a great service by bringing this critical information out into the open. He and you have accomplished far more in a single year of publicizing the facts about our world's vast petroleum reserves than I have been able to do in the 25 years since publishing *Fuels: A New Theory* (1975). It is very exciting that so many people are beginning to learn the true extent and nature of these hydrocarbon fuels since that first publication explaining the reasons: their abiogenic origins and intimate relationships. Even after a quarter century, most scientists still cling to the antiquated fossil fuels concept. Can you believe that a breakthrough of such tremendous import could be stymied for 25 years—and credited to the wrong person? What has happened to our news media?

Thomas Gold's concept, erroneously predicated on the Big Bang/Accretion concept, was published some five years later than my internal nucleosynthesis-polymerization theory of abiogenic origins of hydrocarbon fuels. Even as late as 1985, when Gold presented his theory in the AAAS Meeting in Los Angeles, he believed, and so wrote, that petroleum came from outer space. At that time, he did not understand the crucial, inseparable connection between Internal Nucleosynthesis of atoms and their subsequent linkages into all the molecules comprising these fuels and all other matter of Planet Earth. Even though my concept was at least ten years ahead, I was not permitted to be on the same program with him. Gold received big coverage from the news media, and thus the abiogenic concept erroneously became known as Gold's theory. As the record shows, Gold's beliefs in the BB and that petroleum came from outer space proved his undoing when the drilling in Sweden came up dry, causing the abiogenic concept to suffer a severe setback.

Some credibility was regained via the accurate predictions made in my books, *Undermining the Energy Crisis* (1977 & 1979), which foretold vast fuel reserves and why they are continuously being replenished from below via internal nucleosynthesis/polymerization, consequently bringing low energy prices in the $1.00 per gallon range in the 1980s. In spite of its definitive nature and substantial irrefutable evidence, the IN-E concept has yet to be acknowledged by advocates of the fossil fuels hypothesis. In 1995, this revolutionary concept came full circle via the discovery of the elusive solution (1980-1995) to the Fourth Law of Planetary Motion explaining how the planets attained their orbital spacing around the Sun. Together with Kepler's Three Laws, the Four Laws of Planetary Motion (FL) gave closure to the FL-IN-E model initiated by the Energy Fuels concept in 1973-1975. I was happy to hear that after reading my book, *The Spacing of Planets: The Solution to a 400-Year Mystery,* chemical engineers at DuPont proclaimed it the most logical explanation they had ever read. It does seem only a question of time before the new concept inevitably displaces the antiquated fossil fuels hypothesis.

Minus this valuable underpinning, other misconceptions, including the Big Bang, cannot long survive. We are on the verge of a number of exciting breakthroughs to a genuine understanding of origins and evolution. For the sake of our future, you, Caruba and I should continue promoting this irrefutable concept until the full truth becomes more firmly established.

Memo: Resources and Ramifications: A perspective of the 1980s

Most scientists recognize that matter was, and is indeed, transformed from energy particles. But it may be some time before they agree that Earth created its own atmosphere and crust with atoms made in its own internal nuclear furnace. And in forming crust, Earth created (and is still creating) its own energy fuels, beginning with atoms of carbon and hydrogen that combine to create methane, then polymerize into higher gases and petroleum, some of which inundated lowlands, then cross-linked and solidified into coals.

Since these events occurred worldwide, it is reasonable to expect that such fuels can be found almost anywhere in the world if one drills deeply enough in the right places. And since these reserves of energy appear plentiful, and nuclear fusion sources are on the horizon, it is quite possible that mankind will become extinct long before energy fuels are exhausted.

In view of the new concept of creation of hydrocarbon fuels, whole new vistas have been opened to mankind. The probability that these energy fuels might not have the limitations of finite "fossil fuels" has worldwide ramifications. When more widely recognized, the EFT will bring new perspectives to fuel suppliers and to energy-dependent industries, while greatly altering the world's economic and political outlooks for many centuries.

The notion that our supply of energy fuels may be nearly limitless is a certain deathblow to the stranglehold exercised by OPEC in the 1970s and to unreasonably high fuel prices during that unwarranted energy crisis.

In the long-term view, the most popular sources of energy should be natural gas and nuclear fission or fusion. Coal and petroleum, both less plentiful and more slowly renewable than gas, should not be burned, but should be utilized for more sophisticated purposes: the manufacture of chemicals and the assurance of transportation needs.

Certainly, the fear of running out of fuels should never again be a factor in precipitating an energy crisis. The notion of "fossil fuels" is indeed a fossil of antiquated thinking, the relic of the outdated concept of the 1830s.

Why Earth May Never Run Out of Oil

To: APG-Executive Committee, c/o Editor, AAPG Explorer. **August 25, 1998**

The purpose of this letter is to set the record straight on the concept of abiogenic origin of hydrocarbon fuels and to update you on the latest developments pertaining to this fascinating subject.

The story begins in 1935 with my boyhood realization that a lump of coal is simply solidified petroleum containing pockets of blue-flaming gas, and that dying plants on the farm begin turning into dust at the end of their life cycle—making live encapsulation absolutely essential for live imprints to be found in coal (Logan, 1830s). This inspired a lifetime of curiosity to learn everything possible about how all things came into being. After sufficiently researched evidence finally developed into a viable alternative to the fossil fuels hypothesis and removed all doubt from my mind, I was forced to publish *FUELS: A New Theory* (1975) and *Undermining the Energy Crisis* (1977 & 1979) at my own expense during the energy crisis of the 1970s. In these publications, I accurately predicted the vast stores of hydrocarbon fuels (especially methane) in Earth's crust (and the reasons) and the current low price range of $1.00 per gallon of gasoline when experts were predicting $5.00 per gallon and the depletion of petroleum reserves by 1990.

Unfortunately, some facets of my Energy Fuels Theory (EFT) became known as Gold's theory, which--after Sweden's drilling episode—became unacceptable to the scientific community because Gold's belated efforts failed to present an accurate and total picture of the EFT. Long before the Sweden episode, Gold had refused my offers to work together on the full story to convince the science community of its scientific validity. To compound the problem, my presentations at science meetings thereafter were neither permitted beyond Poster Sessions nor permitted news releases through AAAS and AGU.

Cogent evidence of the abiogenic origin of hydrocarbon fuels and their evolutionary stages from energy to atoms to gas to petroleum to coal, in full compliance with all natural laws, led to the realization that planetary evolution (E) is possible only via an internal source of energy (a.k.a. Internal Nucleosynthesis) (IN) to drive it forward. In any active sphere, IN and E are inseparable; one cannot exist without the other. From this understanding came another realization—corroborated by ever more powerful evidence—that planetary evolution occurs in five stages common to all planets. To observe the five stages, we have only to look, with insight, to the Sun and planets of our Solar System (SS).

Through the years, these realizations became entrenched in much corroborative evidence. But if planets began as nebulous energy masses, how did they attain their orbital spacing?—a mystery that first eluded Kepler in 1595. Inevitably, curiosity led to the mathematical solution (1980-1995) to the proposed Fourth Law of Planetary Motion detailing how the planets attained their original geometrically-spaced orbits around the Sun and their current, somewhat altered, positions in the SS. Together with Kepler's First Three Laws of Planetary Motion, the Four Laws (FL) bring the FL/IN/E paradigm full circle.

Outer space discoveries continually corroborate the FLINE paradigm; all new evidence fits precisely into the new concept. Examples: the findings on Mars and the data from exo-planets are readily explained in its perspective. If a fundamental flaw does exist in the paradigm, it remains elusive. Yet, since its completion in 1995, it has not been able to get past peer-review advocates of prevailing beliefs, even for Poster Sessions and news releases.

I am confident that the concept's supportive evidence is incontrovertible and that the new paradigm will continue to stand on its merits under the closest scrutiny. In comparisons with prevailing beliefs, the paradigm is clearly superior at explaining planetary origins and evolution—including the abiogenic hydrocarbon fuels--and the role of the inner core in Earth's processes, the generation of the geo-magnetic field and the thermal evolution of Earth (and other planets, moons, etc.). It is a factual, mathematically-proven, definitive and viable alternative to current speculative beliefs about the origins and evolution of planetary systems. Another abstract (herewith) on the subject has been submitted for the AGU Fall Meeting in December under the title of *The FLINE Paradigm: Critical Insights into the Origins of Solar Systems and the Orbital Spacing and Subsequent Evolution of Planets.* After three years of rejections, I suspect that my skepticism of the objectivity of reviewers will again be confirmed.

But since this definitive paradigm is crucial to understanding the origin and evolution of hydrocarbon fuels (and all planetary matter) via Earth's internal processes, can you suggest how we might work together, in the best interest of science, to have these findings brought into open discussions among AAPG members? Its potential benefit to science appears unlimited.

To: Glenn Strait, Science Editor, The World & I, Washington, D.C. 03-21-99

Being familiar with my work of the past 26 years, a local geologist, Grady Traylor, sent me a copy of the exciting article, *It's No Crude Joke: This Oil Field Grows Even as It's Tapped (Wall Street Journal, WSJ,* April 16, 1999) by Staff Reporter, Christopher Cooper. The results described therein add more absolute proof of the *Energy Fuels Theory* (EFT) that has been the backbone of my copyrighted writings since 1975, the year of the initial publication entitled *Fuels: A New Theory.* First named the TIFFE (*Theory of Internal Formation of Fuels and Elements)* in my book, *Undermining the Energy Crisis* (1977, 1979), it was re-titled *The Energy Fuels Theory* in 1980, and led inevitably to the concept of the five stages of evolution of planets, which, in turn, led to my pending *Fourth Law of Planetary Motion* (1980-1995) explaining how the planets attained their orbital spacing around the Sun.

Together, the Four Laws of Planetary Motion (FL), Internal Nucleosynthesis (IN) and Evolution (E) form the FLINE paradigm (1973-1996) of the origins of solar systems and the evolution of planets. Thus this new concept, 23 years in the making, consists of three chronological, inseparable and ongoing realities of Nature. The *WSJ* article is the first recognition by scientists of the initial phase of this revolutionary concept that openly predicted and fully explains its findings. The rest of the story is even more exciting.

Cooper's article is timely in that it comes on the heels of the discovery of a multiple-planets solar system in the Upsilon Andromedae system (AAAS news release, April 15, 1999). The findings in both articles add powerful corroborative evidence to the FLINE paradigm—which predicted and explains these discoveries. Although many facets of this revolutionary concept have been presented in a number of national science meetings (including the AAAS in Anaheim on January 23, 1999) and submitted in news releases, the FLINE paradigm remains relatively unknown within the scientific community. Now beyond the shadow of a doubt, it is the crucial key *to understanding the origins and evolution of planets.*

As with many historical recordings, the facts in Cooper's article need some revision to set the record straight. Some five years after my initial publication (*Fuels: A New Theory*), I read *The Deep-Earth Gas Hypothesis* by

Thomas Gold (*Scientific American*, 1980). I wrote and offered to work with him on the subject; he refused. In the 1985 AAAS Meeting in Los Angeles, Gold and I clashed on different versions of the abiological origins of these hydrocarbon fuels. The record will show that Gold believed at the time that Earth's petroleum was brought from outer space in meteorites, as opposed to my nucleosynthesis version (1975) of its internal creation. He did not delve into the origin of coal, the final product of evolution of hydrocarbon fuels. Because he did not understand the full IN concept, Gold failed in future efforts to find oil in Sweden; consequently, the concept lost credibility—until now. Cooper's article restores this credibility—and that is why it's historical and exciting. I believed that I would not live long enough to see this happen. In spite of my ceaseless efforts and Gold's publicized failures, the *WSJ* and others understandably and erroneously credit Gold with the concept of abiogenic fuels.

In view of these corroborating discoveries, the FLINE paradigm becomes even more irrefutable. Now more than ever, it appears destined to displace prevailing dogma about origins of solar systems and evolution of planets and fuels. Eventually, it should rank high among the most monumentally significant discoveries of all time. And it guarantees us a vast supply of hydrocarbon fuels for a very long time.

To: Dr. John Chappell, Jr., San Luis Obispo, CA 93406. April 22, 1999

The enclosed article by Christopher Cooper appeared in the *WSJ* two days after presentation of my paper on the FLINE paradigm at the NPA-SWARM Meeting in Santa Fe. If published a few days earlier, it would have been used as additional corroborative evidence vindicating this revolutionary concept that began with the EFT in 1973 and came full circle with the pending Fourth Law of Planetary Motion in 1995. For sure, it strengthens my faith in every aspect of the FLINE paradigm. I consider its supporting evidence incontrovertible. But this was always the case, since my research during the past quarter century concentrated on the separation of the facts and myths of science.

The paradigm both predicted and now explains the anomalies of Eugene Island 330's "oil field that grows even as it's tapped." The sub-headline states: "odd reservoir off Louisiana prods petroleum experts to seek a deeper meaning." To get the final answers spelled out in the paradigm, they will indeed have to go all the way into the energy core to understand the origins and evolution of planets and all planetary matter, including the hydrocarbon fuels. Their eventual understanding of Internal Nucleosynthesis finally will dispel the erroneous belief in the Big Bang concept in which all elements were created in the very beginning—an added bonus that will please Bill Mitchell.

In all probability, the next breakthrough can come only if scientists are permitted to learn about the FLINE paradigm. My only hope to reach them is through the next AAAS Meeting in 2000 and a subsequent news release that is permitted to be picked up by the news media. Otherwise, I know of no editor willing to stick out his/her neck to risk that challenge to prevailing beliefs. Do you think that Stephen Gould could be influenced to permit the proposed symposium to be on the AAAS program in Washington in 2000?

I'm enclosing some additional information that is self-explanatory.

To: Dr. Floyd E. Bloom, Editor-in-Chief, Science Magazine. April 28, 1999

In response to your challenging editorial *Think Ahead (Science,* April 1999), I submit herewith the information concerning a revolutionary concept that predictably will guide scientists to total knowledge of the origin, orbital spacing and evolution of planets. I predict that by 2050, every planetary anomaly will have been solved, conclusively proven and taught as standard textbook material—all within the realm of the FLINE paradigm.

This revolutionary concept began in 1973 with substantive evidence against the Fossil Fuels Theory—the same evidence that pointed precisely to a more logical theory strongly supported by natural laws of physics and chemistry. First published in 1975 under the title, *Fuels: A New Theory,* it was presented in Poster Sessions at the AAAS Meeting in Washington, DC in 1982 and again at the AAAS Meeting in Detroit in 1983. Its key feature was, and remains, the evolution of hydrocarbon fuels via Internal Nucleosynthesis and polymerization: from energy to atoms to molecules of methane that, via polymerization, evolve into higher gases, light oil, crude petroleum and finally into coal. The rapidly mounting corroborative evidence was presented at various science meetings during the next decade, but to no avail. In 1985, Thomas Gold was permitted a full-day symposium on the subject in the AAAS Meeting in Los Angeles. My offer to work with him to get the full message across in a convincing manner was refused. As you know, the consequences were devastating; the concept lost credibility.

The *Wall Street Journal* article, *It's No Crude Joke: This Oil Field Grows Even as It's Tapped* (*WSJ*, 16 April 1999), is the most recent corroborating evidence that should serve as the final proof of the Energy Fuels Theory of 1973 that both predicted and explained this anomaly.

The mounting evidence inevitably led to the realization that all planets are self-sustaining entities that create their own compositional matter via Internal Nucleosynthesis, thus accounting for their five common stages of evolution. Planetary evolution would not be possible without an internal source of energy to drive it forward. Simply put: no energy source, no evolution. The corroborating evidence continues to mount. Every anomaly of planetary systems is traceable back to its source of internal energy.

This awareness inevitably led to the proposed Fourth Law of Planetary Motion (1980-1995) explaining how the planets attained their orbital spacing around the Sun. Now complete, the FLINE paradigm is comprised basically of three chronological, inseparable and ongoing realities: the Four Laws of Planetary Motion (FL), Internal Nucleosynthesis (IN), and Evolution (E). Every anomaly of solar systems can be understood within the realm of this revolutionary concept. Yet, for inexplicable reasons and to the detriment of science, it remains relatively unknown within the greater scientific community. But by 2050 this scenario of the unknown will be, of necessity, a thing of the past.

To: Thomas McMullen, Georgia Southern University, Statesboro, GA. 05-15-99

Thanks for the interest expressed in your letter of May 12 concerning the background of my Energy Fuels Theory (EFT) of 1973. I'm grateful for this opportunity to set the record straight. It is a long story, but I will be as brief as feasible.

The initial sparks can be traced to 1935 when my seventh-grade teacher introduced us to the fossil fuels concept that gas, oil and coal came from plants and animals. This contradicted my observations that plants and animals on the farm quickly turned to dust after dying. And when we burned coal in fireplaces in those tough times, the small blue jets of flames were recognized as emanating from trapped pockets of gas. The rainbow colors found inside lumps of coal seemed symptomatic of oil. And the tarry matter therein was chewed to whiten the teeth. That was the year I decided to concentrate on science, specifically chemistry, to learn the truth about these fuels and eventually, of how everything came to be as it is. My curiosity would not be denied.

Immediately after graduation from UGA (BS in Chemistry), I was drafted into the army, and subsequently married, earned another degree (Chemical Engineering at Georgia Tech) and went through the usual struggles to earn a living while raising three children. In March of 1973, the emerging energy crisis brought my thoughts sharply back into focus on the true nature of the hydrocarbon fuels erroneously known as fossil fuels. All the pertinent knowledge and answers that had been accumulating in my subconscious since 1935 suddenly came together in a flash: these fuels were intimately related; coal was solidified petroleum, which previously had polymerized from basic methane gas formed deep within Earth.

These thoughts, backed by more research findings, were first published in pamphlet form in 1975 under the title, *Fuels: A New Theory,* followed by an updated version later in the year. In 1977, the concept was expanded to the 40-page booklet, *Undermining the Energy Crisis.*

The next expansion was a 140-page hardback book in 1979 under the same title, *Undermining the Energy Crisis.* Since I was unable to obtain any financial aid, or interest any editors, all were copyrighted, published, advertised and sold at personal expense throughout the USA, primarily to libraries. A 1979 copy was given to Ronald Reagan's presidential campaign manager during their swing through Louisiana that same year. After Reagan's election in 1980, his first edict to the oil companies was, "Produce, produce, produce." They did, and the energy crisis soon abated. The experts' predictions of running out of fuels by 1990, based on finite resources of the fossil fuels hypothesis, proved erroneous (as predicted), while all other predictions of the EFT have proven true—yet the EFT has gained little scientific credibility.

In June, 1980, *Scientific American* published an article, *The Deep-Earth Gas Hypothesis,* by Thomas Gold and S. Soter. While I suspected that they had gotten the idea from one of my books sold in the Ithaca, NY area, I had no proof of it. I wrote Gold about my publications in this field, and offered to work together to get the full concept of the EFT into the scientific literature. He did not respond. My first presentation of the completed fuels concept to an international audience of scientists occurred in 1982 in a Poster Session at the AAAS Meeting in Washington, D.C. Backed by my employer, Ralph Howard, President, Kleen-Tex International here in LaGrange, we distributed pamphlets entitled *New Concepts of Origins,* which contained a brief synopsis of each of its three compositional subjects: *Birth of the Solar System, Evolution of Planets,* and *Evolution of Gas, Oil, and Coal* (the EFT). This was repeated the next year at the AAAS Meeting in Detroit. In 1985-1986, Ralph sponsored a research project at Georgia Tech confirming the EFT's three-layer system of hydrocarbon fuels in Earth's crust. As the corroborating evidence continually mounted, facets of the concept were presented to various organizations of scientists, military and civic groups during the next 13 years. You mentioned the last meeting in Anaheim in your letter. As the Chinese would say, "Very long time, very little headway."

In 1985, Gold had a full-day symposium on the subject of origins of fuels at the AAAS Meeting in L.A. Again, I offered to work with him; again he refused. While he made the headlines, the concept was set back more than 14 years when the oil well Gold convinced Sweden to drill through granite came up dry. But the abiogenic concept became known erroneously as Gold's theory. In reality, the only point on which he and I agreed was the polymerization of gas into petroleum. For a reason I did not comprehend, Gold believed that petroleum came to Earth on meteorites, one of which cracked the granite drilling site in Sweden. This was the reason for drilling in that specific spot. To my knowledge, he never did delve into an explanation for coal or into the deeper subject of internal creation of the gases. He did not appear willing to grasp the full significance of the EFT, and remained aloof from it while clinging to facets of the prevailing beliefs—perhaps to minimize possible adverse reactions from other scientists, or possibly to avoid direct conflict with my earlier copyrighted material.

My hydrocarbon fuels concept in the 1970s was first called the TIFFE (Theory of Internal Formation of Fuels and Elements). My 1986 and 1991 publications called it the ITEM (Internal Transformation of Energy to Matter) theory. In the 1996 book, *The Spacing of Planets,* it was changed to FLPM/IN (Four Laws of Planetary Motion/ Internal Nucleosynthesis). This change came about to include the manner in which the nebulous planetary masses of Sun-like energy were spaced in orbits before beginning their five-stage evolution into the planets we see today. This was made possible by my geometric solution to the proposed Fourth Law of Planetary Motion (1980-1995), which brought the original 1973 concept full circle. Since then, it has evolved into the FLINE model of planetary origins that includes the five-stage Evolution (E) of planets, which is the inevitable result of Internal Nucleosynthesis (IN). The two (IN & E) are inseparable and ongoing, along with the Four Laws of

Planetary Motion (FL). Although the FLINE model was detailed fully in the 1996 book, this full acronym only came along soon after that publication.

Meanwhile, efforts to obtain a grant and to get the concept, or facets thereof, into scientific publications during the past 24 years always proved futile. But many of my abstracts did make it into published science programs, which gave some assurance that these crucial discoveries would be preserved for posterity. My primary concern was to not let them die on the vine before someone in an influential position would be willing to examine their substantial corroborating evidence. The main problem, as I saw it, was that my work was being rejected by peer-review advocates of prevailing beliefs that would be seriously challenged by these new ideas. Who could blame them? But through the years, the reviewers were never able to offer a valid rebuttal or a valid reason for the rejections.

The solution to the proposed Fourth Law of Planetary Motion was submitted to The Royal Astronomical Society in London in mid-January of 1998. Their only response has been the acknowledgment of its receipt and the assignment of the ID number qy016. The *AAAS Journal* quickly rejected it in 1998, again without a valid reason. However, the manuscript, *The FLINE Paradigm: Definitive Insights into the Origins of Solar Systems and the Orbital Spacing and Evolution of Planets,* requested by the science editor of a magazine in Washington, D.C., was recently mailed to him. Hope springs eternal. If published, I believe it will have an impact at least equal to that of the ideas of Copernicus and Kepler.

Thanks again for your valued interest. If I can be of further assistance, please let me know.

Re: News Release (Via the AAAS; Rejected). **For Release: May 19, 1999**

On consecutive days in April, two exciting and highly significant news items were published. The initial item, a news release via the AAAS, revealed the discovery of the first multiple-planets solar system other than our own. The second news item was headlined *It's No Crude Joke: This Oil Field Grows Even as It's Tapped (Wall Street Journal,* April 16, 1999). Strange as it seems, the two discoveries are intimately linked and that linkage, coincidentally, had been presented in a science meeting in Santa Fe on April 14 by Alexander Scarborough in a paper titled *The FLINE Paradigm: Definitive Insights Into the Origins of Solar Systems and the Orbital Spacing and Evolution of Planets.* Scarborough, age 76, is a retired research chemist now with Ander Consultants in LaGrange, Georgia.

The news release about the distant solar system tells of three giant planets orbiting a star. The innermost planet contains at least three-quarters of the mass of Jupiter and orbits only 0.06 AU from the star. The middle planet contains twice the mass of Jupiter and orbits at 0.83 AU from the star. The outermost planet has a mass of at least four Jupiters and orbits at a distance of 2.5 AU. "The gaseous nature of the planets and their orbital positions provide powerful evidence against current beliefs about how solar systems form. These findings will encourage scientists to alter their beliefs about such formations occurring from dust, gas, planetesimals, comets, etc.," according to Scarborough.

Conversely, the findings fit precisely into the FLINE model of planetary origins featuring Scarborough's mathematical solution to the proposed Fourth Law of Planetary Motion, explaining the spacing of planets in solar systems. The masses, velocities, distances and gaseous nature of the three huge Jupiter-like planets account for their orbital positions and their being simultaneously in the second stage (gaseous) of the five common stages of planetary evolution. "The FLINE model consists of three chronological, inseparable and ongoing realities: the Four Laws of Planetary Motion (FL), Internal Nucleosynthesis (IN) and Evolution (E). In every solar system, each reality is dependent on the other two," Scarborough stated. Although single and multiple-planets systems apparently are abundant throughout the Universe, solar systems with nine planets that include an Earth-like planet and three distinct Asteroids belts like ours should be very rare. This is assured in the FLINE model by the strict prerequisites imposed by mathematics and natural laws on the manner in which these systems are created.

The exciting article about the oil field that grows even as it's tapped describes how very hot oil is being forced under very high pressure up through the bottom of the oil field it is refilling—precisely as predicted and explained by Scarborough's Energy Fuels Theory (EFT) (1973-1979). The hot petroleum, a polymerization product from virgin methane made deep within Earth, continuously seeks the upward paths of least resistance in its journey towards the surface. "This crucial facet of the FLINE model, supported by a large amount of incontrovertible evidence garnered during the past 26 years, exposes the obsoleteness of the antiquated fossil fuels hypothesis that originated in the 1830s, and this new finding is a powerful addition to the many corroborating facts comprising the EFT," Scarborough added. "The abiogenic origin and evolution of fuels is an example of the typical manner in which all planetary matter was, and is, created via Internal Nucleosynthesis within every planet."

These findings, along with the discoveries about Mars' tectonic plates, magnetic stripes and its surface features, were predicted, and are explained readily, by the FLINE model. This paradigm clears the path to understanding planetary origins and evolution, and it guarantees us a vast renewable supply of fuels (gas, petroleum, coal) for a very long time. More scientists are beginning to interpret their findings in its perspective. "It is a factual, non-speculative, easy-to-understand concept that makes a lot of sense," according to Dr. Morton Reed, a former professor of thermodynamics at Auburn University. "It provides answers to scientists who rethink current beliefs about planetary origins and evolution."

To: Dr. William Mitchell, Institute for Advanced Cosmological Studies, Carson City, Nevada 89702.

June 12, 1999

Thanks for the tip on the article, *Why We'll Never Run Out of Oil (Discover,* June, 1999) by Curtis Rist. I finally was able to get it downloaded from the local library's computer. Rist did an excellent job in describing the vast quantities of petroleum reserves that have continually increased ever since the headlines in 1935 first warned of running out of oil by 1960, and his information closely agrees with my early predictions and publications (1970s) on the subject. His reasoning of improved technology for continually discovering new sources and salvaging oil from abandoned fields does ring true; however, it is only a small part of the big picture.

In the latter part of the article, Rist makes the mistake of going along with the antiquated Fossil Fuels Hypothesis (FFH) to explain the source of gas, petroleum and coal. Perhaps that is the reason he did not mention the discovery in the early 1990s of the world's largest oil field located in central Arabia. The field contains only lightweight oil slightly heavier than gasoline, and is close to being suitable for use in automobiles. This gigantic field of almost-refined oil could not possibly have been created via the FFH. At the same time, it amounts to incontrovertible evidence against prevailing beliefs about creation—not only via the FFH, but also via the Accretion aspect of the Big Bang.

Simultaneously, the lightweight oil field in central Arabia is incontrovertible evidence for the FLINE model of creation in which all matter evolves via Internal Nucleosynthesis (IN) from hot to cold and all planets evolve through five common stages of Evolution (E). The latter is strongly backed by the Four Laws of Planetary Motion (FL); all are backed by a vast amount of substantive evidence accumulated and self-published during the past 26 years. More recent confirmations of the FLINE model and its accurate predictions include the findings on Mars and the discovery of the oil field that grows even as it's tapped. If any valid argument could be made against this FLINE model, shouldn't someone have been able to find at least one during the many years of its evolving facets?

While Rist is correct in figuring that we will never run out of oil, where was he in 1973-1980 when I needed support for the fledgling Energy Fuels Theory (EFT) that accurately predicts and definitively explains the source of these apparently inexhaustible hydrocarbon fuels? For scientists to continue promoting current dogma on planetary and universal origins in apathetic opposition to the overwhelming evidence favoring more factual models does a great disservice to science, to students and to the public.

Just today, I received another annual rejection from Dr. Carlyle Storm, Director, Gordon Research Conferences, concerning my application to attend the 1999 Conference on *Origins of Solar Systems*. Can you believe such a Research Conference could have a valid reason to refuse to critically examine the solution to the greatest mystery of our Solar System: the long-sought solution to the proposed Fourth Law of Planetary Motion explaining the orbital spacing of planets around the Sun? Copernicus was fortunate not to be living in these times. It is tragic that scientists are denied the right to debate issues crucial to the advancement of science, even sadder that they are denied opportunities to hear the rest of the fascinating story whereby they could interpret their findings in more sensible perspectives. But as Max Planck concluded, science advances one funeral at a time.

To: Grady Traylor, 245 Baywood Circle, LaGrange, GA 30240. 08-05-99

Thanks for your letter of August 4 in which you posed three questions. I prefer to answer them in writing (for the records) rather than calling, and in the chronological order submitted.

1. Concerning my letter of April 21, 1999 to Glenn Strait, Cooper, Hyne, Anderson, Whelan and you, there has been no response other than your letter today. But this is no surprise. Through the years of correspondence on the subjects of abiogenic origins of hydrocarbon fuels (1975-1999) and the FLINE model of planetary origins and five-stage evolution (1995-1999) (to scientists, editors, etc.), the rare responses have remained apprehensive, while they prefer not to get involved. I believe the reasons include: (1) the definitive threat these ideas pose to prevailing beliefs and their advocates; (2) the findings, based strictly on scientific facts minus speculations, leave no room for rebuttal—or else advocates of prevailing beliefs would have quickly and gladly pointed out a flaw of some type; (3) the two concepts put reviewers on the hot spot: they can find no fatal flaw, and they cannot afford to go against the established beliefs of their peers, which, in turn, would jeopardize their continued access to grant money and their favorable standing with their cohorts. Too much is at stake—which is understandable.

One example: my Fourth Law manuscript submitted to the Royal Astronomical Society, London has been held in review (or abeyance) now for 19 months. As the world's top scientific organization, they are truly on the hot spot: damned if they do, and damned if they don't, upset the apple cart of established beliefs about planetary origins.

2. Yes, I definitely am interested in sending information to any and all of the interested parties, if you get the addresses. The best and most comprehensive material can be found in my latest book, *THE SPACING OF PLANETS: The Solution to a 400-Year Mystery.* The contents are cumulative findings since the concept began in 1973 with the abiogenic origin of the hydrocarbon fuels. Herewith are two pieces of literature.

3. George Brawner was a resident and former coworker in the chemical lab during the early 1970s. We later corresponded about my first book, upon which occasion he wrote the statement about the immense ramifications of the new fuels concept. He also wrote that the discovery and proof are worthy of the Nobel Prize. The last I heard, he had accepted a job in Chicago.

Thanks again for your interest and assistance. Maybe some day soon the truth will come out.

Powering the Next Century

The excellent article, *Powering the Next Century,* by Richard Stone and Phil Szuromi *(Science* 30 July 1999) contains a statement that warrants some comments. Speaking about the "unexpectedly" cheap oil prices in the United States, they erroneously state that such low prices were "impossible to foresee in the immediate aftermath of the [energy] crisis" of the 1970s.

The authors apparently remain unaware of my publications in the 1970s that accurately predicted the current price range of $1.00 per gallon of gasoline. These predictions and the logical reasons that made them possible can be found in *Fuels: A New Theory* (1975) and *Undermining the Energy Crisis* (1977, 1979). This revolutionary Energy Fuels Theory (EFT) has been continually updated and expanded in later books to include the mounting evidence that periodically comes with relevant discoveries.

Understanding the abiogenic origin of hydrocarbon fuels led inevitably to understanding how all planetary matter came into being in like manner via the processes of Internal Nucleosynthesis (IN) (the transformation of energy into matter). Inevitably, this led to a paramount issue: the mysterious spacing of the planets around our Sun. The solution (1980-1995) to the enigmatic Fourth Law of Planetary Motion that first eluded Kepler in 1595 details how the planets attained their orbital spacing around the Sun before beginning the five ongoing stages of evolution (E) from energy masses to the planets we see today, all in full accord with size. Combining the Four Laws of Planetary Motion (FL), Internal Nucleosynthesis (IN) and Evolution (E) gives us the revolutionary FLINE model of origins of solar systems and the evolution of planets. These three chronological, inseparable and ongoing realities of the new model remain interactive until, one by one, they reach their ending. Thus, from energy mass to inactive sphere, each planet is a self-sustaining entity that creates its atomic compositional matter via Internal Nucleosynthesis throughout its first four stages of evolution.

Understanding the FLINE model enables one to make accurate predictions concerning planetary anomalies other than those of the hydrocarbon fuels. One example: Early in 1994 the prediction that the collisions of the 21 fragments of Comet SL9 on Jupiter would be the most powerful explosions ever witnessed in our Solar System was in stark contrast to the much weaker predictions of the best supercomputers of the time. The laughter at this "wild" prediction faded quickly after the powerful explosions at Jupiter's cloud-tops fulfilled all predictions of the FLINE model, while simultaneously casting doubt on the snowball concept (of comets) that served as the basic model for the computer predictions.

If the understanding of all planetary anomalies is to be accomplished, it must be done within the realm of the FLINE model of planetary origins and evolution. The confidence expressed in this conclusion is directly proportional to the quantity and quality of evidence for the conclusion. The new concept meets all validation criteria: (1) it makes successful predictions; (2) it is based on observations that could, in principle, refute them, but have not; (3) there is a comparable competing concept that is faring worse (one that fails to meet the validation criteria). An a priori test would present no problem with it.

R. Kerr's *USGS Optimistic on World Oil Prospects (Science* 14 July 2000) is the latest in a series of articles on the status of the world's petroleum reserves. As in every article since 1935, the reserves have been grossly underestimated, causing false and dire predictions about the exhaustion of world reserves. The cause of these inaccuracies resides in the erroneous beliefs of the biogenic origins of hydrocarbon fuels (gas, petroleum, coal). How these antiquated beliefs survive in the presence of today's voluminous and powerful substantive evidence against them remains a mystery, especially since the substantive evidence proving their abiogenic origins is even more voluminous and powerful. According to Nobel laureate, Irving Langmuir, "Pathological science is the science of things that aren't so."

Since the Copernican/Galilean era, history has taught that the progress of science is linked inexorably to the openness of minds to new ideas. Otherwise, adherence to the status quo at all cost ultimately proves too high a price. Should science risk adherence to antiquated beliefs that can lead only to pathological science? Shouldn't we be willing at least to examine, with open minds, all evidence supporting the abiogenic origins of hydrocarbon fuels, the origins of planets other than via accretion, the origins and true nature of comets and asteroids, the sources of planetary waters (not from outer space!), the nature of planetary quakes, the most basic causes of lightning and other weather phenomena, the accelerating expansion of the Universe and even the Big Bang itself? Valid answers to these anomalies have their taproots deeply embedded in two inexorably linked concepts now known to, and accepted by, only a few scientists: the FLINE paradigm and its partner, the Little Bangs (LB).

The FLINE model was brought full circle via the solution to the proposed Fourth Law of Planetary Motion (1980-1995), explaining how the planets attained their orbital positions around the Sun. Other than offering definitive solutions to anomalies of the planets of our Solar System, the FLINE model details how each of the known (40+) extra-solar planets formed so close to its central star (not possible via accretion!). In close conjunction with the FLINE model, the LB model of universal origins explains the accelerating expansion of the Universe without the necessity of a cosmological constant.

The out-of-hand rejection of new ideas that challenge the status quo is perhaps the most effective way to impede scientific progress. The choice is clear: Either we adhere to outmoded beliefs that lead only to costly pathological science, or we can examine newer alternatives that offer definitive solutions to the mysteries of the origins of solar systems and the evolution of planets and moons by means of natural laws. Can science really afford the high price of ignoring the lessons of history?

To: Dr. Richard K. Preston, S.T.A.R. Foundation, Washington, D.C. 20024. January 21, 1982

It was a pleasure meeting you at the AAAS conference earlier this month and discussing my presentation of the new theory of the origin of our Solar System.

As you requested, a copy of the original manuscript, including the six diagrams illustrating the geometric derivation of the creation of the system, is enclosed. In appreciation of your interest, I am enclosing as a bonus a copy of the manuscript *Evolution of Gas, Oil and Coal*. Each theory enhances the other; both are supported by preponderant scientific evidence. *Birth of the Solar System* challenges the current gas-dust-cloud condensation theory of formation. It contains the underlying patterns of perfect geometric proportions and Bode's Law-patterns that make sense out of myriad observations of the clockwork precision and predictability of the solar system, patterns that are logically self-consistent and hold true in any number of tests. I am still discovering new information from the diagrams.

Evolution of Gas, Oil and Coal challenges the current fossil fuels theory in the same scientific manner. It utilizes the natural laws of energy transformation into matter and the subsequent formation of gases, oils and coals by known natural processes. Its underlying patterns include the original three observations of Dr. William Logan plus a number of more recent scientific facts that reveal the truth about the origin of these fuels.

On the basis of the preponderance of evidence, the statements of these patterns fully qualify them as scientific theories. Their truths seem irrefutable and destined to withstand the test of time.

To: Don Fehr, Publisher Smithsonian Books, Washington, D.C. 20560-0950. **04-20-02**
April 20, 2002

Re: Manuscript, ORIGIN and EVOLUTION of PLANETARY SYSTEMS: Example: The Hydrocarbon Fuels (Gas, Petroleum, Coal). Introduction to the New Fourth Law of Planetary Motion and Its Ramifications

I write to you as a scientist retired from a career in industrial R&D, and one with a special continuing interest in origins and evolution. For the past 30 years, my research into the myths and facts pertaining to the planetary and astronomical sciences has produced startling results that remain ahead of their time.

Lacking a valid rebuttal, these well-substantiated findings have been recorded periodically and cumulatively in nine self-publications (1975-2002), and duly presented in national and international science meetings since 1982. They pose a serious challenge to the conjectural (Hawking terminology) Big Bang/Accretion (BB/Ac) hypotheses of Edgar A. Poe (1848) and Simone Laplace (1796), respectively, but have yet to break through the well-guarded barriers of these antiquated beliefs.

By fulfilling the requirements specified by Thomas Kuhn in *The Structure of Scientific Revolutions* and by solving a plethora of anomalies of our Solar System (e.g., the spacing of planets, the planetary angles to the ecliptic, etc.), the new Fourth Law of Planetary Motion that first eluded Kepler in 1595 brings the Pythagorean-Copernican Revolution full circle.

The manuscript, subtitled *The Hydrocarbon Fuels* (*Gas, Petroleum, Coal*), illustrates the thoroughness and scientific validity of each of the facets in my latest book, *New Principles of Origins and Evolution,* on www.authorhouse.com or through Ander Publications (see letterhead). Its contents appear destined to displace the BB/Ac myths (Sagan terminology) in the foreseeable future, which points to a lucrative long-term future for this definitive book as the standard textbook, along with an exciting future for science.

I now consider the evidence supporting the new LB/FLINE model incontrovertible. I recognize that Smithsonian Books could quickly render more justice to these significant discoveries than my limitations permit How can we work together to publish and promote what many readers and I consider one of the six most significant discoveries in the history of science? Now age 81, I am open to sale of the copyrights and all contents, etc.

Memo
Waiting in the Wings: A Comprehensive View of Origins and Evolution

(An open plea to the AAS, AAAS, AGU, GSA)

(Sep 11, 2002)

Twenty-seven years have passed since *Fuels: A New Theory* (1975) (researched & formulated 1935-1975) was introduced to the scientific community. By 1979 a plethora of well-substantiated evidence that remains incontrovertible had put this definitive concept of abiogenic origins beyond controversy; yet today, this critical knowledge remains unknown to most scientists.

Additionally, seven years have passed since my final geometric solution to the enigmatic Fourth Law of Planetary Motion (1980-1995) that first eluded Kepler in 1595, and now brings the Copernican Revolution (a definitive understanding of the origins of solar systems and the evolution of planets) full circle. Embracing the works of Copernicus, Kepler, Newton, Descartes, Dutton, Einstein and a host of modern-day scientists, these concepts form a solid and powerful basis for a comprehensive view of how everything came (and continues to come) into being—thereby definitively revealing how and why everything evolves. This crucial discovery also remains unknown to a host of scientists who would benefit tremendously from it and its ancillaries.

As each and every relevant discovery continues to interlock precisely into the big picture, these concepts continually grow in substance and scientific validity. However, as they become more comprehensive, rejections of abstracts—to the detriment of science—become more common, but always without the mention of a valid reason or specific flaw; if one exists, it has yet to be discovered.

The time is overdue for advocates of prevailing beliefs to recognize and act upon the sagacious insights of Kuhn, Planck, Sagan, and Hawking, all of whom have recognized and expounded upon outmoded, unsubstantiated beliefs. The time to abandon antiquated hypotheses and animosities has come and gone; it's time to move forward into more valid scientific concepts underpinned with a plethora of substantiated evidence—concepts that predict a truly exciting future for science! We need to stage fair and full consideration of viable alternatives that have waited too long in the wings without the help warranted from organizations whose creed is to investigate new discoveries and disseminate them to the greater scientific community for vetting and beneficial purposes—because it is the right thing to do.

As to rejections and delays, posterity will ask the inevitable question: Why?

Memo

To: Joint Technical Program Committee, GSA, Seattle, 2003. November, 2002

In keeping with the criteria for Pardee Keynote Symposia, I submit this proposal for consideration for the Symposia scheduled in the Seattle 2003 Meeting: *New Principles of Origins and Evolution: Bringing the Copernican Revolution Full Circle.*

Obviously, its contents and message are on the leading edges in its many scientific disciplines; it addresses broad, fundamental problems that are interdisciplinary, and focuses on global problems involving all disciplines. In these respects it at least matches, if not surpasses, the ramifications of the 1543 Copernican concept of our heliocentric Solar System. This paper utilizes the works of Copernicus, Kepler, Newton, Descartes, Einstein, Dutton, and other astronomers, philosophers, and geologists to form a valid scientific concept that brings the Copernican Revolution (a definitive understanding of solar systems) full circle. The ramifications are immense.

The abiogenic origins and evolution of the hydrocarbon fuels are used as examples of the manner in which all planetary matter is brought continually into being in strict accord with the natural laws of physics, math, and chemistry; i.e., the solution to every planetary anomaly has its taproot deeply embedded in these principles. Beginning with Logan's three critical discoveries in the 1830s, a plethora of well substantiated evidence forms a solid foundation that leads to the Universal Law of Creation of Matter, which, in turn, leads to the obvious five-stage evolution of planets and their moons, simultaneously revealing the sophistry underpinning today's "conjectural hypotheses". (Hawking's terminology, 1999).

This revolutionary LB/FLINE model is brought full circle by my new Fourth Law of Planetary Motion (1980-1995) detailing, beyond question, how the planets attained their spacing around our Sun. With Kepler's First Three Laws, the Four Laws definitively reveal the dynamic origin of our Solar System (and all solar systems), and guides us toward a non-conjectural, definitive understanding of how everything came into being. These revolutionary discoveries predict a truly exciting future for science!

Your policy (*GSA Today*) is highly commendable: "The goal...is to provide presenters the best possible opportunity for communicating new scientific information rather than to dictate what can or will be presented." It raises the bar for other organizations. As you know, such advantage was not available to Copernicus when he initiated this revolution to understand our Solar System - and it yet remains rare.

Re: ORIGIN and EVOLUTION of PLANETARY SYSTEMS.
Example: The Hydrocarbon Fuels (Gas, Petroleum, Coal, Peat).

This manuscript presents a crystal clear, definitive understanding of the intimate relationship of the hydrocarbon fuels, and the beautiful continuity in which they were, and are, made abiologically by means of natural laws. Its continuity includes 15 indisputable facts, along with a plethora of supportive evidence that includes a definitive research project (1985-1986) by Georgia Tech research personnel.

The conclusive results are products of 30 years of research into the myths and facts concerning the origin and evolution of hydrocarbon fuels. They pave the way for definitive understanding of the fundamentals of origins and evolution of all planetary and universal matter—from energy to atoms to molecules—which was made possible by my discovery of the Fourth Law of Planetary Motion (1980-1995) that, in turn, made it possible to unite definitively the great discoveries of the past 2500 years. Relative to origins & evolution, the Fourth Law is the most significant discovery since Einstein's $E=mc^2$.

I now consider my facts and conclusions incontrovertible and indispensable to the fundamental understanding of geological processes.

P. S.—"Scarborough leaves no holes in his persuasive argument. His impeccable logic, accurate facts and sound evidence rewrite dogma...." Dimitri Veras, *Yale Scientific*

P. S.—"...the most brilliant and fascinating work I have seen in many years."
Brig. Gen. Gordon C. Carson III (Ret), Conyers, GA

To: Science News **December 2, 2004**

Edgar T. Lynk's letter *(Science News*, Nov 20, 2004) concerning "the controversy that Thomas Gold engendered when he first postulated abiogenic origins of earthly hydrocarbons"...and Alexandra Goho's response to the letter, warrant further comments regarding historical facts behind this revolutionary controversy.

The idea behind the non-biological origins of earthly hydrocarbons indeed has been sustained by a number of Russian scientists since Mendele'ev first suggested it in 1877. However, the first substantiated concept of their close relationship and their ongoing abiogenic origins via nucleosynthesis and subsequent polymerization was first self-published in my *Fuels: A New Theory* (1975). This was followed with two books, both titled *Undermining the Energy Crisis* (1977, paperback & 1979, hardback). A copy of the latter was given to Ronald Reagan during his Presidential campaign. Upon his election, President Reagan issued an edict to oil companies: "Produce, produce, produce!" They complied, and the severe energy crisis of the 1970s quickly abated.

Entitled *Evolution of Gas, Oil and Coal* (1975-1983), my paper finally was presented at the Miami International Conference on Alternative Energy (6th), and published by T. Nejat Veziroglu, Clean Energy Research Institute, University of Miami. This was two years before Thomas Gold and I clashed at the AAAS Meeting in 1985, the year the abiogenic origins erroneously became known as Gold's theory. Is there a question of scientific integrity here?

Truth is stranger than fiction, and far more interesting!

P. S. - The most recent book in my cumulative Energy Series (1975-2002) is *New Principles of Origins and Evolution: Revolutionary Paradigms of Beauty, Power and Precision* (ninth edition). Both the eighth and ninth editions feature the new Fourth Law of Planetary Motion that definitively explains the spacing, ecliptic angles, etc., of planets.

Memo: Oil Crisis

In a *LaGrange Daily News* article published June 13, 2004, *Oil crisis may fuel solutions,* Will, a columnist with *Washington Post* Writers Group tells of the world's vast reserves of hydrocarbon fuels: specifically oil and coal. Since 1935, there have been periodic predictions of near-term exhaustion of fuel supply sources. However, these dire warnings have always proven extremely shortsighted. Quoting Will: "But today known reserves are larger than ever,"—a periodic news headline at least since 1960. Quoting further, "However, the rate of discovery has been declining for several decades." [One must wonder why.]

Therein lies the reason for energy shortages and higher fuel prices, and their huge effect on inflation and the economy. According to the LB/FLINE model there is little, if any, danger of hydrocarbon fuels being exhausted in the foreseeable future. Earth is a self-sustaining entity that continually creates its compositional matter, including the hydrocarbon fuels—a fact clearly explained by the LB/FLINE model of origins and evolution. One example: in recent times, an exhausted oil well was observed to be refilling itself from below—exactly as the new model predicts and definitively explains.

To: Jane Selverstone, 2004 Technical Program Chair **September 7, 2004**
Subject: GSA Annual Meeting Abstract , Ref: Abstract No. 73966

Title: ORIGIN and EVOLUTION of PLANETARY SYSTEMS: Example - The Hydrocarbon Fuels (Gas, Petroleum, Coal, Peat)

This is the third consecutive year that a definitive abstract of mine has been rejected without a valid reason by the GSA. These abstracts represent 31 years of in-depth research into the myths and facts of science. I find the Big Bang/Accretion hypotheses without an iota of substantiated evidence: the myth that Sagan called it, and now an albatross around the neck of science. There is a viable alternative: one that Thomas Kuhn predicted as essential for a scientific revolution capable of displacing prevailing beliefs.

The revolutionary LB/FLINE model of origins and evolution provides critical fundamentals for definitive solutions to all planetary anomalies. Its example of origin and evolution of the intimately related hydrocarbon fuels provides a plethora of substantiated evidence that I now consider incontrovertible. The new alternative unites all relevant scientific discoveries of the past 2500 years in a beautiful and well substantiated continuity of ongoing relationships in which everything is connected to everything else. Space discoveries continually strengthen all facets of the concept.

I regret that GSA members are denied the opportunity to evaluate and benefit from this inevitable change in direction of scientific thought by accurately interpreting their findings via the new perspective. If my abstracts cannot breach the wall of resistance, I will continue to add the evidence of future discoveries, along with the reactions of scientists—rejections, etc.—into the record for posterity. Rejections retard the progress of science, but they cannot stop it; substantiated truth has a way of surviving.

A question to ponder: Has science lost its objectivity?

From: Peter F. Erickson, 5555 E. Evergreen Blvd., #217, Vancouver, WA 98661 **Aug. 18, 2006**

Two years ago at an NPA conference, you made a copy of one of your earlier papers on the non-biological origin of oil.

Early this month I attended the 24th annual meeting of Doctors for Disaster Preparedness in Portland, Oregon. Two of the speakers presented theses similar to yours. One of them was Professor S. S. Penner of the University of San Diego.

I told him I knew of a man (yourself) who presented the same thesis back in the seventies. He said he would like to see that paper.

I am enclosing a copy of his paper. You can send him your paper, if you wish.

P.S. I am also sending you a copy of my new book, *Absolute Space, Absolute Time, & Absolute Motion.*

To: Peter Erickson, 5555 E. Evergreen Blvd., #217, Vancouver, WA 98661 August 22, 2006

Thanks for your book, *Absolute Space, Absolute Time, & Absolute Motion.* (I agree with all three, and with what I've read so far), and the timely paper by S. S. Penner received today concerning the origin of petroleum. I'm sending to Penner a copy of my paper, *Evolution of Gas, Oil and Coal* (1975-1983), presented in a Miami Conference in 1983, and published in *Alternative Energy Sources VI* by the University, edited by T. Nejat Veziroglu.

Two years later, Thomas Gold and I clashed at the AAAS meeting. I had the full picture; Gold had only a small part of it, just enough to do serious damage to the concept's scientific validity for the next two decades. *The Hot Deep Earth* by Thomas Gold may involve plagiarism (page 6) more so from my books than from the modem R-U theory. I did send him a book on the Energy Fuels concept in the late '70s, and I did warn him about its publication.

My initial publication was *Fuels: A New Theory* (1975), followed by *Undermining the Energy Crisis* (1977, 1979). These were followed by *New Concepts of Origins* (1986), *The I-T-E-M Connection* (1991), *The Spacing of Planets* (1996), and *New Principles of Origins and Evolution* (2002, authorhouse.com). All are of a cumulative nature, as new discoveries continually added validated evidence to the evolving LB/FLINE model of universal origins.

The R-U findings do add more details of molecular reactions of the atomic elements formed in Earth's nuclear core, later to be polymerized into hydrocarbon fuels in Earth's crust—a crucial factor in the LB/FLINE model. Penner's references dated 2002, 2004, and 2006 (page 9), must have occurred after my paper including the Energy Fuels Theory briefly describing how abiological hydrocarbon fuels formed via Internal Nucleosynthesis was presented (in absentia) at the International Congress in 2000 in St. Petersburg, Russia, July 3-8, 2000. However, the three separate papers by Kenney, Herschbach, and Hoffman (page 5) do add more confirmation to the scientific validity of the new LB/FLINE model.

Initiated by Robert Heaston at the 2006 NPA meeting, my Fourth Law of Planetary Motion has been revised into the Fourth and Fifth Laws. The two new laws have led to a definitive answer to the current question of how to define planets, while removing any reasonable doubt about how solar systems form—not by accretion of gaseous dust-clouds or by planetesimals! A relevant page is included herewith.

At the bottom of page 3 of his paper, Penner draws an accurate conclusion via logical reasoning: comets, asteroids, and planetary atmospheres other than Earth's reveal an abundance of organic compounds, not of plant origin. Oil [land, water, other planetary matter] have been produced by processes [and] reactions not involving plants. [Those processes are Internal Nucleosynthesis of atomic elements, followed by subsequent polymerization, cross-linking, and chemical reactions: the INE phase of the revolutionary model].

One last comment: Much substantiated evidence in the LB/FLINE model assures that hydrocarbon fuels are a renewable resource via Internal Nucleosynthesis; time is the remaining question.

Enclosures, c: Prof. S. S. Penner, University of San Diego

To: Dr. Scarborough **09-06-06**

Thank you for sending me a copy of your 1983 paper. I am sorry that I missed this manuscript in discussing the abiogenic vs. biogenic discussions. It is, of course, gratifying to find that a long-term error in the accepted Western literature is in the process of being corrected. Hopefully, practical results of this assessment will be progressively deeper drilling with large new resource discoveries. The just announced Chevron, etc., findings in the Gulf of Mexico are a happy first step in changing accepted doctrine.

S. S. Penner

U.S. News and World Report **April 12, 2004**

The discovery of marsh gas on Mars (*U.S. News and World Report*) by C. W. Petit confirms another accurate prediction of the LB/FLINE model: methane is one of the most common products produced by Internal Nucleosynthesis (IN) during the active stages of planetary Evolution (E). The wisp of methane in the thin atmosphere of Mars is no exception; it is indeed non-biological, made in the same manner as Earth's tremendous volumes of methane were (and are still being) created.

Mars is exiting its fourth (rocky) stage of evolution, and entering the fifth and final inactive stage; its nuclear core is either on the verge of depletion, or, in all probability, already depleted—only time can tell. Because of its small size, Mars is one or two billion years ahead of Earth in its evolution; it now portends the future barren stage of Earth. Its small size, and consequently its rapid pace through the rocky fourth stage (Earth's status), perhaps did not permit enough time for forms of life to develop beyond primitive stages.

Water on Mars

July16, 2004

In his article, *The Proof Is In: Ancient Water on Mars (The Planetary Report*, May/June 2004), M. H. Garr records definitive evidence that findings by the rover, *Opportunity Meridiani Planum,* "confirmed the presence on Mars of standing bodies of water at some time in the past."

These findings confirm earlier predictions of the LB/FLINE model of planetary origins and evolution, which definitively explains the water's source (clearly not from outer space!), the processes that enabled it to leave clear evidence of its former presence there, and its disappearance from the dried lake and river beds.

The new model goes much further in providing definitive answers to other anomalies of the Red Planet; e.g., its surface features, most notably Olympus Mons, its past and current atmosphere, the magnitude and cause of its climate change, and why there is almost no evidence of chemical weathering in the orbital data.

The revolutionary LB/FLINE model would be of tremendous benefit in understanding the excellent data being collected via the space probes to Mars, other planets and moons. The book, *New Principles of Origins and*

Evolution, by the author is available on authorhouse.com at a very reasonable price. Its definitive explanations of the anomalies of Saturn, including its rings and moons, are an added bonus.

An Historical First-Paper Publication

The new hydrocarbon fuels theory (Energy Fuels Theory) was first presented orally to the American Institute of Chemical Engineers in Atlanta on February 10, 1983. The first bulletin on three theories was presented at the AAAS meeting, Washington Hilton Hotel in D.C., January 3-8, 1982.

The last few pages in this chapter are repetitive, and are recorded for historical purposes and additional information only. It is the first paper of the author's Energy Series (1973-2007) to be presented orally to, and published (1983) by, a scientific organization.

ALTERNATIVE ENERGY SOURCES VI VOLUME 3

Wind/Ocean/Nuclear/ Hydrogen

Edited by
T. Nejat Veziroglu
Clean Energy Research Institute, University of Miami

HEMISPHERE PUBLISHING CORPORATION
Washington New York London

DISTRIBUTION OUTSIDE NORTH AMERICA
SPRINGER-VERLAG
Berlin Heidelberg New York Tokyo

NOTE: For the historical record: After the first four self-publications (1973-1979), *Evolution of Gas, Oil and Coal* (1975-1983) was the first full paper published in the establishment's scientific literature (*Alternative Energy Sources VI*, pp 337-344), copied above.

CHAPTER II

PLANETS, EXPANDING EARTH, EARTHQUAKES, MASS EXTINCTIONS, FLINE AND BIG BANG

Origins of Universal Systems:
From Myth to Reality in 12 Giant Steps for Mankind
The 2006 John Edgar Chappell Memorial Lecture

Abstract This paper melds together 12 significant discoveries crucial to understanding origins and evolution of universal systems. Beginning around 500 B. C. with discoveries and beliefs of the Pythagoreans, the presentation progresses briefly through the principal works of Aristotle, Copernicus, Kepler, Newton, Descartes, Olbers, Einstein, Hubble, Chladni, and Scarborough, right into the 21st century. Interlocked via sound logic and substantiated evidence, these discoveries nudge science away from the Poe/Laplace Big Bang/Accretion hypotheses, and toward a new direction of revolutionary scientific thought: the Little Bangs (LB)/FLINE model of universal origins and evolution.

About John Chappell

Before beginning this paper, I would like to comment on who and what made it possible for me to be here today. John Chappell was a giant man, both physically and mentally. We first met in the mid-to-late 1990s in the Poster Session at a meeting of the American Association for the Advancement of Science (AAAS). He commented on my display and paper on the new Fourth Law of Planetary Motion, and then invited me to join the fledgling Natural Philosophy Alliance (NPA), where I could finally get on the main programs, as opposed to being forced into participating only in ineffective Poster Sessions at meetings of AAAS, AGU, GSA, etc.

I liked the fact that the NPA was always open to new ideas rather than limited only to Big Bang beliefs. Fortunately, John convinced me to switch allegiance, and that is why I am here today to present this paper, *Origins of Universal Systems.* It is soundly based on the revolutionary Fourth Law of Planetary Motion that brought John and me together at the AAAS meeting where he convinced me to join the NPA. Without that meeting, I doubt that the new Fourth Law or any of the critical avenues it opens for understanding and correctly interpreting relevant scientific discoveries would have ever been published in scientific literature.

He was indeed a giant among scientists. We are all indebted to him for his good work in organizing and promoting the open-minded NPA where new ideas are encouraged to grow into valid scientific theories.

Introduction

Since there was no choice other than Edgar A. Poe's 1848 Big Bang (BB) hypothesis and Laplace's 1796 Nebular (Accretion) hypothesis, scientists in the early 20th century combined the two concepts into one that gradually became entrenched in the scientific literature as the BB. They seemed a perfect match.

Despite billions of research and education dollars, and almost 90 years later, no substantiated evidence supports the BB/Ac hypotheses. They remain in a state of "truthiness"—a new word coined in 2005—defined as "the quality of stating concepts one wishes or believes to be true, rather than the facts."

Before his untimely death from cancer, Carl Sagan called the BB a "myth". In 1999, Stephen Hawking stated, "Cosmology's hypotheses are conjectural when pressed to the limit." Many scientists have expressed similar views; some even recognize the BB as the major obstacle to scientific progress.

In the 16th century Copernican era, people were still bound by Ptolemy's edict in 150 A.D. that Earth was the center of the Universe—a belief that had lasted 14 centuries before Copernicus in 1543 placed the Sun at the center of our Solar System (SS) to initiate the Copernican Revolution.

In his book, *The Structure of Scientific Revolutions* (1956), Thomas Kuhn makes it crystal clear that no scientific revolution can occur unless there is an alternative to replace prevailing beliefs. Scientific literature now provides powerful evidence underpinning a new LB/FLINE model of universal origins that fulfills Kuhn's prerequisite for a valid alternative to the BB. The new model is soundly based on the Five Laws of Planetary Motion. The First Three Laws were firmly established by Johannes Kepler in the early 17th century. However, the Fourth and Fifth Laws remained enigmas for four more centuries, which is the fundamental reason science had no alternative to the BB until their recent discovery by the author.

A Brief History of Reality in 12 Giant Steps and 2500 Years
Objective: A Change in Direction of Scientific Thought

August, 2005

1. Around 500 B.C. Pythagoras worked out the Pythagorean Theorem that led to the Golden Ratio, the two crucial keys to Nature's secrets of beauty, power and precision. Because of their belief that Earth revolves around our Sun, the Pythagoreans were killed by enraged neighbors and much knowledge was lost.
2. Aristotle, 384-322 B.C., was first to realize that sea and land can change positions. "Where there is now land was once sea, and where sea, land." He also preserved some of the knowledge of the slain Pythagoreans.
3. In 1543 Nicolas Copernicus placed the Sun at the center of our Solar System, thereby initiating the ongoing modem-day Copernican Revolution in scientific beliefs.
4. Johannes Kepler in the early 17th century discovered the first Three Laws of Planetary Motion governing the birth and functioning of our Solar System. He wrote, "Geometry has two great treasures: one is the theorem of Pythagoras; the other, the division of a line into extreme and mean ratio [the Golden Ratio]. The first we may compare to a measure of gold; the second we may name a precious jewel."
   The failure of Kepler to utilize this valuable information resulted in his failure, beginning in 1595, to discover a Fourth Law governing the geometric spacing of planets around our Sun, and a Fifth Law governing the relatively small displacements of planetary orbits during the five billion years since the dynamic birth of the Solar System.
5. Isaac Newton (1642-1727) found that the force of universal gravitation makes every pair of bodies in the

Universe attract each other. This tells us that gravity is inherent in all forms of energy and matter, right down to the fundamental cosmic strings comprising all things universal.

6. René Descartes invented analytic geometry, and in 1644, described Earth's interior as "Sun-like", a fact that would be revealed in the late 20th century as a vital key to understanding the five-stage evolution of planets and moons, all in full accord with Einstein's formula $E = mc^2$: the energy-matter relationship leading to the atomic bomb (that - in all probability - saved the life of this former combat engineer who was boarding an invasion ship in Italy at the time of the first gigantic explosion in Japan).
7. In 1802 Heinrich Olbers correctly identified asteroids as fragments of a disintegrated planet. The FL, along with much other evidence and sound logic, clearly verify this fact about the former fifth planet of our SS. Contrary to BB beliefs, the Asteroids can never coalesce into their former planet.
8. Albert Einstein published his famous 1905 formula, $E = mc^2$, revealing the interchangeable relationship between energy and matter, a crucial key to understanding Evolution (E) of universal spheres via Internal Nucleosynthesis (IN), the transformation of energy into atomic matter, followed ultimately by subsequent polymerization into gases, liquids and solids. FL, IN and E are inseparable; IN drives all E.

   During their transformation from energy via IN, atomic and molecular matter undergo extreme temperatures and huge expansions, resulting in extremely high internal pressures that are transmitted throughout Earth's interior, forcing the planet to expand and slowly increase in size. Volcanic actions and other pressure-releasing outlets, including an active baseball-type seam (discovered in mid-20th century) that circles a globe scarred with multitudes of cracks, attest to our planet's pressurized internal system initiated by its nuclear energy core. Watching an inflating balloon with its surface dots moving steadily apart illustrates long-term surface movements of an ever-expanding Earth as it grows steadily from within. Sea-floor spreading further attests to these phenomena.

   Radioactivity is a regular, but much cooler, product of nuclear reactions - not a source of Earth's internal heat. Such thermonuclear reactions in stars, planets, etc. continuously produce all points forming Planck's curve, including the short-lived 2.7 K microwave background radiation that's erroneously claimed to be a leftover from the BB some 14 billion years ago. Short-lived 2.7 K microwave radiation is produced continuously by thermonuclear spheres (e. g., stars, planets).

   Even a probe escaping the effects of our SS would never be able to escape the "noise" of other suns (stars) undergoing transformation in the normal course of evolution. Starting with our own planet, it can be heard as the "noise" of nucleosynthesis - the hum of creation - that does indeed pervade throughout the Universe. As long as there is an active, expanding Universe, the hum will not cease.
9. In 1929 astronomer, Edwin Hubble, announced his observation that our Universe is expanding, and its galaxies are receding from us. In 1998 astronomers found the rate of recession is proportionate to distance, with the farthest away traveling the fastest, an inherent factor in the LB/FLINE model of universal origins. This tells them that our Universe is expanding at accelerating rates, just the opposite of what is expected in the Big Bang version. The BB's questionable solution to this dilemma is that something is either pulling or pushing the galaxies along, and perhaps the driving force is dark energy or dark matter, neither of which has ever been seen.
10. Ernst Chladni in 1794 correctly recorded the cosmic origin of comets as relatively short-lived fireballs [from hundreds to a few thousand years, depending on size of the fiery mass] and the generic connection between comets, meteorites and meteors. However, in 1950 scientists accepted Fred Whipple's handy suggestion that comets are dirty snowballs, primordial remnants of the BB from which planets accreted.

11. Other than atmosphere, land and water, the best examples of Earth's ongoing production of elements via Internal Nucleosynthesis are the hydrocarbon fuels (gas, petroleum, coal), as first published in 1973 and 1975 by Alexander Scarborough. Their ongoing origin and intimate relationship are underpinned with substantiated evidence too voluminous for this paper. Suffice it to say that their origins serve as critical evidence for the five-stage evolution of planets, and that this knowledge would be of considerable value in critical energy situations if widely known.
12. By utilizing Pythagorean geometry, Alexander Scarborough discovered the enigmatic solution to the Fourth Law of Planetary Motion (1980-1995) that first eluded Kepler in 1595, and now is revealed as the crucial key to understanding origins and evolution of universal systems. By governing the progressive geometric spacing of ten or more energy masses around our Sun during birth of our Milky Way galaxy, the Five Laws (FL) tell us that all solar systems began as small stars that eventually evolve(d) continuously through the five stages of planetary evolution, each of which is discernible today—all at rates proportionate to size (mass).

Understanding the Solar System presents us with information about its mother galaxy and consequently, about the billions of similar galaxies. When exit energies strike the dark, absolute zero-cold Space at the perimeter of our speed-of-light expanding Universe where all atomic motion ceases, they collapse to form small black spheres (a.k.a. black holes), one of the densest forms of energy, that grow in size while extending the perimeter ever outwards: an expanding Universe. Gravity very gradually slows these fast moving, spinning black energy masses enough to trigger each into releasing from its two poles billions of less-dense white-hot masses we call stars, buried in voluminous gaseous dust-clouds, all in beautiful pinwheel fashion. Scientists have verified the presence of a small specific amount of black mass (0.5% of the orbiting mass) remaining at each center to hold the galaxy together; i.e., no assistance via either dark matter or dark energy is needed.

Each star will accrete the dust-clouds only in its immediate vicinity, presenting the illusion that accretion alone creates stars. Among the conglomerate of stars will be a large number of solar systems, primarily binary and trinary types, but very rarely a multiple-mass system that will evolve into multiple planets like ours. The reason: it is very difficult to fulfill the stiff prerequisites demanded by the Five Laws of Planetary Motion (FL).

When properly visualized on a spherical expansion basis, this scenario will be seen to fulfill requirements of the accelerating expansion of the Universe without the need for assistance from the cosmological constant that Einstein called his greatest blunder. One might wonder whether BB stands for Big Bang or Big Blunder.

Incorporating all of the above, the new LB/FLINE model beautifully melds these historical findings into a valid scientific concept capable of displacing the prevailing Big Bang myth that served its time well. All discoveries, large and small, appear to fit readily and precisely into the new concept, and at this time, no anomaly seems capable of escaping a valid scientific solution.

These 12 giant steps, selected from a plethora of strong supportive evidence underpinning the revolutionary LB/FLINE model, appear capable of moving science away from costly BB myths, and back into reality, finally bringing the work of the Pythagoreans and the Copernican Revolution full circle.

But is the time ripe for a change in direction of scientific thought? Can Kuhn's predictions be circumvented?

In fulfilling Einstein's profound prediction that "...the right answers will be simple and beautiful", each giant Step remains crucial for understanding the beauty, power and precision of universal origins and evolution. In the absence of any one of them, such comprehension might not be simple. The only recourse would be continued dependence on the costly Big Bang myth (Note Steps 7, 9 & 10), against which a plethora of valid scientific evidence underpinning the 12 Steps is overwhelmingly persuasive.

Understanding must begin with the inseparable 1st and 12th Steps that have given science the Fourth and

Fifth Laws of Planetary Motion (1980-1996) governing the geometric spacing of the planets around our Sun, while simultaneously solving the centuries-old mystery of Bode's Law. Together with Kepler's First Three Laws, the Five Laws definitively reveal the dynamic birth of our Solar System some five billion years ago and (with the aid of Bode's Law) the changes it has undergone since then. They also reveal vital clues to formation of galaxies that, in turn, led to the Little Bangs (LB)/FLINE model of universal origins noted in Step 12.

This knowledge of origins makes it possible for scientists to decipher the definitive history of our Solar System, and to accurately predict future changes in orbits and velocities of its planets—and to understand why Uranus and Neptune are the only two planets drifting outward from the Sun, etc., etc.

The evidence underpinning Steps 7 & 10 sharply clarifies the roles of asteroids and comets: Contrary to BB beliefs, they can't and don't coalesce into planets.

Step 8, Einstein's $E = mc^2$, reveals the forever ongoing relationship between energy and matter, making it possible to understand Step 11: the origin and intimate relationship of the hydrocarbon fuels (gas, petroleum, coal) via ongoing Internal Nucleosynthesis (IN) via a nuclear energy core that drives all planetary Evolution (E) (1973-1977). (IN and E are inseparable.) This led to understanding the five-stage evolution of planets (1973-1992), which explained Step 2 (Aristotle's interchangeable positions of land and sea) in full accord with the laws of thermodynamics and other natural laws—a vital link in the LB/FLINE model of universal origins (1973-1996) noted in Step 12.

Conclusion

When the BB myth becomes displaced by the LB/FLINE model (1973-1996), scientists will be pleased by the rapidity and ease of solving the great number of anomalies that now plague current beliefs. In a recent paper on the new model, 41 examples of solved anomalies were listed.

One example is the definitive explanation of the slight outward drifting of Uranus and Neptune, while the other planets are drifting inward toward the Sun: a mass/distance relationship. Another example is the definitive explanation of the geometric spacing of planets in our SS: the new Fourth Law of Planetary Motion (1980-1995). A third example: the accurate prediction (and explanation) of the powerful explosions of 21 remnants of Comet SL9 at Jupiter's cloud-tops in1994—a prediction, in writing, that stood alone in its accuracy. Computer predictions based on BB beliefs were puny by comparison.

The LB/FLINE model appears to be the only pathway to definitive solutions to these (and a great many other) anomalies. The first crucial step is an in-depth understanding of the Five Laws of Planetary Motion. Everything else then begins to make sense by simply interlocking into place in the scheme of things—much as it did when Copernicus placed the Sun at the center of our SS. The new model brings the Copernican Revolution full circle.

The new LB/FLINE model appears capable of withstanding the close scrutiny it warrants, and thereby changing the direction of scientific thought away from conjectural myth and more toward substantiated evidence. There is no better time than now to begin the change.

A Change in Direction of Scientific Thought: Introduction to the LB/FLINE Model of Universal Origins

June 30, 2006

To preserve and build upon scientific innovation, scientists must first determine whether a change in direction of scientific thought on origins and evolution is warranted. A few probing questions with known answers might be a good way to begin.

After almost a century of discoveries and untold billions of dollars in cost, do we have even one substantiated fact in support of prevailing beliefs about universal origins and evolution? No! Do data from space probes ever fit precisely into current beliefs? No! Would they fit better into another concept? Yes! Is too much money being wasted in vain pursuit of evidence supporting the Big Bang myth? (Sagan terminology) Yes!

Since myth begets myth, shouldn't we realize the futility of pursuit of knowledge about universal origins and evolution in this perspective? And about why the Universe is expanding infinitely—just the opposite of what the BB predicts?

If the right answers are not to be found within the BB perspective, doesn't this tell us that the BB concept must be the major problem in this scenario? Is science wrong in clinging to the BB in spite of its incapability of providing accurate interpretations to the many brilliant discoveries of space probes—interpretations likely to provide even more questions than answers? Is there an alternative in which new and old data can be interlocked precisely to provide answers to questions that may never be answered via prevailing beliefs?

Fortunately, there is an alternative that warrants full consideration: the LB/FLINE model of universal origins and evolution. The new model has been carefully constructed with factual evidence during the past 33 years (1973-2006). Findings of space probes continually fulfill its predictions, and verify its scientific validity. No anomaly appears capable of escaping its proclivity for solving the mysteries of universal origins and evolution.

To understand the revolutionary LB/FLINE model, one must first understand the Five Laws of Planetary Motion (FL):

I. The planets travel in elliptical orbits, with the Sun at one focus. (Kepler, 1609)

II. For every planet, in equal intervals of time, equal areas are swept out by a line drawn from the Sun to the planet. (Kepler, 1609)

III. Comparing the orbits of the several planets, it is found that the times of revolution are related to the distances from the Sun by a proportionality: the squares of the periods of revolution are proportional to the cubes of their mean distances from the Sun. (Kepler, 1619)

IV. Comparing the orbits of the several planets, it is found that the distances between the orbits of any pair of planets have a common proportionality: the original orbit of a planet relative to the one on either side of it is defined by the classic Greek golden ratio. (Scarborough, 1995, modified 2006)

V. In additional comparisons of the orbits of the several planets, it is found that gradual changes in their distances and velocities have a common proportionality: all are functions of a distance-mass-velocity-time relationship. (Scarborough, 1995, modified 2006)

This knowledge and pictorial evidence provide powerful clues to the dynamic origins of solar systems consisting of fiery masses of energy (stars of all sizes) embedded in gaseous dust-clouds, all of which form the beautiful center and spiral arms of pinwheel galaxies: ejecta from the two poles of a spinning black hole (a.k.a. sphere). Scientists have discovered a black-sphere remnant (0.5% of the ejected mass) remaining at the galactic center to hold the system together. Such explosive actions from black spheres into brilliant white galaxies are known as Little Bangs (LB)—never ceasing, forever ongoing at the spherical perimeter of the Universe.

As the newborn cloud-embedded energy masses (stars) are hurled outwards, many of them inevitably interlock in various types of solar systems, most commonly, in binary systems, and in lesser numbers of trinary systems. Multiple-planet systems like ours are very rare, but they can and do happen: geometrically spaced (Fourth Law of Planetary Motion), and functioning in accord with the five laws of planetary motion.

In turn, this knowledge and pictorial evidence lead to the realization that the smaller fiery masses soon begin their five-stage evolution into planets and moons as they revolve around larger masses of energy (stars, suns).

The five stages are readily discernible; e. g., nuclear energy (Sun), gaseous (Jupiter), transitional (Uranus), rocky (Earth), and inactive (Mercury), all evolving at rates proportional to mass.

The knowledge of planetary Evolution (E) via Internal Nucleosynthesis (IN), in sync with the five motion laws (FL), will enable scientists to decipher the precise history of solar systems, and to predict future changes in planetary orbits arid velocities as functions of time, and to solve definitively every anomaly of planets and SSs; e.g., why Neptune and Uranus are drifting outward away from the Sun, while all other planets and asteroids are drifting inward toward the Sun (Fifth Law of Planetary Motion).

Understanding the formation of galaxies at the ever-expanding perimeter of the spherical Universe will enable scientists to determine how and why the dense black spheres form via buildup of attractive forces: electromagnetic, gravitational, and other energies (e. g., exiting light) at the absolute-zero interface with outer Space where all atomic motion ceases. Is gravity inherent in electromagnetism (EM)? Probably, else why are all universal bodies attracted to all other bodies of every size at all stages of their evolutionary existence, beginning at the superstrings level? Gravity and EM must be one and the same.

Definitive answers will bring closure to questions about an ongoing continuous source of energy that runs our Universe: strings of electromagnetic energy, perhaps the most fundamental substance.

In summary, the five laws (FL) of planetary motion provide sufficient evidence for a definitive understanding of the origin of solar systems and the evolution of their planets and moons via Internal Nucleosynthesis (IN)—the transformation of energy into atomic matter—that drives all planetary Evolution (E). This revolutionary FLINE model readily explains the dynamic origin of solar systems and the geometric spacing of planets (Fourth Law) before beginning their five-stage evolution into the currently observable stages. Every planet and moon is a self-sustaining entity that creates its own compositional matter via IN. This should bring closure to the prevailing question: What is a planet?

This knowledge provides clues to the formation of black holes, one of the densest forms of energy, and the source of galaxies, leading to an understanding of the origin, nature, and function of a black hole as it spews out a pinwheel galaxy from the two poles of its spinning mass. Scientists will learn definitively how these multitudes of black holes form at the perimeter of the ever-expanding Universe to supply our universal energy needs via galaxy formations, that they are the beginning point for explaining the accelerating rate of expansion of the Universe without the need for dark energy or dark matter, and why galaxies are organized in vast sheets measuring in hundreds of millions of light years: inherent factors in the revolutionary LB/FLINE model.

Science needs definitive reasons the Universe expands, how the forces of Nature are unified, why and how everything in the Universe is connected to everything else, the true nature of planetary quakes, the abiogenic origin and source of the hydrocarbon fuels, and how quantum mechanics and gravity can be reconciled. If these needs are to be fulfilled, the BB must be forsaken and more feasible concepts of universal origins and evolution must be examined.

In sync with relativity theories, the LB/FLINE model of universal origins fulfills these needs. When examined closely, one finds two famous illusions to be inherent factors:

(1) With all other galaxies receding from it, our Milky Way galaxy appears to be at the center of the Universe. However, this illusion also holds true for all other visible galaxies when viewed from any one of them.

(2) The receding galaxies create the illusion of a static Universe.

Definitive solutions to universal mysteries via the LB/FLINE model predict a truly exciting future for the relevant sciences. The time is right for the BB to step aside and permit the vetting of younger ideas: a change in direction of scientific thought, one that bodes a really bright future for science for generations to come.

Corroborative Evidence for Nucleosynthesis within Planets and Moons (1975-1980)

One of the best illustrations of how planetary matter came into being via the processes of Internal Nucleosynthesis and subsequent polymerization is the creation of hydrocarbon fuels (i.e., gas, petroleum, coal) (Chapter I), erroneously called fossil fuels. In the beginning, the countless numbers of atoms of carbon and hydrogen were (and still are) created via nucleosynthesis in Earth's energy core. One carbon and four hydrogen atoms join readily together to make one molecule of methane gas. When two of these molecules link together via the process known as polymerization, ethane is produced. Further polymerization produces propane gas composed of three methane molecules, followed by butane gas consisting of four methane molecules.

When five molecules of methane combine, they form pentane, the lightest weight of the oils found in lightweight petroleum. As a variety of other elements become involved in the mix, the polymerization and cross-linking continue onward into the crude petroleum stage of evolution. Scientists have found that every small sample of crude petroleum contains tiny particles of coal—conclusive evidence of the ongoing evolution of petroleum into coal.

When huge volumes of petroleum were forced onto Earth's surface, they inundated the low-lying swamplands and encapsulated the plants and animals in a sea of oil that began cooling, polymerizing and cross-linking into thick beds of coal. In the 1830s, a scientist named William Logan discovered the imprints of these live plants in coal. From this and similar evidence, Logan initiated the Fossil Fuels Theory (FFT) by concluding that coal was made from plants. However, he overlooked the fact that such imprints of live plants can be preserved only through sudden encapsulation—in this case, by the liquid petroleum that solidified into coal. Those imprints of live plants in coal, including tree trunks, had been preserved only by their sudden encapsulation by the gushing petroleum. The tree trunks became carbonized masses, in a manner analogous to that of petrified trees.

According to this Energy Fuels Theory (EFT) of 1973, these hydrocarbon fuels should exist in an irregular pattern of a three-layered system throughout Earth's crust—and they do (Fig. 4). Coal is found on or very near the surface, petroleum is found at moderate depths, while the vast majority of the huge volumes of gas exist at deeper levels. Of course, noticeable overlapping of the three layers does occur as the gas and oil are forced upward through paths of least resistance—but not enough to discount the fact of a three-layered system of fuels. The varying mixtures of gas and petroleum at moderate levels, and the fact that every lump of coal contains traces of methane and petroleum add corroborating evidence to this EFT, Nature's way of making these hydrocarbons.

The huge volumes of methane hydrates discovered throughout Earth's crust during the 1980s were predicted in the 1970s, and are explained, by this EFT. And at the peak of the energy crisis in the mid-to-late 1970s, the knowledge that petroleum was not made from fossils allowed me to predict confidently in print the 1980s price of around $1.00 per gallon of gasoline, as detailed in Chapter I.

Much more powerful corroborating evidence for the EFT can be found in the scientific literature, while the powerful evidence against the FFT continues to mount. The EFT is a typical example of how all planetary matter was, and is, created via Internal Nucleosynthesis, polymerization and cross-linking—the natural processes of evolution.

The processes used by Nature to create all matter are reversible by mankind. For example, coal can be converted back into oil, which can be changed into molecules of gas from which the atoms can be isolated and then split to release the original powerful energy with which they were created.

The FLINE Paradigm of Planetary Origins and Evolution (1)

With this information, we can put together the FLINE paradigm of planetary origins and evolution consisting of three chronological, inseparable and ongoing realities:

1. The Five Laws of Planetary Motion (FL)
2. Internal Nucleosynthesis (IN)
3. Evolution (E).

These three basic principles of the FLINE paradigm apply to all solar systems, except that single-planet systems disregard only the Fourth Law. No solar system like ours is possible without undergoing these inseparable realities. In any such system, the three realities remain interactive until, one by one, they reach their ending. They reveal that planetary evolution in any solar system is not possible with any type of core other than that hypothesized by Descartes as "Sun-like"—a concept now corroborated by abundant substantiated evidence. Such evidence mounts with each relevant discovery of planetary anomalies, both of Earth and of space probes. From energy mass to inactive spheres, every planet is a self-sustaining entity that creates its own atomic compositional matter via Internal Nucleosynthesis throughout the first four stages of its inevitable evolution. The essential relationship of Internal Nucleosynthesis to the evolution of stars, planets, moons, etc. (in which one cannot exist without the other) (1) poses a serious challenge to the idea that all light (and some heavy) atoms were created solely in a Big Bang. (2)

This FLINE model has an excellent record of accurate predictions. For example, the discoveries in the 1980s of vast quantities of methane hydrates in Earth's crust had been predicted and explained in the 1970s by the Energy Fuels concept. Further, the FLINE model accurately predicted that the crashes of Comet SL9 fragments on Jupiter in 1994 would produce the most powerful explosions ever witnessed in our Solar System. The competing super computer predictions proved to be puny in comparison with the actual explosions. For example, Fragment G (Fig. 5), exploding upon contact with the thin cloud-tops of Jupiter, released at least 6 million megatons of energy. (One megaton is the equivalent of one million tons of TNT.) Each of the 21 crashes was far more powerful as nuclear explosions than the small one that hit Tunguska in 1908. However, both situations reveal the true nature of fiery comets: nuclear energy. There is much substantive evidence in the scientific literature in strong support of this conclusion.

The dramatic and costly explorations on Mars have revealed nothing that had not been predicted and explained by the FLINE model. The same statement holds true for the explorations of other planets and moons of our Solar System and for the findings about the 30 extra-solar systems.

The rapid progress being made in understanding the Sun's plasma is encouraging. (Ref: SN, 3/27/99, p 200). The results, when eventually applied to planetary cores, will corroborate Descartes' belief in Earth's "Sun-like" core that explains the five stages of planetary evolution, thus bringing this revolutionary concept full circle.

FIVE STAGES OF PLANETARY EVOLUTION.

Figure 3

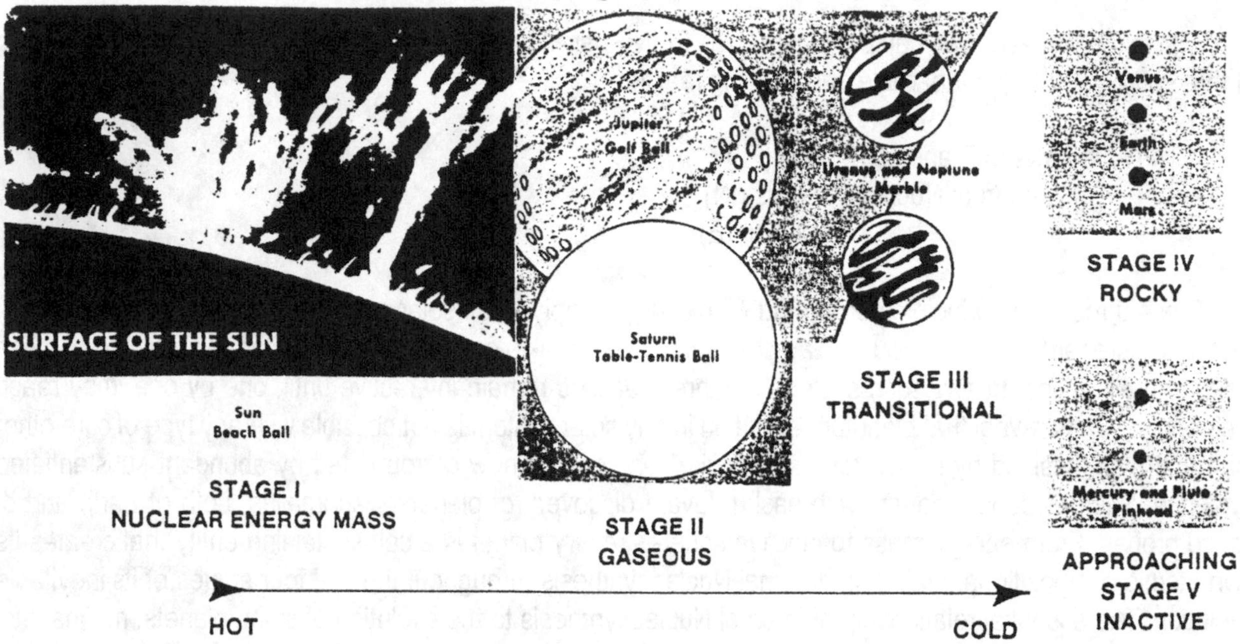

SURFACE OF THE SUN

Sun
Beach Ball

STAGE I
NUCLEAR ENERGY MASS

Jupiter
Golf Ball

Saturn
Table-Tennis Ball

STAGE II
GASEOUS

Uranus and Neptune
Marble

STAGE III
TRANSITIONAL

Venus
Earth
Mars

STAGE IV
ROCKY

Mercury and Pluto
Pinhead

APPROACHING
STAGE V
INACTIVE

HOT → COLD

HOW PLANETS EVOLVE THROUGH FIVE COMMON STAGES.

The relationship between the size and rate of evolution from energy masses to gaseous planets (Jupiter, Saturn) to transitional combinations (Uranus, Neptune) to rocky planets (Earth, Venus, Mars) and finally to the last stage of inactive (now nearly inactive) crustal spheres (Mercury and Pluto) — (Can include our Moon.).

A. All planetary bodies evolve through these five stages, IAW the laws of thermodynamics and all natural laws.

B. In any solar system, the smaller the sphere, the more rapid its evolution through these stages.

How Planets Evolve Through Five Common Stages

The relationship between the size and rate of evolution from energy masses to gaseous planets (Jupiter, Saturn) to transitional (Uranus, Neptune) to rocky planets (Earth, Venus, Mars) and finally to the last stage of inactive (now nearly inactive) crustal spheres (Mercury and Pluto) — (Can include our Moon.)

A. All planetary bodies evolve through these five stages IAW the laws of thermodynamics and all natural laws.

B. In any solar system, the smaller the sphere, the more rapid its evolution through these stages.

Planet Earth has always been in flux, and will be until its core's energy transformation into atomic and molecular matter is completed. In the new *National Geographic Atlas of the World*, there are more than 15,000 listed changes since its last edition five years ago. Two examples: Mount Everest is seven feet taller, while the Dead Sea is somewhat lower.

Such changes are attributed to the power of Earth's nuclear energy core that continuously alters our lands, oceans, and atmosphere, and is responsible for earthquakes, volcanism, and all other disasters, all of which are factors contributing to our planet's ongoing, constant flux. But without that source of energy that also drives all evolution, Earth would be barren: a dead and lifeless planet. So we must learn to accept the bad along with the good that comes from within Earth and all other universal spheres.

Conclusion

Embracing the great ideas of Copernicus, Galileo, Kepler, Newton, Descartes, Buffon, Dutton, Einstein *et al.*, the revolutionary FLINE model of planetary origins and evolution was brought full circle with the solution to the Fourth Law of Planetary Motion explaining the spacing of planets around our Sun. The original 12 orbits of our Solar System, along with Kepler's First Three Laws, verify both the dynamic origin of solar systems and the correct interpretation of Olbers in 1802 that Asteroids are fragments of disintegrated planets. Further, the acknowledgment by astronomers that each giant gaseous planet in extra-solar systems is too close to its central star to have formed there via accretion of dust, gases and planetesimals, adds powerful evidence to the FLINE model that argues strongly against the accretion concept of planetary origins. Thus, this new model seems destined to displace the modified Kant-Laplace Accretion concept of planetary origins early in the 21st century. This change in direction of scientific thought will be expedited with the eventual confirmation of the true nature and sources of comets (ref. example, Comet SL9) and asteroids.

The inseparable connection between Internal Nucleosynthesis and Evolution will lead to understanding the five stages of evolution of planets and moons, and why all active spheres are self-sustaining entities that create their own compositional matter. A prime example of the manner of creation of all planetary matter by means of natural laws is the origin and evolution of hydrocarbon fuels (erroneously called fossil fuels). With its foundation solidified by the Five Laws of Planetary Motion, the FLINE paradigm definitively connects past, present and future discoveries about all planetary systems. The recent discoveries about Mars, entering its fifth (inactive) stage, and other planets and moons continually corroborate and fulfill predictions of the FLINE model, thereby enhancing its position as a valid scientific theory that warrants closer examination.

At this writing, the new Fourth Law explaining the spacing of planets has been in peer review by The Royal Astronomical Society in London for two years and four months.

REFERENCES

1. A. A. Scarborough, *The Spacing of Planets: The Solution to a 400-Year Mystery*. Ander Publications, LaGrange, Georgia, USA, 1996.
2. W .C .Mitchell, *The Cult of the Big Bang: Was There a Bang?* Cosmic Sense Books, Carson City, Nevada, 1995.

Note: Such original works usually do not require or use many references other than commonly-known theories.

Excerpts from Author's New Concepts of Origins (1986)
Orbital Fluctuations in the Outer Planets

Tapered Sizes of the Planets

Another example of the beautiful symmetry and precise order of the solar system can be observed in the tapered sizes of the planets. As would be expected, Jupiter, at the geometric mean of the system, is the largest of the planetary bodies. Between the Sun and Jupiter lie Mercury, Venus, Earth, Mars, and the Asteroids, with Earth being the central and largest of the five masses. On either side of Earth lie Venus and Mars, the second and third largest of this group. The two smallest masses, Mercury and Asteroids serve as bookends for this beautifully tapered sequence of inner planetary bodies. Of course, the Asteroids are no longer planets, but the evidence is strong that they were once a group of planets with moons, totaling a mass perhaps close to that of Mercury (or perhaps much larger).

On the other side of Jupiter, the four outer planets almost form another tapered sequence. Beginning with Saturn (next to Jupiter in size and position), then Uranus, Neptune and finally Pluto, the sizes progressively decrease—except that Neptune is very slightly larger than its predecessor, Uranus. (See the Solar System Table for a comparison of sizes of the solar system bodies.)

Solar System Table

Planet	Equatorial Diameter (Kilometers)	Inclination to Ecliptic ° ' "
Sun	1,392,000	--------------------
Moon	3,476	--------------------
Mercury	4,870	7 00 15
Venus	12,100	3 23 40
Earth	12,756	0
Mars	6,790	1 50 59
Jupiter	142,800	1 18 17
Saturn	119,300	2 29 22
Uranus	47,100	0 46 23
Neptune	48,400	1 46 22
Pluto	5,900	17

Such tapered size patterns could only have been established by a balanced set of forces of natural laws of physics and mathematics during their formation as a planetary system. Some day all of these forces will be more fully identified and understood by mankind.

Ecliptic Inclinations of Planetary Orbits

The ecliptic is an imaginary plane that runs between the center of Earth and our Sun. That great circle of the celestial sphere is cut by the plane containing the orbit of Earth. None of the other planets of our Solar System lie in this plane; all of them are inclined, by various degrees, to the ecliptic. (These degrees of inclinations are shown in the Solar System Table.) Previously, the cause of their differences has remained a genuine mystery—without a clue.

Some important clues, however, can be gleaned from the SS Table. For example, Mercury's ecliptic inclination of 7° 00' 15" is almost identical to the Sun's axial inclination of 7° 15'. Under the GBSST this would be expected, since the cometary mass passed the Sun at that angle and accelerated the equatorial mass of the Sun accordingly. Mercury, the closest and first planetary mass to break out, naturally retained the same angle.

The Venus inclination is slightly less than one-half of Mercury's angle, placing it approximately half-way between the angles for Mercury and Earth (0). One can visualize the vector forces between the Sun, each planetary mass at birth, each of the preceding orbital masses and the mother mass. The resultant vector force accounts for the different ecliptic angle for each of the planets. Again, a pattern of harmony and precision can be seen in these figures, viewed in this perspective.

Other than Mercury and Venus, the only planets with angles greater than two degrees are Saturn and Pluto. The strong combination forces of Jupiter's huge mass and Saturn's large size could account for the larger angle of Saturn. Pluto's extreme angle (17 degrees) is explained by the small planet's severe drop below the critical momentum point (CMP) during birth, leaving only the relatively weak forces of Neptune and the too-distant Sun, to coax it into a tenuous orbit. (This is shown more clearly in the Solar System Diagrams, especially in SS-3).

Again, some day the precise mathematical relationships among all the forces involved in the birth of our Solar System will be formulated and more clearly defined. The ecliptic inclinations will be a vital part of these solutions.

Differences in Speeds of Rotation—Why?

Everything in nature rotates—from the tiniest quark particles to the universe itself. Why? Why not? Spin, analogous to that imparted to water going down a drain, transferred to each mass as it broke from the mother mass to go into orbit around the Sun. The amount of spin was determined by size, angle of break-out and the forces acting on the mass.

A comparison of Jupiter and Earth can be used to illustrate how a history of the speeds of rotation may be visualized. Jupiter spins rapidly in just under 10 hours per cycle while Earth rotates once in 24 hours. The mass of Jupiter is 318 times the mass of Earth while its density is less than 25% of Earth's. In the beginning stages, Earth may have been spinning almost as rapidly as Jupiter. Scientists know that Earth is gradually slowing in rate of rotation. This may be attributed to the gradual expansion of the sphere as energy converts to matter, much as ice-skaters slow their spins by extending their arms outward.

Once Jupiter begins forming its crust and enters into a gradual expansion phase, as Earth has been doing for a few billion years, its rate of rotation should be expected to decrease accordingly toward a more Earth-like speed, though perhaps not quite as slowly as 24 hours per cycle.

Summary

Any valid theory of formation of our Solar System should satisfactorily explain: (1) the speed and directions of rotation and revolution of the planets; (2) the slow, reverse spin of Venus; (3) why Bode's Law holds true for the first eight planetary orbits and not for Uranus and Neptune; (4) if and how moons formed contemporaneously with planets; (5) the tapered sizes of the planets; (6) the huge size of Jupiter; (7) the highly elliptical orbit of Pluto; (8) if and why Pluto is the final planet of the Solar System; (9) why the planetary orbits are inclined to the ecliptic; (10) the anomaly of the asteroids; (11) the physical and chemical differences among planets and moons; (12) if and how the Solar System is based on precise geometric configurations; (13) the huge discrepancy in angular momentums of the Sun and the planets.

Any concept that can explain and offer substantial proofs of these observations must be considered a valid scientific theory of the origin of our Solar System. The Cometary-Mass-to-Planets Theory (CMPT) and its Seven Diagrams of the Solar System (© 1980) clearly supply such answers and necessary evidence. It seems destined to obsolete current theories (e.g., the condensation of gaseous dust-clouds) that do not offer such clear, logical and provable explanations.

The new concept states that a huge cometary mass of energy passed close by and partially around the Sun. As the decelerating mass sped away from the Sun, critical interactions of forces pulled huge pieces from the mother mass and placed them into planetary orbits in accordance with natural laws. The Seven Diagrams of the Solar System clearly reveal step-by-step how these forces geometrically placed into orbit each and every planetary nebula which later transformed into the planets of our Solar System.

Creation of Binary and Other Solar Systems
(1982)
Kant's Ideas Expanded

Immanuel Kant (1724-1804), an 18th century German philosopher, gave science a new concept of creation: he believed that the Universe was formed from the center outward.

He wrote: "The sphere of developed nature is incessantly engaged in extending itself. Creation is not the work of a moment....Millions and whole myriads of centuries will flow on, during which always new worlds will be formed after each other in the distant regions away from the center of nature, and will attain to perfection.... While nature thus adorns eternity with changing scenes, God continues engaged in incessant creation in forming the matter for construction of still greater worlds."

Kant's brilliant ideas are proving to be accurate. The incessant creation of matter does indeed occur—the majority of it at the spherical perimeter of our ever-expanding universe. This matter forms as all matter does: by the strict natural laws of physics and chemistry, directly from extremely hot nuclear energy, as proven by Einstein's formula, $E = mc^2$. The question becomes, where and how does this energy originate?

Most scientists recognize that energy is the substance of which all matter is composed. Energy cannot just exist—it <u>must be created</u> before any matter can be formed. This can be accomplished only by the laws of physics, operating under the ideal conditions of absolute-zero temperature in an absolute vacuum. Under such conditions, the incessant creation of energy does indeed occur, taking the form of a black mass (or black hole).

Scientists are coming ever closer to the realization that when a massive, spinning black sphere of newly created, tightly-packed energy reaches a critical point, it explodes violently into the billions of masses of expanded, white-hot, nuclear energies that form galaxies. The new energies race madly outward in accordance with the pattern established by the exploding forces of the black sphere, thus creating the several types of galaxies observed by astronomers. As these nuclear energies race outward (and somewhat parallel) along the spiral arms of each rapidly expanding, new-born galaxy, billions of close encounters and interactions occur among the gaseous plasma and fireball masses.

It is at this stage of galactic development that nearly all multi-bodied systems are formed. These include binary and multi-star systems and single-star planetary formations such as our Solar System. One theory, proposed first by the Frenchman, Buffon, states that planets are formed as the result of close encounters between passing stars. This concept is the only old theory of our Solar System formation strongly supported by available scientific evidence, natural laws of physics, and studies reported in this text.

Our Solar System began its formation during the early stages of our Milky Way galaxy. It is not unlike the many solar systems that exist throughout all galaxies. As one Russian poet put it, "We must learn to understand our world first; then we can expect to understand other worlds."

Extra-solar Systems: How and Why They Differ From Our Solar System

The Five Laws of Planetary Motion reveal the absolute necessity of a great momentum of a smaller, faster mass interacting with a larger mass as it sped past our Sun to create the elliptical planetary orbits with geometric clockwork precision. Further, the solution to the Fourth Law reveals that in order to create a multiple-planet solar system like ours, in which the smaller planetary masses break from the speeding mother mass at the GR points along the way, the two initial masses must meet precise prerequisites of relative velocities, relative masses, and a specific distance apart to thereby effect a critical angle of diversion and subsequent breakup around a central star. Since it is a rarity for two masses to meet these exact prerequisites, the vast majority of solar systems will contain only one intact giant planetary mass circling its central star. At this time, astronomers have discovered about 30 of these exo-planetary solar systems, all at very great distances from our SS, and each with a planetary mass that failed to meet the prerequisites specified in the solution to the Fourth Law of Planetary Motion. As astronomers have acknowledged, each of these giant gaseous planets is much too close to its central star to have formed there via the prevailing accretion concept of planetary origins!

All solar systems comply with Kepler's First Three Laws. Whether or not the solution to the Fourth Law applies solely to our Solar System remains to be seen. But as we shall see, the sizes and gaseous nature of all the giant Jupiter-like planets, as well as the sizes and nature of all smaller planets and moons in our Solar System, can be explained by a new concept of planetary evolution that is based on the Five Laws of Planetary Motion.

How Planets Evolve

The FLINE Paradigm: Definitive Insights into the Origins of Solar Systems and the Orbital Spacing and Evolution of Planets (1973-1995)

Abstract

Three fundamental and inseparable realities of nature comprise a revolutionary paradigm of planetary origins and evolution. The first two realities are of a geological nature: planetary Evolution (E) and Internal Nucleosynthesis (IN). In any sphere, neither E nor IN can exist without the other. These realities provide the connecting links among all the discoveries and mysteries of Earth's surface features and its internal characteristics. The same IN and E principles are applied to the evolutionary cycles and characteristics of all planets and moons. The critical connection is made between planetary evolution and the essential source of energy that drives planets through five stages of evolution common to all such spheres until each planetary core is depleted of energy, leaving only an inactive (dead) sphere (e.g., Mercury) in orbit. The supporting evidence embraces thirteen clues to the true nature of planetary cores: Sun-like energy, the taproot of the mechanisms of magma replenishment into caldera magmatic systems and other fiery outlets. This paradigm provides an understanding of the role of the inner core in Earth's processes, including the generation of the geomagnetic field and the thermal evolution of Earth. The Internal Nucleosynthesis-Evolution (IN/E) concept leads inevitably to the mathematical solution to the proposed Fourth Law of Planetary Motion (1980-1995) that first eluded Kepler in 1595. Together with Kepler's First Three Laws of Planetary Motion, the Five Laws (FL) (the third fundamental reality) bring the FL/IN/E paradigm full circle by definitively revealing how nebulous planetary masses, including exo-planets, attain their orbital spacing around their Sun. Prevailing theories fail to make these connections between SS origins, orbital spacing and the evolution of planets, moons, etc. Discoveries of the space probes to Mars, Venus, Europa, Jupiter, etc. continually add corroborating evidence.

Introduction to Planetary Evolution

All planets began as relatively small masses of fiery energy: small stars, each placed in orbit in full accord with the Laws of Planetary Motion. Each is destined to obey these Laws as long as it remains in orbit around its sun.

The transition of our planets from nuclear energy to matter has required some five billion years of evolution to reach the current stages observed in the nine planets and the asteroids of our SS.

The Phi geometry of the SS in Chapter III reveals how and where each planet was placed in orbit, and how much each has been displaced from its original orbit. Knowing this, we can uncover many clues that divulge the secrets of how each planet evolved by means of natural laws to its present stage of evolution. This is possible within the realm of the natural laws of physics and chemistry, and can be accomplished without speculation or assumptions that too often mislead.

How is it possible for planets and moons to evolve from nuclear fireballs into the spheres of the SS as observed today in their geometrically-spaced orbits? Our Sun furnishes significant clues to the evolution of planets by natural laws. An atomic force, it is readily recognized as one in which fission-fusion reactions release energies in the form of heat and light from atomic nuclei undergoing constant changes. For example, four hydrogen nuclei fuse together to create one nuclei of helium and almost 1% of the original weight or mass of the hydrogen material is changed into heat and light.

The glowing hot gases of our Sun are made of the lighter elements that comprise the huge gaseous planets and are found in the crust, oceans, and atmosphere of Earth. The most common elements in the Sun are hydrogen, helium, calcium, sodium, magnesium, and iron; the number of other elements includes oxygen, nitrogen and carbon. The main point here is that atoms are created in situ from the energy of nuclear masses under extreme conditions of temperature and pressure. The big question becomes: How does a nuclear mass (a small sun) metamorphose into a planet or moon?

The first phase of transformation (see Figure 3, Chapter I) begins with the accumulation of countless numbers of atoms, ions and simple compounds that eventually form an enveloping spherical blanket high above their source (the energy mass, now identifiable as the core). Simultaneously, similar matter, as with the Sun, accumulates on the extremely hot "surface" of the core. During billions of years, virgin matter pours from the manufacturing plant below, slowly filling the huge space between the energy mass and its layered spherical blanket to form clouds of chemical vapors throughout the towering atmosphere. Scientists see the results in Jupiter and Saturn, both in the second stage of planetary evolution, and classify them as gaseous planets.

In the second phase of transition, the voluminous clouds of towering atmospheric matter gradually close in on the hot interface, and precipitation of its chemical vapors is initiated. The precipitates are instantly repelled by heat; the gigantic battles between the evaporation and the condensation forces rage on for eons. The atmospheric matter is trapped between the unbearable heat of the energy mass and the super coldness of outer space. In time, matter gains a toehold in the form of the first tenuous liquid layers, eventually followed by thin, tenuous crustal formations. Uranus and Neptune are into this transitional third stage of evolution, although their outward appearances may or may not justify their classification as gaseous planets.

During the third phase in the evolution of a planetary sphere, shallow seas cover the entire surface, cooling it sufficiently to permit formation of a more permanent crust. The crust thickens as a function of time, a continual process for billions of years. Newly created matter from volcanic outpourings of lava, elements, water, gases, and compounds of great varieties continually build and alter atmospheric and land systems. Planet Earth is an excellent example of this fourth (rocky) stage of planetary evolution.

With these dramatic, often violent, and persistent environmental changes, species, where feasible, have their beginnings; they come, they flourish, and they vanish into extinction as the ever-changing environment dictates. They adapt or they perish. The era of the dinosaurs is a prime example—perhaps the most popular of all time. Their history offers powerful evidence that Earth is a self-sustaining entity creating its own systems from within, always maintaining control of the creation and extinction of its species. Other prime examples include the life that exists around the "black smoker" vents discovered on deep ocean floors, along with outpourings of virgin materials that never cease. Even more voluminous outpourings of virgin materials occur from the baseball-like seam circling the globe, as often viewed on TV screens.

There is no need to look to outer space for answers to the origin and extinction of species. As it continuously thickens over time, Earth's multi-layered crust offers clues to events affected by virgin ejecta, including the iridium layer at the K-T boundary. Although other layers of iridium have been found in other eras, this particular iridium layer has been interpreted erroneously as coming from outer space and wiping out the dinosaurs. In reality, it is a layer of ejecta matter created within Earth via nucleosynthesis under extreme, specific conditions of temperature and pressure—as were the other iridium layers.

A similar example of a puzzling heavy-metal layer was discovered recently on Venus (*Science*, 5 Jan 1996). The loftiest parts of the planet, like the highest peaks on Earth, are covered by a perpetual frost—a coating of the lustrous, silvery white element, tellurium. This explains why the highlands of Venus appear so bright in radar images of the surface. Tellurium has just the right properties to explain the radar brightness of Venus, and just the right melting point to coat its highlands, but not its plains.

Planetary scientists now argue that the metal, nearly as rare as gold, could be spewing from volcanoes as metallic vapors that would freeze out onto the cooler highlands. This is in full agreement with the 1973 idea that has developed into the FLINE concept during the past 28 years, and is powerful confirmatory evidence that the iridium layers on Earth were made in the same manner via the nucleosynthesis that exists in both (and all) planets. Also, it is powerful evidence of the ever-changing planetary environments that spawn and eventually kill the species that cannot adapt to these changes: the true cause of the demise of the dinosaurs.

As planets evolve from tiny, fiery stars into crusty old spheres with ever-diminishing nuclear cores, startling things happen. In Earth's case, the hot, humid environment spawned the first small wonders of life. Bacterial life spawned in temperatures of a few hundred degrees before the surface cooled sufficiently to allow the formation of eubaryotic cells. It was only a matter of time and additional cooling before various forms of life could evolve and be sustained.

Dinosaurs: The Reason for Their Extinction

Eventually, Earth's average temperature cooled to the mid-100s, with high humidities, heavy rains and vicious thunderstorms prevailing: ideal conditions for plush tropical forests worldwide. The steamy forests were fully capable of sustaining huge life forms—a real paradise for giant animal life. Dinosaurs thrived on the lush plant life and smaller animals during the 140 million jungle years of the Mesozoic Era, often referred to as the Age of the Dinosaurs.

During this rich time span, environmental conditions were changing ever so imperceptibly. Temperatures gradually declined as Earth's crust thickened, its insulation propensities ever-increasing to block out the heat emanating from within. The food supplies of Nature's lush forest began to diminish gradually, while the high oxygen content of the atmosphere declined proportionally over the ages.

Vicious, fiery eruptions from Earth's hot interior were commonplace—the order of the day—as evidenced by the multiple layers of ejected materials that together with sedimentary layers comprise the crust. With each passing millennium, the composition of the erupting materials changed in accordance with the prevailing internal conditions of both temperature and pressure as functions of time that determined the type of end products added as new crust. Each new layer of material contributed its effects, good or bad, to the changing surface and atmosphere. These crustal layers now serve as clocks and books from which mankind extracts their recorded history.

With the passing of time, the dinosaurs adapted as best as they could, while slowly declining in numbers as the environment became ever more hostile. Gradually, Nature began delivering its "coup de grace" from the bowels of the mother Earth that had spawned and nurtured the giants through almost 140 million years, and now was dictating their rapid demise, relatively speaking. Internal conditions combined to emit a series of violent eruptions, spewing the iridium that is now found worldwide in a well-defined layer containing concentrations from 10 to 100 times the normal levels of this rare element.

Analyzed sediments from this K-T boundary layer that marks the end of the dinosaur era revealed that the iridium enrichment, along with the other chemical anomalies found there, was deposited over a period of 10,000 to 100,000 years or more.

These anomalies are more consistent with volcanic rather than meteoritic origins. They are significant confirmatory evidence favoring the IN/E facet of the FLINE concept pertaining to eruptions of virgin materials that persistently alter the surface and the atmosphere of Earth, and thereby control the destinies of all its creatures. In this case, the dinosaurs and 90 percent of all genera of protozoan and algae, along with 60 to 75 percent of all species, disappeared from Earth because of the changes wrought via nucleosynthesis within the planet.

During the past few years, a debate has persisted over the source of the iridium in the K-T boundary layer. Some scientists believe that the element came from outer space, brought in by a crashing asteroid that cloaked Earth with a cloud of dust, resulting in darkness, suppression of photosynthesis, the collapse of food chains and ultimately, mass extinction. However, such happenings do not require anything from outer space; Earth is even more capable of stirring up its own mess.

The K-T boundary at Gubbio, Italy was re-sampled in 1990 in detail for both iridium content and magnetostratigraphy by a team that included both the terrestrial and impact proponents. The results confirmed that the iridium anomaly covers about three meters of the vertical section, representing about 500,000 years of deposits. There now seems little doubt that the iridium was not deposited there by the impact of a large asteroid. The evidence clearly shows that the deposits are the result of volcanic actions, thereby adding more powerful support for the FLINE concept.

Many scientists still attribute this demise to an extraterrestrial crash of a huge meteorite or comet. But the evidence against this belief grows stronger. Recently, Nicola Swinburne and her co-workers found glass spherules and high concentrations of iridium within relatively young 61-million-year-old rocks in West Greenland. When researchers detect such evidence in rocks of the K-T boundary age, they often misinterpret it as a sign of an impact.

However, these materials were found in volcanic rock, increasing the probability that an eruption created the spherules and iridium layer. The Greenland rocks also contain large chunks of nickel-iron metal, a principal component of some meteorites. These findings fit precisely into the FLINE concept in which all of these materials are simply ejecta created via nucleosynthesis within Earth's nuclear core.

Ice-core drillers at Russia's Vostok Station, atop the great ice sheet of East Antarctica, recently passed 3000 meters—a depth at which the ice is about 300,000 years old. Analyses of air bubbles trapped in the ice have confirmed that levels of carbon dioxide and methane were higher between glacial periods than during them. These findings illustrate how our atmosphere changes over time as a result of subtle changes in Earth's internal conditions that determine the types and quantities of elements (matter) created at any given time within the nuclear core. The creation and expelling of significantly less carbon dioxide and methane to the atmosphere resulted in eras of drastic cooling, now identified as glacial periods (just the opposite of the current warming period).

The most obvious and critical factor determining the rate of planetary evolution is size. The smaller the planet, the more advanced its stage of evolution. Scientists observe Earth and Venus, and classify them as rocky planets (the fourth stage of planetary evolution).

In the fifth and final stage, planets and moons become inactive spheres. Core energies become depleted to the point that no new material can be created within. Outpourings and seismic activity cease. Electromagnetism may or may not be detectable. Mercury, Mars, Pluto and Moon are examples of smaller, nearly inactive spheres. Each has only traces, if any, of seismic activity and electromagnetism remaining.

For example, Moon's very faint seismic activity is revealed in tiny moonquakes, indicating that its nearly depleted core is still capable of creating the weak outgassings observed in craters. Moon once generated its own magnetic field, which may have been nearly twice as strong as the present-day magnetic field of Earth, according to S.K. Runcorn and colleagues, who used magnetized lava rock from Moon as evidence. This verifies that our Moon's small nuclear core originally was larger than Earth's energy core is today (and both are being depleted).

The proof of the declining strength of Moon's magnetic field is a strong indication that such magnetism is of nuclear-energy origin rather than of a steady-state iron core, rock or other origin. This large decline is attributable to the dwindling size of the core as its energy transforms into matter. An iron core, or any other type core, would not dwindle.

From this discussion of the transition phases of planetary evolution, two conclusions can be drawn:

1. All planets and spherical moons evolve through five common stages of evolution via nucleosynthesis by means of natural laws.
2. In any SS, the smaller the mass, the more rapidly it evolves through the five hot-to-cold stages.

One last comment on current beliefs about the origin of our Moon (the giant impact scenario and its most plausible alternative, co-accretion) seems appropriate: In the words of one scientist, "They are more a testament of our ignorance than a statement of scientific knowledge."

Update on Dinosaurs' Extinction

Chemical reaction killed dinosaurs? New clues point to toxic vapor, an AP article appearing in the Atlanta newspapers on December 18, 2000, made the startling announcement: "The rock and dust kicked up by an asteroid impact 65 million years ago was not enough to kill the dinosaurs, according to researchers—but the debris may have sparked a deadly chemical reaction in the atmosphere."

"New studies show the Chicxulub impact crater on the coast of Mexico's Yucatan Peninsula is smaller than once thought, making dinosaur extinction difficult to explain completely. Researchers presented those findings Sunday at the American Geophysical Union's fall meeting." These findings agree with the arguments of the FLINE model during the past two decades; they move scientific progress much closer to its teachings. The article continued: "Since 1980, research on the dinosaurs' disappearance has focused on the 125-mile crater and the [alleged] 10-mile-wide asteroid believed to have created it. Dust from the impact was thought to have blocked sunlight for years." Here, the FLINE model remained steadfast against these assumptions, arguing that the crater is of an ejecta or sinkhole origin rather than via an asteroid impact.

Quoting from the article again, "Now, however, drilling around the Yucatan crater indicates the presence of carbonates and sulfate rocks. The new theory is that these were vaporized by the [alleged] asteroid impact, a process that would have released chemicals that produce sulfur and the greenhouse gas carbon dioxide." The FLINE model goes a big step further by explaining the origins of the carbonate and sulfate rocks, then explaining the clues that substantiate the ejecta origin of the crater.

Quoting again, "The sulfur compounds would have been especially toxic," Sharpton said. "They do nasty things. They form little globules that persist in the atmosphere for some considerable time—decades to a hundred years. They also mix with water in the atmosphere and produce sulfuric acid." However, the major problem here is that dinosaurs died out over a very long period of time, estimated at 10,000 to 100,000 to 500,000 years by a number of researchers. Rather than the short period mentioned in the article, this long extinction time clearly reveals that the demise of the huge animals was the result of gradual changes, wrought by Internal Nucleosynthesis, in an environment to which the dinosaurs could not adapt sufficiently to survive. These gradual environmental changes via volcanism are accurately explained solely by the two inseparable principles of the FLINE model: Earth's Internal Nucleosynthesis that drives its evolution through the five common stages of planetary Evolution. Once again, we can state with confidence that all planets are self-sustaining entities in which their evolutionary anomalies can be explained without the need to look to outer space for answers.

Quoting again, "How do you initiate the global crisis? It had to be atmospheric chemistry of some sort," Sharpton said. "That's the only way you can transport the effect globally of something that [doesn't] dump the majority of its energy into a single spot on the Earth's surface." In agreement with the FLINE model, Sharpton is correct only in that it had to be atmospheric chemistry of some sort. However, the world-wide atmospheric chemistry changes and transport assumptions in his interpretations that are based on a single impact crater are puny in comparison with the effects of worldwide volcanism. Certainly, the latter scenario of the well-substantiated FLINE model—e.g., the several separate layers of iridium-- can be explained and understood more clearly than with any other known concept.

It is encouraging to think that the article's wide-spread circulation might convince more people of the fallacy of the extinction of dinosaurs via an asteroid impact 65 million years ago, one that simply was not enough to kill the giant animals. After all, they did survive many such impacts during the 140,000,000 years of the Mesozoic era. A better understanding of the actual processes responsible for their demise will be gained by scientists who are willing to investigate the basic principles of the FLINE model's Internal Nucleosynthesis-Evolution connection (formerly known as Internal-Transition-of-Energy-to-Matter).

References: Previous books in this energy series by the author.

NEW CONCEPTS OF ORIGINS: With White Fire Laden, 1986.

THE I-T-E-M CONNECTION: How Planet Earth and Its Systems Were Made, 1990.

The SPACING OF PLANETS: The Solution to a 400-Year Mystery, 1996.

The World of National Geographic's "Volcano", a spectacular video of volcanic eruptions that occur worldwide, was featured on GPTV (December 26, 2000 at 9:00 p.m.). The statement was made that some 1500 volcanoes are active at various times, but the film did not venture into either the source of the voluminous magmatic outpourings or their virgin atomic nature. The illustration of Earth's interior still depicted its core as made of iron rather than of a nuclear-energy nature.

While the surprising statement was made that all atoms in human bodies came from within Earth, it did not venture into either the origin of these atoms or how they were created via the processes of nucleosynthesis within the planet's energy core. But the statement does move the FLINE model a step closer to realization and acceptance; the final step will come with the inevitable acknowledgment of the manner in which these human atoms, along with all atoms comprising planetary matter, have taproots deeply embedded in the planet's nuclear energy core. The same principles apply to all planets throughout their five stages of evolution.

The video clearly reveals the status and fallacies of current beliefs about the origin and evolution of Planet Earth (and all other planets).

On the evening of Nov. 2, 2006, TV's station NGEO documentary brought viewers a step closer to reasons for ancient extinctions of species. The emphasis was on volcanic-type outpourings of lava and gases worldwide, using the voluminous Deccan Traps in India as one of the best examples.

Each explanation revealed vital details that moved viewers closer to the FLINE model of universal origins developed through the years since 1973. [Ref.: *New Concepts of Origins: With White Fire Laden* (1986), and *The I-T-E-M Connection (1991)]*.

However, the documentary adhered to the Big Bang belief that a comet hit Earth, and triggered the long series of eruptions. It stopped short of explaining the fundamental source of internal heat and explosive power driving the eruptions throughout Earth's history: a nuclear energy core (1973). Without such a power source, no planet would be capable of evolving through five evolutionary stages via Internal Nucleosynthesis (IN).

Meteorite Clues

Richard Greenwood of the Open University in Milton Keynes, UK, and colleagues found that "the ratio of isotopes was identical in every meteorite from the same asteroid, regardless of what part of the asteroid they came from." (*New Scientist*, 18 June 2005). They concluded that at one time the asteroids had been molten enough for these isotopes to be distributed evenly throughout. (*Nature*, Vol 435, p 916).

These findings are powerful evidence of the conclusion of Heinrich Olbers in 1802 that asteroids are fragments of a former planet that disintegrated. "These pieces have a very large range of sizes, indicating disintegration as the cause of the catastrophe... Much fusion occurred, and many heavy metals such as iron and nickel, if not already in existence in the hot core, were formed... The stony irons could fit the theoretical half-metal, half-rock composition of some planetary layer between core and crust. These compositions are probably similar to the interior portions of our planet, either as it is now or as it will evolve..." (*Undermining the Energy Crisis*, 1977, A. A. Scarborough).

The idea of a nuclear energy core (1973) in each evolving planet led eventually to a definitive solution to the Fourth Law of Planetary Motion. Together, the Four Laws detail how the fiery energy masses attained their orbital positions around our Sun before beginning their five-stage planetary evolution (from energy to gaseous (Jupiter), to transitional (Uranus), to rocky (Earth), to inactive (Mercury)). They did not form from gaseous dust-clouds or comets, all of which are forms of ejecta from larger masses of energy.

NASA Prepares Mercury Probe

NASA's Messenger spacecraft was launched August 1, 2004 on a five-billion-mile trip to Mercury, almost 30 years after the last visit by Mariner 10. Scientists hope that the innermost planet will illuminate fundamental questions about the formation of the early Solar System. "What we are really chasing is the processes that led to the formation of Venus and Mars and Earth, and why they produced such different outcomes," said Sean Solomon, Messenger's principal investigator.

Mercury's density of 5.44 is close to Earth's 5.5; its diameter is 4878 km, somewhat larger than our Moon's diameter of 3456 km. Their crater surfaces make the two spheres look like identical twins, a powerful indication of the manner in which both were made: via Internal Nucleosynthesis (IN), exactly as all planetary spheres were, and are, made. Both are in their final stage (fifth, inactive) of evolution.

The differences in all planets and moons are attributed to size (mass) differences and location with respect to the Sun and to any other large source of heat (e. g., the early Earth/Moon and the Earth/Venus situations) that would affect their rates of evolution, surface features, and their types and quantities of atomic and molecular matter produced via IN, the processes now being sought by the Messenger spacecraft, and which already are definitively explained by the LB/FLINE model of origin and evolution of planetary systems (1973-2002).

The differences in magnetic fields of universal spheres are discussed in another section.

Four Original Clues to Planetary Nuclear Cores

A number of observations and facts contribute to the conclusion of the 1970s that a nuclear mass, rather than rock (then, the prevalent theory) or iron (the current prevalent belief), exists as Earth's core. First, the three-layered system of energy fuels throughout the crust made it obvious that a central source of nuclear energy was essential for creation of elements and compounds comprising these fuels: gas, petroleum, and coal. The predicted vast stores of deep gas, composed of tremendous volumes of the two elements, carbon and hydrogen, strongly indicated only one possible source for these building blocks comprising methane gas—the starting point for creation of hydrocarbon fuels.

The second and most common observations are the tremendous outpourings of lava, gases and other matter from volcanoes and rifts throughout Earth's history. Examples include the lava flows that played major roles in building the continents of North America, Europe and India (Deccan Traps), the mile-high multi-layer Grand Canyon and numerous other layered systems throughout the crust.

The third clue is the radioactivity within Earth's crust—known to be a natural by-product of fission-fusion reactions in nuclear masses. The obvious interpretation of this powerful evidence remains in direct opposition to the persistent belief that credits radioactivity as the source of internal heat, rather than the by-product of the interior reactions. However, the fact that fission-fusion reactions inside nuclear masses (e.g., nuclear bombs) do produce radioactivity (as well as heavy elements 99 and 100) positions this as the most logical reason for its presence in the crust. One must question why the radioactivity found in crustal matter is always cold. And how would radioactivity get inside Earth without the presence of a bona fide source to produce it?

The fourth original clue indicative of nuclear cores inside planets and moons can be observed daily: the high mountains and uplifted plateaus found on Earth, Mars, Moon, and more recently, as predicted by this concept, Venus. The presence of such similar characteristics on these spheres is strongly indicative of a common cause: the creation and uplifting of their surface features by very powerful forces within each one. Only nuclear power, in conjunction with isostacy, seemed capable of wielding such magic. These observations and interpretations made it easy to predict that other planets and moons have undergone similar processes. Time and space probes have proven the accuracy of this prediction.

Volcanoes and Extinctions

The article, *Did Volcanoes Drive Ancient Extinctions? (Science,* 18 August 2000, p. 1130), presented powerful evidence of the connections between volcanism and extinction events. Jozsef Palfy and Paul Smith "reported that they have improved the dating of both the extinctions and the accompanying large volcanic outpouring. They gathered the latest argon-argon and uranium-lead dates, recalculated them to account for uncertainties peculiar to each, and found a peak in this so-called Karoo-Ferrar volcanism at 183 million years +/- 2 million years."

"The rough coincidence of a large volcanic outpouring and a sizable extinction event brings to five the number of examples of apparent volcanic-extinction correlations including three of the big five mass extinctions. As Olsen pointed out recently *(Science,* 23 April 1999, p. 604), increasingly abundant and reliable radiometric dating techniques have linked in time three of the largest mass extinctions—the Cretaceous-Tertiary 65 million years ago, the Triassic-Jurassic 200 million years ago, and the Permian-Triassic 251 million years ago—with three of the largest flood basalts: the Deccan Traps of India, the Central Atlantic Magmatic Province of northeastern South America, and the Siberian Traps, respectively. And deep-sea extinctions and a turning point in mammal evolution 55 million years ago at the Paleocene-Eocene boundary *(Science,* 19 November 1999, p. 1465) coincide with the massive lavas laid down when Greenland and Europe parted tectonic ways. The coincidences are within a million years or so, as tight as current dating allows." Congratulations to the researchers on their volcanism-extinction dating, a powerful confirmation of the scientific validity of the FLINE model.

"The very big flood basalt provinces are remarkably correlated with extinctions," says Paul Renne of the Berkeley Geochronology Center in California. With these findings, scientists have speculated that episodes combining climatic warmth, massive volcanic eruptions, oceanic anoxia, and bursts of methane may lie behind major extinctions—a giant leap toward vindication of the revolutionary FLINE paradigm. During the past quarter century, the new model has argued persistently and with logical reasons in favor of such volcanism as the basic cause of all extinction events, including that of the dinosaurs. The FLINE model goes a step further by explaining with incontrovertible evidence the fundamental causes of volcanism: Internal Nucleosynthesis, the force that drives all planets through five common stages of evolution.

All facets of the evolving concept have been presented to various scientific organizations during the past quarter-century, but it remains to be examined, vetted and accepted by the greater scientific community. Thus, to preserve it for posterity, a cumulative set of Energy Series publications was initiated in 1975 with *Fuels: A New Theory.* Specific volcanic-extinction connections were explained in *New Concepts of Origins: With White Fire Laden* (1986), then expounded in *The I-T-E-M Connection: How Planet Earth and Its Systems Were Made by Means of Natural Laws* (1991). This was followed by *The Spacing of Planets: The Solution to a 400-Year Mystery (1996)* featuring the revolutionary Fourth Law of Planetary Motion, explaining how the nebulous planetary masses attained their orbital spacing around the Sun. These writings inexorably link the Five Laws of Planetary Motion (FL), Internal Nucleosynthesis (IN) and Evolution (E) via volcanism—three inseparable principles of Nature that apply to all planets. They explain the specific volcanism-extinction connections used in the subject *Science* article.

The FLINE concept makes it easy to understand the fundamental connections between volcanism and extinctions, as well as the basic causes of all other planetary anomalies. It eliminates the need for speculation about asteroids and/or comets being involved in extinctions. There is no need to look to outer space for answers that can be found right here, inside Planet Earth. As scientists dig ever deeper into the FLINE model, their understanding of the new concept will open new avenues to easier solutions to all planetary anomalies.

Recent Clues to Nuclear Cores in Planetary Spheres

A number of clues indicative of nuclear cores have been discovered during the last few years. Foremost among the predicted evidence is the discovery in 1987 that Earth's center is hotter than the surface of the Sun. The inner core has a temperature of about 12,420°F, scientists from the University of California at Berkley and California Institute of Technology reported in the April journal, *Science.* They calculated temperatures of 11,900°F for the boundary between the inner and outer cores, and 8,640°F for the outer core-mantle boundary.

In comparison the surface temperature of the Sun is about 10,000°F.

The researchers based their experiments on the assumption of an iron core at Earth's center and a pressure of 49 million pounds per square inch. Their finding was surprising to them because it suggests that the core, not the enveloping mantle, is the source of much [actually all] of the internal heat.

"Thus, the forces that drive the plates and give rise to earthquakes and volcanoes have their origins in the Earth's core," said Thomas Ahrens. "This provides us strong insight into how Earth works."

While this insight is a giant leap forward from the rocky core concept of a few years ago, it falls far short of the true situation inside Earth: a nuclear core with temperatures ranging into the millions of degrees—perhaps into the 100-million-degrees range believed to be a requisite for creating uranium, Earth's heaviest stable element.

The core temperatures calculated by Ahrens *et al.* are far too high to have been generated by the slow decay of radioactive elements. Further, such calculations are invalidated by the assumption of an iron core and 49 million pounds pressure, which result in temperatures far below reality.

Viewing these factors in the light of the second law of thermodynamics, one must conclude that the extreme heat was present from the beginning over five billion years ago, even before the Earth began forming its crust. This evidence adds dramatically to the FLINE model of an original nuclear mass transforming into matter that formed our atmosphere and crust.

The second important piece of new evidence supporting this concept is the discovery made in the 1980s by seismic tomographers exploring the interior of our planet via sound waves. By slicing open Earth with tomographic techniques and modern computers, geophysicists have uncovered features from the crust down to the core. Topographic maps prepared by scientists at Galtech and Harvard show "blobs" of hotter material rising from the core-mantle boundary, while cooler masses are sinking from the upper mantle into the interior.

Contradictory to the serene, onion-layer concept, the outer core does not have a smooth, bland surface; rather it consists of alternative deep valleys and mountains. The scale of these deep depressions and elevations is estimated to be between five and ten kilometers—greater than Mount Everest—in some locations. These findings remind one of Sun flares, conforming to the visualized violent images of a nuclear mass encapsulated in the mantle of hot, molten matter of its own making.

The third piece of predicted and confirmatory evidence favoring the new concept came with the discovery in the 1980s of the simultaneous increases in sea levels and polar ice sheets. According to the 1985 National Academy of Science report, sea level is rising about one-tenth inch annually, but scientists don't know why. They have been unable to put the responsibility on the greenhouse effect or on the polar ice sheets.

However, when viewed in the perspective of the FLINE model, the answer becomes obvious: the internal creation of virgin water, as with observable land increases, is an ongoing, never-ending process.

The best examples of virgin water can be seen in the outpourings from deep-sea hydrothermal vents. The Juan de Fuca ridge in the Pacific Ocean has hydrothermal jets spewing jets of mineral-laden water at 662°F in a fairly continuous stream. The minerals are making their surface debut as new matter from Earth's transformation factory.

Scientists believe that mega plumes (large columns of warm water) come from fields of these vents. However, the mega plumes represent an explosion of fluids, like a giant underwater burp. Multiplied many times throughout the World, the result is a steady increase in sea levels. For virgin water and other new matter, there are, of course, many other outlets: volcanoes, rifts, etc.

From as far back as September, 1981, government scientists began analyzing metal-bearing chunks spewed out of an undersea volcano 270 miles west of Oregon. H.E. Clifton, Chief of the USGS's Pacific-Arctic Branch of Marine Geology, reported, "New earth crust is actually being made" by the volcano.

Other strong arguments for creation of virgin materials that form Earth's crust include the "black smokers" discovered in 1979 by a team of American, French, and Mexican investigators. They observed turbulent black clouds of fluid billowing up from chimney-like vents, much like factory smokestacks (which, in reality, they are). The venting fluid, a metal-rich hydrothermal solution, was measured at 350°C. Mixing with the ambient seawater causes copper-iron-zinc sulfides to precipitate as fine black particles suspended in the plumes.

The grade of the metals is comparable to that of many ancient massive piles of sulfides on land: 31% zinc, 14% iron, 1% copper, plus small amounts of silver and gold. On the island of Cyprus some 90 large deposits of copper-iron-zinc sulfides occur as saucer-shaped bodies up to hundreds of meters in diameter. They fill depressions in volcanic lavas that erupted on the sea floor some 85 million years ago. Still further back in geological time, similar hydrothermal convection systems were active 2.7 billion years ago in rocks of the Archean period, now exposed in the eastern Canadian field.

Further verification of an active energy core came on December 5, 2000. Ocean researchers, sponsored by the National Science Foundation, discovered huge, eerie spires of a deep-sea hot-water oasis that dwarfed all known sea-floor vents. They are the largest and oldest formations of their kind yet observed in any hot water vent field.

The researchers were stunned by the enormous sizes of the skyscraper-like structures with multiple white-capped towers soaring as high as 18-story buildings, with large diameters of up to 30 feet. The area, at least as large as a football field and submerged 3,200 feet below the surface of the cold North Atlantic Ocean, was appropriately named "Lost City". It actually is the construction work of a system of hydrothermal vents and fissures through which extremely hot water from within Earth's crust is forced upward into the cold sea-water, thereby building up deposits of virgin minerals in the same manner as the younger black smokers described above.

Minerals and lavas, water and land, air, gases, oils, sulfides and other chemicals, etc., constantly pouring from an internal source over eons of time. Such observations continue to add credibility to the theory of Internal Transition of Energy to Matter (I-T-E-M) theory, now identified by more recent terminology as the Internal Nucleosynthesis (IN) concept. In combination with the new Fourth Law of Planetary Motion, it has evolved into the all-encompassing FLINE concept that explains our planetary origins and evolution from the beginning to the present day.

The reason for these continuous outpourings is a crucial point in the concept. The transformation of energy into molecules of matter entails a tremendous expansion, relatively speaking. In such transformation, each molecule expands into far more space than had been occupied by the energy from which it was made. This forced expansion of uncountable molecules is what creates even more pressure and, consequently, ever greater temperature within the nuclear energy core. These dramatic increases in the demand for space, and in temperature, account for the capability of Earth to produce all heavy elements up to uranium.

Simultaneously, the expansions result in a steady, but imperceptible increase in the size of Earth. As it expands, the crust cracks, giving the surface an appearance similar to the cracked shell of a hard-boiled egg.

The fourth piece of predicted and confirmatory evidence for the IN concept can be illustrated by events occurring in Lake Nyos in Cameroon, West Africa. One night in 1986, Nyos experienced a large burst of carbon dioxide from the lake depths. The spreading gas snuffed out the lives of 1700 people. Two years earlier the same type gas had burst from Lake Monoun, killing 17 people. The other 37 lakes in Cameroon posed no immediate danger, although they, too, are nestled in volcanic craters. Scientists remain puzzled by the processes involved and the true source of the gas.

Nothing but a warm, mineral-laden subsurface spring seemed capable of delivering the type of ions identified as increasing quantitatively. The composition and warm temperatures of the bottom water point to a hot, deep spring feeding it from below. Much like deep-sea hydrothermal vents and our familiar volcanoes, the volcanic craters holding the lakes simply serve as outlets for the virgin gases. Worldwide and on other planets and moons, many similar outlets exist for such outgassings.

Methane and hydrogen sulfide are examples of gases discovered on a number of SS bodies besides Earth. Hydrogen sulfide, found for the first time (1989) outside our planet, is present on the surface and in the atmosphere of Io, one of Jupiter's moons. More recently and seemingly stranger, French astronomers reported the presence of hydrogen sulfide in the comets, Austin and Levy. These crucial discoveries confirm that gases are consistently present in all SS bodies suspected of being powered by nuclear interiors. Comets, as we shall see later, are no exception.

A good example to consider here is the gaseous atmosphere of our nearest neighbor, Venus. Known for some time, its bright envelope is 96% carbon dioxide with a substantial admixture of argon. Such an atmosphere argues strongly for outgassing from a hot interior capable of selectively creating such voluminous gases. Further, recent mappings of the surface of Venus reveal that its unique surface characteristics, conforming to predictions, are readily explainable by principles of the FLINE model discussed previously. Its excessive atmospheric heat contributes to its uniquely flowing surface features.

More Evidence Favoring Earth's Energy Core

An article on *The Globe inside Our Planet (Science News,* Jul 25, 1998) states that Earth's inner core consists of solid iron. In the accompanying article, *Solid Core Proof Ends Half-Century Search,* researchers have observed shear waves passing through the inner core, which they interpreted as evidence of a solid inner core. To quote one statement: "Seismologists have not doubted that the inner core is solid because several lines of indirect evidence point toward that conclusion." This conclusion of the two articles raises some key questions: Aren't the electromagnetic fields of Earth and our Sun and stars all produced in identical manner? If Earth's electromagnetism is created via an iron core, does this mean that our Sun's powerful electromagnetism that stretches for billions of miles also is produced by an iron core inside this huge sphere? In view of the very short ranges of magnetism created by mankind's iron-core magnets, how can electromagnetism generated via an iron core remain effective over billions of miles of space? Do all stars possess iron cores that generate their powerful electromagnetism?

Serious consideration of these questions points to a more powerful source of electromagnetism residing inside every active sphere. Rather than a solid iron core producing magnetic fields, a much more powerful source is inherent in the core of every active sphere. That source is nuclear energy. *Without this source of energy to drive Evolution via the processes of Internal Nucleosynthesis (IN), in which the 92 atomic elements of Earth's atmosphere, land and water are created, Evolution (E) of our planet (and all planets, moons, Sun and stars) would not be possible. In any active sphere, IN and E are inseparable; one cannot exist without the other. When that source of energy is depleted to the point of becoming a solid core, evolution ceases and the planet or moon in question enters its fifth and final stage of evolution: a dead (inactive) sphere with little, if any, electromagnetism.* A plethora of supportive evidence is detailed in *THE SPACING OF PLANETS: The Solution to a 400-Year Mystery* by the author (1996).

The final piece of the enigmatic puzzle of how our Solar System came into being is the recent solution to the proposed Fourth Law of Planetary Motion detailing how the planets—while still in their original Sun-like energy stage—attained their orbital spacing around the Sun. Thus, the planets began their evolvement through the five stages of evolution common to all planetary spheres—from hot to cold stages. We have only to observe the Sun, Jupiter, Uranus, Earth and Mercury to see examples of these five stages of evolution—all in full accord with size and with all natural laws.

The strong supportive evidence furnished by the Four Laws of Planetary Motion (FL) that properly spaced the energy masses around the Sun, together with the subsequent close relationship of IN and E, provide a valid, testable and definitive scientific concept of planetary origins and evolution via natural laws: the FLINE paradigm.

One remaining question: Until the results of shear waves passing through a dense energy core are known, how can researchers be certain that the results of shear waves passing through Earth's inner core can be interpreted as clues to an iron core rather than to a dense energy core? A comparison of such results, if ever obtained, would add the fourteenth clue to the FLINE paradigm's evidence of the true nature of Earth's core: Sun-like energy—just as Descartes first predicted.

One of the best illustrations of how planetary matter came into being via the processes of Internal Nucleosynthesis and subsequent polymerization is the creation of hydrocarbon fuels (i.e., gas, petroleum, coal), erroneously called fossil fuels. In the beginning, the countless numbers of atoms of carbon and hydrogen were (and still are) created via nucleosynthesis in Earth's energy core. One carbon and four hydrogen atoms join readily together to make one molecule of methane gas. When two of these molecules link together via the process known as polymerization, ethane is produced. Further polymerization produces propane gas composed of three methane molecules, followed by butane gas consisting of four methane molecules.

When five molecules of methane combine, they form pentane, the lightest weight of the oils found in lightweight petroleum. As a variety of other elements become involved in the mix, the polymerization and cross-linking continues onward into the crude petroleum stage of evolution. Scientists have found that every small sample of crude petroleum contains tiny particles of coal—conclusive evidence of the ongoing evolution of petroleum into coal.

When huge volumes of petroleum were forced onto Earth's surface, they inundated the low-lying swamplands and encapsulated the plants and animals in a sea of oil that began cooling, polymerizing and cross-linking into thick beds of coal. In the 1830s, a scientist named William Logan discovered the imprints of these live plants in coal. From this and similar evidence, Logan initiated the Fossil Fuels Theory (FFT) by concluding that coal was made from plants. However, he overlooked the fact that such imprints of live plants can be preserved only through sudden encapsulation—in this case, by the liquid petroleum that solidified into coal. Those imprints of live plants in coal, including tree trunks, had been preserved only by their sudden encapsulation by the gushing petroleum. The tree trunks became carbonized masses in a manner analogous to that of petrified trees.

According to this Energy Fuels Theory (EFT) of 1973, these hydrocarbon fuels should exist in an irregular pattern of a three-layered system throughout Earth's crust—and they do. Coal is found on or very near the surface, petroleum is found at moderate depths, while the vast majority of the huge volumes of gas exist at deeper levels. Of course, noticeable overlapping of the three layers do occur as the gas and oil are forced upward through paths of least resistance—but not enough to discount the fact of a three-layered system of fuels. The varying mixtures of gas and petroleum at moderate levels, and the fact that every lump of coal contains traces of methane and petroleum add corroborating evidence to this EFT, Nature's way of making these hydrocarbons.

The huge volumes of methane hydrates discovered throughout Earth's crust during the 1980s were predicted in the 1970s, and are explained by this EFT. And at the peak of the energy crisis in the mid-to-late 1970s, the knowledge that petroleum was not made from fossils allowed me to predict confidently in print the current price of around $1.00 per gallon of gasoline and its bountiful supply of the 1990s *(Undermining the Energy Crisis,* 1977, 1979). You may recall the news headlines of near-panic fears of the 1970s of running out of these fuels by 1990, while the price of gasoline would soar above $5.00 per gallon.

Much more powerful corroborating evidence for the EFT can be found in the scientific literature while the powerful evidence against the FFT continues to mount. The EFT is a typical example of how all planetary matter was, and is, created via Internal Nucleosynthesis, polymerization and cross-linking—the natural processes of evolution.

The processes used by Nature to create all matter are reversible by mankind. For example, coal can be converted back into oil, which can be changed into molecules of gas from which the atoms can be isolated and then split to release the original powerful energy with which they were created.

Earth Story - A TV Series

On September 20, 1998, the *Earth Story* series was initiated on TV Station TLC. The first two parts of the program's six segments were shown from 9:00 to 11:00 pm, EST. The first part of the video was a brief discussion of some important geological discoveries (Wegener, Dutton, Hall, Vine, Sykes, etc.) in chronological order. Conspicuously absent was the concept of isostacy, first recognized by Aristotle and later identified and expounded by Dutton.

Unfortunately, the entire TV series apparently will be based on the erroneous Big Bang (BB) hypothesis in which the overwhelming evidence against this explosive concept far outweighs the speculative beliefs and misinterpretations of data offered in its support. In turn, through no fault of their own, advocates of the BB are forced into misinterpretations of their findings about the planets of our SS—which, in every case, can be interpreted definitively via the FLINE paradigm of the origins, orbital spacing and evolution of planets.

The formation of planets via the condensation/accretion hypothesis of the origin of our SS is an excellent example of the situation in which the Four Laws of Planetary Motion and the principles of evolution are blithely ignored. This is borne out by the fact that planetary evolution is not possible without Internal Nucleosynthesis (IN), the engine that is absolutely indispensable for driving planets through their five stages of evolution. The vast amount of supportive evidence, in which no relevant discovery can remain outside its realm, assures that this basic principle—the First Principle of Evolution (FPE)—is both indispensable and indisputable.

According to the FPE, planetary evolution would not be possible if Earth's center consists of either a rocky core (1980s) or an iron core (1990s). The existence of any type of core other than a nuclear energy mass in any planet is possible only in inactive (dead) spheres in which evolution has reached the end of its cycle. Certainly, the prevailing idea of an iron core at Earth's center, as emphasized in the TV series, needs to be re-examined if we are to truly understand planetary evolution.

Wegener's concept of continental drift contains much verifiable truth—as far as it goes. However, it needs to explain that the days of continental drifting, per se, have ended; the continents have been locked into their positions by Earth's ever-thickening crust, while the planet imperceptibly expands in size. Although it still prevails over the expansion concept of the FLINE paradigm, the theory of subduction and recycling of the crust is an idea whose time has come and gone. The continual creation of virgin matter via IN accounts for this ongoing expansion, which, in turn, accounts for the illusion of continental drift and for the ubiquitous cracking of Earth's crust—and its "plates"—and is a key factor in earthquakes and in the cause of Dutton's isostacy.

While *Earth Story* is interesting and helpful, it does leave wide gaps in the understanding of planetary origins and evolution. A series based on the FLINE paradigm, underpinned by the Little Bangs Theory (LBT) of the origin of the Universe, could fill all of these gaps.

Ringing Earth's Bell: What makes our planet constantly quiver? (Science News, 7-4-98)

In mid 1998, two teams of Japanese researchers reported discovery of the faintest of reverberations, an inaudible incessant pulse coming from Earth itself. "This planetary ringing has set seismologists around the world searching for an explanation... Unlike the bursts of vibrations from earthquakes, the mystery vibrations appear to resonate continuously... In the past, seismologists had thought that these slow planetary stirrings came only with energetic quakes. But researchers are now starting to tune into extremely feeble oscillations that ring all the time." They have reached a provocative conclusion: "The observed 'background'-free oscillations represent some unknown dynamic process of Earth."

Logic, tests and calculations have eliminated all suspected causes (earthquakes, wind, tectonics, etc.) of the ringing. Quoting the article again, "At this early point in the investigation, however, researchers must also consider geological forces as a source of the vibrations. 'It would be exciting if these observations lead to the discovery of some slow, deep process,'" says Kanamori.

These observations will indeed lead to the "discovery" of a slow, deep process that is the heart of creation of all planets: a nuclear energy core. Even more exciting is that this "discovery", when added to the 13 previous (1973-1995) clues to the nuclear nature of Planet Earth's core, should be sufficient to convince skeptics of the scientific validity of the FLINE paradigm of planetary origins. One of those clues is the recent discovery that our Sun also rings like a bell; from this fact, we can conclude that all other stars and nuclear masses also ring like a bell.

An exciting aspect of this turn of events is the probability that all planetary spheres of the Universe pulsate in accord with their stage of evolution—until they enter the fifth and final stage of evolution when their cores are depleted. Thus, such observations can serve as a tool to determine whether the core of any planetary sphere is active or depleted.

The results and conclusions raise some interesting questions. Was the project inspired by previous knowledge of the already-proven nature of Earth's nuclear energy core? Were the researchers already aware of the "unknown dynamic process of Earth", and, if so, did they design the project in order to be able to reach their "provocative conclusion"?

In any case, their results and conclusions add corroborative evidence of the nuclear energy nature of the dynamic core of Earth (and all active planets): the basic foundation of my Evolution-via-Internal-Nucleosynthesis (IN/E) concept of 1973-1977 that inevitably led to the Fourth Law of Planetary Motion (FL) to become the FLINE paradigm (1995) of planetary origins and evolution.

Why Electromagnetic Field Strengths Vary Among Planets (1994)

One of the most puzzling anomalies in science concerns the origins of the electromagnetic fields embodied in the spheres of our SS. Why do they exist and why do their strengths range so widely between very weak and very strong from sphere to sphere? Do they remain constant for billions of years?

A comparison of the field strengths of Venus, Earth, and Mars should offer some insight into the reason for the wide range of electromagnetic strength from sphere to sphere. Venus is almost as large as Earth, but Mars is only one-seventh as large. The fields of Venus, a very slowly rotating planet, and of Mars, a very small planet, are much weaker than Earth's. At this writing, NASA scientists have concluded that if Mars does have an intrinsic magnetic field, it is not of any consequence. Likewise, although nearly the size of Earth, Venus has been identified as a planet with little, if any, electromagnetism. Additionally, our Moon, being small and having only one rotation for each revolution around Earth, shows no detectable magnetic field.

Further, the highly tilted and offset magnetic field of Uranus, a midsize sphere, is the strangest one among planets. The offset from the planet's center causes its magnetic field strengths at the surface to vary by a factor of ten between the north and south magnetic poles. Adding to the evidence, the two largest planets, Jupiter and Saturn, both with faster rotational speeds than Earth's, can be expected to have much stronger EM fields. And they do. Mercury, the smallest planet, has a field-strength of only one percent of Earth's, and a very thin atmosphere. Both clues are indicative of the presence of a small, active remnant of a nuclear core.

There are two basic principles of electromagnetism: Electricity in motion produces a magnetic field, and a magnetic field in motion across an electrical field produces an electromotive force. Combining these principles with the above observations, one can make a general rule: The electromagnetic field strength of each SS sphere is a function of its core size, speed of rotation, (and perhaps its velocity of revolution) and the angle of inclination of the axis.

The electromagnetic field strength (EFS) rule can be true only in the concept of nuclear cores. Thus, another new confirmation of this revolutionary IN idea appears to have been established. If, and when, proven, it will further verify the validity of the total FLINE concept.

In contrast to the new concept, prevailing beliefs dictate the existence of various types of cores in individual spheres, which serve as generators of their magnetic fields. For example, Earth's core consists of iron, either in the molten state or a crystallized solid. Using Earth as an example, one of the main problems with such non-nuclear cores can be seen in a comparison of potential strengths between a field created by an iron core versus a field created by a nuclear energy core; e.g., the Sun. The powerful, extensive magnetic field created by the Sun's nuclear mass reaches many billions of miles to the edge of the SS—billions of times greater than the size of the Sun.

Experiments with iron magnets and electric generators quickly reveal their very limited ranges: only relatively small multiples of their sizes. In this perspective, the choice of a nuclear core rather than an iron or a generator type appears more logical. Thus, the fifth piece of predicted and confirmatory evidence (the ninth clue) favoring this revolutionary idea of a nuclear core in each of the bodies comprising our SS becomes more firmly established.

The Mysteries of Earthquakes

The abstract (1992 Western Pacific Geophysics Meeting), *Infraplate Earthquakes and Crustal Horizontal Temperature Differences in Europe* by I. Stegena, told of three series of tests that correlated the numbers of earthquakes in specific time frames with the geothermal temperature gradients in those areas.

In the Pannonian basin, temperature differences in 1 km depth are compared to earthquakes that occurred between 1859 and 1958. It was found that 95% of the earthquake energy in that period occurred on that half of the basin where horizontal geothermal gradients are large.

In the West European area, 84 of the 93 quakes between 1901 and 1955 occurred in that third of the area where the horizontal gradients of heat flow density are the most abrupt.

A third study carried out in East Europe gave a similar result in which 81% of the total quake energy between 1901-1973 burst out on only 20% of the area (where the horizontal changes of heat flow density are most abrupt).

The concordant results of these investigations, in which the epicenters are lying mostly on places of large horizontal temperature differences, suggest that the sporadic infra-plate quakes are generated by thermal stresses and relaxation. The significant conclusion reached by Stegena is that there is no expressed correlation between tectonics and epicenters: the earthquakes of the area are not tectonic quakes *sensu stricto*.

At the same meeting in 1992, the abstract by Mary Ann Glennon, on the subject of deep-focus earthquakes beneath the island of Sakhalin, reached the conclusion that at least part of the observed pattern residuals is due to path effect away from the source, not that of subducted slab.

The so-called "rim of fire" virtually surrounding the Pacific Ocean conveys the notion to many minds that earthquakes and volcanic actions should occur only along the boundary lines of two or more adjourning tectonic plates. However, there are too many exceptions to this assignment to make a general rule. For example, during the week ending July 7, 1995, a *Chronicle Feature* map pinpointed 14 significant earthquakes that had occurred during the previous seven days. Of this number, ten were inland, well away from known tectonic plate boundaries. Five of the ten quakes occurred within the USA.

In October, 1992, Georgia Tech scientists announced that new analyses of molten, pressurized rock, minerals and other materials making up the mantle deep in Earth may provide insights into deep quakes, volcanoes and even the formation of the planet.

Scientists, studying the mantle, which extends from about 60 miles to 1,800 miles below the surface, say the region consists of two distinct layers containing different properties of key minerals. The difference between the two layers could help account for certain deep quakes and volcanoes that cannot be explained by conventional theories.

As far as they go, these findings and those of Stegena and Glennon are in complete agreement with the FLINE concept. They, too, indicate the necessity of taking another look at current beliefs about the nature of earthquakes. The next step should be to question the source of the molten magma and its specific role in quakes. The answers will point the way to a better understanding of the nature and relationship of quakes and tectonic plates.

Most scientists now interpret their findings about quakes strictly in the perspective of plate tectonics in which rock slippage at plate boundaries cause the earthquakes. In turn, these beliefs are linked to the Accretion Disk hypothesis, a supportive branch of the Big Bang theory.

However, when viewed in the perspective of the FLINE model, each booming epicenter of every quake pinpoints a powerful explosion that rocks the magma that causes the vicious or gentle shakings, cracking, sinking and uplifting of the crust. These events and the cracked eggshell-type surface they create worldwide attest to the powerful forces constantly at work deep within Planet Earth. This compelling new cause-and-effect version offers genuine opportunities for understanding the true nature of earthquakes—and all other planetary quakes.

Although not consistent in their warning signals, earthquakes do emit pre-quake clues: sudden changes in emissions of gases, electrical signals, and swelling of the ground under tremendous pressures. All of these signals can be traced back to the nucleosynthesis processes in Earth's nuclear energy core.

The evidence for precursory warning signals continues to mount. An amateur who studies radio signals had warned that an earthquake capable of devastating damage would hit California within three days. Jack Coles, a former stereo salesman with no college degree, operates the Early Warning Earthquake detection network out of his home in San Jose. The fax he issued on a Saturday in January, 1994 said he had received reports of "increased radio signals, magnetic anomalies and many cases of electrical problems." He warned that the results could mean an earthquake measuring more than 6 on the Richter scale.

The quake that hit two days later in Southern California measured 6.6. Reporters had ignored Saturday's message.

Coles had started developing his theories a few years back when a radio he was repairing started making strange noises. "About four hours later, we had a quake, and I wondered if there was a connection," he said.

Scientists at the USGS in Menlo Park find little to back Coles' theories. "We sent three scientists over there some time ago to look at his stuff. We couldn't make any sense out of it," stated one scientist (an iron core advocate).

A study published in September, 1992 suggests that the eruption patterns of an "Old Faithful"-type geyser (near Calistoga) in California could give warning of impending large quakes. In the report in the journal, *Science*, scientists said they entered into a computer the record of the geyser eruptions since 1971, and then compared the eruption pattern for months around major quake events.

They found that three major quakes within 155 miles of the geyser occurred within one to three days after the geyser's eruption pattern underwent abrupt and dramatic changes. A mathematical examination of the Calistoga record eliminated both chance and the effects of rainfall as an explanation for the abrupt pattern changes preceding the quakes.

The Kobe Earthquake Signals

In January, 1995, as aftershocks of the destructive Kobe earthquake continued to rumble beneath the region, reports of aurora-like flashes just before and after the deadly tremor were announced by a Japanese professor. Tamenari Tsukuda of Tokyo University said one of the most intriguing sights was a flash that streaked from east to west about eight feet above the ground shortly after the quake.

Phenomena like this are believed to be due to electrical and magnetic waves "generated by the grinding of Earth's crust", when interpreted in the "rock slippage" version of the cause of quakes. In the IN version, such phenomena are generated before, during and after the huge explosion at the quake's epicenter, while rock movements are the <u>results</u> of the powerful blasts.

Precursory signals that forewarn of pending quakes are very real, although not always consistent and not yet dependable. All signals can, and some day will, be traced to their source: Earth's nuclear core.

Radon concentration in groundwater increased for several months before the Kobe earthquake. From late October, 1994, the beginning of the observation, to the end of December 1994, radon concentration increased about fourfold. On January 8th, nine days before the quake, the radon concentration reached a peak of more than ten times that at the beginning of the observation, before starting to decrease. These radon changes apparently were precursory phenomena of the disastrous quake.

Chloride and sulfate ion concentrations of groundwater issuing from two wells located near the epicenter of the Kobe quake fluctuated before the magnitude 7.2 events on 17 January 1995. The samples measured were pumped groundwater packed in bottles and distributed in the domestic market as drinking water from 1993 to April, 1995. Analytical results demonstrated that the concentration of both ions increased steadily from August, 1994 to just before the quake. Water sampled after the quake had much higher ion concentrations. The precursory changes in chemical compositions apparently reflect the preparation stage of a large earthquake.

These precursory changes in concentrations of radon and other chemicals do not appear to have any connection with strain buildup in rocks. Rather, they appear to lead in a direct path to Internal Nucleosynthesis as the source of the changes that account for the basic cause of earthquakes.

Depth as a Safety Factor

On June 8th, 1994, a powerful earthquake sent panicked Bolivians, Chileans and Brazilians into the streets, and was felt as far away as Canada. Power outages occurred in parts of Chile and Bolivia. Although it was perhaps the biggest deep-focus quake of the century, there were no reports of major damage or casualties.

Unusually deep at 400 miles beneath Earth's surface where solid rock does not exist, the widely felt quake caused very little damage. The lack of rocks there allowed the shock waves to travel through the magma and far into the distance, dissipating the explosive energy quickly and safely.

Scientists have learned that our Sun has radical pulsations that make the solar surface contract and expand like a ringing bell. In addition, like a bell, Earth has its own natural frequencies—or normal modes—which start ringing if the globe is hit hard enough—as was the case here. The most persistent of these modes causes the planet to expand and contrast every 20 minutes, almost as if it were breathing. This mode can be detected even three months after a great quake. "This is really going to change the level of our information about the deep Earth," one seismologist said.

Contrary to this magnitude 8.2 quake, the much smaller, shallower quake of a magnitude 6.7 in Northridge, California killed 61 people and caused at least $20 billion in damage.

Scientists have yet to unravel exactly the cause of deep earthquakes. According to one prominent, but speculative theory, deep tremors occur when increasing pressures cause minerals in the ocean crust to undergo a sudden structural transformation!

Such speculation is not necessary; the cause is obvious. In any size quake, the deeper the explosion's epicenter, the less the danger at Earth's surface. This holds especially true when deep explosions occur where no rocks exist. If no rock exists there, then rock slippage cannot be the cause of the quake. If not the cause, crustal rock movement must be the result of a powerful explosion at the epicenter of each quake, no matter how deep or shallow. If below the level of solid rock, Earth's bell will ring until the shaken magma settles down again.

Why the Explosions?

In April, 1992, an explosion packing the power of an earthquake ripped open an underground propane gas pipeline, killing one person, flattening nearby mobile homes and shaking buildings more than 140 miles away. It registered 3.5 to 4.0 on the Richter scale—as strong as an earthquake that could cause slight to moderate damage.

"It was just a big bang, a tremendous bang," stated one survivor. One child was killed, and at least 18 persons were injured, three critically.

The explosion occurred in a rural area seven miles south of Brenham, Texas, a community of 12,000 about 70 miles northwest of Houston. Officials suspect gas at a liquid petroleum storage and pumping facility collected in a ravine, and was ignited by a car or a pilot light in a home.

A most interesting point here is the comparison of this powerful explosion to a 3.5 to 4.0 quake. One must wonder how different or how much alike the two events actually are.

The Tiny Mystery of Polonium Halos: Creationism, Big Bang or the FLINE Model?

In the book, *Creation's Tiny Mystery,* Robert V. Gentry presents a good argument for Creationism, using polonium halos found in granite as evidence supporting this belief in opposition to the evolutionary viewpoint of science. The book was sent to me in June, 1995 by Glenn C. Strait, Natural Science Editor of *The World & I* magazine. His cover letter asked how the presence of these tiny halos in granite could be explained by the FLINE concept.

Under the microscope, these halos show a tiny radioactive particle at the center of concentric ring patterns in the granite, much like the bull's eye at the center of the rings. Because of their radioactive origin and their halo-like appearance, these microscopic ring patterns became known as radioactive halos.

After reading the interesting book, the essence of my reply can be summed up in two short paragraphs: While I do not agree with the conclusion that his discoveries of polonium halos in granite support Creationism, per se, I am excited about two aspects of these findings. First, they offer strong support for the evolution of planets via Internal Nucleosynthesis (IN), which, in turn, is powerful evidence against the Accretion Disk hypothesis of the origin of the SS (a vital factor in the Big Bang theory).

Second, the short half-life of radioactive polonium gives solid assurance that they could not, by any stretch of the imagination, have been formed in distant supernova explosions (as claimed in the Big Bang Theory) and survived the eons that supposedly elapsed before they became a part of Earth's crust. Thus, in the prevailing beliefs about planetary origins, it is impossible for polonium to be a primordial constituent of Earth's granite.

The evidence clearly shows that the halos had to have been formed in situ—and they were. Only in the FLINE model would this be possible! Both the origins of Precambrian granite and the polonium halos therein <u>can be explained readily in the new perspective</u>—a concept firmly structured on the solid foundation of the SS's geometric origin that proved crucial in deriving the Fourth Law of Planetary Motion on the spacing of planets.

Granite is the foundation rock of Earth's ever-thickening crust. It was among the first layers to solidify atop the molten mantle worldwide to form tenuous sections, and eventually a permanent foundation on which layers gradually built—and are still building. How did the compositional elements and compounds form into granite with polonium halos inside?

Polonium is a radioactive metallic element belonging to the uranium decay series. It occurs naturally in pitchblende as a decay product of radium, and can be produced artificially by bombarding bismuth with neutrons. Its most stable isotope has a mass number of 210. Polonium, with a half-life of about 138 days, decays into an isotope of lead by giving off alpha rays. One polonium isotope is the product of the radioactive decay of radon, a common gas that still emanates continually from deep within Earth.

The short half-life of polonium-218 of three minutes means that every three minutes, one-half of its remaining mass will decay. If created along with other elements within Planet Earth, it would be no surprise that traces of any and all polonium isotopes are found in granite. The isotopes formed in situ during cool-down of the granite mix containing the polonium and other elements.

The textures and composition of granite give important clues to its formation processes. Scientists know that these foundation rocks have coarse-grained, crystalline textures, which are only found in rocks that cool slowly from a hot molten mass. The process can be observed by crystallizing compounds in the laboratory; these experiments always form crystalline textures.

These clues point to the only apparent explanation: the elements and compounds comprising Earth's hot mantle-like, thick liquid surface some 4.6 billion years ago came together and slowly cooled to form granite. Radioactivity was at its peak surface performance; huge numbers of large particles of radon, polonium, etc., continually bombarded the coagulating granite; particle entrapment was common, followed by decay that left their halo marks in the granite.

Through the eons the newly-mixed materials from the hot mantle continuously contributed to the ever-thickening crust. The process continues today, and will do so at diminishing rates until all the energy of Earth's core is expended in the processes known as nucleosynthesis.

Thus, to the previous nine clues to Earth's nuclear core, we can add one more: the tiny mystery of polonium halos in granite. While adding much support for the FLINE concept, these tiny clues to Earth's origin clearly aim their arrows at both the hearts of Creationism and the Big Bang Theory.

Galileo's Stunning Probe into Jupiter

A stunning weather report from the probe launched into Jupiter's atmosphere (7 Dec 1995) by the Galileo satellite hit the news media. The report revealed that the skies were hotter, windier, drier, and clearer than forecasters predicted. In fact, the preliminary data from the probe is so shockingly different than expected that it inevitably will lead planetary scientists to rethink not just the meteorology of the gaseous planet, but its very origins. According to the understatement of one Galilean project scientist, the data "doesn't fit very well. In fact, it's darn uncomfortable."

Contrary to this situation, the report's results are very exciting in that they are precisely what one would expect when interpreting the data in the perspective of the FLINE model. The best way to verify this statement is by comparing the interpretations of the data in the two different perspectives: the current view of Jupiter's origin versus the FLINE view.

The temperature at Jupiter ranged from -144°C at the top of the ammonia-cloud covered atmosphere to +152°C (306°F) at only 600 km (360 miles) into the thick blanket of clouds covering the giant planet whose diameter measures 142,980 km (86,000 miles). Pressures ranged from 400 millibars to 22 bars over the same descent path, compressing the gases to densities up to 100 times greater than previously postulated. The extremes of temperature and pressure created a vertical convective motion in the atmosphere, stirring up turbulent winds more than 50 percent stronger than predicted. These winds were constant throughout the probe's descent.

In theory, the probe should have passed through a region where wind speeds drop to zero. However, it never reached such a point. Contrary-wise, the wind speed increased with depth, which led several investigators to speculate that the energy source driving the circulation of the Jovian atmosphere is probably escaping from the interior. The answer moves us a giant step closer to truth.

In the perspective of the FLINE concept, one would predict conclusively that the engine driving Jupiter's powerful winds is the internal heat source: its nuclear energy core. The core also accounts for the thick, extremely dense blanket of gaseous clouds and the high temperature at the depth of only 600 km. Jupiter's core temperature, even though well-insulated, was recorded by Pioneer 10 in the 1970s at 30,000°C, but this will prove to be a very conservative figure when more accurate measurements can be made.

The probe's helium abundance detector recorded that the outermost regions of Jupiter now contain much less helium than the planet started with, a figure calculated from the helium-to-hydrogen ratio in the Sun. Current beliefs go with the suggestions "that Jupiter's helium is now condensing into droplets under the deep interior's megabar pressures; the droplets then fall even deeper into the planet. So, the gravitational energy released as heat by the fall of the helium raindrops must, in fact, be fueling Jupiter's infrared glow, which is brighter than anything the solar energy reaching the planet could account for."

A more sensible explanation is that Jupiter's core, in strict compliance with natural laws of evolution from energy to matter, simply is producing a smaller percentage of helium as it gradually evolves into the stage that produces less helium and/or more heavy elements. Rather than being produced by helium raindrops, the heat fueling Jupiter's infrared glow emanates directly from the nuclear core, just as predicted by the FLINE model.

Further, the probe revealed Jupiter to be much drier than anticipated and relatively free of condensation. The observation that its atmosphere contains water concentrations equal to that of our Sun left investigators wondering. However, both the helium and water concentration are in line with predictions of the FLINE concept in that they reveal the close relationship of the two masses. They are clear indications that Jupiter has evolved into the second stage of evolution from the energy stage by covering itself with the initial blanket of chemicals that still match the Sun's production. As the huge planet evolves toward the rocky stage, its production of chemicals will gradually lean more and more toward the heavier elements, and its higher-than-expected radiation doses will gradually decline as more and more of its core energy is transformed into encapsulating matter.

Scientists can learn a great deal about planetary evolution through the realization that planets actually do evolve from energy masses through five common stages. Jupiter is an excellent specimen to study for details of evolutionary changes as the giant planet progresses through the second stage.

As for water clouds, mere wisps of water particles were found in the atmosphere. Even many of the heavier elements—carbon, oxygen, and sulfur—as well as neon were found to be at lower-than-expected concentrations. Here again, these findings are revealing great details about planetary evolution. There simply is no need to use the excuse that Galileo's probe dropped into a very rare "hot spot" to explain these undoubtedly stunning results. With winds at 531 km/hr, temperatures would tend to level out any such spots. It is a safe bet that the results are truly representative of the atmosphere on Jupiter.

Just as Galileo himself drove the dagger into the heart of the Ptolemaic version by promoting the Copernican idea of a Sun-centered Solar System, the Galileo orbiter has plunged a dagger into the heart of the prevailing dogma about planetary origins and evolution.

The Revolutionary FLINE Model: Creation by Natural Laws

All efforts since the initial attempts of Johannes Kepler in 1595 to solve the mystery of how the planets attained their orbital positions around our Sun have failed to produce a definitive solution to this enigma. To understand the mystery of the origins of solar systems and the evolution of planets, one must first explain, beyond doubt, the critical relationship between the current spacing of our planets and the forces that power the evolution of these celestial spheres. Prevailing concepts fail to explain this inseparable connection.

The key to understanding this relationship resides in Kepler's First Three Laws of Planetary Motion and the solution to the proposed Fourth Law of Planetary Motion. Together, these Laws offer a valid explanation of the dynamic manner in which our SS came to be, while providing a solid foundation for understanding the forces that drive the evolution of planets. The complete mathematical solution is detailed (Chapter III).

A valid solution must provide a solid foundation for supplying definitive answers to all anomalies of our SS; e.g., how and why planets attained their current orbital positions, why planets differ in size and composition, how and why the atmospheric and surface features of every planet progressively undergo evolutionary changes, and what forces drive planetary evolution. Why does Earth contain the full range of elements from hydrogen to uranium, while only the lighter elements can be found on Jupiter, Saturn, our Sun, extra-solar planets, etc.? How and why did the known extra-solar gaseous giant planets form so close to their central star? Why will water and signs of early forms of life likely be found on each of our nine planets and many moons?

The corroborating evidence is both substantive and substantial. A definitive and testable alternative, the revolutionary FLINE model of planetary origins and evolution is brought full circle by this solution to the proposed Fourth Law of Planetary Motion, now undergoing peer review. The new model reveals the crucial relationship between the spacing of planets and the forces that power their ongoing stages of evolution.

The solution to the 400-year mystery proved to be the final link in a revolutionary model of the origin of our SS and the evolution of planets—a new concept that evolved during the research years of 1973-1995. This geometric solution that first eluded Kepler in 1595 proved to be the key to bringing the decisive discoveries of the past millennium full circle. It all began with the realization in 1973 that hydrocarbon fuels (gas, petroleum, coal) could not have been created from fossils. The encapsulated imprints of live plants and animals found within coal - the last stage of polymerization of these three fuels—could have been preserved only via a sudden encapsulation of the live plants and animals by petroleum gushing from the earth and inundating lowlands, then polymerizing and cross-linking into solidified coal via these very common processes of Nature. Substantive evidence since then has confirmed beyond doubt that these fuels were created—as was all matter comprising Earth—via these natural processes of Nature, and not from fossils.

The ongoing creation of atomic matter comprising all planets would not be possible without an internal source of energy that keeps the planet active until the energy source is depleted. These internal processes eventually push each planet through five common stages of evolution. Earth is in the fourth (rocky) stage; Mercury and Pluto are in the fifth and final (inactive) stage.

Scientists know that all stars are driven by these internal processes called Internal Nucleosynthesis (i.e., energy into matter, in accord with Einstein's formula, $E = mc^2$). Many scientists also realize that all planets of the Universe are driven via Internal Nucleosynthesis, often referred to as the heat within. The Five Laws of Planetary Motion (FL) reveal the fiery, dynamic origin of our SS. This internal heat, properly called Internal Nucleosynthesis (IN), is the driving force behind all planetary Evolution (E). We have only to look to the skies to observe each of these five stages of ongoing planetary evolution: all at rates in full accord with size.

Fusion-fission processes occur in all spheres of the universe; in making the stars shine, they are the universal powers that drive forward the evolution of all spheres. Rene Descartes, brilliant in his own right, first recorded in 1644 that Earth's center is "Sun-like". His writings on the subject appeared 30 years after Johannes Kepler had discovered the last of his Three Laws of Planetary Motion, but had failed in efforts to discover the fourth law explaining the geometric spacing of the six then-known planets. Soon afterwards, Newton discovered gravity holding the SS together. These great discoveries, along with those of a number of geologists, form the solid foundation of a revolutionary definitive concept of planetary evolution.

The discovery of the mathematical solution to the Fourth Law during the years of 1980-1995 proved to be the final link in the new model of the origin of our SS and the evolution of planets through five stages of evolution via Internal Nucleosynthesis. The Five Laws clearly reveal the dynamic origin of the SS and how the planets attained their orbital spacing. The processes of Internal Nucleosynthesis (IN) in full accord with Einstein's formula clearly are the forces that drive all planetary Evolution (E) forward until the energy-core source is depleted. These three interlocking and inseparable principles serve as a solid foundation for the revolutionary FLINE model of the planetary origins. This new definitive concept provides crystal-clear explanations that are continually corroborated by every relevant discovery about Earth and all other planets, including the recently discovered exo-planets. The solution to every anomaly of every planet and moon has its taproot deeply embedded in the FLINE model—a statement already proven true by its record of accurate predictions.

The new model remains poised to displace the prevailing, but highly speculative, Accretion concept that has survived via speculative modifications since its introduction in 1796 by Pierre Simon Laplace in his *Exposition of the System of the Universe.* Einstein's work defining the energy-matter relationship that earned him the title of *Man of the Century,* along with Descartes' insight into the nature of planetary cores, Kepler's Three Laws of Planetary Motion and the new Fourth and Fifth Laws, stand as beacons capable of guiding scientists to a fully proven understanding of the origins of solar systems and the evolution of planets—and perhaps to understanding our Universe—in the 21st century.

Sadly, the disbelief in Darwin's animate evolution concept has an equally tragic parallel in the inanimate evolution of all things universal. The two are intimately intertwined; both stem from the same taproot of origins: energy. Nature's greatest universal truth can be summarized in two words: *everything evolves.* Evolution—animate or inanimate—is not possible without an energy source to drive it forward. Every planet is a self-sustaining entity that creates its own compositional features via the Internal Nucleosynthesis of atomic matter and subsequent polymerization, etc.—including any form of life existing thereon. We do not need to look to outer space for answers; they have always been right before our eyes and under our feet.

Whether speaking of the animate or the inanimate, the evidence for evolution is indisputable, incontrovertible, overwhelming and conclusive. To the two revolutions Freud designated as paramount (the ideas of Copernicus and Darwin), we must add another one that is destined to displace prevailing beliefs: the FLINE model of planetary origins and evolution. Had Kepler succeeded in his quest (initiated in 1595) for the Fourth Law of Planetary Motion, there would have been no need for scientists later to establish the speculative, but erroneous, concept of planetary accretion from gases, dust, asteroids, comets, etc. The solution (1980-1995) to the elusive Fourth Law was the final link in the FLINE model that inevitably, as with the ideas of Copernicus and Darwin, will force scientists to rethink their beliefs about inanimate and animate origins. In view of its factual, non-speculative structure and vast scientific potential, why must this new model face the same hurdles encountered by those revolutionary ideas?

The definitive FLINE model of planetary origins (1973-1999) consists of three chronological and ongoing realities of Nature: (1) the Five Laws of Planetary Motion (FL); (2) Internal Nucleosynthesis (IN); and (3) Evolution (E). No solar system is possible without undergoing these inseparable realities that remain interactive until, one by one, they reach their ending. They reveal that planetary evolution is not possible with any type of core other than that hypothesized by Descartes as "Sun-like"—a concept now corroborated by abundant substantiated evidence. Such evidence mounts with each relevant discovery of planetary anomalies, both of Earth and of space probes. From energy mass to inactive sphere, every planet is a self-sustaining entity that creates atomic compositional matter via Internal Nucleosynthesis throughout its first four stages of evolution.

During its quarter-century evolution, the revolutionary FLINE model has compiled an impeccable record of accurate predictions. One example best illustrates this point: Its deadly accurate prediction of the extremely powerful explosions of the 21 fragments of Comet SL9 on Jupiter in 1994 stood alone as a giant among the many puny predictions of others, including those of the world's best supercomputers. Recognition of the true fiery nuclear energy nature of comets made it no contest against the erroneous belief that comets are dirty snowballs (Whipple, 1950), the basis used in the computer models.

The 1975 pamphlet, *Fuels: A New Theory,* the initial publication of the revolutionary concept, led to a number of accurate predictions of fuel reserves—gas, petroleum, coal—that always remained ahead of all other relevant predictions; e.g., the tremendous volumes of methane hydrates were predicted and explained before being discovered during the past two decades. By understanding how these closely related fuels were created via nucleosynthesis and polymerization, one can understand how all planetary systems were, and are, made in ongoing processes that adhere strictly to the laws of physics and chemistry. Their entrapped fossils played no vital part during the creation of these hydrocarbon fuels, nor do they play a vital role in their ongoing creation.

Every planetary sphere is a self-sustaining entity that creates its own compositional matter. The definitive solution to every anomaly of planetary systems has its taproot deeply embedded in these five principles of the FLINE model—a fact that soon should bring the Copernican revolution full circle.

The latest substantiated support of the FLINE model's five-stage evolution of planets came on December 5th, 2000, via the news media three days before its scheduled publication in the journal, *Science. Researchers find evidence of lakes on ancient Mars* (*LD News* via AP) reported that photos from a satellite orbiting Mars suggest the Red Planet was once a water-rich land of lakes (Malin Space Sciences Systems). The photos, taken by the Mars Global Surveyor spacecraft, show massive sedimentary deposits with thick layers of rock stacked in miles-deep formations. The pictures show clear views of horizontal deposits of rock, a characteristic of sedimentary layers, in the walls of craters and chasms cut into the surface of Mars. Like Earth's Grand Canyon, such formations are possible only in the presence of water. Further, the splitting of planetary crust via either expansion or spotty sinkhole contraction, as well as volcanic actions, must be taken into account as probable causes of other formations.

"We have never before had this type of irrefutable evidence that sedimentary rocks are widespread on Mars," said Michael Malin. "These images tell us that early Mars was very dynamic [as with all active planets—a basic principle of the FLINE model] and may have been [was] a lot more like Earth than many of us had been thinking." Such incontrovertible conclusions, all readily explained by the revolutionary model, continually move scientists ever deeper into the FLINE paradigm, and perhaps, ever closer to its long overdue acceptance as a valid scientific concept.

The anticipated findings on Mars add to the impeccable list of accurate predictions of the past two decades of the evolving FLINE model. This smaller planet is entering, if not already in, its fifth and final stage of evolution (inactive, dead), bleak and barren, just as Earth is destined to become after its current fourth evolutionary stage completes its cycle in a few billion years. Such depletion of the energy core in each planet and moon defines the boundary between the fourth and fifth stages of planetary evolution. These substantiated findings clearly corroborate the fact that planets undergo the five-stage evolution at rates in full accord with size.

The FLINE paradigm goes a giant step further in explaining how the planets evolve through their five stages of evolution, via Internal Nucleosynthesis from energy into atoms that form the molecular matter comprising all atmospheres, lands, waters, life, etc. Understanding this new model opens the floodgates to full comprehension of these predicted discoveries on Mars, and eventually, to irrefutable solutions to all planetary anomalies. But in the process, scientists must re-examine a number of ancient myths, especially the modified Laplace accretion hypothesis of 1796, allegedly explaining the origin of planets; they do not accrete from gaseous dust, planetesimals and comets, all of which are by-products of creation processes involving the intimate relationship among black holes, quasars, galaxies, solar systems, stars, etc.

The findings on Mars, Earth, exo-planets and all similar spheres shout out the dynamic origins and on-going evolution via Internal Nucleosynthesis of planets and moons. At the current pace of such discoveries of irrefutable evidence on other planets, can it be only a question of time until the shouts are heard and heeded to the point of displacing the antiquated myths about origins of solar systems and the evolution of planetary spheres?

Thirteen Revelations of the FLINE Model

- Reveals the beautiful geometric origin of our solar system, thereby enabling scientists to understand its origin and history, and to predict precisely the future displacement and fate of each of the planets.
- Details how each planet became spaced in its present orbit.
- Reveals how the precise geometric clockwork of the solar system came into being.
- Describes how planets evolve through five common stages of evolution via Internal Nucleosynthesis.
- Details how Earth became Earth: the origin of its atmosphere, land and water—not from outer space!
- Explains the anomalies of all recently discovered extra-solar planets.
- Describes the origin and evolution of Earth's contemporary partner, the Moon.
- Reveals the ten clues to the true nature of Earth's core: nuclear energy.
- Explains the explosive power of the epicenters of earthquakes and their precursory signals.
- Details factually the non-biological origin of fuels (gas, petroleum, coal)—not from fossils!
- Reveals the source and true nature of fiery comets.
- Explains the powerful explosions of Comet SL9 at Jupiter's cloud-tops.
- Details why species come and go: provable reasons dinosaurs became extinct.

Now after 23 years of research into the myths and truths of science, the solutions to some 41 anomalies of planetary systems intermesh in a continuity of powerful, incontrovertible evidence—as if in a giant jigsaw puzzle with every piece locked into place. The FL-IN-E concept is a definitive and testable alternative to current beliefs about our planetary origins and evolution—a genuine paradigm shift ready to be examined more closely by scientists.

Solutions to Anomalies via the LB/FLINE Model

The Four Principles (LB-FL-IN-E) comprise the Rosetta stone, the crucial key for solving anomalies of universal systems. They supply answers to many problems of both the Big Bang and the Static models of our Universe. Listed below are 41 examples (out of many) that illustrate the capabilities and potential of the new model.

1. Why planets obey Kepler's First Three Laws of Planetary Motion.
2. The complete geometry of the Solar System's origin.
3. The original and current spacing of planets.
4. The valid explanation of the enigmatic Bode's Law of the spacing of planets.
5. Why Uranus, Neptune and Pluto do not obey Bode's Law.
6. Why planetary orbits are inclined to the ecliptic.
7. Why the displacement (AU) of each planet from its original orbit.
8. The highly elliptical and large inclination of the orbit of Pluto.
9. The speeds and directions of rotation and evolution of planets.
10. The slow spin of Venus.
11. The huge size of Jupiter (the Geometric Mean of the SS).
12. The tapered sizes of the planets.
13. How large moons formed and evolved contemporaneously with planets.
14. Why the Sun's equatorial mass has a 7° inclination to the ecliptic.
15. Why the Sun's equatorial mass rotates faster than the remaining mass.
16. The huge discrepancy in angular momentum of the Sun and the planets.
17. The nature of Earth's core: nuclear energy (ten clues).
18. The abiogenic origin and evolution of hydrocarbon fuels (1973-1980).
19. The origins of surface features: water, land, salt mines, etc.
20. The expanding Earth: increases in sea level and land.
21. How the Asteroids came into being: Olbers was right.
22. The craters on moons, asteroids, comets and planets: Mills was right.
23. How planets evolve through five common stages in accordance with size.
24. The physical and chemical differences among planets and moons.
25. Why species come and go: the extinction of the dinosaurs.
26. Why Earth and other planets are self-sustaining entities.
27. Plate tectonics: from highly mobile to barely movable.
28. Why the electromagnetic strength of each planet is different.
29. The unexpected powerful explosions of Comet SL9 on Jupiter.
30. The origin and evolution of comets.

31. The origin and fate of planetary rings.
32. The cometary moons of Mars: Phobos and Deimos.
33. How the discoveries of space probes fit into the FLINE concept.
34. The 2.7 K radiation throughout the Universe: not from the Big Bang.
35. The erroneous notions of the Big Bang.
36. The too-high Hubble constant revealed by the Hubble Space Telescope.
37. The new conflict between the oldest stars and a younger Universe.
38. How and why the Universe expands at its spherical perimeter—forever.
39. The 51 Pegasi system discovered by the HST.
40. The large extra-solar planets: 70 Virginis and 47 Ursae Majoris.
41. The origins of known extra-solar systems (50+) and binary systems.

The Three Fundamental Infinities

To understand the LB/FLINE model of universal origins, we begin with the three plausible infinities: Space, Energy, Time (SET). Space remains totally dark at absolute-zero temperature until exposed to energies, usually in the form of light, an event that is happening continually at the spherical perimeter of our Universe. The exiting energies, moving at the speed of light, crash into the absolute-zero coldness of space where all atomic motion ceases. The result is the collapsing of energies into the densest form of energy: tiny black holes that grow ever larger by accretion of more and more energies as they move outward at the speed of light on the perimeter of our spherical Universe. (6) Meanwhile, Time else moves onward into Dark Space, forever creating spherical Space-Time.

Eventually, due to the pull of universal gravity, each black hole (BH) is slowed somewhat, warmed slightly and triggered into a continuous ejection of dust and brilliant stars from the two poles of the spinning black mass, creating a beautiful pinwheel galaxy. This process continues until it reaches a balance between each black hole and its surrounding masses of billions of stars buried in gaseous dust-clouds. (6) As determined recently by Carl Gebhardt *et al.*, this balance occurs when each black hole is reduced to a mass equal to 0.5% of the ejected mass. The remnant of the original BH remains at the center of each galaxy to hold the system together; no other forces such as dark energy or dark matter are needed to hold the masses in orbit around the BH. (6)

This ongoing dynamic formation of speeding galaxies explains the amazing discovery that the farther out astronomers look in any direction, the faster the galaxies are moving away. The LB/FLINE model, when visualized in the manner explained above, needs no cosmological constant or other forces to explain this phenomenon. (6)

According to the new model, nuclear energy cores of all active spheres of the Universe consist of quark-gluon plasma (qgp); it is the energy source of their Internal Nucleosynthesis: the ongoing transformation of nuclear energy into all forms of radiation and atomic/molecular matter found in varying quantities in all stars and all active planets, moons, comets, etc. It is what makes the stars shine, and spheres evolve. It is key to understanding the interior of atomic nuclei and gravity; in all forms of atomic energy and matter, gravity is an inherent property that holds everything in orderly fashion. (6)

But rather than this continuous process in the ongoing, expansive universal creation in which all data, both old and new, fit precisely into the new model, Big Bang advocates interpret qgp as a one-time primordial product produced at the birth of the Universe some 13 billion years ago—a belief that puts undue limits on our expanding, evolutionary Universe—and forces scientists into speculation and misinterpretations of their excellent discoveries and subsequent data. (1) (2)

All universal systems continually evolve from energy into matter; nothing is as it once was, and nothing is where it was before—an impossibility within the rigid confines of the BB concept that places undue limits on the ongoing creation of atomic elements comprising universal systems. Evidence clearly shows that in the absence of Internal Nucleosynthesis (IN) via qgp in planets, Evolution (E) of planetary systems (comprised of virgin atomic matter) via IN would not be possible: qgp, IN and E are inseparable; one cannot exist without the other. (3) (4) (5) (6) The ongoing, never-ending source of the qgp via an ever-expanding Universe is key to understanding universal origins and the integrated functions of universal systems. (6) Perhaps recognition of the LB/FLINE model will come when scientists learn the true nature and sources of asteroids, comets and the cosmic microwave background (the 2.7 K radiation) early in the 21st century (4) (6) (8) (9)—findings that will definitively expose the sophistry of the BB/Ac myth.

References

(1) Thomas Kuhn, *Structure of Scientific Revolutions*, (1956).

(2) William Mitchell, *The Cult of the Big Bang. Was There a Bang?* (1995). (Cosmic Sense Books, Carson City, Nevada 89702).

(3) Alexander Scarborough, *Bringing the Copernican Revolution Full Circle via a New Fourth Law of Planetary Motion*, *Journal of New Energy*, Fall 2003. Proceedings, National Philosophy Alliance, International Conference, June 9-13, 2003.

(4) Alexander Scarborough, *The Spacing of Planets: The Solution to a 400-Year Mystery*, (1995), 8th Edition, Energy Series. (Ander Publications, LaGrange, Georgia 30241).

(5) Alexander Scarborough, *Origin and Evolution of Planetary Systems*, *Journal of New Energy*, Fall 2003. Proceedings, NPA, International Conference, June 9-13, 2003.

(6) Alexander Scarborough, *New Principles of Origins and Evolution: Revolutionary Paradigms of Beauty, Power and Precision*, (2001), 9th Edition, Energy Series. (Ander Publications, LaGrange, Georgia 30241).

(7) R. A. Karam, J. M. Wampler, Neely Nuclear Research Center, Georgia Institute of Technology, Atlanta, Georgia. *Characterization of Natural Gas, Oil and Coal Deposits with Regard to Distribution Depth and Trace Elements Content*, (1986).

(8) R. G. Lerner, G. L. Twigg, *Encyclopedia of Physics*, Second Edition, VCH Publishers, Inc., NY.

(9) N. S. Hetherington, *Encyclopedia of Cosmology*, (1993). (Garland Publishing, Inc., NY).

Origins of Solar Systems and the Evolution of Planets

FL-IN-E Model vs.	ACCRETION Model
1. Structured solely with facts; non-speculative	1. Remains highly speculative; not factual
2. Explains the spacing of planets in all solar systems	2. Cannot explain the spacing of planets in any SS
3. Explains planetary evolution via Internal Nucleosynthesis	3. Cannot explain planetary evolution via metal cores
4. Explains planetary energy sources that drive evolution	4. Does not explain these energy sources in planets
5. Explains compositional differences in all planets	5. Speculates on these compositional differences
6. Explains differences in sizes of all planets	6. Does not explain differences in sizes of planets
7. Explains the sources of moons around planets	7. Speculates on origins of planetary moons
8. Explains the origins of planetary rings	8. Speculates on origins of planetary rings
9. Capable of explaining all planetary anomalies of our SS	9. Incapable of explaining our planetary anomalies
10. Explains origins and differences of all moons	10. Speculates on origins and differences of all moons
11. Explains the sources, nature, anomalies of comets	11. Speculates on these sources, nature, anomalies
12. Explains sources, nature, anomalies of asteroids	12. Speculates on these sources, nature, anomalies

13. Incorporates three inseparable principles of Nature	13. Ignores these three inseparable principles of Nature
14. Explains electromagnetism via natural energy laws	14. Attributes electromagnetism to metallic cores
15. Based on works of Descartes, Buffon, Kepler, Olbers	15. Based on antiquated beliefs of Kant, Laplace, *et al.* Ignores works of Descartes, Buffon, Kepler, Olbers
16. Obeys the Second Law of Thermodynamics	16. Violates the Second Law of Thermodynamics
17. Fully explains all anomalies of extra-solar systems	17. Can't explain any of these anomalies; speculates
18. Explains why our SS layout is so rare among many SSs.	18. Cannot explain the rarity of our SS's layout.
19. Predicts and explains all anomalies of probes to planets	19. Cannot predict or accurately explain these results
20. Has an impeccable record of accurate predictions	20. Findings are usually surprising and puzzling
21. Supported by powerful corroborative evidence	21. Not supported by corroborative evidence

Conclusion

Embracing the great ideas of Copernicus, Galileo, Kepler, Newton, Descartes, Buffon, Dutton, Einstein *et al.*, the revolutionary FLINE model of planetary origins and evolution was brought full circle with the solution to the Fourth Law of Planetary Motion explaining the spacing of planets around our Sun. The original ten orbits of our Solar System, along with Kepler's First Three Laws, verify both the dynamic origin of solar systems and the correct interpretation of Olbers in 1802 that Asteroids are fragments of disintegrated planets. Further, in view of the acknowledgment by astronomers that each giant gaseous planet in extra solar systems is too close to its central star to have formed there via accretion of dust, gases and planetesimals, adds powerful evidence to the FLINE model that argues strongly against the accretion concept of planetary origins. Thus, this new model seems destined to displace the modified Kant-Laplace Accretion concept of planetary origins early in the 21st century. This change in direction of scientific thought will be expedited with the eventual confirmation of the true nature and sources of comets (ref. example, Comet SL9) and asteroids (Chapter IV).

The inseparable connection between Internal Nucleosynthesis and Evolution will lead to understanding the five stages of evolution of planets and moons, and why all active spheres are self-sustaining entities that create their own compositional matter. A prime example of the manner of creation of all planetary matter by means of natural laws is the origin and evolution of hydrocarbon fuels (erroneously called fossil fuels). With its foundation solidified by the Four Laws of Planetary Motion, the FLINE paradigm definitively connects past, present and future discoveries about all planetary systems. The recent discoveries about Mars, entering its fifth (inactive) stage, and other planets and moons continually corroborate, and fulfill predictions of the FLINE model, thereby enhancing its position as a valid scientific theory that warrants closer examination.

At this writing, the new Fourth Law explaining the spacing of planets has been in peer review by The Royal Astronomical Society in London for three years.

References

1. A. A. Scarborough, *The Spacing of Planets: The Solution to a 400-Year Mystery,* (1996). (Ander Publications, LaGrange, Georgia, 30241).

2. W.C. Mitchell, *The Cult of the Big Bang: Was There a Bang?* (1995). (Cosmic Sense Books, Carson City, Nevada, 89702).

<u>Note:</u> Such original works usually do not require or use many references other than commonly-known theories.

Exploring Earth's Mysteries
Global Expansion and Continental Drift

As the internal energy of our planet undergoes transition into matter, the continuous increase in the volume of hot molten mantle causes a slow, steady expansion of the crust. The newly-created matter must find more room along the paths of least resistance. Something must give, and it does. The crust cracks apart as it slowly expands, and new materials quickly fill the gaps and holes. The subsequent gains in size of the crust are offset by equivalent losses in the core; thus, the planet's mass remains constant, while its density slowly decreases.

If necessary, the new matter makes room for itself by moving older matter out of the way. When this happens on a large scale, the result may be a new mountain or a range of mountains or a huge plateau of land pushed straight up. The Grand Canyon plateau is an excellent example of flat land pushed a mile higher by these circumstances. Other examples include the spectacular tilted mountains common all over the globe.

Other than the elevated land masses, the same powerful uplifting force, in conjunction with isostasy, created the deep oceans and their underwater ranges. The crustal expansions of lands and waters on Earth's surface caused separations of land masses, much like dots separate on a slowly inflating balloon. The cracking played a role in creating the "plates" comprising Earth's floating crust. Such happenings were violent and extensive during the young stages of Earth's evolution, but have decreased geometrically through the eons. One might question how large a role these factors play in the popular Continental Drift (CD) theory.

The theory of continental drift was first published in 1915 by Alfred Wegener, a German meteorologist. He taught that land masses on Earth were originally one vast continent (Pangaea), which gradually split into several smaller continents that may still be drifting apart. In spite of its supportive evidence, the theory was ridiculed and declared impossible by colleagues. Wegener died an intellectual outcast in 1930. Now, his idea is widely accepted by geologists under the popular name of "plate tectonics".

The CD theory teaches that the crust of our planet consists of about eight large plates and more than a dozen smaller ones, all floating on the molten mantle, traveling in various directions like conveyer belts, at speeds between one-half-inch-to-six inches per year. At the junctions where plates separate, collide or shear, earthquakes occur. In the past, huge mountain ranges and volcanoes were created by collisions of plates.

Continents do float on the molten mantle. They can and do drift to some small degree, but not nearly as much as in the beginning before the crust thickened sufficiently to stabilize itself via solidification, locking the continents in their current positions. Our crust can be visualized as resembling the shell of a moderately cracked hard-boiled egg. Some movements can occur in each shell segment (plates), but all segments remain secure in their relative positions.

Yet, too many critical questions remain unanswered in the current CD theory. How can it account for the gigantic chain of underwater mountain ranges that encircle our globe in a pattern resembling the seam of a baseball? Were these mountains created by collisions of plates? Were the mountains on Moon, Mars, Venus and other spheres created by collisions of plates? And how could the plates on the other side of the globe drift if squeezed in the vice of Pangaea's scattering plates?

The CD theory, no pun intended, appears to be on shaky ground when viewed in the perspective of current beliefs. It cannot answer these questions except in the perspective of the FLINE concept. First, the chain of underwater outlets is the exact pattern one would expect from a steady global expansion by extrusions from within. The steady flow of virgin matter from the underwater baseball-like seam (as from "hot spots") around the globe consistently drives the continents apart, like dots on an inflating balloon, as the circumference expands.

Mountains, as discussed previously, were pushed up by the powerful combination of internal nuclear and isostatic forces on all spheres on which high or elevated peaks exist. While mountain building via internal power is obvious in volcanic actions, it is not readily discernible in other types, unless one views them in the perspective of the FLINE Model. Then, that uplifting force mentioned by Dutton stands out more clearly than when he studied the landscape without the benefit of today's knowledge of nuclear power.

In the original ITEM concept, the global crust formed fairly evenly under a blanket of shallow seas worldwide. This crust-under-a-blanket-of-seas theory was first proposed by Wegener to help explain why marine fossils can be found embedded in high peaks. The two theories are in total agreement on this point. Powerful expansive forces from within lifted whole plateaus and mountain ranges from beneath the seas to form large land masses that became our continents. Simultaneously, the sinking of sea floors, via the process known as isostasy, formed the first deep-water oceans.

In future measurements of movements of continents, the interpretations of any increases in distances between them should be considered in the perspective of a slowly expanding globe in which virgin materials are positioned as new crust.

Birth of an Ocean

Discounting volcanic outpourings, it was necessary to wait only six years for the most startling and exclusive evidence of the ITEM concept in action. The article by Uwe George first appeared in GEO (Vol. 1), 1979, and was reprinted in *Reader's Digest*, September, 1979. In seven spellbinding color photographs that accompanied the article, *An Ocean Is Born*, the German writer/photographer recorded a rare and momentous event: a geological spasm along East Africa's Rift Valley, where an entire continent is slowly splitting apart to produce a new ocean.

The process is similar to the one that split Africa and the Americas into three continents some two hundred million years ago. At that time Earth's crust separated in the manner previously described, breaking the super-continent apart. As the three parts slowly separated, the Atlantic Ocean grew ever wider. Even today its growth continues as the globe expands with new materials.

The present distance between Africa and the Americas is harmonious with the amount of expansion planet Earth has undergone since that time. In accordance with the ITEM concept (formerly called TIFFE), virgin matter has continually extruded from the rift (now the mid-ocean seam) for two hundred million years while pushing the continents apart. As described previously, the planet was, and still is, expanding in the manner of a balloon being slowly inflated. The results have been continental drift via creation of more and more waters and lands, along with disappearances of large land masses such as Atlantis via isostatic forces. (More on Atlantis later)

The George photographs recorded the actual extrusions of various materials from Earth's interior. Most notable among these were spirals of tons of boiling sodium carbonate surging, swelling and spurting to the surface. Thousands of miniature soda geysers found their way to the surface through a network of dry cracks. A few grew to sizable spirals, hundreds of yards across.

These extrusions illustrate the manner in which all shafts of pure or mixed materials were formed from Earth's interior. For example, the bottomless shaft of pure white marble in Tate, Georgia, described earlier, gives all indications of having been formed in this manner, except that calcium carbonate, instead of sodium carbonate, was extruded. Slight differences in internal conditions dictated whether sodium or calcium would be made in each case.

Imagine the excitement of seeing the George photographs less than two-and-one-half years after publication of descriptive details of such phenomenon in the third edition of this book. In the same edition, it had been predicted that many years would pass before sufficient evidence would become available to prove the ITEM concept. In view of this evidence meshing so well with all other cited evidence, the theory appears irrefutable. The passage of time is adding more and more supportive evidence, while opening doors to long-sought answers to creation of planet Earth and its systems.

More on Earthquakes: Understanding the Causes

Earthquakes are fascinating and deadly clues to the inner working of Earth, while planet quakes and moonquakes present opportunities to learn more about internal functions of all planetary spheres. Current beliefs guide research into the mechanics of faulting, or rock mechanics, as the basic cause of earthquakes. In this perspective, a number of earth science disciplines have combined forces to accomplish much work in basic physics of rock friction, crack propagation, fault zones, seismotectonics, hazards and predictions.

The prevailing theory of the cause of earthquakes teaches that tectonic plates grind against each other, resulting in build-up of stresses and strains that eventually are relieved by sudden movements, or cracking of rocks. The mechanics of the stresses include pressures created by the effects of heating and cooling and the sheer weight of the rocks.

However, some problems have been encountered with this concept, requiring much ingenuity to squeeze its ill-fitting pieces into the jigsaw puzzle. All earthquakes don't fit into established patterns of expectancy: many occur in central sections of plates (e.g., Hawaiian Islands in the middle of the Pacific plate and Charleston, SC in the North American plate), rather than at the intersections where grinding takes place, and many occur at depths of 400 miles of more, where rocks are too malleable or molten to fracture. Further, the large, vertical movements of land during a quake often exceed its lateral movements; e.g., large sections of land rose as high as 50 feet in North America's greatest earthquake in 1964. No lateral movements were reported in this explosion that shook the entire state of Alaska.

On March 27, 1964, North America's greatest earthquake, with its epicenter at the head of College Fiord, 75 miles north of Chenega, Alaska, shook the entire state, extending as far as 800 miles. It registered an 8.5 magnitude on seismographs around the world. Some 100,000 square miles of land lifted up or subsided: large sections rose as high as fifty feet; others dropped by more than six feet.

For the next two days, there were 55 aftershocks exceeding 4.0 on the Richter scale, including one of a 6.7 magnitude. Aftershocks continued over the next 18 months. In spite of its viscosity and because of several favorable factors, the disaster killed a relatively low number of people—114 total.

In March, 1986, a series of 50 small earthquakes rattled the Lake Sinclair area in central Georgia. The largest measured only 2.5 on the Richter scale, so no structural damage occurred. One resident living directly above the center of the quake stated, *"It feels like a tremendous amount of dynamite going off, an explosion kind of thing."* His description was deadly accurate.

In the perspective of the ITEM concept, these and other anomalies are much easier to understand: they become key pieces that fit precisely into the completed jigsaw puzzle. All earthquakes have an epicenter, or central point of greatest explosive force, from which P and S waves radiate, and are duly recorded on seismographs around the world. The intensity of the quake's epicenter is measured on the well-known scale of 1-10.

What was the source of the super powerful force capable of moving massive land sections like straws in the wind, lifting some areas as high as 50 feet straight up? There should be no doubt about its explosive nature, more powerful than several shiploads of hydrogen bombs. The extended pattern of aftershocks indicates a long series of smaller explosions. Yet, the severe violence on the surface pales to insignificance by comparison with the unimaginable violence of these underground explosions and magma stirrings that fed the aftershocks.

The dearth of significant lateral movements of land, plus the factors described above, challenges the concept of plate slippages as the cause of such earthquakes, while strongly supporting the conclusion that all planet quakes and moonquakes are caused by internal explosive forces. As published in the 1986 edition, fault slippages are **results** of such explosiveness—not the **cause** of them.

Georgia Tech seismologist, Leland T. Long, reached the same reversed conclusion in 1989. Further, faults denote zones of weakness that can actually disappear after a quake, according to Long, leaving only healed scars for geologists to interpret. Where shallow quakes might be attributed to hydrostatic pressure on water-porous crustal structures, deep quakes denote more serious hydraulic movement.

Long's beliefs mesh into the ITEM concept, except for the explosive nature at the epicenter of quakes. Where Long attributes it to hydraulic pressures, as do many geologists, the ITEM concept points to nuclear energy in the form of solar flares or huge jets of gases from the energy core as the source of powerful and numerous explosions that rock Earth and its magma with some one million quakes every year.

These sun-like actions of Earth's core should give clues to their strong influence in creating earthquakes, and apparently they do so. Research teams recently reported finding unusual pre-quake crustal signals: electromagnetic bursts and elevated gas emissions. One report published in NATURE, Sept. 27, 1990, describes how two Japanese scientists, Kozo Takahashi and Yukio Fujinawa, found that on several occasions anomalous electromagnetic changes preceded, by a few hours, several quakes and an undersea volcanic eruption that shook the central eastern coast of Japan in July, 1989. Their monitoring system detected the electromagnetic bursts in the very low frequency ranges, between 1 and 9 kilohertz.

Just hours before the October 17, 1989 California quake, an ultra low-frequency radio receiver set up for other purposes, accidentally recorded unusual magnetic signals in the 0.01 to 10 hertz range near the epicenter-to-be. The radio background noise shot up to 30 times its previous level during the three hours before the quake. The recordings were discovered when scientists happened to check their data several weeks after the earthquake.

Since U.S. scientists have tended to ignore electromagnetic measurements for a long time, realization of their significance remained elusive. However, such recordings in the future should prove to be indispensable tools in early warnings of pending earthquakes.

Some scientists think anomalous soil-gas levels may be good quake predictors. For example, 32 hours before the October 17th quake, a soil-monitoring station on the San Andreas Fault recorded a significant increase in helium levels at Stone Canyon, 37 miles southeast of the eventual epicenter. This helium increase contradicts ten years of data that show decreases in helium preceding San Andreas quakes of magnitude four or greater.

Scientists also have observed a gradual increase, followed by a sudden decrease, in concentrations of radon before some Hawaiian temblors. Although scientists can't identify the specific processes behind the helium and radon fluctuations, they believe any unusual changes in soil-gas levels may signal a future quake. *"Gases really should play a part in earthquake prediction,"* Says G. Michael Reimer, a USGS geologist in Denver, thereby moving a step closer to the ITEM concept. Geochemist Donald M. Thomas of the University of Hawaii's Center for Study of Active Volcanoes in Oahu warns, *"Until you understand the process, it's going to be very difficult using gas geochemistry…to predict earthquakes."*

Both electromagnetic bursts and elevated gas emissions are symptomatic of sun-like masses of nuclear energy. When viewed as such, a better understanding of the process is assured. In the perspective of the ITEM concept, these symptoms should prove to be the best clues for predicting earthquakes at least as accurately as scientists forecast tornadoes. One can't help but wonder how many lives might have already been saved if the ITEM concept, offering a better understanding of earthquakes via its nuclear core evidence, had been taken more seriously as far back as its introduction in 1975.

Source of Iridium Layers: Volcanism or Asteroid Impacts?

The final decision in the debate on volcanism vs. asteroid impact rests in identifying the true source of the iridium found in the K-T boundary that marks the end of the dinosaur era. As is true with all other substances that comprise Earth's crust, the iridium came from within the planet as typical massive amounts of ejecta material—usually via vast volcanic outpourings. According to the ITEM concept, and with the assistance of sedimentary actions, Earth's crust was built, layer by layer, in this manner.

In this factually-based perspective, any assumption of an asteroid impact to bring the iridium from outer space seems superfluous. Besides, would the massive momentum of such a high-impact projectile bury its component parts deeply within the crust—or would it explode and scatter the iridium around the world as efficiently as volcanism is capable of doing?

For further thoughts on the subject let's read a recent (1991) letter from an expert on dinosaur fossil evidence:

"I used to believe the Alvarez impact theory of dinosaur extinction. It was appealing because it was a simple, dramatic and tragic explanation of the Cretaceous extinctions, which have puzzled people for several decades.

Last summer I spent two weeks in Montana excavating fossils from rock strata straddling the Cretaceous-Tertiary (K-T) boundary. The fossil evidence in the Hell Creek (Cretaceous) and Tullock (Tertiary) formations shows a decline in dinosaur populations and diversity at the end of the Cretaceous. Thirty-three dinosaur species present 100 million years ago gradually decline to 13 species just below the iridium layer. Contrary to the predictions of the impact theory, 13 dinosaur species appear just *above* the layer. They gradually disappear over the next 500,000 years. The fossil record shows that no dinosaur species go extinct across the boundary. It takes another 500,000 years until dinosaur fossils are absent from the Tertiary strata. The fossils we uncovered were not abraded, which rules out the possibility of Tertiary erosion and redeposit of Cretaceous fossils."

The outpouring of massive amounts of mantle material from the Deccan Traps eruptions in India over a period of 500,000 years injected large amounts of sulfur aerosols into the atmosphere. In a large enough scale, these emissions would cause acid rain, reduction in pH of the ocean surface, global atmospheric temperature changes and ozone layer depletion. The Deccan Traps eruptions, as Alvarez notes in the *Scientific American* article, "*represent the greatest outpouring of lava on land in the past quarter of a billion years.*" In addition to the flood basalt volcanism in India, there is evidence of substantial explosive volcanism at the K-T boundary over an extensive area from the South Atlantic to the Antarctic.

The iridium layer is most often cited as evidence of an extraterrestrial body. Although iridium is rare in Earth's crust, it is more plentiful in the mantle. Modern samples taken from the Kilauea volcano in Hawaii show a prolific enrichment of iridium in deep-mantle lavas (about 100,000 times) over other Hawaiian lavas.

Contrary to Mr. Chapman's assertion that we *"know what caused most K-T extinctions: an impact,"* the paleontological record is ambiguous. The controversy over the impact versus eruption theories is science at its best. It has stimulated the close examination and interpretation of the geological record from a variety of disciplines. Although chastised by Chapman, *Scientific American* did its reader a service by presenting both sides of the debate and letting readers draw their own conclusions.

I challenge Mr. Chapman to get out and dig (literally) into the fossil record in Montana (or anyplace else where the K-T boundary is exposed). Although lacking in drama, terrestrial volcanism lasting over several hundred thousand years is a more probable cause of the Cretaceous extinctions.

-- Carol A. Schneier, Marietta, Georgia

The case for the Alvarez impact theory grows weaker as time permits more scientists to ponder the evidence on both sides of the issue. In the end, all beliefs based on impact theories rather than on ejecta concepts will be proven wrong. This, of course, includes the current belief that our Moon formed when a giant asteroid struck a glancing blow on Earth, taking a chunk of our planet with it as it bounced into orbit to form into the spherical beauty we admire from afar.

When viewed in the perspective of the LB/FLINE concept, all such impact theories will seem comically illogical. Earth and Moon initially formed as a double-planet system, evolving simultaneously. In slowly drifting away from Earth, Moon is in compliance with the Fifth Law of Planetary Motion.

An important lesson can be learned from the debate on volcanism vs. asteroid impact: Earth is indeed a self-sustaining entity, an autopoietic Gaia that can supply answers from within itself; one does not need to look to outer space phenomena for explanations of its anomalies.

Why Species Come and Go (1986)

Throughout the evolution of planets from small fiery masses into crusty old spheres with ever-diminishing nuclear cores manufacturing elementary building blocks for all things animate and inanimate, change is the only constant factor. In Earth's case, the hot and humid environment spawned bacterial life that eventually formed into self-replicating cells that, with the aid of chemical evolution and self-assembly processes, multiplied into life forms. As in the beginning, the ever-decreasing synergistic energy, in synchronization with the products of the powerful internal forces, still wields its powerful influence in establishing the cycles and destinies of all species.

The evidence for at least four other mass extinctions of species exists. Similar iridium layers have been found at the end of the Eocene, at the end of the Permian, between the Middle and Upper Jurassis, and toward the end of the Devonian.

The evidence for repeated mass extinctions led in 1984 to a theory of cyclical cataclysms. All the evidence on reliably dated, documented extinctions over the last 200 million years was computerized. Analysis revealed that the rate of extinctions was not constant; rather, it rose and fell through 12 peaks. Eight of the peaks fit closest to a 26-million-year cycle. Because no Earthly cycles were known to take so long, many scientists again looked outward into space instead of inward for the answer.

Comets seemed to supply the best answer, but this called for Nemesis, a hypothetical, unseen companion star to the Sun. It was necessary to speculate that the unknown Nemesis might exert enough gravitational pull on comets from the speculative Oort cloud of icy masses to disrupt their theoretical distant orbits around our Solar System, and send some of them careening toward the Sun! Then, once every 26 million years or so, one or more comets might hit Earth! And, thereby, a tangled web based on pure speculations was woven.

However, some sanity did prevail. Dewey McLean of Virginia Polytechnic Institute used available evidence to theorize that instead of a sudden catastrophe, the late Cretaceous experienced a gradual increase in the rate of species' extinctions, which merely peaked at the K-T boundary and then slowly declined to a more normal rate. His belief, now supported by many scientists, that extinctions came gradually because of changes in volcanic eruptions, meshes well with the ITEM concept.

Severe periodic eruptions throughout Earth's history have indeed controlled the cycles of all species. Scientists need not look to outer space for origins of species or cause of their extinctions. Earth creates, sustains and perishes its own, simply by changes induced by violent eruptions that create while they destroy. These changes include global warming (greenhouse effect) and cooling (ice ages). Species that cannot adapt their reproductive endocrinology to the temperature, chemical, vegetation, atmospheric and other changes soon become extinct. The question of time is relative to the severity of the changes. The severity of change is proportional to the severity and types of global outpourings. Such outpourings were worldwide initially, but have become more and more localized as the crust continued to thicken as a function of time.

McLean's discovery of a link between air temperature and extinctions and reproductive endocrinology is a vital clue to the extinctions of the dinosaurs and other species. The iridium layers are vital clues to the cause of temperature variations and other critical changes. Violent eruptions spewed the virgin iridium from deep within Earth's internal manufacturing plant—its nuclear energy core—and set in motion the changes in environmental conditions necessary for extinctions of species and rebirths of others. Even today, Nature continues to dictate its control over the destinies of all current and future species, although at perhaps a slower pace and with smaller-size species.

Earth's Ubiquitous Water (1991)

Water, water everywhere. It covers most of the surface of our planet, and is found underground almost any place, no matter how deep the drilling. Just as it does today, this ubiquitous compound of hydrogen and oxygen dominated the scene throughout all, except the very earliest stages, of Earth's evolution. It played critical roles in extinguishing fires and in transporting elements and compounds that comprise the crust. Even the hydrocarbon fuels, gas, petroleum and coal, do not escape the presence of water.

The ubiquity of water makes it obvious that hydrogen and oxygen were two of the most abundant elements created from the energy of Earth's core. This fact, along with the great abundance of carbon, offers a valid explanation for the tremendous volumes of methane trapped in underground waters, as described in an earlier chapter. The ITEM concept suggests that the two elements, along with the abundant carbon and lesser quantities of nitrogen, sulfur, helium, metals and other atoms found in fuels, formed into steam and gases that escaped along the easiest routes possible until trapped by impervious layers of rock. Subsequently, partial polymerization of the gases into petroleum formed the contaminated pockets of gas-oil-water combinations familiar to all geologists. Additional contaminants were picked up along the upward routes, and continued to accumulate to the extent now found in surface coals. Water, thereby, played a role, good or bad, in the formation of its companion energy fuels.

However, we have evidence today that many compounds were created and deposited in situ without the presence or assistance of water. Examples discussed previously include salt, marble and other chemical deposits. Such pure deposits apparently were made directly from the gaseous products of the thermonuclear source without utilizing water as a carrier. Heavy metals and rare metals should fit into this category. Such flawless depositions from the gaseous state explain the high purities of metals like gold, and of other uncontaminated substances like salt mine and white marble shafts and sodium carbonate and other chemical outpourings.

Thus, the ITEM concept vindicates Aristotle's belief that metals are deposited directly from the gaseous state.

The evidence for internal creation of elements and subsequent compounds seems incontrovertible. The new concept appears destined to open many avenues of research for the scientific community, while altering the direction of scientific thought on creation of matter and systems universal.

Lightning: A Shocking Revelation

Like fish in the sea, we live in an ocean of electrical energies. And like fish needing water, we could not survive without these energies to which we owe the very composition and existence of our bodies and minds, and perhaps souls. In the absence of these energies within and around us, existence would be impossible. These energies make it possible for us and our worlds to exist.

The beauty, power, and variety of lightning are well known. The most familiar of these instantaneous energy discharges are ground-to-cloud (erroneously called cloud-to-ground) strikes. Other lightning occurs as intra-cloud, air discharge, eye of the storm, and the rare ball and ribbon displays.

Intra-cloud lightning occurs when electrons drift upwards under milder conditions of updraft, lower moisture content of the air, and lower concentrations of these energies en route. Discharge occurs when sufficient positive potential is accumulated in the clouds.

When atmospheric conditions are such that the concentration of electrons near the ground receive no updraft and have no clouds low enough to reach, these energies take the form of either ball or ribbon lightning. These rare, fascinating displays just above ground level tend to linger longer than the GC flashes that move at one-third the speed of light. But their potential for moving equally fast remains dormant as long as the environment offers no conditions for these energies to flash upwards.

Every atom of your body and every atom in the air you breathe and the food you eat consists of powerful nuclear energies from which they were made. We experience the magnetic field forces of Earth when we see a compass point to the north. The jolt of a shock when we reach for a doorknob after treading across a nylon carpet is a sharp reminder of the electricity that can accumulate in your body by this simple act of walking. The powerful discharge to the doorknob can amount to thousands of volts of electricity, easily accumulated by you from the environmental ocean of electrical energies. In reality, the jolting discharge is an extremely small lightning bolt from your fingertip to the doorknob—too tiny to electrocute you. But large lightning bolts do kill people regularly. Each second, over 100 lightning bolts strike from the Earth. One bolt alone can produce ten times more energy than the annual output of a large power company. That's powerful! How and why do over 6,000 bolts occur every minute of every day and night?

"A bolt from the blue" may be the most popular way of defining lightning. But a more accurate description would be "a bolt from the ground", a geological fact familiar to Charles Darwin, an unknown English naturalist. Darwin found his evidence in buried fulgurites—sand that has been fused by lightning into slender limbs. He described what later observations would prove: *"the fulgurites are records of roots of a towering tree of light that bursts from the Earth's surface, stretches into the clouds, and dissipates within one blinding flash—so swiftly that it seems to originate in the clouds and shoot downward."*

In later years, time-lapse photography showed the single flash to consist of several ascending and descending strokes, with the more luminous one beginning at ground level. It terminates in a canopy of tapering branches in the overhead clouds. Thus, what is still identified as cloud-to-ground (CG) lightning is actually the reverse: ground-to-cloud (GC) discharges.

Here again, literature researches uncovered another historical discovery laying dormant and unrecognized for its true significance to science. Even more exciting than the clues it offers to scientific enigmas was its predictability by, and perfect confirmatory fit into, the FLINE concept—as will be seen further along.

Another critical clue to understanding the mechanics of lightning was proposed in 1916 by C.T.R. Wilson, a Scottish physicist. He postulated that thousands of daily thunderstorms around the planet generate a global electrical current that circulates between land and upper air layers, then flashes its way through the troposphere (5-11 miles thick) in the form of lightning exchanges between Earth and clouds. Upon exiting from the top of clouds, the current zips through the stratosphere and into the highly conductive ionosphere some thirty miles up. Herein, it encircles the globe, and begins its slow leakage back down through clear areas, thereby playing a major role in creating the fair-weather electrical ocean described earlier.

Wilson's excellent theory fits perfectly in the FLINE concept, but does not address the issues concerning the source of the electrical charges and how a thundercloud becomes electrified. Scientists have offered several theories that appear to contribute to solving the enigma of the actions of electrical charges during thunderstorms. Most efforts have been concentrated on the causes of buildup and release of positive and negative charges in clouds and on the ground. The more advanced theories include the effects of water and ice as critical role players in these reactions.

However, laboratory experiments have failed to substantiate any of these collision, polarization and precipitation theories. Further, the proposed concepts do not account for the lightning strokes (GC) that occur at low levels of, or in the absence of, precipitation, or for the polarization of clouds that never produce a drop of rain. As to convection theories in which updrafts are credited with creating charge differentials in clouds, the most intensive lightning discharges occur in shallow winter snowstorms over Japan in which updraft convection is practically non-existent. So the true origin of lightning remains elusive. What does cause lightning?

The ocean of electrical matter in which we exist is susceptible to forces that forever alter it. As illustrated by the example of doorknob shock one receives after walking on nylon carpet, atomic charges of the electrical matter can be readily separated and rapidly accumulated under relatively mild conditions—and they can be discharged through a finger even more quickly to a receptive doorknob, or to a stream of tap water.

Before offering an explanation in the perspective of the FLINE concept, let's examine more clues. Sometimes, just before a storm, a person can feel a tingling sensation and his hair will stand on end. Too frequently, we read where animals or people sheltering under a tall tree are killed by lightning. Ballplayers and farmers in open fields often get hit by bolts. What happens in these instances, and what clues do they offer? How do these powerful bolts of electrical energies select their targets? What is their source?

With all these facts in mind, we can look inward to the nuclear core for the source of the electrical matter that forms our ocean of atmospheric energies. Ionized atoms and electrons accumulate at and below the surface, and forever seep outward and upward, adding their potential to that of the Earth's magnetic field of force. When atmospheric conditions dictate the formation of clouds as receptive matter (as with doorknobs, or streams of tap water), the potential for lightning storms is created.

When sufficient electrons have saturated the localized area to the hair-raising level, the difference in charge between cloud and ground has built to a high level—in some cases, to as high as 100 million volts. The sudden discharge equalizing the electrical difference between the two is seen as a streak of lightning. This process can be repeated many times within the stormy region until all differences in potential are essentially equalized.

Many of the GC bolts leave their trademarks as evidence of the violent discharges from the ground or from tall objects protruding from the ground. If from the ground directly, the brilliant flashes may leave trademarks as the fulfurites in sand do. If from taller objects, they leave their more visible trademarks of destruction on the objects through which the charges escaped from the ground. An important point here is to avoid being identified with, or as, the taller object, especially if in direct contact with the ground.

To better understand lightning, two key words are helpful: concentration and coagulation. The buildup of electrical energies must be concentrated sufficiently to permit a discharge to be initiated. During the discharge, coagulation of electrical energies creates its brilliant path. High-energy particles from the immediate area are instantaneously and irresistibly drawn toward it, thereby creating a flash made of several high-current pulses called strokes. Each stroke lasts about a millisecond and the separation between them is several tens of milliseconds. Thus, the flash often appears to flicker because our eyes can just resolve the individual pulse produced by each stroke as the surrounding energies join in parallel to the primary stroke. These consecutive strokes can number up to twenty, giving the flash a broader flickering appearance.

Coagulation also accounts for the tributary-type flashes, or channels, that coalesce as jagged branches zipping instantaneously upward and inward toward the cloud end of the main flash in GC lightning, thereby forming spectacular three-dimensional multi-branched displays over a large area. Many times, these branches can be seen emanating directly from the ground to join the primary flash. Such multi-channels can exit simultaneously from areas more than five miles away. Thus, to continue activities in an open field when lightning can be seen in the distance is very dangerous. All other forms of Nature's most spectacular lightning displays can be understood better when viewed in this concept of GC discharges.

In this perspective, distance of clouds above the ground is an important factor in determining the type of discharge, both visible and invisible. Higher clouds and drier conditions result in long, vertical, viciously powerful single strikes. Lower clouds with usually wetter conditions normally display more jagged, branched and crooked flashes. Cloud-to-cloud lightning occurs under conditions in which the electrons have concentrated higher up in the atmosphere before discharging with no visible (only invisible) means of ground connections.

Artificially initiated lightning can occur during man-made events. Examples are the strikes to Apollo 12 and to the Atlas-Centaur 67, and accounts of strikes to aircraft flying in clouds that were not producing lightning. Such damage during spacecraft launches could be reduced by a British instrument that predicts lightning strikes. The system uses aerials at three ground stations spaced more than 60 miles apart to detect electrical discharges that indicate a buildup in electrical potential difference, the cause of lightning. It can predict when lightning will strike up to 30 minutes beforehand and within a few miles of where it strikes—simply by recording the increasing concentration of electrical energies emanating from within planet Earth.

This perspective of electrical energies can enable scientists to better understand the role updrafts play in the intense winter lightning over the warm waters of the Gulf Stream. Further, the reasons for the prevalence of lightning over volcanoes and lightning on all planets with atmospheres become more obvious.

Thus, mankind will be able to better understand and predict Nature's powerful, brilliant displays that have their origins within planets with energy sources: nuclear cores.

What the Findings on Mars Will Tell us about Planet Earth

On July 4, 1997, the Pathfinder with its six-wheel Sojourner landed on Mars to investigate the surrounding Martian terrain. Sojourner's spectrometer has provided analyses that identify the compositions of the soil and a few rocks. The first rock, Barnacle Bill, contained quartz and resembled andesite, a volcanic rock found on Earth in areas of explosive volcanism. The processes needed to create andesite were not thought to have existed on Mars. The whitish appearance of other rocks hinted at the probability of sedimentary rocks that might contain ancient life. The dry river beds and basins on Mars were confirmed. These findings raise some critical questions: Why are the rocky surfaces of Mars and Earth so similar? Was Mars once a thriving planet that supported life? And if so, why is it now bleak and barren? The answers to these three questions provide deep insight into the origins and evolution of Planets Mars and Earth and inevitably into the geometric origin of our Solar System. The findings on Mars corroborate a continuity of cogent evidence; altogether they provide definitive answers to the three questions. Everything is brought full circle by the geometric solution to the proposed Fourth Law of Planetary Motion (FL) (1995) that explains how the nebulous masses attained their orbital spacing around the Sun and evolved to their present stages of planetary evolution via Internal Nucleosynthesis (IN). This new perspective—the FL/IN/E paradigm—opens new vistas of research into planetary origins and evolution.

On January 12, 2000, the question, "*Did Life on Earth Come from Mars*?" was posed and discussed by a team of international astronomers at the national meeting of the American Astronomical Society. "Infra-red spectra readings of short-lived, carbon-rich stars that are engulfed in clouds of gas and dust show that the clouds are rich in some of the most advanced organic molecules ever detected in outer space," astronomer Sun

Kwok said at the AAS meeting. "There is no doubt now that such complex molecules exist and the stars are able to make them with no difficulty." "Such chemicals would eventually be ejected into interstellar space," he said, "which makes it possible that they could end up on planets such as Earth where, under the right conditions, life could have evolved." Among the chemicals detected was acetylene, a building block for benzene and other aromatic molecules that, in turn, can form complex hydrocarbons, the chemical stuff of life. The key point here: stars make and eject complex molecules into space, the opposite of the acclaimed accretion processes of star formation.

In contrast to the stated complicated process for life, a simpler one comes with the realization and proof that any planet, under the right conditions, can and does create its own complex hydrocarbons and amino acids so essential for life. Earth did not need to depend on sources from outer space to create and evolve its atmosphere, land, water and life. The same situation holds true for all planets as each one evolves through a common five-stage cycle of evolution. At some point along the way, if time permits, conditions become ideal for some form(s) of life to evolve. Thus, the FLINE model predicts that every planet, at some time in its evolutionary cycle, will spawn some form(s) of life.

Spacecraft Pioneer 11

Spacecraft Pioneer 11 (1979) and the two Voyagers in succeeding years furnished much information about Saturn and its beautiful rings. In early July, 2004 the Cassini spacecraft flew by Saturn's giant moon, Titan, and furnished views of its atmosphere that were too murky for scientists to distinguish any surface features of the huge sphere, according to news reports. Since the close-up views taken earlier by Voyager 1 of Titan's surface had shown just cloud, cloud and more cloud, no attempt had been made by Voyager 2 to repeat this event.

"That haze is kind of an organic goo much like the smog one might see in Los Angeles, composed of hydrocarbons, and not allowing us to see through to the surface," said Linda Spilker, the Cassini deputy project scientist.

However, the images of Titan did appear to show linear features that suggest tectonic activity. "Scientists believe Titan could have chemical compounds much like those that existed on Earth billions of years ago before life appeared."

The three earlier spacecrafts produced definitive evidence that proved essential to the development of the LB/FLINE model of origins and evolution of planetary systems. The Cassini probe already is revealing corroborating evidence that Saturn and perhaps Titan are indeed models of earlier stages of evolution through which Earth has already passed; in Saturn's case, billions of years ago.

In every case, the Internal Nucleosynthesis (IN) of their nuclear energy cores is the driving force behind their five-stage Evolution (E) into planet and moon. Because of its much smaller size, Titan is much further advanced in its evolution into the fourth (rocky) stage, while huge Saturn remains in its second (gaseous) stage, where it will continue evolving for some one or two billion more years before losing most of its rings and moons, and entering the third (transition) stage of evolution.

Meanwhile, Jupiter's rings and moons will grow in numbers, and will eventually surpass Saturn's dwindling numbers.

NASA's Messenger spacecraft was launched August 1, 2004 on a five-billion-mile trip to Mercury, almost 30 years after the last visit by Mariner 10. Scientists hope that the innermost planet will illuminate fundamental questions about the formation of the early Solar System. "What we are really chasing is the processes that led to the formation of Venus and Mars and Earth, and why they produced such different outcomes," said Sean Solomon, Messenger's principal investigator.

Mercury's density of 5.44 is close to Earth's 5.5; its diameter is 4878 km, somewhat larger than our Moon's diameter of 3456 km. Their crater surfaces make the two spheres look like identical twins, a powerful indication of the manner in which both were made: via Internal Nucleosynthesis (IN), exactly as all planetary spheres were, and are, made. Both are in their final stage (fifth, inactive) of evolution.

The differences in all planets and moons are attributed to size (mass) differences and location with respect to the Sun and to any other large source of heat (e. g., the early Earth/Moon and the Earth/Venus situations) that would affect their rates of evolution, surface features, and their types and quantities of atomic and molecular matter produced via IN, the processes now being sought by the Messenger spacecraft, and which already are definitively explained by the LB/FLINE model of origin and evolution of planetary systems (1973-2002).

The differences in magnetic fields of universal spheres are discussed in another section.

Messy Findings

Quoting from *Messy Findings (Science News,* Oct 23, 2004), "Rocky planets such as Earth are born through countless acts of violence—the collision and merging of many smaller bodies." An infrared survey of 266 youthful stars revealed that 71 of them have disks of dusty debris.

"The disks are a sign that newborn planets are being clobbered by asteroids and that the asteroids—planet-formation leftovers—are banging together and making dust that glows at infrared wavelengths. The only way to produce as much dust as we are seeing in these older stars is through huge [recent] collisions."

Here again, belief in the Big Bang/Accretion myths has misled scientists to erroneous conclusions about the formation of dust rings around stars and planets. Evidence for the alternative concept—the LB/FLINE model of origins and evolution—clearly reveals that huge volumes of gaseous dust-clouds are ejecta from huge masses of energy. One source is a spinning black hole (the densest form of energy) that spews from its two poles billions of energy masses embedded in gaseous dust-clouds—energy masses we call stars, all forming the common pinwheel-shape of a galaxy.

In like manner, even smaller energy masses embedded in dust-clouds are ejected at later times from the various-size stars; most are destined to become planets via processes of Internal Nucleosynthesis and polymerization. In turn, these planets form their own rings of ejecta consisting of dust and various-size chunks of matter circling their equators.

Improbable collisions of chunks of matter orbiting in a cold vacuum, and all in the same direction, is not a likely way to produce dust rings. All rings consist of ejecta from the energy mass in which they were made before being forcibly shot into orbit. Ejecta are common characteristics of powerful energy masses of all sizes—from the largest to the smallest.

Extra-solar Planet News

Extra-solar Planet News (Science News, Nov 27, 2004) tells of new observations of an oddball planetary system 150 light years from Earth that "may force astronomers to rethink the textbook definition of a planet and the accepted idea about how such a system forms." The Swiss team announced that a body at least 17.4 times as heavy as Jupiter orbits the sun-like star, HD 202206, at a distance equal to only 0.82 times the Earth-Sun distance. A second unseen body at least 2.4 times as heavy as Jupiter orbits the same star at a distance equal to only 2.55 times the Earth-Sun distance.

In accordance with the current standard definitions, astronomers classified the large mass as a brown dwarf and the smaller mass as a planet. Both masses are in a curious synchrony: the heavier inner body revolves around its star five times for each single orbit of the lighter outer planet. The team concluded that the two bodies were born in the same way and at the same time, and that the heavier body wouldn't be a brown dwarf after all, but the heaviest planet known.

While posing another dilemma for the Big Bang/Accretion hypotheses, all the observations fit precisely into the LB/FLINE model of origins of solar systems and the evolution of planets. Such oddball solar/planetary systems inevitably form when billions of fiery masses are spewed from the two poles of a spinning black hole during the creation of every beautiful pinwheel galaxy. Such ubiquitous formations are generally known as binary and trinary systems. On very rare occasions, very specific conditions will result in a multiple planetary solar system like ours, well-governed by the Four Laws of Planetary Motion.

The article, *NASA to Announce New Class of Planets* (*LDN*, Aug 31, 2004), tells of the discovery of four new planets in a week's time, "...some appear to be noticeably smaller and more solid—more like Earth and Mars—than the gargantuan gaseous giants identified before." They are described as a "new class" of exo-planets—those orbiting stars other than our Sun. At least two of them "probably are comparable in scale to intermediate-sized planets in our solar system, like Neptune and Uranus, which are about 14 times the mass of Earth."

These findings give a tremendous boost to the scientific validity of the LB/FLINE model (1973-1996) that foretold and explained how such smaller exo-planets came into existence as energy masses that simply evolved into gaseous bodies before evolving into solid masses at rates in accord with size: the smaller the mass, the sooner it passes through each of the five stages of planetary evolution.

The LB/FLINE model gives astronomers two tremendous advantages:

(1) Pre-knowledge of the reasons for existence, composition, etc., of smaller exo-planets: exactly what to look for;

(2) The findings are easier to understand and interpret accurately via the revolutionary model.

Thermonuclear Fusion

Aug.12, 2003

On April 6, 2003, at a meeting of the American Physical Society in Philadelphia, Sandia researchers reported a third method for the successful thermonuclear fusion of hydrogen into helium and tritium nuclei. These thermonuclear reactions were accomplished at temperatures exceeding 10 million degrees, and each blast in the Z machine yields some 10 billion neutrons.

Thermonuclear fusion takes place when substances (energy or matter) become so hot under pressurized conditions that violent collisions force atomic nuclei to fuse together. Such processes always involve the radiation of X-rays. In this third method of approaches that achieves thermonuclear-fusion reactions in a lab, scientists are getting ever closer to Nature's method of achieving nuclear-energy-to-atomic matter within all active spheres of the Universe. But ever higher temperatures and pressures, duplicating the conditions within all active spheres, must be achieved in future trials before ever heavier atoms can be created via thermonuclear fusion. For example, uranium, the heaviest element, is produced in Earth's nuclear core under extreme conditions of temperature and pressure.

What Are Planets?

January 13, 2006

Planets are self-sustaining spheres, generally revolving around a larger central mass while evolving through five observable stages of Evolution (E) via Internal Nucleosynthesis (IN) at rates in accord with size, and in full compliance with natural laws of planetary systems; e.g., the laws of planetary motion, and the ongoing energy-matter relationship expressed as $E=mc^2$.

Other than different rates of evolution and subsequent variations in compositional matter, spherical size is immaterial, while IN and E are crucial: one cannot exist without the other during the billions of years of transformation of their nuclear energy cores into planetary matter.

A Fuller Definition of Planets

Dec. 26, 2006

An accurate definition that covers all sizes, types, locations and activities of planets would be of boundless benefit to the future of science. But such a definition must avoid speculation, and be factually correct.

A candidate that meets these specifications is offered for consideration to the scientific community:

Planets are universal spheres, inestimable in numbers, and active in countless quantities of galaxies. Usually (but not always) locked in orbit around a central star, they form countless solar systems ranging from one to a dozen orbiting masses of various sizes, masses that do not shine of their own accord. Size is irrelevant, but all obey Kepler's three laws of planetary motion.

All planets evolve through four observable stages of evolution at rates in accord with size: (1) Gaseous; (2) Transitional; (3) Rocky; (4) Inactive. Examples are: (1) Jupiter; (2) Venus; (3) Earth; (4) Mercury, respectively.

In addition to Kepler's three laws, the planets of our Solar System obey the author's new fourth and fifth laws of planetary motion. The fourth governs the geometric spacing of planets, while the fifth law governs orbital displacements versus time (in all solar systems).

In summary, a planet is a sphere that does not shine of its own accord, while evolving through four observable stages of evolution via Internal Nucleosynthesis (IN) as it revolves—with exceptions—around a central star in accord with the laws of planetary motion. Size, or mass, is irrelevant. When revolving around a larger planet, the smaller mass is called a moon.

These facts, along with many others, spell doom for the Big Bang hypothesis. Could this be the reason BB advocates railroaded Pluto out of its rightful position—just as Kuhn indirectly predicted they would in his *Structure of Scientific Revolutions* (1956)? Or, could it be a case of not being aware of, or receptive to, the recent spate of new facts? Truth has a way of surviving; let us move forward by not allowing Planet Pluto to die in a vain attempt to salvage the BB hypothesis.

More Evidence of Water

In an AP report, *Mars Rovers Find More Evidence of Water (LDN,* Aug 19, 2004), the Spirit rover "found indications water had altered an outcropping of bedrock dubbed Clovis... Sulfur, chlorine and bromine found inside the rock were in much greater concentrations than in rocks on the plain. Those elements are commonly emitted from volcanoes and could have combined with liquid water or water vapor."

After finding ripples in sedimentary rocks that indicated a pool of salt water once existed at the landing site in the vast Meridiani Planum, Opportunity rover rolled into Endurance Crater and found rippled dunes. "Tiny ripples in a rock dubbed Millstone are clear signs that it had contact with flowing water." In April, Spirit found evidence that limited amounts of water had deposited minerals in a volcanic rock.

These findings, along with the dry canyons and river beds discovered via other means, clearly confirm the past presence of water on Mars, just as predicted and definitively explained by the LB/FLINE model of origin and evolution. Mars is now in its fifth and final (inactive) stage of evolution. Its energy core is depleted.

And its barren surface is an example of Earth's destiny in the far distant future when its energy core becomes depleted.

A Few Brief Comments

Life on any sphere can evolve only during the Third (Transitional) and Fourth (Rocky) stages of its evolution. The more favorable the other conditions, including sufficient time, the more advanced the life forms. In Earth's case, the conditions remain favorable in its current Rocky stage, and will remain so for another three-to-five billion years, or until the energy core is depleted and nothing remains to sustain further evolution. In other examples, Mars, Mercury and our Moon, because of their small sizes, are close to (or already may have entered) the Fifth and final stage: inactive (dead) spheres, with depleted, or nearly depleted, energy cores. Note that since the pace of evolution is a function of mass, the smaller the mass the more rapidly each planet passes through each of its five stages of evolution. Obviously, the smaller masses do not have sufficient time for evolution of life (if any at all) to advance beyond the most primitive stages of bacteria. Earth is fortunate to be the ideal size and the proper orbital distance from a Sun of the proper size for the highest forms of life to evolve—factors that will remain so for a few billion years more. The solution to the new Fourth Law, in revealing why and how these fortunate circumstances of our multi-planetary system came into being (in full compliance with all natural laws), clearly explains why this rare event was no accident.

In reference to the Big Bang, there exists a more viable concept of the origin of the Universe: the Little Bangs Theory in which the evidence allegedly supporting the prevailing belief can be interpreted more realistically, yet has been unable to get past peer-review advocates of the Big Bang. Suffice it to say that the arguments put forth by a number of name scientists against the Big Bang are powerful and conclusive, and they continue to mount with each relevant discovery—which often is poorly force-fitted into this hypothesis. The most complete account of these cogent arguments can be found in *The Cult of the Big Bang: Was there a Big Bang?* by William Mitchell. The evidence overwhelmingly confirms that the Big Bang hypothesis is the grandest and costliest scientific scheme ever perpetrated (probably naively and unintentionally) on an unsuspecting public.

All matter in the Universe evolves in full compliance with natural laws, and when the proper conditions are met, life itself evolves. But in each and every case, there must be a local source of energy to drive that evolution. However, neither life nor anything else can evolve on Fifth-stage (dead, inactive) spheres in which their local sources of internal energy that supplies the atomic building blocks for all planetary matter are depleted.

The FLINE paradigm makes it crystal clear that atoms form via Internal Nucleosynthesis in all active spheres of the Universe. For example, atomic elements ranging from hydrogen to iron have been observed on the Sun's surface, while all 92 atomic elements comprise Earth's crust. In both cases, the atomic elements were created via nucleosynthesis within each of these active spheres; they did not simply fall together from outer space. Without Internal Nucleosynthesis, Evolution of the two spheres would not be possible. Since the same principles apply to all spheres of the Universe, it is obvious that atoms are being created continually within the nuclear masses of all active spheres rather than having been created in the first second after the alleged explosion of the Big Bang, later to coalesce into planetary systems and systems universal.

If scientists are to make a great leap forward in the planetary sciences, we must forsake the Big Bang and the coalescing planets hypotheses, and seriously examine the valid alternatives.

Impact or Ejecta?

The article, *Scientists: Giant meteorite struck Wisconsin millions of years ago (LaGrange Daily News,* April 26, 2004), describes an ancient catastrophic event 450 million years ago that created a massive hole in a four-mile area called Rock Elm about 70 miles east of Minneapolis.

The report said, "The impact at Rock Elm released more than 1,000 megatons of explosive energy, lifted the earth at the center more than 1,650 feet and sent shock waves through the rocks, crushing them. ...the rocks here have always appeared different than those just a few miles away. They're tipped at an angle in many places, reflecting the damage inflicted millions of years ago."

Based on this evidence, scientists stated that a 650-to-700-foot meteorite had crashed into the earth at speeds up to 67,500 mph.

But their conclusion raises a question about the true nature of the catastrophic event. The evidence of explosive energy is more indicative of a powerful ejecta force that lifted the center more than 1,650 feet, tipped rocks at an angle in many places, and sent shock waves through the rocks, crushing them. Powerful ejecta events were more common 450 million years ago than in more recent times.

The best example of evidence created by a crashed meteorite can be seen in The Great Meteor Crater of Arizona; its evidence is in stark contrast to the evidence presented in this article.

The Pluto Controversy

The recent meeting of 2000 astronomers for the purpose of definitively defining planets finally achieved its primary goal. Pluto was demoted from its rightful position as the ninth planet in our Solar System.

Substantiated evidence clearly reveals the Big Bang myth as sophistry tightly trapped in "truthiness". Since myth begets myth, the definitive definition of planets remains impossible until viewed from a more factual standpoint.

Fortunately, the new LB/FLINE model of universal origins provides the proper foundation of sufficient evidence to define planets definitively.

PRESS RELEASE

Planet Pluto: Science at the Crossroads:

To Be or Not To Be --That Is The Question!

Alexander A. Scarborough

September 20, 2006

Stripping Planet Pluto of its popular name and scientifically validated status as the ninth planet in our SS is a sham that has no place in science. Accomplished politically by a small consensus, it is akin in principle to placing Earth once again at the center of our SS, a giant step backward into the Ptolemeic-Copernican-Galilean era when consensus of powerful leaders ruled the day, in spite of contradictory facts. When politics and science are mixed together, the results will be wrong, often catastrophically, as in this case.

The Pluto controversy places science squarely at the crossroads of choosing substantiated facts or speculative consensus. Such decisions should be made solely on the basis of all known facts—not on selected changes in criteria or by consensus: a simple poll on speculative beliefs whenever the right answer is unknown or not available. The Pluto issue stands as a symbolism for what is happening in the sciences based on BB beliefs that must always remain speculative. Fortunately, there is a way out of the tangled web.

The five laws of planetary motion clearly reveal, beyond any reasonable doubt, the status of Pluto as the original tenth planet, and now the ninth planet in our SS (fourth law). The change in orbital position occurred when Planet Asteroid disintegrated into countless thousands of asteroids—as correctly interpreted by Heinrich Olbers in 1802. Three sets of geometric diagrams reveal the precise, but relatively small, changes that have occurred in the SS during its five-billion-year existence. Each of the three sets corroborates the other two.

These findings led to the fifth law of planetary motion governing orbital changes: a matter of mass-distance-velocity-time relationship; e.g., why the orbits of Uranus and Neptune are the only two drifting outward from the Sun, and all others, including Planet Pluto, are drifting closer to the Sun. While not yet proven, any matter beyond Pluto should be drifting ever farther away. The possibility of another planet or two beyond Pluto must be further investigated. If validated beyond doubt, it would simply be revealing an extension of the SS and the five laws of planetary motion that govern it.

To: Editor, Discover Magazine, New York, N.Y. October 30, 2006

In *War Over the Worlds* (*Discover*, Nov. 2006, p 41), Alan Stern of the Southwest Research Institute stated it best: "The IAU's planet definition resolution is an example of science at its worst—it's sloppy, internally inconsistent, and is designed backwards, that is, to produce a desired result that only eight planets are in our solar system. I am embarrassed for the IAU."

Many other scientists, including myself, are deeply embarrassed for the IAU and for science in general. The fact that Pluto is the original tenth, and now the ninth, planet of our solar system is clearly revealed by the Five Laws of Planetary Motion (FL), supported by a plethora of other substantiated evidence. The IAU committee may or may not be aware of the new Fourth Law governing the mathematical spacing of the planets, and the new Fifth Law governing the small changes in orbital positions during the past five billion years, while giving astronomers the capability of predicting similar changes in the future.

The numerical change in Pluto's orbit occurred when Planet Asteroids in the fifth orbit disintegrated into countless thousands of asteroidal fragments—as first correctly recorded by Heinrich Olbers in 1802. Those asteroids can never re-form into a planet via accretion predicted by the Big Bang hypothesis.

The above evidence clearly reveals that all forms of matter beyond Pluto, regardless of shape, size, or nomenclature are leftovers from the dynamic birth of our Solar System some five billion years ago. As extensions of our Solar System, some may meet specifications for a planet—a spherical object obeying the laws of planetary motion, regardless of size, while evolving through five observable stages of planetary evolution via Internal Nucleosynthesis (IN) at rates in accord with mass (size) and the new LB/FLINE model of universal origins.

Neither asteroids nor comets play a major role in the formation of planets; both are products stemming from much larger masses.

Response to Why Planets Might Never Be Defined

November 27, 2006

In *Why Planets Will Never Be Defined* (*ad Astra*, National Space Society, Winter 2006), R. R. Britt presents valid reasons for his titled conclusion in the Great Pluto War. As based on the BB hypothesis, Britt's reasoning is sound, his conclusion accurate: a valid definition of "planet" will always remain impossible.

But there is hope. When based on the revolutionary Little Bangs (LB)/FLINE model (1973-2000) of universal origins, a valid scientific definition for "planet" becomes implicitly relevant to our small Solar System, and remains equally relevant when applied across all universal galaxies and other worlds several times the mass of Jupiter floating freely in space without a host star.

Brown dwarfs 70 times as massive as Jupiter are definitively explained via the new model that's soundly based on the five laws of planetary motion that, in turn, offer definitive clues to the dynamic origins of the billions of galaxies.

The new model clearly explains the fact that nearly 25 percent of the known extra-solar planets are in binary or multiple-star systems.

Without a thorough knowledge and application of the five laws of planetary motion, an accurate definition of a planet will indeed remain impossible via the BB concept that must ignore them for the sake of its own survival.

To: LaGrange Daily News **Aug, 2006**
Re: Pluto News, plus Manuscript: A Really Bright Future for Science

The current controversy about Planet Pluto bodes a truly exciting future for science, but only if followed through until truth via substantiated evidence is definitively established and accepted. For now, it is forcing science into the resistance-to-change phase predicted by Thomas Kuhn in his *Structure of Scientific Revolutions* (1956), a phase somewhat akin to the era of Copernicus and Galileo.

Big Bang beliefs today are the major obstacle to defining planets, making the task impossible to accomplish definitively. While the new definition is good, it does remain incomplete.

What makes the future of science potentially bright is the beautiful continuity in the revolutionary Little Bangs (LB)/FLINE model of ongoing evolution of universal systems, all in sync with Einstein's relativity theories. The definition of planets is inherent in the new model, in which the Five Laws of Planetary Motion (FL) assure its scientific validity and the fact that Pluto—beyond any reasonable doubt—is the ninth planet (originally the tenth planet) in our Solar System, a fact firmly established by the new Fourth Law of Planetary Motion. A valid explanation of all matter beyond Planet Pluto is derived via the new Fifth Law of Planetary Motion explaining and governing the ongoing changes in planetary orbits since their dynamic origin five billion years ago.

The five laws furnish a solid foundation for the new model, governing the origin, movements, spacing, and subtle changes in orbital positions for the past five billion years. They are powerful arguments against the BB hypothesis in which no explanation, beyond speculation, for these activities can be found. Could this be a reason behind the Pluto controversy?

The definitive definition of a planet: a sphere that does not shine of its own accord, and usually (but not always) revolves around a larger mass of nuclear energy (star, sun). Size is irrelevant to its planetary status, but when revolving around a larger planet, the smaller sphere qualifies as a moon. In all cases, planets and moons evolve through five observable stages of Evolution (E) via Internal Nucleosynthesis (IN) at rates in accord with size, and in which gravity is an inherent factor and the reason behind the spherical shape of universal spheres. In this manner, planets are self-sustaining entities, each evolving its own planetary matter via IN-E, while governed by FL, all made possible by LB.

To: Letters Editor, LaGrange Daily News, LaGrange, Georgia 30240. **August 28, 2006**

Congratulations to Pam Duke's second grade students at West Georgia Christian Academy for their efforts to preserve Pluto's status as a planet. They are more right than they now realize. The controversy over whether Pluto is or is not a planet is one of the most pivotal questions in the history of science. It now appears destined to unwittingly trigger a change in direction of scientific thought about origins of universal systems in the near future. (Add: Sep 5, 2006; Congratulations to NMSU for your protest rally.)

The removal of Planet Pluto from its rightful place in our Solar System only prolongs the 2500-year struggle for truth in the astronomical sciences, initiated in 500 B. C. by the Pythagoreans. It gives the impression of a diversionary tactic in a vain attempt to preserve the Big Bang myth's precarious position as long as possible.

The five laws of planetary motion firmly underpin the fact that Pluto—beyond any reasonable doubt—was originally the tenth planet, and is now the ninth planet in our Solar System. The change occurred when the original fifth planet disintegrated into uncountable thousands of asteroids, correctly recorded by Heinrich Olbers in 1802, and now firmly established by the newly modified (May 2006) fourth and fifth laws of planetary motion governing the geometric spacing and ongoing changes in planetary orbits, respectively. In moving almost 10% (4.68 AU) closer to the Sun during its existence, Planet Pluto's highly elliptical orbit is a prime example of the new fifth law (May 2006) in action.

The five laws (FL) and their ancillaries present powerful arguments against the Big Bang hypothesis in which no explanation, beyond speculation, for these activities can be found. Could this be one reason behind the unwarranted Pluto controversy, and a fulfillment of Thomas Kuhn's prediction in his book, *The Structure of Scientific Revolutions* (1956)?

Definition of a planet: a sphere that does not shine of its own accord, and, with rare exceptions, revolves around a larger mass of nuclear energy (star, sun) in compliance with the laws of planetary motion, while obeying all laws relative to Internal Nucleosynthesis. Size is irrelevant to its planetary status, but when revolving around a larger planet, the smaller planet is called a moon.

Historically, prevailing beliefs are never easily displaced by scientific revolutions, but before the time Pam's students enter college, their protests should be fully vindicated.

Reference: ***New Principles of Origins and Evolution: Revolutionary Paradigms of Beauty, Power and Precision.*** **Alexander Scarborough, 2002. Available: authorhouse.com, dubuissonk@bellsouth.net**

c: Pam Duke, Bernie McNamara, Al Tombaugh (address unknown; please give Al a copy).

Letters and Memos

To: Glenn Strait, Science Editor, The World & I, Washington, D.C. 20005. 05-04-98

Beginning at 6:00 p.m. on May 3, I watched *Dinosaurs: Inside and Out, and Then There Were None* on the TV Discovery channel. The film first dealt with the impact of a massive asteroid as the cause of their alleged short-term demise, as described by Bob Barker. This was followed by the new concept of a longer-term extinction via the decreasing oxygen content (from the original 35%) in the atmosphere which gradually became insufficient to support the huge dinosaurs (as explained by a Dr. Rigby).

As you may recall, this latter version deals with a vital aspect of the FLINE paradigm that goes a big step further by explaining the taproot cause of the decrease in oxygen content: Earth's nuclear energy core that persistently alters the content of our atmosphere and surface features during the ongoing evolvement of Earth through five stages of evolution common to all planets. The section that gives the details can be found in my writings as far back as 1986 *(New Concepts of Origins: With White Fire Laden*, pp. 53-55). Updated versions are detailed in my later books (1991, 1996) and letters written since that time. All the reasons species come and go point directly to the irrefutable fact that Evolution and Internal Nucleosynthesis are inseparable: one cannot exist without the other. Further, all evidence used in support of the theories of Barker, Rigby *et al.* is even stronger support of the new paradigm.

While I am happy to see other scientists contributing supportive facts to the FLINE paradigm, it is ironic that they can get aspects of the concept published so readily via various media outlets while I have struggled for the past 25 years to put these factual aspects together into what now appears to be an irrefutable and flawless paradigm—and cannot get published. I am beginning to doubt the value of copyrights—even though they have been strengthened in recent decades—and to doubt the objectivity of planetary science in general.

After 3 ½ months, I have not yet received an evaluation of my manuscript, *On the Spacing of Planets: The Proposed Fourth Law of Planetary Motion,* submitted to the Royal Astronomical Society. During that time, I have learned that the Royal Astronomer is Martin Rees, while the Isaac Newton Chair is Stephen Hawking; both are avid supporters of prevailing beliefs about planetary origins. I did write them last week for a status report. The final decision on the proposed Fourth Law and the FLINE paradigm should be very interesting, and should make history one way or the other.

Thanks again for your interest and support.

To: EOS, AGU, AAAS, Strait, Astronomical Journal, Science News, UGA, Dubuisson, Herndon.
Examining the Overlooked Implications of Natural Nuclear Reactors (Herndon, AGU, Eos, 9-22-98)

During the past two decades beliefs about the composition of Earth's core have undergone significant changes. In the early 1980s scientific papers concentrated on how and why a rocky center formed in our planet. Later in the decade the emphasis was placed on an iron core as the more logical composition to generate a magnetic field. The latest concept (Herndon, Eos, Sep 22, 1998) promotes the idea of uranium from outer-space missiles as the core ingredient that furnishes the internal heat via fission alone.

However, all of these theories have problems; all are speculative and contain fatal flaws. For example, a uranium core decaying via fission alone as the source of the heat within would fall far short of the energy necessary to drive planetary evolution, and would provide no valid source for Earth's electromagnetism. We have only to observe our Sun to understand that the lighter elements, primarily hydrogen and helium, created via Internal Nucleosynthesis (IN) in nuclear masses (the true source of heat and electromagnetism in all active spheres) are the initial building blocks for fusion reactions into all atomic elements of which planetary spheres are composed—often including the last and heaviest one, number 92, uranium. Theories based on this specific IN perspective offer more logical solutions to all planetary anomalies.

The most logical concept of Earth's core was first defined by Descartes as being "Sun-like". When one ponders all possible solutions to the mysteries of planetary evolution, this old idea takes on the reality of truth. Some historical information on the subject will fill in the missing pieces of Herndon's article, and perhaps help maintain an accurate record. My initial publications in 1975 on the powerful supportive evidence for nuclear cores in active planets include three titles: *Birth of the Solar System, Evolution of Planets, and Evolution of Gas, Oil and Coal.* These subjects were first presented in poster sessions at the AAAS Meeting in 1982 in Washington, D.C. and again in the 1983 Meeting in Detroit, Michigan.

The subject was expanded to booklet form in 1977 under the title, *Undermining the Energy Crisis,* then to book form in 1979 under the same title—always emphasizing the critical role of nuclear energy cores in the evolution of planets via Internal Nucleosynthesis of the atomic matter comprising their atmospheres, crust, fuels, etc. During the ensuing years, these revolutionary concepts were presented at various meetings, including the

AAAS, the AGU, the AGS, other professional groups and civic organizations. Many abstracts appear in the records of these meetings. However, my manuscripts and news releases continued to be rejected, and on occasions, were handled unfairly by officials in charge of the programs.

So, I was forced to continue writing and publishing in ever-expanding book form as fast as new discoveries were made (and finances permitted it). In all situations, the new knowledge fitted precisely into the ever-evolving concept. These books include: *From Void to Energy to Universe* (1980) (manuscript copyrighted, not published), *New Concepts of Origins: With White Fire Laden* (1986), *The I-T-E-M Connection: How Planet Earth and Its Systems Were Made by Means of Natural Laws* (1991), and *The Spacing of Planets: The Solution to a 400-Year Mystery* (1996). The additional confirming evidence was continually submitted to various organizations and editors, but to no avail except in one case: *Evolution of Gas, Oil and Coal* was published in *Alternative Energy Sources* (Vol. 3, pp 337-344) by the Clean Energy Research Institute, University of Miami (1983). Unfortunately, a partial version of my *Energy Fuels Theory* was published in the *New York Times* the same day, and thereby became erroneously known as Gold's theory. However, Gold's incomplete concept did not include the nuclear energy source of the elements comprising hydrocarbon fuels—his source was missiles from outer space.

To: Martha M. Hanna, Letters Editor, Physics Today, College Park, MD. 06-14-00

Thank you for the belated response to my letter of June 14, 1999 in response to articles by Burbidge *et al.* in the April 1999 issue of your fine magazine.

Just as the Copernican idea of a Sun-centered Solar System revolutionized how mankind views our planetary system, the solution to the new Fourth Law of Planetary Motion (1980-1995) explaining the spacing of planets around our Sun presents a powerful, definitive and irrefutable argument that is destined to change current views in a similar revolutionary manner. This conclusion was reached after more than two decades of researching the truths and myths of science and presenting the evolving factual facets of the FLINE paradigm to various scientific and civic organizations.

During the five years of presentations and submissions since the solution to the new Fourth Law as the final phase of the new concept of origins and evolution, not even one scientist has been able to give a valid scientific reason for its continual rejection. The real reason, of course, is that there is no valid reason other than that its revolutionary nature threatens, and inevitably will displace current beliefs about how solar systems came into being and evolved into their different stages of evolution.

Over the past twenty years, the FLINE model has an impeccable record of now-verified predictions. Further, it provides a solid foundation for definitively understanding every anomaly of all planets, including the more than three dozen gaseous (second stage of evolution) giants discovered in extra-solar systems. As you know, these giants are much too close to their central stars to have formed there via the accretion concept. Simultaneously this anomaly was predicted, and is explained beyond any doubt, via the new FLINE model of origins and evolution. The five stages of planetary evolution leave no doubt that Evolution is possible only via Internal Nucleosynthesis that drives it forward: the two are inseparable—one cannot exist without the other.

In comparison, the one-year delay in response to my last letter is not too bad; my manuscript, *On the Spacing of Planets: The Proposed Fourth Law of Planetary Motion*, has been in peer review at the Royal Astronomical Society, London, for two years and five months. Any comment here would seem superfluous. Meanwhile, I'm continuing to successfully present the FLINE model to scientific and civic groups. The most recent talk was given to a science group at the University of Connecticut on June 5, 2000, and was well received. Further, I'm sending a copy of this letter and my manuscript on the subject to Editor Stephen G. Benka for his evaluation as a publishable paper.

As in the Copernican era, it is time to move onto newer ideas if we are to learn all there is about our planetary systems. When definitive ideas are developed, antiquated ideas, no matter how admired, must inevitably surrender their hallowed places in science. The time for change has arrived.

To: F. Todd Baker, Physics and Astronomy Dept., UGA, Athens, GA 30605. Aug 28, 2000

The existing model for planetary origins is not consistent with the recent discoveries of nearly 50 extra-solar planets orbiting stars like our own, as described in the scientific literature *(Discover,* March 2000). Therefore, I would ask your indulgence for a few moments while I introduce some elements of an alternative model that is consistent with the unorthodox positions of these gaseous giants of which most, if not all, are too close to the central stars to have formed there.

The *Discover* article states that "orbits in which [our planets] were born some 4.6 billion years ago have remained the same ever since. Until recently that was the accepted scenario. But now the detection of extra-solar planets has forced astronomers to re-examine such notions, because they present us with a paradox. Many [if not all] are so monstrous in size, and hug their stars so closely, that they could not have formed in their present positions. The searingly hot stars around which they circle would have melted their rocky cores before they got started. Instead, it's assumed that they coalesced some distance away, then barreled inward over millions of years. And if such chaos characterizes the birth of extra-solar planets, could not similar disorder have reigned closer to home?" However, as previously admitted, these giants could not have formed very far out because of the thinness of the formation substances at greater distances.

This proposed scenario presents more problems than it solves. However, there is an alternative that warrants consideration, one in which the data do fit precisely without the need of assumptions. Such extra-solar planets were predicted and explained by this revolutionary FLINE concept before these amazing discoveries were made. In it there is no need for chaotic disorder in solar systems. The new theory eliminates both the need for gaseous giants to form at much greater distances from their central stars and the need to sling planets out of their orbits in a chaotic manner. It is strongly reinforced by the laws of planetary motion, which include my solution to the enigmatic Fourth Law of Planetary Motion, detailing how the planets attained their orbital positions around our Sun. The solution to this revolutionary law has been submitted to the Royal Astronomical Society in London, where it has been under peer review for more than 2 1/2 years—perhaps a record?

While it is exciting to see these extra-solar systems adding much to the credibility established in past years by the impeccable record of accurate predictions of the FLINE concept, it is discouraging that mainstream scientists have been unwilling to attempt, or simply unable, to point out any fatal flaw; perhaps none exists. Researched and developed during the past quarter-century, this new theory does not require the speculative assumptions so necessary in other concepts.

I would like to discuss these new findings with you and any associates at your earliest convenience, with emphasis on determining whether or not a flaw can be found in the new concept.

To: The Royal Astronomical Society, Burlington House, Piccadilly, London. Jan.12, 2001
Re: qy016; On the Spacing of Planets: The Proposed Fourth Law of Planetary Motion

January 12, 2001 marks the third anniversary of submission of my manuscript for review and publication of the proposed Fourth Law of Planetary Motion detailing how the planets attained their spacing around our Sun—the enigmatic solution that first eluded Kepler in 1595. In conjunction with Kepler's three laws, the four laws have proven to be the final link in the revolutionary FLINE model that reveals - beyond any doubt - the dynamic manner in which our Solar System and all solar systems came into being and subsequently evolved into their current orbital positions and evolutionary stages.

Researched over a 27-year time frame, this factually structured, non-speculative model continues to open new avenues of research that inevitably lead to definitive solutions to planetary anomalies that cannot be comprehended fully in the perspective of any other known theory about the origins of solar systems and the evolution of planets. The most recent example, dated Jan 10, 2001: "Two 'clearly bizarre' planetary systems found in the orbits of distant stars are puzzling astronomers and raising new questions about how planets form." (See enclosed ref: *Mystery planets intrigue experts* via the AAS national meeting.) To fully comprehend the two bizarre (and all other) planetary systems, an in-depth understanding of the FLINE model that explains the eccentricities of all planets and answers the "new questions about how planets form" is crucial. That is why I urge you to add these findings to your long prestigious record of published historical discoveries.

One has only to read *The Spacing of Planets: The Solution to a 400-Year Mystery* to become convinced of its scientific validity and become an advocate of the revolutionary FLINE model of origins and evolution. Herewith are a few testaments from knowledgeable readers. Additionally, information about the book can be found on the Internet at www.iuniverse.com via Bookstore and typing in the author's name.

As with the findings of Copernicus and Kepler, these new concepts are destined to revolutionize the manner in which scientists view the origins and functions of solar systems and planets. Delay of its review and publication denies scientists the opportunities to benefit from its revelations. In view of the tremendous import of the manuscript's contents and in the best interest of scientific progress, I respectfully urge you to expedite the procedure so that scientists may have the opportunities to comprehend their findings in its perspective.

Please advise the status of the manuscript and whether or not there are problems or questions in need of answers that might help in your esteemed evaluations.

To: Alexander A. Scarborough, LaGrange, Georgia 30241. 2001-02-20
From: Amanda McCaig, Editorial Assistant, Astronomy & Geophysics, The Journal of the RAS
Re: On the spacing of planets: The proposed fourth law of planetary motion - our ref. qy 016

I'm writing to let you know that, regrettably, we have to find a new referee for your paper. We have done our best to get a report back from the referee with whom we first placed it, but to no avail. I can only keep on apologizing, but this, of course, cannot make up for the quite unacceptable delay in dealing with your work. I do hope you are prepared to bear with us a little longer while we find a more reliable, although just as eminent, referee. However, we would quite understand if you preferred to withdraw your paper and place it elsewhere.

To: Dr. Amanda McCaig, Leeds LS18 4NJ, UK. March 1, 2001
RE: Ref. qy016 - On the Spacing of Planets: The Proposed Fourth Law of Planetary Motion

Thank you for your letter of 2001-02-20 advising the status of the subject manuscript submitted in January, 1998 for peer review and publication. Your efforts to get a report back from the referee with whom it was first placed and your efforts now to find a new referee for the paper are deeply appreciated.

You probably are aware of the daunting task you face to accomplish this crucial mission. History teaches that any major breakthrough that poses a threat to prevailing beliefs unfortunately must undergo much resistance from advocates of those beliefs. Perhaps the referee found the paper to be in this category, and was unwilling to commit either a negative or a positive response because of some apprehension of obvious consequences

either way.

Revolutionary concepts inevitably run into the major problem of lack of an expert capable of properly judging it, simply because it is a new concept with which no one has become familiar enough to make an accurate assessment because no one is yet an expert in the new arena—a chicken-and-egg situation. Neither has anyone been able to offer a valid rebuttal to any of its intermeshed facets.

I realize its potential impact, and do understand the problem being faced by the manuscript in spite of its factual nature. It was submitted to the world's most prestigious science institution because of your history of properly handling such historical breakthroughs, and I remain prepared to bear with the Royal Astronomical Society as long as it takes to find a more reliable, although just as eminent, referee. A correct assessment of the manuscript demands an unbiased and very courageous person with a strong inclination to understand and accurately judge it; one who realizes that his/her decisive action either way will be recorded throughout the history of science. The concept's impact on the understanding of solar systems should remain as huge and as memorable as the crucial discoveries of Copernicus and Kepler.

Because of its factual nature, mathematical evidence, intermeshed principles, beautiful continuity and continual verification via the ongoing space discoveries, the scientific validity of this 27-year researched concept seems assured. Because of its tremendous potential for enabling scientific discoveries to be accurately interpreted without the oftentimes fairy-tale aura, I encourage and urge you to push it through the system in the manner that discoveries of this import warrant. The definitive solutions to all planetary anomalies have their taproots deeply embedded in the FLINE paradigm which, in turn, is based solidly on the Four Laws of Planetary Motion and the principles of the inseparable Nucleosynthesis-Evolution connection; only within its realm can these solutions spring forth crystal clear and beyond any doubt.

Thank you again for your efforts in this highly significant undertaking.

From: Gordon Rehberg. **April 22, 2001**

Received your latest correspondence and all I can say is how completely disappointed I am in the reply from England. It borders on disgusting. You and I will die and disappear from this life and never have the satisfaction of someone realizing the core essence of your work. It's one thing to think people stupid, and another for them to offer a reply such as you've received and confirm the fact.

Was watching a program on *A&E or Discovery* the other evening about the planet and oceanographic research. Long session on Methane Hydrates. Bought a 1973 Oceanographic textbook two weeks ago for one dollar at the Dalton Library book sale. Been plowing through it. Find many questions in the book published in 1973 have since been answered with research in the past 20 years.

Enclosed are a small series of web pages I have printed out dealing with Methane Hydrates. The web pages go onto infinity with research findings on methane hydrates. Comments like, "Found on the ocean floor, methane seeping from the floor in small discharges along with OIL." Unexplained radioactivity levels at the ocean floor, etc., etc., etc. Unusual life, new undiscovered life, evidently feeding and growing on methane as a food.

I have concluded that most of the people on this planet are not interested in thinking about anything. Most of the scientists cannot question their foundations of knowledge based on theories. Rather, they are secure in thinking these theories are not questionable at all and represent "truths" rather than "theories". It just would drive them out of their comfort zone of basic knowledge, and this they do not question.

I believe the methane hydrate knowledge only further confirms the nuclear core of all planets. It reinforces the concept of the core, the shell, the pressure vessel, the polymerization of compounds, etc.

I am not versed in the Chemical Engineering foundations to expand these facts, but know methane hydrates confirm your theories as well as other physical evidence surrounding us. The shear numbers of people that have been exposed to your ideas and the absolute rejection by all of them is confirmation of their primitive mental capacity. I thought two years ago I would have some luck at helping spread your ideas, but must recognize nothing has happened of importance.

Always appreciate your sharing these comments with me. All the best of luck, and hope your health is well and continuing to improve. All well here, and Nicholas is close to reaching "normal", and has decided he would like to teach high school mathematics in Columbia, SC. He will hopefully be able to start this profession in August. Really think he likes the idea of only working 9-10 months a year. Sounds good to me, only 40 years too late.

To: Glenn Strait, Science Editor, The World & I, Washington, D.C. 20002. 04-18-01

Thanks for the interesting and valuable information in the e-mail of March 29, March 30, and April 10, 2001. It's always exciting to get new information on planetary discoveries that seem always to blend so easily into the FLINE model. I would like to comment on the four issues.

The plume activity on Io is as predicted by the new model: it is definitive evidence of the young age of Io, now in the early cycle of the fourth stage of evolution through which all planets evolve. If this moon and Jupiter were the same age, Io would now be an inactive planet much like Mercury. This was discussed in more detail in a previous letter to you.

I'm always amazed at the logic used in such press releases as *Turning Stars into Gold.* To wit: "Many common elements, such as oxygen and carbon, are known to be made in stars and distributed through the Universe when a star explodes as a supernova. This is the origin of most of the material that makes up Earth [many questions here]. It is becoming clear, however, that normal stars cannot make enough of the heavy elements, such as gold and platinum. Thus, the origin of gold and platinum—on Earth and throughout the Universe—remains a mystery [?]. ...When binary neutron stars are close enough to collide, they soon create the most powerful explosions in the Universe." Dr. Rosswog "has found that a large quantity of gold and platinum is made and thrown into space. Relatively small seed nuclei, made of elements like iron, collect neutrons and build themselves up to become heavy elements such as gold and platinum." Rosswog *et al.* "have shown that the relative amounts of elements formed in his computer models of colliding neutron stars match those seen in our Solar System. This provides strong evidence that most of the gold and platinum on Earth was formed in the violent collisions of distant stars." [logic?]

You recall the puny predictions of supercomputers that did not come even close to predicting the super powerful explosions of Comet SL9 fragments on Jupiter in the mid 1990s, thus casting serious doubt on the validity of computer findings and conclusions. The FLINE model alone accurately predicted these powerful explosions, simply because its initial assumptions of comet compositions were accurate. Certainly, heavy metals are formed in powerful explosions, as was illustrated in the explosion of the atomic bomb that created traces of elements 99 and 100. This is indicative of how the full range of elements can be created within the confines of an encapsulated nuclear mass like those found in all active planets that have advanced into their fourth stage of evolution: the reasons reside in the energy core's gradual increases in the internal conditions of extreme temperatures and pressures in which quantities of specific atoms are created at any specific time, eventually to become crustal matter. All planets are self-sustaining entities that create their own compositional matter via Internal Nucleosynthesis and polymerization. [Ref.: *Fuels: A New Theory* (1975), *Undermining the Energy Crisis* (1977, 1979)].

One basic problem with prevailing theories is illustrated in the news release of April 10: *Giotto's heritage: the past and future of comet exploration.* To wit: "...ESA's Giotto spacecraft made history by obtaining the first close-up pictures of a comet's black, icy nucleus." In reality, close examinations of these photos clearly reveal the true nature of comets: fiery masses of energy encased in a carbonized black shell of its own making, spewing out separate trails of virgin ionized gases and dust through several separate "portholes" of its own making. If comets do not crash or explode first, they eventually become burned-out shells; examples are Phobos and Deimos, the two moons of Mars. Recent findings reveal that comet tails and the gigantic comae remain with comets after they move far beyond the warming effects of the Sun and into the nearly absolute-zero temperatures of distant space—exactly as predicated by the new model. Only fireballs have the capability to accomplish these feats.

Glenn, I understand your trepidation about promoting the new model; I felt that way until the data became—and is still becoming—overwhelmingly persuasive. During the 28 years of piecing its volumes of persuasive data together—finalizing it with the solution to the new Fourth Law of Planetary Motion—and presenting the findings to scientific personnel, not a single valid counter-argument has prevailed against it. Nor does one seem likely to do so.

I fully agree with the mainstream model of stellar and supernova nucleosynthesis, but I do not agree with the accretion hypothesis: an illogical concept that cannot be proven, but can be indirectly disproven. I have ventured one giant step further by recognizing the true nature of planetary Evolution (E) via Internal Nucleosynthesis (IN) and polymerization. IN and E are inseparable to the end; one cannot exist without the other. Planetary evolution is absolute proof of Internal Nucleosynthesis; it is crucial that every active sphere have an internal nuclear engine to drive its evolution. Only in this perspective can all planetary anomalies be understood beyond any doubt; it is the crucial key to understanding origins and evolution. The FLINE model is simple, beautiful, and easy to understand: the type of concept preferred by Einstein, who gave us the dynamic formula $E = mc^2$ that reveals the energy-matter relationship, the most fundamental principle of the IN-E concept that eventually led to the solution to the enigmatic Fourth Law.

In 1985-86, a project at Georgia Tech's Nuclear Division confirmed the three-layer system of the hydrocarbon fuels within Earth's crust, as predicated by the new IN-E model. I discussed IN with Dr. Karam, the Director, who gave no reason to believe that it was not feasible. But I believe his position in the hierarchy prevented his committing to a final answer. The same results came from my visit with Dr. Todd Baker at the University of Georgia's Physics Department, and from the Royal Astronomical Society and the American Astronomical Society with regard to my manuscript on the Fourth Law. This reaction appears to hold true throughout the system with other academic personnel; they prefer not to step outside prevailing beliefs, no matter the data being overwhelmingly persuasive. I understand their position, one much akin to that of the Copernican-Galilean era. Thus, science has another classic case of an irresistible force of incontrovertible facts meeting an immovable object of highly speculative, unproven, but well-entrenched, beliefs. Who knows the answer to this dilemma? Time, the FLINE's best ally, will tell—but only after the true nature of comets is "discovered" and confirmed. Meanwhile, must students be forced to continue learning beliefs only?

To: Kathy Sawyer, News Reporter, Washington Post, Washington, D.C. 20071. 04-30-01

Your article, *Astronomers detect birth of planets (Science,* news reprint in the *Atlanta Journal-Constitution,* April 27) was interesting information well done. But some helpful comments seem in order.

The prevailing perspective—in which interpretations of the new scientific data discovered by the researchers are unduly restrictive—is based on beliefs rather than on factual science. The belief that stars and planets accrete from gaseous dust, grains, etc., was initiated by Simon Laplace in 1796. Scientists do not have any definitive evidence of the accretion process, which does have a nice convenience about it. But it raises more questions than it answers: why can't it explain all anomalies of planetary systems?

But there is a definitive alternative that does explain them, one that's based solidly on the Four Laws of Planetary Motion (FL), clearly revealing how the planets attained their orbital spacing around our Sun before beginning their five-stage evolution at rates in full accord with size. Definitive evidence can be ascertained simply by looking to the skies to observe the various stages of evolution of our planets at rates relative to their sizes. From there, it is a simple step into the reality that planetary Evolution (E) is not possible without a core engine to drive it forward. It is an irrefutable fact of science: Evolution and Internal Nucleosynthesis (IN) are inseparable; to the end, one cannot exist without the other.

In reality, planets (and all other spheres) must obey the laws of physics and chemistry; beliefs without substantiated facts, and based on misinterpretations of data, are always illusory.

I am writing while under the impression that you are the same young lady who worked for the *Gwinnett News* and interviewed me some ten years or so ago about the erroneous fossil fuels hypothesis. Since then, the fast-mounting evidence finally led to the enigmatic solution to the revolutionary Fourth Law (1980-1995) that brought the LB/FLINE concept full circle. (Herewith are some pages of more details.) In spite of the stubborn resistance against it, this new concept is destined to displace current beliefs early in the 21st century; truth based strictly on facts has a way of prevailing in the long run.

One major question to pose to the researchers: How can smoke particles and sand-sized grains—typical products of fiery comets—circling a star be interpreted as proof of accretion, especially in view of the powerful evidence that shouts so loudly against their conclusion and in favor of the definitive conclusion that all rings around spheres are composed of ejecta powerfully propelled into orbits around each equator near the end of the second stage of evolution? Saturn, being further along than Jupiter's early second stage, is the prime example of being at the peak of such processes.

The erroneous fossil fuels hypothesis—a costly contributor to beliefs about energy crises—still survives after 170 years as a detriment to scientific progress; unbelievable in this age of enlightenment.

To: Letters Editor, AAA Science, Washington, D.C. 20005. June 26, 2001

The article, *Infrared Gleam Stamps Brown Dwarfs as Stars (Science* 15 June 2001), concludes that free-floating brown dwarfs form like stars rather than like planets. The conclusion was made based on the belief or assumption that if the dwarfs formed from contracted clouds like a star, a warm, dusty disk should orbit the dwarf and radiate additional infrared light. The additional IR light, if found, would be interpreted as evidence for the presence of a disk.

After finding 63 dwarfs that showed such evidence, Charles Lada *et al.* concluded that since an oversized free-floating planet formed by agglomeration would not have a disk, these dwarfs must have formed the way stars do. But questions remain: Could these brown dwarfs have the same type of hard-to-detect rings as those recently discovered around Jupiter? Shouldn't the principles of agglomeration remain identical in all sizes of accreted spheres?

The belief that an oversized free-floating planet formed by agglomeration won't have a disk appears to contradict the consensus belief that all planets and stars form via accretion, usually, but not always, with leftover material from their formation forming into disks. In these cases, how could size make a difference?

Further, "the surprisingly bright IR light from the 63 brown dwarfs in the nearby Trapzium star cluster is helping to make the case that the free-floating brown dwarfs are failed stars and not stray planets." Would the bright IR light from Jupiter and from comets strengthen the case—or would this tend to weaken it? Could these infrared lights be interpreted as coming from a source of hot energy similar to the 30,000°C core of Jupiter (as measured by the Pioneer 10 spacecraft instruments)?

Concerning nuclear energy sources, should size make any difference in the principles involved, whether the smallest source on Earth or inside any active planetary sphere and its moon(s) or comprising the largest stars? Can we be absolutely certain that accretion is the best and only way to create universal spheres? If not, then why not examine other probabilities that meet the criteria of staying strictly within the realm of natural laws that require no speculation? In such cases, size as a function of time readily accounts for all the differences attributed to free-floating brown dwarfs. They were, and are, made in the same original manner as all other spheres of the universe: via the transformation of energy into matter at rates in full accord with size and in full compliance with all natural laws; i.e., stars, brown dwarfs, planets, moons, etc., all formed, and continually form, and evolve in identical manner.

To: Glenn Strait, Science Editor, The World & I, Washington, D.C. 20002. September 23, 1998

New analysis of data from the Mars Pathfinder Mission has revived a nagging question: Why do the inner planets exhibit different mean densities when presumably they formed from the same material (C1 chondrites)? The new analysis suggests that future modelers of inner solar system origin must account for a set of inner planets with differing elemental compositions. A new perspective on the subject offers some common sense answers.

To begin, three fundamental points can be mentioned briefly: (1) Since the accretion theory ignores the Four Laws of Planetary Motion (FL) and cannot explain the orbital spacing of planets (including the exo-planets), we must question its scientific validity; (2) Planetary Evolution (E) is not possible without the Internal Nucleosynthesis (IN) that drives it forward; (3) Differing elemental compositions of planets are functions of the parameters in which IN occurs.

When viewed in the perspective of this FLINE paradigm in which planets evolve through five stages of evolution, all of the aforementioned problems can be displaced by valid scientific answers.

The density of any planet is a function of its current stage of evolution, its distance from the Sun (i.e., temperature), and its size. These three realities are the critical factors that interact to determine the parameters for the sphere's energy core to produce various quantities of elements ranging from hydrogen to uranium. Mercury's small size and its close proximity to the Sun accounts for its greater amounts and types of lost volatiles and retained heavy materials, resulting in a high density of 5.5 in its fifth and final stage of evolution (inactive, dead). The slightly less density (5.2) of Venus can be accounted for by greater distance from the Sun and lower losses of volatiles than Mercury, (but greater losses and smaller size than Earth), and its thick, heavy cloud cover during its current fourth stage of evolution. Earth's 5.5 density is attributed to its mid-range fourth stage of evolution (rocky), slightly larger size (with less cloud cover), greater distance and a slightly larger, heavier nuclear core than Venus.

By these relative measurements, the lighter density (3.9) of Mars can be attributed to its small size, which accounts for its fifth stage of evolution (inactive, dead), plus its great distance from the Sun (-23°C average temperature), its retention of more types of volatiles, all of which, in turn, account for the accumulation of more lightweight volcanic material throughout its stages of evolution. Note: The last outlet for volcanic material via IN in Mars resulted in the largest volcano in the Solar System. The low density of our Moon is accounted for in similar manner to that of Mars. Their lightweight volcanic compositions are the primary reasons for their lower densities. This evidence enhances the belief that differing elemental compositions of spheres are functions of the parameters in which their IN occurs.

To explain one step further, the low density (1.3) of Jupiter is attributed primarily to its current second stage of evolution (gaseous) in which tremendous volumes of virgin gases consisting of lightweight elements were made via IN (the transformation of nuclear energy into atoms) within the core. The huge volume of gases relative to the heavy core is sufficient to account for Jupiter's very low density. The same reasoning applies to Saturn, which is somewhat ahead of Jupiter in their second stage of evolution—which accounts for the greater ejecta rings of Saturn. But Jupiter's rings will catch up some day. All of the same reasoning can be applied to all other spheres of our SS, as well as to the exo-planets.

Valid scientific evidence supporting the three fundamental points corroborates the FLINE paradigm, which, in turn, offers opportunities to fill the many gaps existing in current theories of planetary origins.

To: Glenn Strait, Science Editor, The World & I, Washington, D.C. 20002. February 27, 1999

Thanks for the information in your e-mail of February 10 (1) and February 18 (4) received while on my week's vacation. I will comment on them in chronological order of date and time clock.

Subject: *Vast Stellar Disks Set Stage for Planet Birth in New Hubble Images.*

Comments: The evidence, as interpreted by advocates of the Accretion concept stemming from the Big Bang hypothesis, can be interpreted more accurately in the perspective of the Little Bangs Theory (LBT). When viewed carefully, the patterns of distribution of the gases, dust, and stars in the Hubble photos reveal them as products of tremendous explosions of gigantic masses (perhaps black holes or quasars) of energy (a generic term yet to be defined by physicists). The patterns clearly show the expanding gas and dust, along with the gravity effects of the encapsulated fiery masses (stars) interacting with these materials. Exchanges of ejecta and infalling material are ongoing for eons of time before the cycles are exhausted. As to planetesimals, space probes soon will prove conclusively that they are remnants of one or more disintegrated planets, just as Olbers first stated in 1802, and as is shown to be true via the FLINE paradigm comprised of the Four Laws of Planetary Motion (FL), Internal Nucleosynthesis (IN) and Evolution (E): three chronological, inseparable and ongoing realities of Nature.

Subject: *Volcanism Had Far Greater Role on Mars Than Was Thought*. [or is now thought].

Comments: A truer statement was never made. To quote, "New photos from the Mars Global Surveyor show that horizontal layers extend deep into the canyons of Mars. The structure and composition of the layers suggest that volcanism played a far greater role in the early geology of the Red Planet than previously believed, scientists report in this week's issue (Feb 18) of *Nature*." This conclusion and its cogent evidence in the news release add powerful (and predicted) support to the three principles of the FLINE paradigm. As a matter of fact, the article sounds as if its evidence stemmed directly from my writings of the past two decades. Certainly, it corroborates that Mars and Earth (as with all planets) evolve in identical manner via IN through five common stages of evolution in full accord with size.

Subject: (1) *CU-Boulder Researchers to Map Polar Ice on Mars.* (2) *News about Sand Dunes on Mars.*

Comment: The evidence revealed in these two releases can be added to the findings in the previous paragraph as powerful (and predicted) evidence of the similarities of Mars and Earth that strongly corroborate the FLINE paradigm.

Subject: *New Climate Modeling of Venus May Hold Clues to Earth's Future.*

Comments: The same things stated about Mars and Earth apply to Venus, as they relate to the FLINE paradigm: all three are evolving via its same three principles. All differences, including compositions, atmospheres, surface characteristics, temperatures, etc., among the three planets are explained in this revolutionary theory by their differences in mass and in distance from the Sun.

The true significance of all past, present and future findings of space probes can be fully grasped only in this new perspective. When its time does arrive, the FLINE paradigm will prove of tremendous value to scientists in interpreting their discoveries. Apparently and hopefully, that time cannot be too far away. My lectures and presentations do seem to be making some headway against the prevailing winds.

Thanks again for keeping an open mind and for the information. When and if you decide to publish the copyrighted FLINE paradigm or any of its facets, I will give you the first magazine publication rights.

To: Mr. David H. Levy, Writer, c/o Parade Magazine, New York, N.Y. 10017. May 31, 1999

An interesting article, *A Planet by Any Other Name...* (PARADE, May 30, 1999), puts the controversy about the status of Pluto in perspective. On the same Sunday night I watched *Asteroid Impact* on TV in which one scientist's prediction that the impacts pf Comet SL9 on Jupiter in 1994 would be the greatest explosions ever witnessed in the Solar System, a prediction laughed at by other scientists. The reasons in making that prediction reside in the true nature and sources of comets, asteroids and planets. I would like to comment on both the status of Pluto and why my accurate prediction of the extremely powerful explosions of SL9 was easy to make without a supercomputer.

We must first realize that Whipple's icy conglomerate model for the cometary nucleus is a highly speculative concept with no basis in fact. The facts clearly reveal that a comet is a fireball with a nuclear nucleus that creates its own encapsulating shell of black carbon, its million-mile hydrogen coma, and its tails of ionized gases and fine particles of matter via Internal Nucleosynthesis. Comets are ejecta from larger nuclear masses, and have relatively short active lives (a few thousand years).

The facts also make it clear that moons and planets evolve in the same manner via Internal Nucleosynthesis through five common stages of evolution; many scientists recognize their original cometary status via still-visible clues. Other incontrovertible evidence is supplied by Kepler's First Three Laws of Planetary Motion and the solution to the Fourth Law (1980-1995) (that first eluded Kepler in 1595) explaining the spacing of planets around the Sun. Understanding the true nature and sources of fiery comets made it easy to predict the powerful nuclear explosions of SL9 at Jupiter's very thin cloud-tops.

The Four Laws of Planetary Motion (FL) and the five-stage Evolution (E) of planets via Internal Nucleosynthesis (IN) clearly include Pluto as the last planet of our Solar System. The small mass and highly elliptical orbit and the surface features of this outermost planet and those of all other planets (including the exo-planets) and moons, are explained definitively in my publications of the past 24 years, culminating in my latest book, *The Spacing of Planets: The Solution to a 400-Year Mystery* (1996). Both the true status of Pluto as a planet and the true nature of fiery comets are backed by a vast amount of incontrovertible evidence carefully researched during the past 26 years. The findings of this long research also confirm that Olbers was partly correct in 1802 when he identified asteroids as remnants of a disintegrated planet (actually remnants of the explosions of three Asteroids planets whose orbits are still well defined). Some burned out shells of comets can be classified as asteroids (e.g., Phobos and Deimos).

Unfortunately, these findings, along with the volumes of evidence by many other authors, clearly undermine the Big Bang concept of origins. As such, they encounter opposition in the form of apathy and disbelief and will continue to do so until (another prediction) the true natures of comets and asteroids are confirmed by future space explorations early in the 21st century. Meanwhile, every relevant discovery of Earth and of the space probes adds corroborating evidence to this FLINE model of planetary origins. The new knowledge tells us that it is time to quit laughing; it is time to re-evaluate the myths of science.

Memo
Findings on Mars
December 5, 2000

The latest substantiated support of the FLINE model's five-stage evolution of planets came on December 5, 2000, via the news media three days before its scheduled publication in the journal, *Science. Researchers find evidence of lakes on ancient Mars* (*LD News* via AP) reported that photos from a satellite orbiting Mars suggest the Red Planet was once a water-rich land of lakes (Malin Space Sciences Systems). The photos, taken by the Mars Global Surveyor spacecraft show massive sedimentary deposits with thick layers of rock stacked in miles-deep formations. The pictures show clear views of horizontal deposits of rock, a characteristic of sedimentary layers, in the walls of craters and chasms cut into the surface of Mars. Like Earth's Grand Canyon, such formations are possible only in the presence of water. Further, the splitting of planetary crust via either expansion or spotty sinkhole contraction, as well as volcanic actions, must be taken into account as probable causes of other formations.

"We have never before had this type of irrefutable evidence that sedimentary rocks are widespread on Mars," said Michael Malin. "These images tell us that early Mars was very dynamic [as with all active planets—a basic principle of the FLINE model] and may have been [was] a lot more like Earth than many of us had been thinking." Such incontrovertible conclusions, all readily explained by the revolutionary model, continually move scientists ever deeper into the FLINE paradigm and perhaps ever closer to its long overdue acceptance as a valid scientific concept.

The anticipated findings on Mars add to the impeccable list of accurate predictions of the past two decades of the evolving FLINE model. This smaller planet is entering, if not already in, its fifth and final stage of evolution (inactive, dead), bleak and barren, just as Earth is destined to become after its current fourth evolutionary stage completes its cycle in a few billion years. Such depletion of the energy core in each planet and moon defines the boundary between the fourth and fifth stages of planetary evolution. These substantiated findings clearly corroborate the fact that planets undergo the five-stage evolution at rates in full accord with size.

The FLINE paradigm goes a giant step further in explaining how the planets evolve through their five stages of evolution via Internal Nucleosynthesis from energy into atoms that form the molecular matter comprising all atmospheres, lands, waters, life, etc. Understanding this new model opens the floodgates to full comprehension of these predicted discoveries on Mars and eventually to irrefutable solutions to all planetary anomalies. But in the process, scientists must re-examine a number of ancient myths, especially the modified Laplace accretion hypothesis of 1796 allegedly explaining the origin of planets; they do not accrete from gaseous dust, planetesimals and comets, all of which are by-products of creation processes involving the intimate relationship among black holes, quasars, galaxies, solar systems, stars, etc.

The findings on Mars, Earth, exo-planets and all similar spheres shout out the dynamic origins and on-going evolution via Internal Nucleosynthesis of planets and moons. At the current pace of such discoveries of irrefutable evidence on other planets, can it be only a question of time until the shouts are heard and heeded to the point of displacing the antiquated myths about origins of solar systems and the evolution of planetary spheres?

To: Stephen Mackwell & Program Committee
Lunar and Planetary Institute, Houston, TX 77058 **01-30-04**

Re: Abstract #I117: Origin and Evolution of Planetary Systems: Bringing the Copernican Revolution Full Circle via the New Fourth Law of Planetary Motion

The reason given in your letter of January 27, 2004 for rejecting my abstract is false and unacceptable; it conveys the impression of censorship brought on by bias for E. A. Poe's Big Bang myth (1848) and Laplace's Accretion conjecture (1796); i.e., the antiquated Big Bang/Accretion (BB/Ac) hypotheses. To quote: "We do not believe that the abstract contains sufficient scientific content [like the BB/Ac does?] for presentation at this conference." This tops all excuses given during three decades of off-and-on rejections and two decades of stymied press releases. Rather than fight progress, advocates should accept the fact that no substantiated evidence can ever be offered in support of their beliefs. But there is a viable alternative underpinned with a plethora of such evidence.

I would be remiss if I failed to assure you and the Program Committee for the 35th Lunar and Planetary Science Conference that this abstract and its papers of the past three decades contain more substantiated content of valid scientific revelations than all the scheduled conference papers combined. This understatement is underpinned by the submitted summary list of 41 substantiated solutions to anomalies that have stumped scientists since the heliocentric idea of Copernicus (1543). Understanding them is a great joy that needs to be shared with all LPI members. If a flaw exists in the content, it has proven elusive during the 30 years of research, development, presentations and copyrighting (1975-2002) of its many facets, and should be disclosed, if found via normal vetting.

Topping this abstract's partial list of answers is the geometric solution (1980-1995) to the enigmatic Fourth Law of Planetary Motion that first eluded Kepler in 1595, and now presents real opportunities to bring the Copernican Revolution full circle. In doing so, this discovery is key to the aforementioned (and future) solutions to anomalies, without which they might have remained unsolved for another very long time, and key to understanding the discoveries on Planet Mars, and how they relate to Earth, as explicitly described in my Energy Series (1975-2002). Due to entrenchment of the BB/Ac hypotheses, these nine editions, apparently too advanced for their time, were self-published at great personal expense as the only option available for preserving this crucial revolutionary knowledge for posterity.

Yes, these findings eventually will force scientists to revise their beliefs about origins and evolution of inanimate worlds. But this is real progress. Inevitably, the well-substantiated LB/FLINE model (1973-1996) will displace the BB/Ac hypotheses that Sagan called myths, and Hawking admitted are conjectural (1999). A start in that direction was made in the TV documentary, *Mars Rocks,* shown on *Discovery* channel on the evening of January 29, 2004. In it, one scientist correctly attributes the reason for the different appearances of Mars and Earth mainly to their difference in size; i.e., the smaller, barren Mars is now what the larger Earth will become in the far distant future: their rates of change are functions of the amounts of heat originally contained and utilized in each planet. This conclusion [Evolution via Internal Nucleosynthesis] is explicit in my publications (1975-2002) and ongoing work. Whether this is copyright infringement, or simply additional confirmation of the LB/FLINE model, or both, remains to be determined.

To understand the fundamentals underpinning the mysteries of Planet Mars, one must first understand the LB/FLINE model embracing the principles of origin and evolution, and growing ever more valid with each relevant space discovery, while bringing the Copernican Revolution full circle. This is the message my abstract intended to convey. My greatest concern, as expressed in an earlier letter, is that the abstract contains far too much scientific content for presentation, unless granted at least four hours. To present one or two facets of the concept would be like giving members small pieces of a jigsaw puzzle, and only hope they could visualize the complete picture—an impossibility. As with pieces of a completed puzzle, every facet is inexorably connected to all other facets: a word picture too monumental even for a poster session. These facts should never be permitted to become irrelevant for the sake of propagating the BB/Ac hypotheses.

I believe the new Fourth Law of Planetary Motion always will rank high among the six most significant discoveries in the history of planetary sciences. Certainly, it is the most significant finding since the great work of 1905. But why should it be forced to go through the same arduous ordeal encountered by the heliocentric idea of Copernicus? If we are to keep closed minds to viable concepts, how can we expect to advance the cause of science?

I deeply regret, as we all should, that the LPI members will be denied the opportunity to learn the intricacies and immense ramifications of the new Fourth Law and its subsequent LB/FLINE model of origin and evolution, now proven beyond any doubt. Sadly, this rejection does raise questions about the objectives, sincerity and integrity of the LPI. Even though I prefer to share these findings, there is one advantage gained by such rejections: they always provide time and incentives to add more facts to the already overwhelmingly persuasive evidence underpinning the revolutionary LB/FLINE model; e.g., the findings on Mars will add much confirmation to the scientific validity of the new model. The timing is perfect for precisely matching these findings with the fundamentals of the new model of origin and evolution of planetary systems. What vast potential this huge scientific content presents for understanding the origin and evolution of inanimate worlds: a truly productive conference!!

If your strategy was to obtain release on my copyrights before rejecting my abstract, I hereby cancel the signed release as having been obtained under false pretense. I much prefer to work openly with the LPI, but I am prepared and willing to fight for both my constitutional rights and my hard-earned copyrights, if need be.

Someone once wrote these words of encouragement to me: "Truth has a way of surviving."

I sincerely hope that your meeting in March will be as productive as ever.

<u>Addendum: Common Sense Insight:</u>

"What people don't know isn't nearly so aggravating as what they do know that ain't so." Mark Twain.

To: Editor, New Scientist, London WC1X 8NS, UK . **December 6, 2004**

In *The Last Word (New Scientist,* Dec 3, 2004), Paul Barrett of the UK poses two questions: Is it theoretically possible that a fault in the Earth's crust on an ocean floor could drain an enormous quantity of seawater? What would be the effect, not only on life on Earth, but on the planet's structural integrity if sea levels underwent a colossal drop?

Answers: The thickness of Earth's crust now averages some 40 miles, and is continually increasing. It is unlikely that any new fault would ever be deep enough to reach the magma below. If by chance it did go deep enough, the tremendous pressure of the magma would instantly provide a solid seal quickly quenched by the neighboring, now-steaming, seawater—effects somewhat analogous to a blend of geyser and volcanic actions.

Earth's crust continually thickens as a function of time, so the possibility of an enormous drainage has passed the point of no return. In the early stages of atmospheric and crustal formations, such contact between seawater and magma was a common event, producing steamy jungles: a suitable habitat for dinosaurs. Inevitably, the environment changed often over time; millions of other species have developed and died out. We view the results of an evolving planet: volcanism, extrusions, black smokers, baseball-like seam of activity circling the globe, etc.—all in the fourth stage of Earth's ongoing five-stage planetary evolution.

Ref: *New Principles of Origins and Evolution: New Paradigms of Beauty, Power and Precision* (2002), featuring the new Fourth Law of Planetary Motion, Alexander A. Scarborough, Ander Publications, 202 View Pointe Lane, LaGrange, Georgia 30241, USA. Available on www.authorhouse.com

To: NPA Members. **August, 2005**

Much substantiated evidence clearly supports some claims in the two L. S. Meyers letters via e-mail:

1. "The nebular hypothesis and the concept of subduction as the mechanism of plate tectonics are shown to be null and void by unequivocal proof the Earth is growing and expanding."

2. "All oceans and waters of Earth have been generated...," etc., etc.

These claims are an integral part of the Little Bangs (LB)/FLINE (1975-1996) model of origins and evolution of planetary and universal systems. The new model's plethora of substantiated evidence further advances these claims by definitively detailing the mechanisms that unite everything into beautiful and simple answers—just as Einstein predicted the right answers would be—about which Meyers suggests NPA members seek unanimity. Certainly, we do not want to follow the Establishment into the speculative and costly Big Bang quagmire.

If the Meyers suggestion can be fulfilled via the facts already available—and speculative concepts abandoned—the future of a factual science looks extremely promising and exciting.

With the realization by more and more people that Earth is expanding and growing, we are on the brink of an epoch-changing direction in scientific thought on origins and evolution: from speculative to substantiated. If successful, this year's NPA meeting could be ranked among the most significant in the history of science.

To: All NPA members.
Subject: Earth's Gassy Gift. **Aug 23, 2005**

Quoting *Earth's Gassy Gift (New Scientist* 6 Aug 2005): "Ever since the Apollo missions brought back lunar soil, the scientists who analyzed it have been puzzling how nitrogen got into it. Turns out the nitrogen may have come from Earth."

"Until now, the best bet was the nitrogen from the solar wind was deposited on the moon. But the amount of lunar nitrogen and the ratios of its isotopes do not match the corresponding values in the solar wind. Now, Frank Podosek of Washington University in St. Louis, Missouri and his colleagues have an alternative explanation."

"If you turn off Earth's magnetic field, the wind from the sun can blow nitrogen from Earth's outer atmosphere onto the moon," says Podosek. And he thinks this is what happened before Earth's field became active. After that, the magnetic field shielded the atmosphere from the solar wind and no more nitrogen was blown away (*Nature*, vol.436, p655).

Unfortunately, this reasoning seriously rivals that of the White Queen in *Alice in Wonderland,* and forces science ever deeper into the fairy tales initiated by the Big Bang of E. A. Poe in 1848.

Earth's magnetic field has been an inherent property of our planet ever since the forces involved in the Four Laws of Planetary Motion geometrically spaced the planetary masses precisely around our Sun, with Earth and its Moon in the third orbit, during the dynamic birth of our Solar System some five billion years ago. Since that epoch-making time, Earth and its satellite have evolved, along with all other planets and spherical moons, into the beautiful spheres we enjoy today.

All heavenly spheres are self-sustaining entities that create and regulate their own planetary matter—gaseous (e.g., nitrogen), solids, and liquids—until their internal energy source is depleted, a process known as evolution. The transfer of nitrogen from one sphere to another is neither likely nor necessary; the Moon's nitrogen is a product of its own making.

To: Editor, New Scientist, 84 Theobald's Road, London, UK. **September 24, 2005**

The ten informative articles in *The World's Biggest Ideas (New Scientist,* 17 Sep 2005) presented very good explanations of each of the subjects. However, due to a number of recent events and more definitive interpretations, two articles, #1, *The Big Bang* (BB), and #9, *Tectonics,* warrant further considerations.

Many scientists openly recognize the BB as conjectural, with no substantiated evidence to support it. In reality, the CMB 2.7 K radiation pervading the Universe, like all other spaces on Planck's thermonuclear radiation curve, is short-lived, and therefore cannot be a leftover from the BB some 14 billion years ago. All such radiation must depend on its constant source of re-supply from ongoing thermonuclear reactions in all *active* universal spheres (e. g., stars, planets), as revealed by the newer LB/FLINE model of universal origins.

The same argument applies to the proportions of hydrogen, helium and deuterium in the Universe: they are the lightest and easiest elements produced quantitatively and proportionately in the thermonuclear reactions in all *active* spheres of the Universe—all in compliance with Einstein's $E = mc^2$: a vital facet of the new model.

The article raises other questions: How did our Universe evolve into the immensely intricate cosmos we see around us? How and why was the Universe set up to expand the way it does? Why does the BB contradict one of the most powerful laws of physics, the second law of thermodynamics?

In reality, these questions and the second law pose no problems for the ongoing Little Bangs (LB/FLINE) model of universal origins and evolution. Researched in-depth during the last 32 years, the answers are simply additional facets comprising the concept, now waiting in the wings for opportunities to become better known to inquisitive readers seeking more definitive science.

Concerning article #9, *Tectonics,* the new concept presents definitive explanations of the underlying basics of plate tectonics: the ongoing energy-matter relationship via Internal Nucleosynthesis that drives all universal Evolution.

Tragically, the two biggest obstacles to comprehension of universal origins are belief in the BB myth and the suppression of newer, more substantiated ideas.

To: Kay Scarborough
From: Hartwig Thim, NPA Member

Thank you for sending me all of these very interesting notes. In the last note entitled "From Myth to Reality in 12 Easy Steps" (August 26, 2005), I'd like to criticize a statement in step 8 in connection with $E=mc^2$.

This expression was published by Hasenoehrl and Abraham in 1904 in the same journal Einstein published his 1905 paper on special relativity (*Annalen der Physik*). Ives has already shown, in 1952, that Einstein's derivation was defective. The reason is that special relativity excludes energy transfer which is responsible for mass increase. Mass increase is therefore a one-way process or absolute effect, and has nothing to do with relative velocities. Both mass increase and $E=mc^2$ can be derived from Poynting's vector product E x H, yielding the mass equivalent of an electromagnetic wave. When such a wave is absorbed by a massive body, its mass increases in addition to its (absolute!) velocity.

Do you agree with me?

The Ongoing, Unlimited Source of Universal Oxygen

The article, "Oxygen Rocks" (*Science News*, Sep 1, 2007), states that "young Earth acquired a healthy portion of oxygen, and credited the change to the rise of cyanobacteria, known as blue-green algae, that produce oxygen by photosynthesis."

A more accurate statement is that oxygen was (and is) utilized by algae, from which it is released into the atmosphere. According to the Big Bang theory, all oxygen atoms were created early in the BB, and no more have been made since then. However, even this concept, if true, could have furnished only a tiny percentage of today's ongoing production of oxygen and all other elements via Internal Nucleosynthesis (IN) in all universal spheres (e.g., stars, moons, planets).

Every active sphere in the universe is a self-sustaining entity that creates its own atomic matter via IN in accord with $E=mc^2$.

The greater question becomes: From whence comes the unimaginable amounts of energy on which the universe depends for its ongoing light-speed, expanding growth and multiple functions?

The endless source of universal energy is a key factor in the LB/FLINE model of origins of universal systems—a simple and beautiful answer: Space-time.

To: All NPA members.
Subject: A Faint Light at the End of the Tunnel. **July 12, 2005**

After 70 years of the ongoing battle for truth in science, including the past 32 years of concerted efforts, and now age 82, I can see a faint light at the end of the tunnel via two recent articles in *New Scientist,* 2 July 2005, (received July 11, 2005). In it, an editorial comments on the status of the Big Bang (BB), *Not the End of the World: Does it matter that some cosmologists* are *finding fault with the BB?* It refers readers to an article on page 30: *End of the Beginning: Is it time to admit that the idea of a BB just doesn't stack* up?

"Look at the facts," says Riccardo Scarpa of the European Southern Observatory in Chile. "The basic BB model fails to predict what we observe in the universe in [at least] three major ways. The temperature of today's universe, the expansion of the cosmos, and even the presence of galaxies, have all had cosmologists scrambling for fixes. Every time the basic BB model has failed to predict what we see, the solution has been to bolt on something new—inflation, dark matter and dark energy," Scarpa says.

Going a big step further, last year the first ever *Crisis in Cosmology* conference members wrote an open letter warning that failure to fund research into BB alternatives was suppressing free debate in the field of cosmology (*New Scientist,* 22 May 2004, p20).

Each of these valid points (and many additional ones) has been argued via my writings of the past 32 years, but to little avail. It is both surprising and a good thing that Scarpa and other Conference members have managed finally to get these points into a prestigious magazine. Perhaps this will be the spark that brightens up the tunnel to the point of moving more and more scientists toward a crucial change in direction of scientific thought.

If this does happen soon, I may yet live long enough to see the LB/FLINE model come to full fruition where solutions to enigmatic anomalies interlock precisely into place, advancing science into a truly exciting future of comprehending origins and evolution of universal systems. If I do not live to see its fruition, I hope that each member will continue the battle for truth in science via the LB/FLINE model.

Evidence uncovered during the 70-year struggle to separate the myths and facts of science clearly exposes the BB as the grandest and costliest myth in the history of science. Yes, it is indeed time to admit that the idea of a speculative BB just doesn't stack up, and to examine closely new ideas that can move science into the epoch-making era of definitive knowledge. From a personal standpoint, it will make the 15-year struggle with the solution to the new Fourth Law of Planetary Motion—the crucial key to the revolutionary LB/FLINE model—even more worthwhile.

As a former member of several of the largest USA science organizations that showed little, if any, interest or faith in new ideas throughout those years of struggle, I now can fully appreciate the open-mindedness of NPA members whose goals include embracing and discussing all new ideas that could lead to more definitive truths in science.

At this time, I would like to acknowledge the persons who played, and are playing, critical roles at crucial times in saving the evolving LB/FLINE model from oblivion: Drs. John Chappell (deceased), Cynthia Whitney, Neil Munch, and the many other patient members of the international NPA science organization, along with Ralph Howard, President of Kleen-Tex and Riccardo Scarpa, as well as my daughter, Kay Scarborough, who always has been there to handle the necessary computer work.

Thank you all!

CHAPTER III

THE FIVE LAWS OF PLANETARY MOTION: Crucial Keys to Understanding Solar Systems and Galaxies

Paper: Intro to the New Fourth Law of Planetary Motion and Its Ramifications
AAAS-NPA Meeting, Denver, April 7-10, 2004

The crucial keys to understanding the origin of our Solar System (SS) were discovered in 500 B. C. by a secret society of Pythagoreans. They rationalized that number was important for understanding Nature, putting much faith in the Pythagorean theorem and the Golden Ratio (GR). And they believed that Earth moved around a central fire. Because of these beliefs, they were killed by enraged neighbors.

In 1543, over 2000 years later, Copernicus confirmed our heliocentric SS, reviving the revolution in scientific thought begun by the Pythagoreans in 500 B. C.

In his book, *The Divine Proportion,* H. E. Huntley wrote, "It was suggested in the early days of the [20th] century that the Greek Letter Phi should be adopted to designate the GR."

The pervasiveness of the GR was expressed best by C. Arthur Coan:

"Nature uses this as one of her most indispensable measuring rods, absolutely reliable, yet never without variety, producing perfect stability of purpose without the slightest risk of monotony... We shall find it flung broadcast throughout all Nature."

Although Bode's Law was not known in his time, Johannes Kepler (1571-1630) realized that spacing of the six then-known planets was geometric. He successfully established the First Three Laws of Planetary Motion, revealing the elliptical orbits and precise clockwork of the revolving planets.

However, Kepler failed in his attempts to solve the geometric spacing of the planets — a potential Fourth Law of Planetary Motion. Although on the right track, Kepler somehow missed the right train of thought: the GR.

Ironically, he wrote, "Geometry has two great treasures: one is the theorem of Pythagoras; the other, the division of a line into extreme and mean ratio. The first we may compare to a measure of gold; the second we may name a precious jewel."

The solution to the Fourth Law, along with Kepler's Three Laws, reveals the absolute necessity of a great momentum, along with geometric precision, to create the ellipses and precise clockwork of our SS.

In every galaxy throughout the Universe, untold numbers of binary star systems have been, and are being, created when two rapidly moving, fiery energy masses (ejecta from a central black hole) lock in orbit. In rare and special cases, two star masses will meet the prerequisites of proper relative sizes, relative velocities, and proper distance apart to effect fragmentary break-outs from the smaller, faster energy mass (SEM) after it swings past and beyond the larger mass at a decelerating pace, while releasing a portion of its mass into orbit at each Phi point along the way.

The solution to the elusive Fourth Law required the graphing of seven geometric Diagrams, each scaled and drawn in full accord with the GR. During the 1980s, the current BL spacing of the planets was plotted in two graphs to illustrate how the inner and the outer planets attained their current positions around the Sun. Modified during ensuing years, both schematics show what happened during the dynamic layout of (1a) the inner planets and (1b) the outer planets. As the smaller mass moved up the y axis, fragments broke from it at GR points along the way, due to the combined gravities of the Sun and the preceding planet at tangent t. Every consecutive pair of orbital spacing complies with both the GR and its reciprocal.

Arc xy' and Tangent xy meet at P to reveal the endpoint establishing line y' Sun, which reveals the 14° angle of diversion. This shifts the base line x to x', and t' to t , an adjustment in accord with the Pythagorean theorem triangle, affirming the correct reading for the angle of diversion of the SEM as it passed the larger Sun.

As the SEM swung past the Sun at the diverted angle of 14°, several things happened. The faster SEM pulled the Sun's equatorial gases to a faster pace and at an angle of 7° to the plane of the ecliptic. This began the erratic distribution of different degrees of inclination of the planets to the ecliptic, each affected by the combined force of gravity of the Sun and of the preceding angled planetary masses. Mercury's angle of 7°, the same as that of the Sun's equatorial gases, is an important clue in verification of this concept.

Initially, the fiery fragment destined to become Planet Mercury broke from the sunward side of the SEM, followed soon by the breakout of the Venus fragment from the SEM's outer edge. Along with the correct Phi spacing, these breakouts account for the slow counterclockwise rotation of Mercury and, via a gear-type action, for the slow clockwise rotation of Venus. In Figure 1a, Mercury's breakout at the first Phi point is 0.618034 AU from the Sun. Abetted by the strong gravity of the Sun and the gravity exerted by the Mercury mass at Tangent t, Venus broke off at exactly 1.0 AU, the second Phi point. Thereafter, at each successive Phi point along the way, gravitational forces of the Sun and of the previous planet, now positioned at Tangent t, combined to pull the next fragmentary mass into orbit, all in full accord with Nature's ubiquitous GR, with each new potential planet rotating in the counterclockwise direction (viewed from above).

Set 2, (Figures 2a and 2b), shows how it happened. The two Diagrams present a clearer picture of the relative motions of the two fiery masses. Figure 2a, the inner planets, and Figure 2b, all planets, show the relative motions of the Sun and the SEM during the dynamic layout of the SS. As the Sun moved to the right along the x-axis at a steady pace, the faster SEM moved up the hypotenuse at a decelerating pace. At each Phi point along the way, gravitational forces of the Sun and the previous planet, then at Tangent t, combined to pull the next planetary mass into orbit (as was shown in Set 1). In 2a, the outer GRL curve is the apparent path of the SEM during layout of the inner planets; the inner BL curve shows the current orbital positions of the inner planets. In 2b, the smooth curve is the apparent path of the SEM; the warped curve shows the current orbital positions of all planets. In both 1a and 1b, the differences are tabulated as changes in distances of the inner and outer planets, respectively, during the past five billion years. This side-by-side comparison of the two curves clearly reveals the warped BL (Bode's Law) curve as a warped version of the original smooth GR layout of the SS. As the Sun reached the right end of the x-axis, Pluto was pulled into the final orbit, closing the system exactly as depicted in the first Set of Diagrams. So we see that Sets 1 and 2 corroborate each other.

Set 3 consists of three Diagrams, 3a, 3b, 3c, each plotted in the GR scale of distance in AU and orbital velocities in mi/sec. Figure 3a shows the current BL orbital positions and velocities of the planets and Asteroids. Figure 3b shows the original GR layout of the SS, a smoother curve.

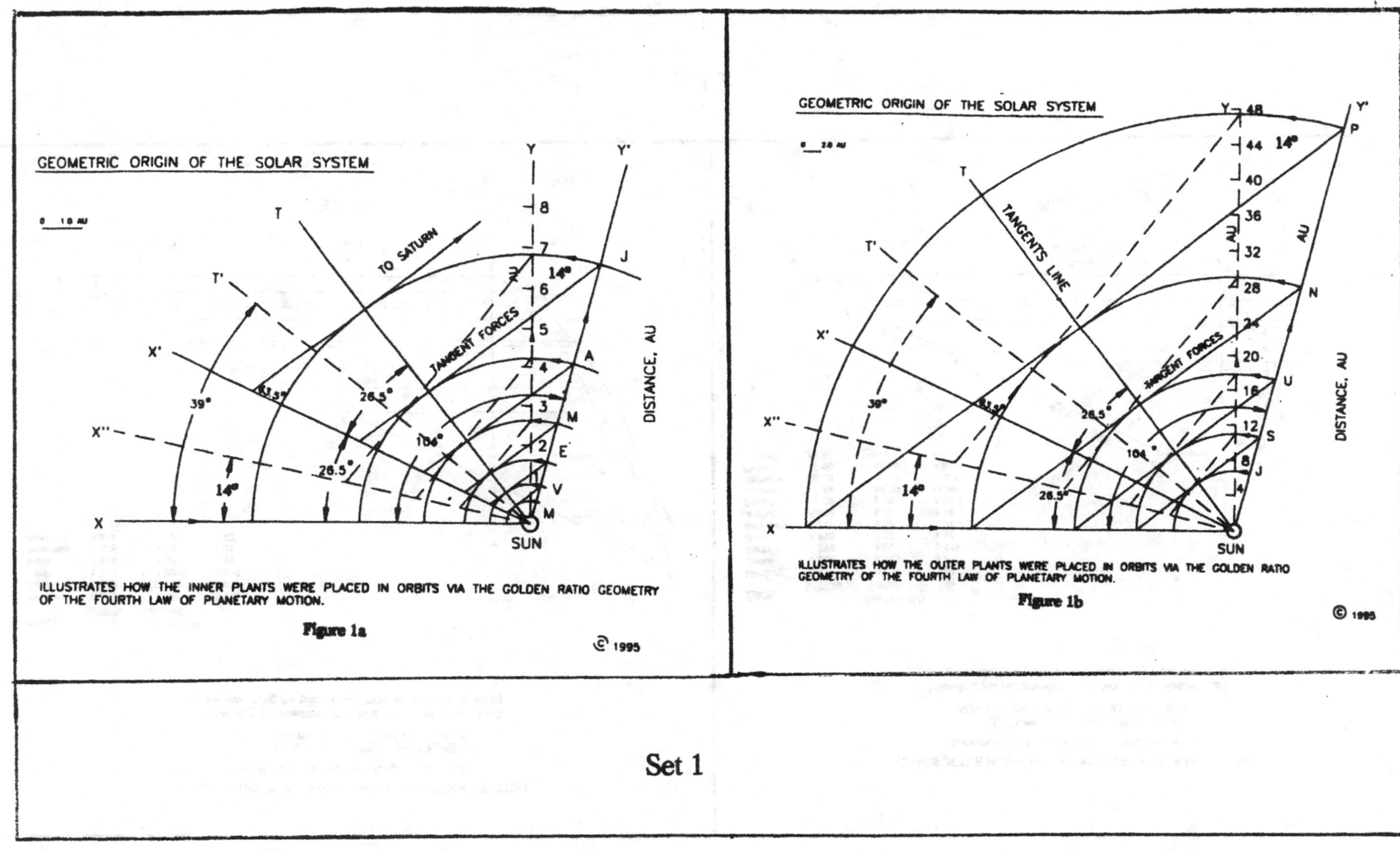

ILLUSTRATES HOW THE INNER PLANTS WERE PLACED IN ORBITS VIA THE GOLDEN RATIO GEOMETRY OF THE FOURTH LAW OF PLANETARY MOTION.

Figure 1a

© 1995

ILLUSTRATES HOW THE OUTER PLANTS WERE PLACED IN ORBITS VIA THE GOLDEN RATIO GEOMETRY OF THE FOURTH LAW OF PLANETARY MOTION.

Figure 1b

© 1995

Set 1

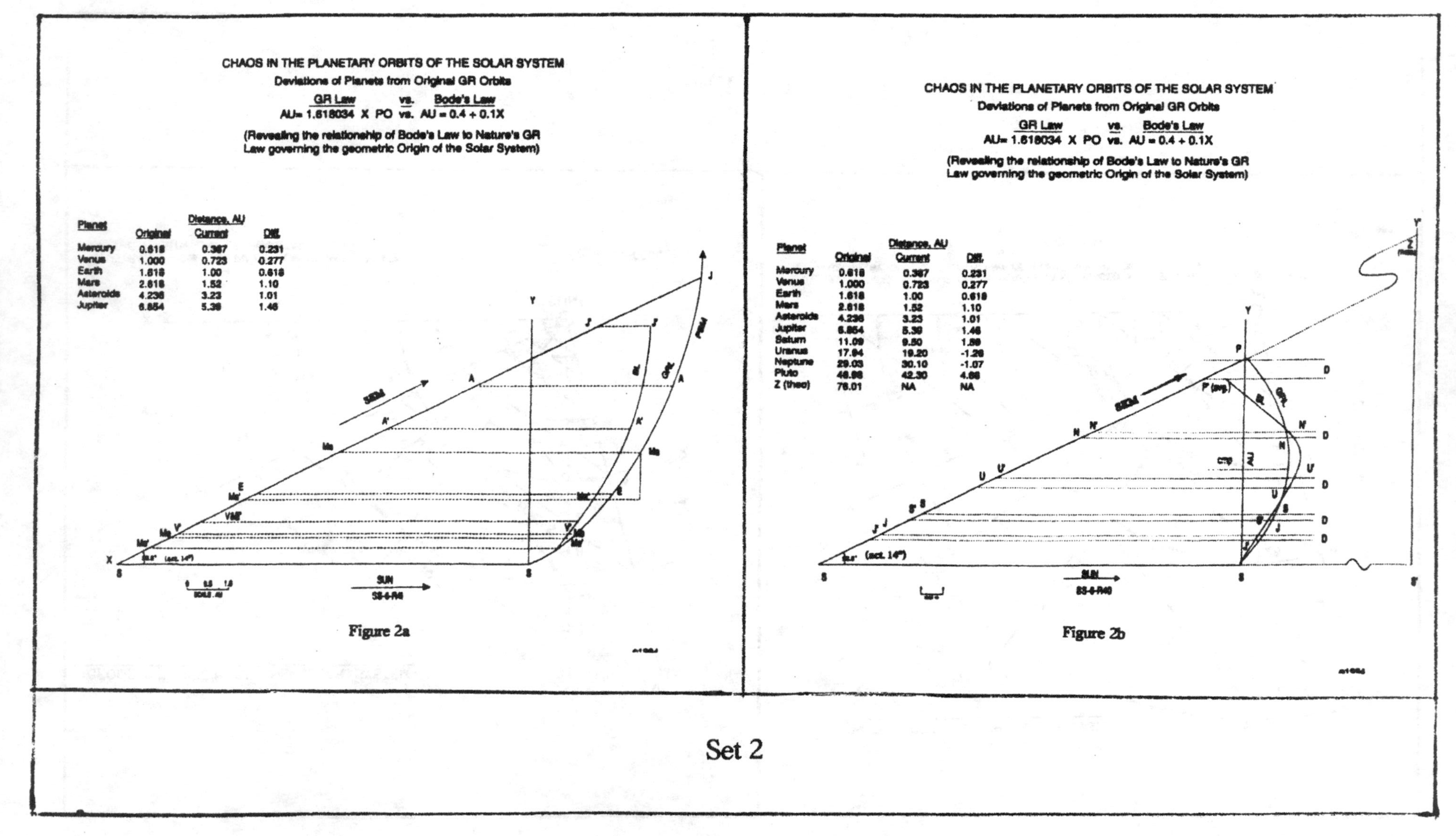

Planet	Distance, AU Original	Current	Diff.
Mercury	0.618	0.387	0.231
Venus	1.000	0.723	0.277
Earth	1.618	1.00	0.618
Mars	2.618	1.52	1.10
Asteroids	4.236	3.23	1.01
Jupiter	6.854	5.39	1.46

Figure 2a

Planet	Distance, AU Original	Current	Diff.
Mercury	0.618	0.387	0.231
Venus	1.000	0.723	0.277
Earth	1.618	1.00	0.618
Mars	2.618	1.52	1.10
Asteroids	4.236	3.23	1.01
Jupiter	6.854	5.39	1.46
Saturn	11.09	9.50	1.59
Uranus	17.94	19.20	-1.26
Neptune	29.03	30.10	-1.07
Pluto	46.98	42.30	4.68
Z (theo)	76.01	NA	NA

Figure 2b

Set 2

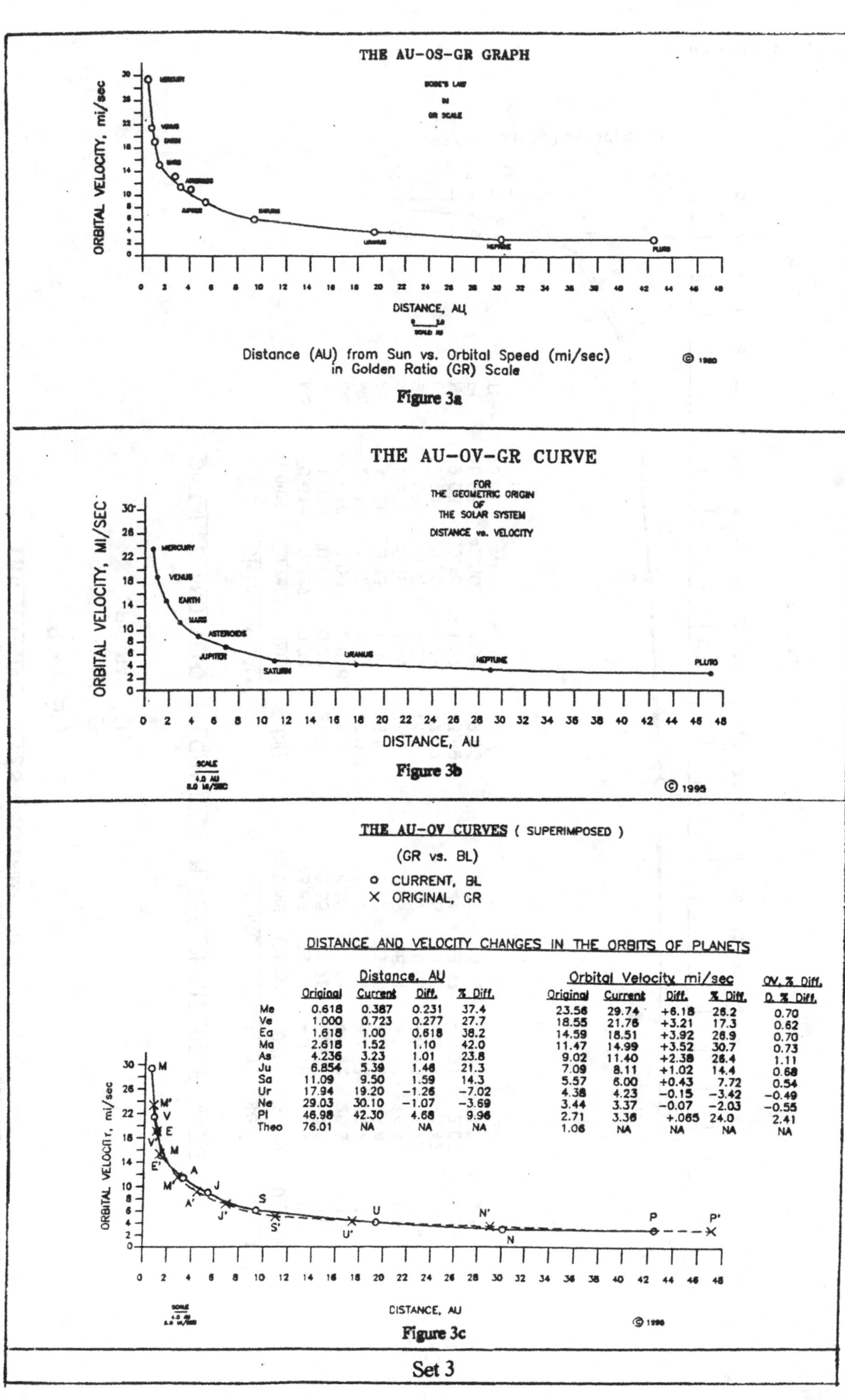

Figure 3a

Figure 3b

DISTANCE AND VELOCITY CHANGES IN THE ORBITS OF PLANETS

	Distance, AU				Orbital Velocity mi/sec				OV. % Diff.
	Original	Current	Diff.	% Diff.	Original	Current	Diff.	% Diff.	D. % Diff.
Me	0.618	0.387	0.231	37.4	23.56	29.74	+6.18	26.2	0.70
Ve	1.000	0.723	0.277	27.7	18.55	21.76	+3.21	17.3	0.62
Ea	1.618	1.00	0.618	38.2	14.59	18.51	+3.92	26.9	0.70
Ma	2.618	1.52	1.10	42.0	11.47	14.99	+3.52	30.7	0.73
As	4.236	3.23	1.01	23.8	9.02	11.40	+2.38	26.4	1.11
Ju	6.854	5.39	1.46	21.3	7.09	8.11	+1.02	14.4	0.68
Sa	11.09	9.50	1.59	14.3	5.57	6.00	+0.43	7.72	0.54
Ur	17.94	19.20	−1.26	−7.02	4.38	4.23	−0.15	−3.42	−0.49
Ne	29.03	30.10	−1.07	−3.69	3.44	3.37	−0.07	−2.03	−0.55
Pl	46.98	42.30	4.68	9.96	2.71	3.36	+.065	24.0	2.41
Theo	76.01	NA	NA	NA	1.06	NA	NA	NA	NA

Figure 3c

Set 3

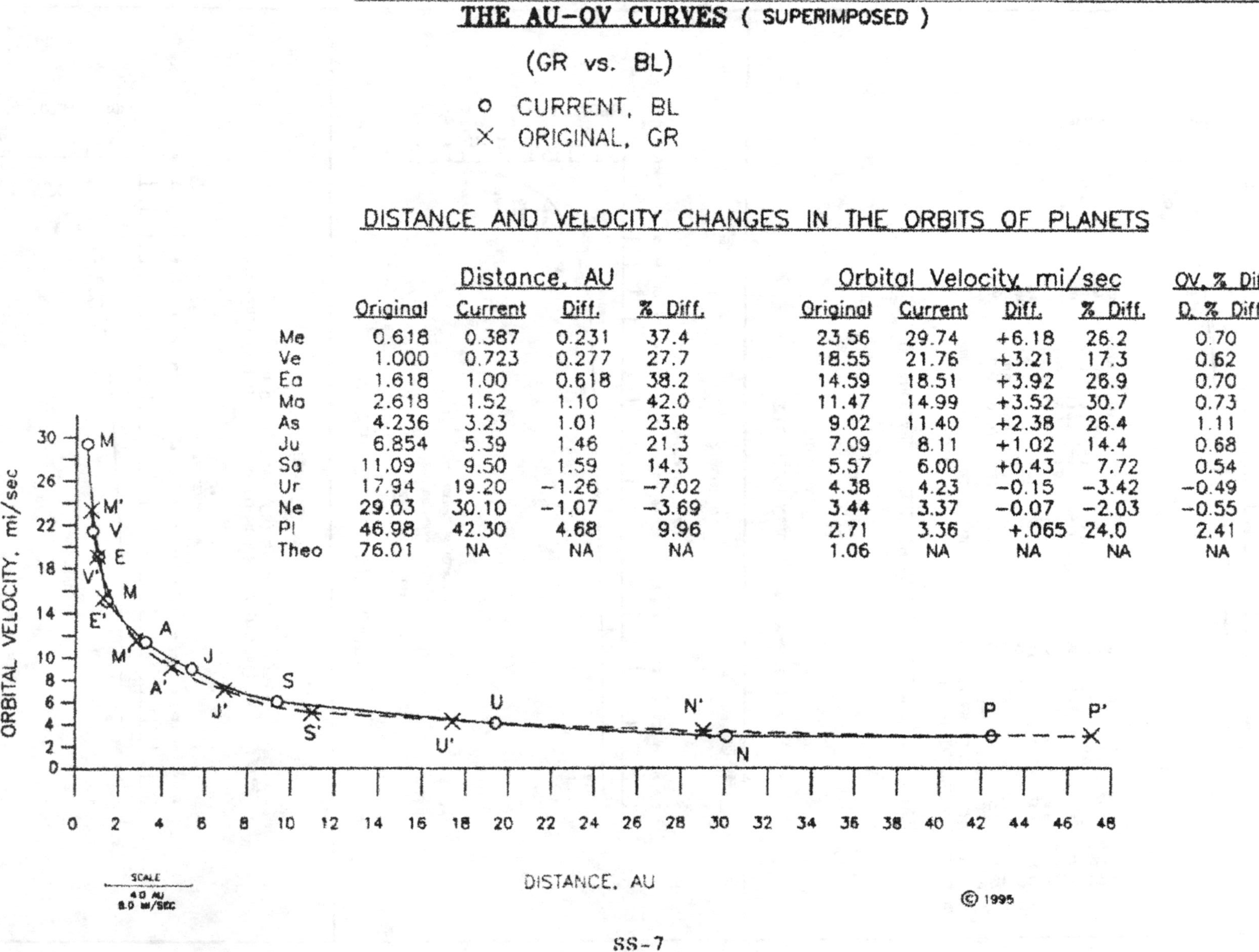

THE AU-OV CURVES (SUPERIMPOSED)

(GR vs. BL)

o CURRENT, BL
× ORIGINAL, GR

DISTANCE AND VELOCITY CHANGES IN THE ORBITS OF PLANETS

	Distance, AU				Orbital Velocity, mi/sec				OV, % Diff.
	Original	Current	Diff.	% Diff.	Original	Current	Diff.	% Diff.	D, % Diff.
Me	0.618	0.387	0.231	37.4	23.56	29.74	+6.18	26.2	0.70
Ve	1.000	0.723	0.277	27.7	18.55	21.76	+3.21	17.3	0.62
Ea	1.618	1.00	0.618	38.2	14.59	18.51	+3.92	26.9	0.70
Ma	2.618	1.52	1.10	42.0	11.47	14.99	+3.52	30.7	0.73
As	4.236	3.23	1.01	23.8	9.02	11.40	+2.38	26.4	1.11
Ju	6.854	5.39	1.46	21.3	7.09	8.11	+1.02	14.4	0.68
Sa	11.09	9.50	1.59	14.3	5.57	6.00	+0.43	7.72	0.54
Ur	17.94	19.20	−1.26	−7.02	4.38	4.23	−0.15	−3.42	−0.49
Ne	29.03	30.10	−1.07	−3.69	3.44	3.37	−0.07	−2.03	−0.55
Pl	46.98	42.30	4.68	9.96	2.71	3.36	+.065	24.0	2.41
Theo	76.01	NA	NA	NA	1.06	NA	NA	NA	NA

SS-7

When the 3a and 3b curves are superimposed in Figure 3c, they almost become one. However, significant differences are noted in the tabulated results for each and every planet. Of the ten original orbits, eight have moved closer to the Sun, while Uranus and Neptune have drifted slightly away. In compliance with Kepler's Third Law, changes in velocities correlate with changes in orbital positions. The negative results of Uranus and Neptune signify slight increases in distances from the Sun correlating with slight decreases in their velocities. Each of the other eight planets (including the Asteroids) shows a significant decrease in distance that correlates with its increase in velocity. The tabulated results show that Mercury, the innermost planet, has gained the most velocity (6.18 mi/sec), while being among the planets losing the highest percentages of distance. Earth is an interesting example of changes since its birth; it has moved 0.618 AU, or 38.2%, closer to the Sun, while gaining 26.9% velocity. This knowledge led to a modified fourth and a new fifth law (2006) of planetary motion.

All changes in displacement and velocity are proportional to the mass/distance/velocity/time relationship of each planet to the Sun and to other planetary gravitational effects. Uranus and Neptune are examples of very distant planets with sufficient mass and distance to resist the inward pull of the combined gravities of the Sun and the inward planets. Each is gradually drifting away while its velocity decreases in compliance with Kepler's Third Law.

Contrary-wise, Pluto is so small that it has been, and still is, greatly affected by these gravities, in spite of its greater distance. This planet's tiny size permits it to be affected by any and all gravitational forces within range, thereby accounting for its highly unorthodox orbit (eccentricity and angle of inclination to the ecliptic), and its large decrease of 4.68 AU from its original Phi position at 46.98 AU.

This third Set of three Diagrams reveals the changes in planetary orbits and velocities since the dynamic layout of our SS some five billion years ago. The results of each of the three Diagrams correlate precisely with results from Sets 1 and 2. Together they give a clear picture of the relative motions of two fiery energy masses of a potential binary star system that met GR prerequisites during its dynamic layout to become a rare type of SS with multiple planets.

The Seven Geometric Diagrams (1980-1995) reveal the true beauty of the GR origin of our SS. The mysterious BL spacing of planets finally is revealed as a warped version of their original GR spacing.

INTRODUCTION TO THE FLINE MODEL

The new model consists of three chronological, inseparable and ongoing realities:

1. The Five Laws of Planetary Motion (FL);
2. Internal Nucleosynthesis (IN) that drives;
3. all ongoing Evolution (E).

In his *Principles of Philosophy* in 1644, Descartes defined Earth's interior as "Sun-like".

In the 17th century, Newton discovered that gravity holds the SS together.

In 1905, Einstein's formula $E=mc^2$ revealed the intimate and interchangeable relationship between energy and matter. Transformations from energy to matter usually occur under extreme conditions of high temperatures and pressures within sources of nuclear energy. Various combinations of these conditions over varying periods of time determine the types and quantities of elements produced via nucleosynthesis within our Sun, stars, moons, planets, comets, and previously, in some asteroids.

As explained in previous papers, planets are self-sustaining entities transforming via nucleosynthesis from nuclear energy masses into the gaseous second stage of evolution, followed in order by the transitional, rocky, and inactive stages. This knowledge provides researchers with the essential fundamentals for definitive understanding of all planetary anomalies; e.g., the findings on Mars were predicted and explained by the evolving FLINE model two decades before.

The non-biological origin of the hydrocarbon fuels—not from fossils!—was proven beyond question during the ten years of 1973-1983. It is an excellent example of the ongoing transformation of all planetary elements from energy to atomic matter. (Details in Chapter III)

NPA 2007 MEETING
University of Connecticut

Modifications to Kepler's Laws of Planetary Motion

Abstract: The Revised Fourth and New Fifth Laws of Planetary Motion

Discussions at the 2006 NPA meeting resulted in a later revision of the Fourth Law of Planetary Motion that governed the dynamic origin of geometrically-spaced orbital positions of the original ten planets of our Solar System. Whether or not the Fourth Law applies in certain principles to other solar systems remains a question for the future.

Dr. Robert Heaston and I worked together in restating the Law to bring it in line with the style used by Kepler in the First Three Laws. The new Fifth Law—originally included as a part of the Fourth Law—is now a separate entity that continues governing the ongoing gradual changes in orbital positions since the dynamic origin of our solar system some five billion years ago. Principles of the new Fifth Law may apply universally to all solar systems.

Simplified geometric diagrams are utilized in the explanations. The Five Laws (FL) offers the capability of deciphering the full history of our Solar System.

Introducing the New Fourth and Fifth Laws of Planetary Motion

To preserve and build upon scientific innovation, scientists must first determine whether a change in direction of scientific thought on origins and evolution is warranted. A few probing questions with known answers might be a good way to begin.

After almost a century of discoveries and untold billions of dollars in cost, do we have even one substantiated fact in support of prevailing beliefs about universal origins and evolution? No! Do data from space probes ever fit precisely into current beliefs? No! Would they fit better into another concept? Yes! Is too much money being wasted in vain pursuit of evidence supporting the Big Bang myth? (Sagan terminology) Yes!

Since myth begets myth, shouldn't we realize the futility of pursuit of knowledge about universal origins and evolution in this perspective? And about why the Universe is expanding at an accelerating rate—just the opposite of what the Big Bang predicts?

If the right answers are not to be found within the BB perspective, doesn't this tell us that the BB concept must be the major problem in this scenario? Is science wrong in clinging to the BB in spite of its incapability of providing accurate interpretations to the many brilliant discoveries of space probes—interpretations likely to provide even more questions than answers? Is there an alternative in which new and old data can be interlocked precisely to provide answers to questions that may never be answered via prevailing beliefs?

Fortunately, there is an alternative that warrants full consideration: the LB/FLINE model of universal origins and evolution. The new model has been carefully constructed with factual evidence during the past 33 years (1973-2006). Findings of space probes continually fulfill its predictions, and verify its scientific validity. No anomaly appears capable of escaping its proclivity for solving the mysteries of universal origins and evolution.

To understand the revolutionary LB/FLINE model, one must first understand the Five Laws of Planetary Motion:

I. The planets travel in elliptical orbits, with the Sun at one focus. (Kepler, 1609)
II. For every planet, in equal intervals of time, equal areas are swept out by a line drawn from the Sun to the planet. (Kepler, 1609)
III. Comparing the orbits of the several planets, it is found that the times of revolution are related to the distances from the Sun by a proportionality: the squares of the periods of revolution are proportional to the cubes of their mean distances from the Sun. (Kepler, 1619)
IV. Comparing the orbits of the several planets, it is found that the distances between the orbits of any pair of planets have a common proportionality: the original orbit of a planet relative to the one on either side of it is defined by the classic Greek Golden Ratio. (Scarborough, 1995, modified 2006)
V. In additional comparisons of the orbits of the several planets, it is found that gradual changes in their distances and velocities have a common proportionality: all are functions of a distance-mass-velocity-time relationship. (Scarborough, 1995, modified 2006)

Although the BL was not known in his time, Johannes Kepler (1571-1630), perhaps history's greatest astronomer, realized that the spacing of the six then-known planets is of a geometric nature. While successful in establishing the First Three Laws of Planetary Motion by sound logic and by solving the relevant mathematics, Kepler failed in his attempts to solve the geometric spacing of planets—a potential Fourth Law of Planetary Motion. Although on the right track, Kepler somehow missed the right train of thought: the Golden Ratio and the Pythagorean Theorem.

Ironically, he wrote, "Geometry has two great treasures: one is the theorem of Pythagoras; the other, the division of a line into extreme and mean ratio [Golden Ratio (GR)]. The first we may compare to a measure of gold; the second we may name a precious jewel."

The Geometric Solution to the New Fourth Law of Planetary Motion

The First Three Laws of Planetary Motion were discovered by Kepler in the early seventeenth century (1609-1618). Dealing with the dynamic precision of the Solar System (SS), they describe the elliptical orbits and precise clockwork of the revolving planets. As will be shown, the solution to the new Fourth Law, along with Kepler's laws, reveal the absolute necessity of a great momentum (from a powerful expulsion), along with a dynamic geometric precision, in order to create the ellipses and precise clockwork of our SS. Imagine a potential binary-star system in which the smaller, faster mass zips past the larger mass at a specific distance apart. If the two masses are of proper relative sizes and have proper relative velocities, the result will be a breakup of the smaller mass into fragments—much like the fragments of Comet SL9 around Jupiter in 1993—that split off at points determined by Nature's ubiquitous GR geometry. With this scenario in mind, one begins to realize why and how the planets were placed in geometrically-spaced orbits around the Sun. By combining this realization with the current spacing of the planets, one can derive a plausible Fourth Law of Planetary Motion that explains the geometric spacing of the ten (or more) original planets around our Sun during the dynamic layout of the SS a few billion years ago.

The solution to the elusive Fourth Law of Planetary Motion required the graphing of Seven Geometric Diagrams, each scaled and drawn in full accord with the Golden Ratio (GR), also known as Phi geometry. In 1980, the current orbital positions, velocities and spacing of the planets were plotted in the GR scale (Fig. 3a). The nearly perfect BL curve easily could (and did) lull theorists into assuming it represents the planetary orbits, velocities and spacing that had not changed since the original layout of the system some five billion years ago. They had not changed in five billion years??!! This false assumption has prevailed for nearly 400 years, and is the primary reason for the lack of understanding the enigmatic mystery of the spacing of the planets. But orbits, velocities and spacing do change over time; nothing can remain in orbit forever. Such changes did, and still do, occur in any solar system. Further, the original relationships had to comply with Nature's ubiquitous Phi geometry. The BL curve for current positions, velocities and spacing of planets had to be a flawed version of GR relationships—a flawed version reflecting changes made by the forces of gravity over the very long time frame of some five billion years. But how could this be verified?

Pluto is so small that it has been, and still is, greatly affected by these gravities, in spite of its greater distance. This planet's tiny size permits it to be tossed around (relatively speaking) by any and all gravitational forces within range, thereby accounting for its highly unorthodox orbit (eccentricity and inclination to the ecliptic), and its large decrease of 4.68 AU from its original Phi position at 46.98 AU. At this rate, Pluto's potential for a collision with another planet in the far distant future seems real.

The large masses and distances of Uranus and Neptune and the small mass and distance of Pluto account for their unorthodox displacements; the disintegration of the Asteroids can be attributed to their vulnerable position, proportionately balanced between the two huge masses of Jupiter and the Sun. The three original Asteroids' masses could not long withstand the battering by these powerful gravitational forces; they disintegrated—exactly as Heinrich Olbers concluded in 1802. Some tens of thousands of planetesimal remnants (asteroids) remain today in three distinct orbital paths, the average of which is beautifully spaced in precise compliance with the warped BL and the GR versions of the original Phi geometry of the SS.

The tabulated results in Diagram 7 (Figure 3c) of Set 3 summarize the changes in orbit, velocity and spacing of each and every planet since the origin of the SS. The displacement of each planet from its original orbital position is tabulated in actual AU distance and as a percentage change. An interesting example to ponder: Planet Earth has moved 0.618 AU, or 38.2% closer to the Sun, while gaining 26.9% velocity.

This third set of three Diagrams (BL and GR graphs, superimposed) clearly shows the changes in planetary orbits that have occurred since the geometric layout of the SS some five billion years ago. The results of each of the three sets correlate precisely with the results of the other two sets. Together they give a clear picture of the relative motions of the two masses of a potential binary-star system that met the special GR prerequisites to become a very rare type of solar system with multiple planets during its geometric layout.

Developed during a time frame of 15 years (1980-1995), the Seven Geometric Diagrams reveal the true beauty of the geometric origin of our SS. The enigmatic BL of current spacing of the planets finally has been shown to be a flawed version of the original Phi spacing of the planets. From this graphical picture with its seven intertwined geometric Diagrams, the plausible Fourth and Fifth Laws of Planetary Motion were stated with a reasonable degree of confidence.

THE GEOMETRIC BEAUTY OF OUR SOLAR SYSTEM

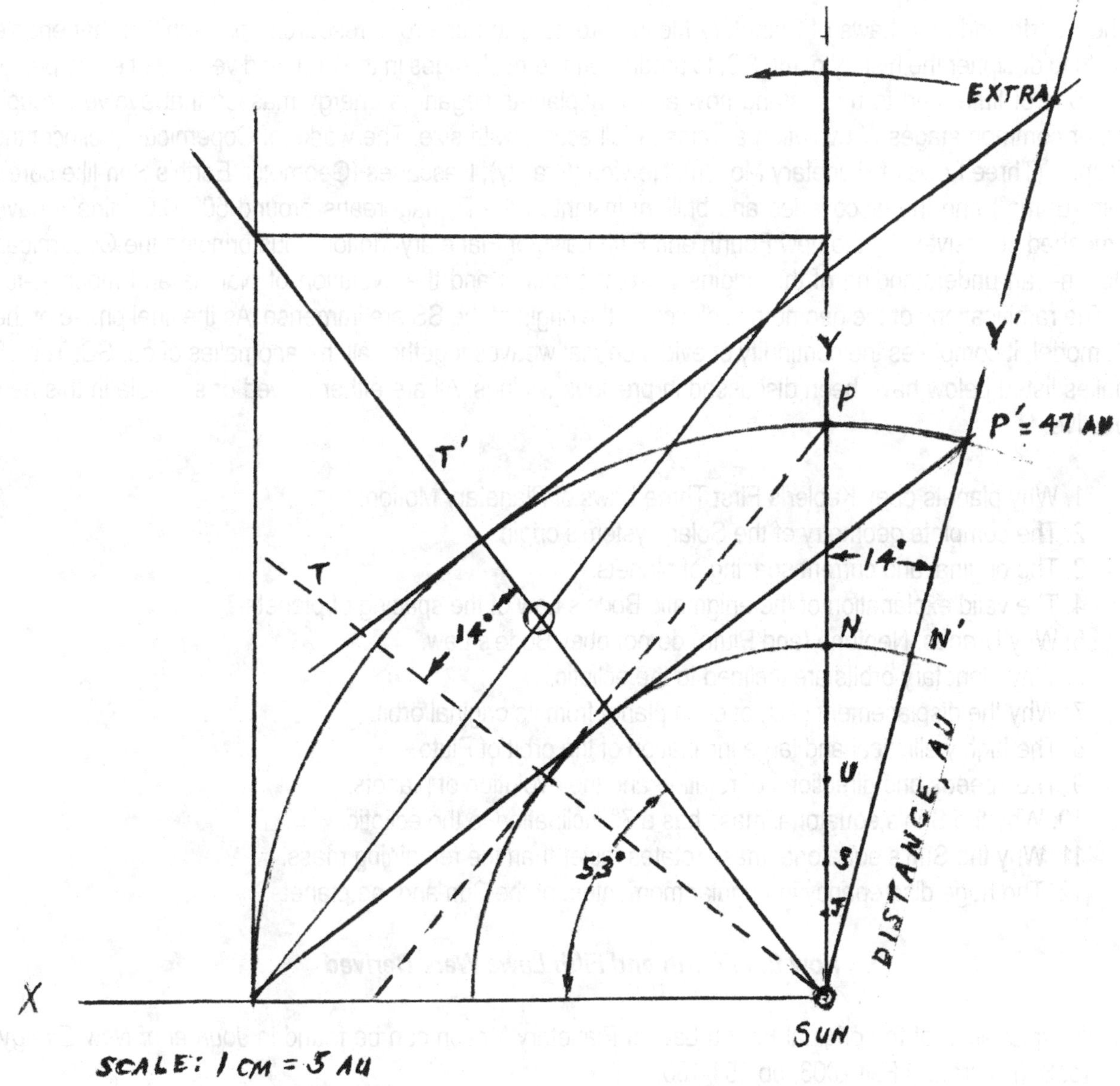

ILLUSTRATION OF THE GEOMETRY OF THE 4TH LAW OF PLANETARY MOTION: THE SPACING OF PLANETS.

NOTE: BOXED IN BY 4 PYTHAGOREAN TRIANGLES.

CAN BE EXPANDED INDEFINITELY FOR PLANETS BEYOND PLUTO, IF ANY, OR FOR LEFTOVER MATTER.

A. A. SCARBOROUGH
(1973-2006)

Conclusions

The Fourth and Fifth Laws of Planetary Motion are keys to numerous research opportunities that enable scientists to decipher the history of the SS, to predict long-term changes in the orbit and velocity of every planet as functions of time, and to understand how and why planets began as energy masses that evolve through four other common stages of evolution at rates in full accord with size. The works of Copernicus (heliocentric SS), Kepler (Three Laws of Planetary Motion), Newton (Gravity), Descartes (Geometry, Earth's Sun-like core), Einstein ($E=mc^2$) and the discoveries and brilliant insight of the Pythagoreans around 500 B.C., finally have been meshed cohesively by the new Fourth and Fifth Laws of Planetary Motion, thus bringing the Copernican Revolution—an understanding of the origins of solar systems and the evolution of planets and moons—full circle. The ramifications of the geometric solution to the origin of the SS are immense. As the final phase of the FLINE model, it completes the continuity of evidence that weaves together all the anomalies of our SS. The 12 anomalies listed below have been discussed in previous writings. All are either solved or solvable in this new perspective:

1. Why planets obey Kepler's First Three Laws of Planetary Motion.
2. The complete geometry of the Solar System's origin.
3. The original and current spacing of planets.
4. The valid explanation of the enigmatic Bode's Law of the spacing of planets.
5. Why Uranus, Neptune (and Pluto) do not obey Bode's Law.
6. Why planetary orbits are inclined to the ecliptic.
7. Why the displacement (AU) of each planet from its original orbit.
8. The highly elliptical and large inclination of the orbit of Pluto.
9. The speeds and directions of rotation and the evolution of planets.
10. Why the Sun's equatorial mass has a 7° inclination to the ecliptic.
11. Why the Sun's equatorial mass rotates faster than the remaining mass.
12. The huge discrepancy in angular momentum of the Sun and the planets.

How the Fourth and Fifth Laws Were Derived

A full explanation of the original Fourth Law of Planetary Motion can be found in *Journal of New Energy*, Vol. 7, No. 3, (illustrated) Fall 2003, pp 154-160.

This current paper presents the modified version of the Fourth Law and a new Fifth Law originally embedded in the Fourth Law. Both Laws now are separate, but connected, entities: the Fourth Law details how the ten original (now nine) planets attained their initial geometric spacing around our Sun. The Fifth Law continues governing the ongoing gradual changes both in orbital positions of the nine remaining planets of our Solar System, and in all extra-solar systems.

Following the NPA meeting in 2006, Dr. Robert Heaston and I worked together in restating both laws to bring them in line with the writing style used by Kepler in his first Three Laws of Planetary Motion.

On January 1, 1980, the author accepted the challenge of solving a 400-year-old mystery that had eluded Kepler in 1595: the geometric spacing of planets around our Sun. Although he was aware that the six then-known planets were geometrically spaced, and knew about the Pythagorean Triangle and the Golden Ratio (GR), Kepler simply got off-track by attempting to place different geometric figures between each pair of orbits.

His model still resides in a museum.

Other than these two clues, the third significant key to solving the mystery was Bode's Law, a series of numbers that came close to fitting the actual spacing between all planetary orbits except those of Uranus and Neptune. A popular belief was that anyone who could explain why the two planets did not obey Bode's Law would have the right answers to the origin of our Solar System. With these three keys, the author could not resist the challenge.

So for the next 15 years, Scarborough periodically worked on his fascinating hobby. Finally, in the summer of 1995, the pieces fell into place. The following diagram should simplify the explanation of how our Solar System came into existence with geometrically-spaced orbits, and why Neptune, Uranus and Pluto do not obey Bode's numbers. (The explanations are in the form of the Fourth and Fifth Laws of Planetary Motion.)

The first Seven Diagrams have been explained previously, and are recorded in the *Journal of New Energy*, Vol. 7, No. 3, Fall 2003. A large copy of each is found at the end of this section.

The eighth diagram is an enlarged version of an explanation of how the Fourth Law was derived, using only the last two planets for simplicity.

A horizontal x-line and a vertical y-line are drawn on graph paper. The distances (AU) of Neptune and Pluto from the Sun are marked in accord with the Golden Ratio (GR). From point Sun, arcs are drawn through N and P to line x. When Neptune is pulled into orbit (by the previous planet, Uranus), and reaches the dotted tangent line t, its gravity, along with the gravity of the Sun, pulls Pluto into orbit. When Pluto reaches the same tangent line, its gravity and the Sun's gravity combine to pull anything that might exist beyond the tiny planet into orbit.

Next, line t' is drawn at a 53° angle to line x. This shifts the tangent points of arcs N and P (and all other planets) 14° while forming four Pythagorean triangles that box in the entire Solar System when one vertical line and one horizontal line are drawn in place. Shifting the tangent line 14° requires extensions of all arcs to points aligned 14° to the right of line y, forming line y'.

These results tell us that the Pythagorean Triangle and the Golden Ratio are crucial keys to explaining how our Solar System came into existence via the momentum and interaction of two energy masses (one large and one small), with the smaller mass moving slightly faster as it passed the larger mass, and was diverted 14° by the gravities of the Sun and the previous planetary mass: the major forces in spacing the planets geometrically around our Sun to become the Fourth Law of Planetary Motion.

The last three of the seven SS Diagrams simply show the original layout of the Solar System in accord with the GR, and the current orbits of planets. When superimposed, the changes in orbital positions during the past five billion years are revealed. The gradual changes in their distances (AU) and velocities have a common proportionality: all are functions of a distance-mass-velocity-time relationship. This is the Fifth Law that explains why none of the planets are in exact agreement with Bode's Law: a warped version of the dynamic GR layout of the Solar System some five billion years ago.

The Judgment of Time

Because the solution to the fourth and fifth laws governing the spacing of planets offers definitive clues to understanding our Solar System, galaxies, and the Universe, time may reveal it as perhaps the most significant scientific discovery of the past 100 years.

Extra-solar Systems: How and Why They Differ From Our Solar System

The Four Laws of Planetary Motion reveal the absolute necessity of a great momentum of a smaller, faster mass interacting with a larger mass as it sped past our Sun to create the elliptical planetary orbits with geometric clockwork precision. Further, the solution to the Fourth Law reveals that in order to create a multiple-planets solar system like ours, in which the smaller planetary masses break from the speeding mother mass at the GR points along the way, the two initial masses must meet precise prerequisites of relative velocities, relative masses, and a specific distance apart to effect a critical angle of diversion and subsequent breakup around a central star. Since it is a rarity for two masses to meet these exact prerequisites, the vast majority of solar systems will contain only one intact giant planetary mass circling its central star.

At this time (Dec., 2000), astronomers have discovered about 50 of these exo-planetary solar systems, all at very great distances from our SS, and each with a planetary mass that failed to meet the prerequisites specified in the solution to the Fourth Law of Planetary Motion. As astronomers have acknowledged, each of these giant gaseous planets is much too close to its central star to have formed there via the prevailing accretion concept of planetary origins! All solar systems comply with Kepler's First Three Laws. Whether the solution to the Fourth Law applies solely to our Solar System remains to be seen. But as we shall see, the sizes and gaseous nature of all giant Jupiter-like planets, as well as the sizes and nature of all smaller planets and moons in our Solar System, can be explained by a new concept of planetary evolution that is based on the Five Laws of Planetary Motion.

About 1,500 galaxies, many only one four-billionth as bright as the dimmest light the unaided human eye can see, can be seen in the spectacular Deep-Field picture. The galaxies are stacked up against one another, so the real challenge is to disentangle them. A high priority is to determine the distances from Earth for as many of the galaxies as possible. Since distant galaxies move away from each other faster than nearby galaxies, astronomers can determine within reason each one's distance from Earth.

According to the LB/FLINE concept, the most distant galaxies will be the most recent ones created at the expanding perimeter of the spherical Universe, and will be moving at nearly the speed of light (details in Chapter V). Thus, to fathom universal and planetary origins, mankind must learn their mathematical relationships and their continuity in the chain of evolution from energy to all forms of matter. The new Fourth Law of Planetary Motion, detailing the beautiful geometric spacing of the nebulous energy masses that subsequently and continuously evolve via nucleosynthesis and polymerization as self-sustaining planetary entities, brings this conclusion sharply into focus.

Letters and Memos

To: Letters Editor, Science Magazine. **January 26, 1998**

In writing about the problem of how the giant gaseous planets got so close to a central star *(Migrating Planets, Science,* Vol.279, p.69) Murray *et al.* based their findings and conclusions on the assumption that there is no known formation method that would allow such large planets to initially form that close. They suggest that a giant planet forms in the outer parts of its system and then migrates toward the star, pushed via instabilities created by planetesimals that linger in the disk after most of the gas has been consumed in planet formation.

Two counterpoints are warranted here: First, the formation of any singular or multi-planet solar system must comply with the basic principles of the Laws of Planetary Motion. These principles did not exist in the article.

Second, the assumption that there is no known formation method that would allow such extra-solar planets to initially form so close to the central star is not an accurate statement. A formation method that also complies with the basic principles of the Laws of Planetary Motion does exist, although it is not yet well-known.

Kepler's First Three Laws of Planetary Motion are familiar textbook material. Together, with the recent mathematical solution to the proposed Fourth Law of Planetary Motion (1995) governing the geometric spacing of multi-planets in solar systems, the Four Laws provide sufficient clues for solving the mystery of how gaseous giants got so close to the central star. This evidence provides a valid scientific alternative to the findings described in the *Migrating Planets* article. Meanwhile, the findings on Mars and of other space probes continue to corroborate the only other choice (the LB/FL/IN/E paradigm).

To: Christine Gilbert, Letters Editor, Science Magazine, Washington, D.C. February, 1998

Thank you for the copy of Dr. Norman Murray's response of February 26 to my letter of reply to his paper, *Migrating Planets (Science,* Vol.279, p 69). His frank response touches on some of the problems that scientists face en-route to the truth about planetary origins and evolution.

Kepler's First Law of Planetary Motion deals with ellipses rather than "the radius vector of each planet with respect to the sun sweeps over equal areas in equal time," (the Second Law). But Murray is correct in stating that "it is unlikely that Jupiter formed at 0.05 AU...Jupiter's orbit has shrunk." Actually, the Four Laws of Planetary Motion clearly reveal that Jupiter formed initially at 6.854 AU and now resides in the 5.39 AU orbit (an inward movement of only 1.46 AU in some 5-6 billion years). Perhaps he meant the Jupiter-size mass in an exo-planetary system.

While these small mistakes can be overlooked, Murray's letter gave the alarming impression that he is unaware of the new Fourth Law of Planetary Motion detailing how the planets of our Solar System attained their current orbital positions around the Sun — the law that first eluded Kepler in 1595. Further, the Fourth Law explains: (1) the unorthodox orbits of Neptune and Pluto; (2) the outward drift of Uranus and Neptune; and (3) the ecliptic angles, etc. Together, the Four Laws clearly reveal how our SS came into being, while clearly disproving the condensation/accretion hypothesis of planetary origins from a planetesimal disk (which accounts for the difficulties in getting it past peer reviewers for publication and for the impossibility experienced by Wetherill in attempts to correctly space the planets via computer trials based thereon).

Thus, the Four Laws make it unnecessary to ponder whether large planets can or cannot form so close to their parent stars via the prevailing hypothesis. But clearly their orbits (original and current) can be any distance from the parent star as long as they obey the Laws of Planetary Motion. The reason resides in the fact that the distance is a function of relative sizes, original velocities and initial distance apart to affect the end result of being in close, medium or distant orbits around the parent star. This information is the basis of my argument against Murray's statement "that there is no known formation method that would allow such large planets to initially form that close." Although little known and less publicized, this provable formation method does exist in the scientific literature—in my latest book, *THE SPACING OF PLANETS: The Solution to a 400-Year Mystery.* A paper on the subject is again in process of peer review for publication; however, its irrefutable arguments against prevailing beliefs about planetary origins might again prove too revolutionary and too unsettling for publication at this time.

I argue against only the basic foundation and conclusion of Murray's math—not his math. Evolution (E), Internal Nucleosynthesis (IN) and the Laws of Planetary Motion (FL) are inseparable: one cannot exist in any SS without the other two. If we are to understand all anomalies of solar system and planetary evolution, we must first understand the relationship among these three tangible facets of creation that are the basic foundation of the provable FL/IN/E paradigm of planetary origins and evolution.

Thank you again for giving us the opportunity to discuss our differences.

To: Glenn Straight, Science Editor, The World & I, Washington, D.C. 20002. 06-02-98

The enclosed article, *Images emerge of suspected planet 450 light-years away* (*Atlanta Journal-Constitution*, May 29, 1998), contains some crucial information pertaining to planetary origins—information predicted and explainable via my proposed FLINE paradigm. To quote Toner's article, "If the newly discovered object, named TMR-1C is confirmed as a planet, it would force scientists to revise their notions that giant planets, like Jupiter, require millions of years to coalesce out of a solar, or stellar dust cloud." Such confirmation and revised notion would indeed add powerful corroborative evidence to my writings of the past 25 years on planetary origins and evolution. It is a giant step in the right direction: planets do not coalesce from gaseous dust clouds—its speed of more than 20,000 miles per hour and its trail of glowing gases are crucial factors that cannot be ignored as clues to a definitive understanding of its origin and nature.

Once past this roadblock, the next advance in planetary origins will come with the acknowledgment that planets do indeed evolve through five stages of evolution: from energy to gaseous to transitional to rocky to inactive, or dead, spheres. Without the original energy core mass that drives each planet through these five stages before its depletion, <u>evolution simply would not be possible.</u> Kepler's Three Laws of Planetary Motion and the proposed Fourth Law of Planetary Motion explain how the original energy masses were placed in geometrically-spaced orbits around the Sun before undergoing evolution via Internal Nucleosynthesis into the planets we observe today—all in accordance with size and in full compliance with all natural laws. Thus, in our Solar System, the Four Laws of Planetary Motion (FL), Internal Nucleosynthesis (IN), and Evolution (E) are inseparable: one cannot exist without the other two.

The article touched on an important key to understanding planetary origins: ubiquitous binary star systems. The geometric solution to the Fourth Law of Planetary Motion reveals that our Solar System is a very special and rare situation in which the two stars of a potential binary system met the specific prerequisites of relative size, relative velocities and distance apart to cause the smaller, faster mass to break up into fiery fragments that were pulled into orbit in full accord with nature's ubiquitous Phi (or Golden Ratio) geometry. The Seven Geometric Diagrams detail this event beyond the shadow of a doubt.

The newly discovered planet is another example of a potential binary-star system that failed to materialize in the usual manner: a "planetary nomad, cast adrift in interstellar space by a newly forming pair of binary stars." It will eventually evolve through the five stages of evolution common to all planets. Binary-star systems and planets originate from powerful explosions (perhaps of quasars or black holes) that scatter the fiery masses along their rapid lines of flight (a facet of the Little Bangs Theory that is beyond the scope of this letter, but is detailed in my book, *THE SPACING OF PLANETS: The Solution to a 400-year Mystery*, 1996).

This important discovery should convince most skeptics that the time indeed has come to revise our notions about the origin and evolution of self-sustaining planets.

To: Glenn Strait, Science Editor, The World & I, Washington, D.C. 20002. 04-20-99

Thanks for the news release about the discovery of the first multiple-planet system around a nearby Sun-like star, which was already known to have one planetary companion. The news had come off the AP wire one day after my return from the National Philosophy Alliance (NPA) meeting in Santa Fe. Had it been known two days earlier, I would have used it as additional evidence corroborating the FLINE paradigm in my paper there. The planetary data fit precisely into the new concept—as will that of any system with any number of planets. The sizes, velocities and gaseous nature of the three huge Jupiter-like planets account for their positions in the system and their being in the second stage of the five common stages of planetary evolution. Simply put, the

FLINE paradigm answers the big question of how such a solar system arose.

Marcy's statement that "no theory predicted that so many huge planets would form around a star" fortunately is not true. Apparently, he is unaware of the FLINE paradigm that both predicted and explains this anomalous situation. It is frustrating to read such erroneous statements while remaining unable to get this viable alternative into the scientific community through normal peer review channels. As long as the planetary accretion concept prevails, erroneous speculation about planetary origins will persist.

Immediately after my presentation to the group of some three dozen international NPA members and visitors, I asked Dr. Roland Hron if he had been able to hear it clearly. His response: "Yes, and it was spellbinding." To make the paradigm crystal clear, I had used 23 transparency illustrations to convey its full message. Apparently, its factual, non-speculative, common-sense nature makes this concept easy to understand and accept whenever it is given the opportunity to be heard in its entirety. If you would like a copy of my paper, *The FLINE Paradigm: Definitive Insights Into the Origins, Orbital Spacing and Evolution of Planets*, featuring the proposed Fourth Law of Planetary Motion, I would be happy to send a fully illustrated copy for editing and publication.

P.S. Dr. George Galeczki (President, Society for the Advancement of Physics, Cologne, Germany) informed me of similar work on planetary spacing being done by Beckmann. He promised to send a copy of the book as soon as it is translated from German into English. Galeczki has one of my books.

To: Sue Bowler, Editor, The Journal of the Royal Astronomical Society, UK. August, 1999

On the Spacing of Planets: The Proposed Fourth Law of Planetary Motion. RefqyO16; Submitted January, 1998

Thank you for your letter of 7 July concerning the above paper and the apology for the long delay in the hands of the referees. At that time, you anticipated knowing the outcome of the refereeing process within the next two weeks.

If our current correspondence hasn't just passed in the mail, please advise the status of the paper and whether there are questions about any unclear points in the article and its illustrations.

Conservatively speaking, the significance of the Fourth Law appears equal to that of Kepler's First Three Laws of Planetary Motion, while its revolutionary impact on the sciences of solar systems should equal that of the Copernican idea of a Sun-centered Solar System. Certainly, it does hold much promise as the crucial key to understanding and proving the origins of solar systems. As such, and in the interest of science, I'm sure you agree that its evaluation and publication warrant a high priority.

Thank you for your valuable time and kind consideration.

Memos

Recent Corroborative Evidence

The Extra-solar Planet with an Earthlike Orbit (Science News, Aug 14, 1999) brings up a good point. Of some 20 known planets outside our solar system, the latest find, announced July 29, 1999, stands out from the group. The planet's mass is at least 2.26 times that of Jupiter, and circles the sun-like star, iota Horologii, at a distance that's 29 million km (20% of the Earth-Sun distance) from its parent star. Its orbit is elliptical. To quote astronomer Geoffrey W. Marcy: "This new planet adds to the suspicion that our solar system with its neat, circular, coplanar orbits, may be the exception rather than the rule."

Definitive reasons for our solar system's being "the exception rather than the rule" are clearly stated in the

FLINE model (1973-1997) of planetary origins and evolution. The geometric solution to the new Fourth Law of Planetary Motion (1980-1995) (in conjunction with Kepler's First Three Laws) explains how our planets attained. their orbital spacing around the Sun. In the vast majority of instances in which two fiery masses interact, the results will be a binary-star system, as described in *The Spacing of Planets: The Solution to a 400-Year Mystery* (p. 12). If the two masses are of proper relative sizes and have proper relative velocities, and the smaller, faster mass zips past the larger mass at a very specific distance from it, the result will be a multiple-planet solar system like ours—truly a very rare event!

When these very specific conditions are not met, as happens in the vast majority of cases, the result will be simply another of many solar systems consisting of one or two extra-solar planets. In all cases, the planets in question eventually evolve through the five common stages of evolution. As a planet gravitates ever closer toward its central star, its orbit should become less and less elliptical.

Marcy's suspicion that our solar system may be the exception rather than the rule is in full agreement with the FLINE model's predictions and explanation of why the event is so rare. To go one step further, the solutions to all solar system and planetary anomalies are deep-rooted in the three basic principles of this paradigm. Only in the FLINE model can fully provable solutions to all anomalies of solar and planetary systems be realized.

Our SS's planetary orbits are somewhat chaotic, elliptical, and multi-planar, just as is revealed by a plethora of evidence; they definitely are not "neat, circular, co-planar orbits!"

Plentiful Planets (Science, 17 December 1999, p 2241)

The article discusses the extra-solar planets discovered during the past four years: "All planets detected so far are gas giants. Moreover, most of them move in very eccentric orbits, unlike the planets in our solar system, leaving theorists scrambling to explain why. Although the string of new data confirms that planets are common, at this stage it seems that planetary systems configured like our own are rare indeed... In 1999 astronomers caught an exo-planet transiting the face of its star, significantly dimming that star's brightness for a few hours every 3.5 days. It is 200 times more massive than Earth, has a diameter significantly larger than Jupiter's, and like Jupiter, is composed mainly of hydrogen and helium."

All the data compiled from these important discoveries were accurately predicted, and are readily explained by the new definitive FLINE model. The eccentric orbits of the larger-than-Jupiter planets, their huge sizes and low densities are critical clues revealing the dynamic origins of the systems whose planets, because of their giant sizes, are still in the early phases of the second (gaseous) stage of their planetary evolution, perhaps a billion years or so behind Jupiter's current second stage of evolution.

Since binary-star systems and the extra-solar systems are created in an ongoing and similar dynamic manner, astronomers can expect to find many more extra-solar planets and very many more binary-star systems. The major differences between the two systems, as with our own planets, can be accounted for by the size differences of the masses in question.

The reason for the rarity of a solar system like ours, as predicted and explained by the FLINE model, resides in the fact that the dynamic layout of our system was accomplished under exacting specifications that are in full accord with the Golden Ratio. These specifications call for two energy masses of precise relative sizes and relative velocities, both masses interacting at a precise distance apart, to effect a layout in full accord with the formula, D=1.618034 x Dpo, where D is the distance in AU of any pair of orbits, and Dpo is the distance in AU between the two previous planetary orbits. These specs and formula are the keys to the solution to the new Fourth Law of Planetary Motion, detailing how the planets attained their orbital positions around our Sun. All of the original 12 nebula planetary masses (including the original three asteroids spheres that later disintegrated, as correctly interpreted by Olbers in 1802) adhered strictly to Kepler's First Three Laws of Planetary Motion as well as to the new Fourth Law.

The short article, *Flat and Happy,* on the same page of *Science* describes how the theory of inflation "holds that a burst of expansion in the first instant of time stretched space almost perfectly flat, so that parallel light rays stay parallel forever. Measurements of distant exploding stars suggested that a mysterious energy in empty space is accelerating the expansion of the universe." Astronomers use the tiny "fluctuations in the microwave background as proof that energy can curve space just as matter can, and the finding suggested that energy might fill in for matter and flatten the universe." Surely, Occam's razor can be utilized here to render a more logical explanation.

Einstein was correct in two relevant statements: (1) space is curved; and (2) his earlier statement that an unknown force (lambda) is acting against the inward forces of the universe was the biggest blunder of his life. When viewed in the perspective of the Little Bangs theory, space is indeed curved and is expanding at its perimeter at the speed of light; no lambda factor is needed to explain the naturally accelerating expansion of the spherical universe. In the vacuum of space, everything is spherically shaped, including the universe itself. Why must everything become more and more complicated via the speculations of a Big Bang origin? When will the tangled web cease to be woven?

News Release June, 1996 for immediate release

Retired Scientist Claims Planetary Motion Discovery

After 23 years of researching the myths and truths of science, a retired scientist in LaGrange, Georgia claims discovery of the Fourth Law of Planetary Motion (FL) that explains **how and why** planets became spaced in their present orbits in our solar system—a crucial key to our planetary origins.

"The next challenge," says Alexander Scarborough, "is to convince the scientific establishment that my findings are correct. The spacing of planets is a 400-year mystery begun in 1595 with Johannes Kepler's efforts to learn why the six known planets were spaced in a geometric progression. His attempts proved futile, and all efforts since then have fared no better," Scarborough said. "But in 1995, my 23-year search ended with the discovery that ties everything together: the Fourth Law of Planetary Motion explaining *how and why* each planet became spaced in its present orbit."

Scarborough believes that the proposed Fourth Law is the most crucial and final link in understanding the true geometric origin and precise clockwork of our Solar System. Further, the new Fourth Law validates his 1973 concept that planets and moons are dynamic, self-sustaining entities in which all systems evolve via Internal Nucleosynthesis (IN), the transformation of energy into atomic matter.

"Along the way to full circle, the solutions to some 40 anomalies of planetary systems simply fell into place, each one intermeshing with all others as if in a giant jigsaw puzzle," added Scarborough.

The revolutionary nature of the concept and the large amount of supportive, intermeshing evidence forced the researcher to publish a book detailing his findings. Entitled *THE SPACING OF PLANETS: The Solution to a 400-Year Mystery*, the fully illustrated book is available from the publisher or in bookstores.

"As we head into the 21st century, this unique treatise—unlike any contemporary book or film on planetary origins—should put the planetary sciences back on the right track established during the Copernicus-to-Newton era," he stated. "Space-probe discoveries are confirming its scientific validity."

The controversial book presents powerful arguments against current beliefs stemming from the Big Bang and Scientific Creationism concepts of planetary origins, according to the author.

"The new FL-IN concept is a definitive alternative to current beliefs—a genuine paradigm shift in planetary origins—ready to be examined by all scientists who would be willing to take the time to learn its factual message rather than rejecting it simply because it challenges textbook beliefs," he said.

Scarborough predicts that his concept should displace current dogma before the turn of the century. "Since the new concept is factual, non-speculative, testable and provable, the change appears inevitable," he stated.

To: John W. Hess, Executive Director, Geological Society of America, Boulder, CO 80301 March 14, 2003

Thank you, and congratulations, for publishing the insightful perspectives in *Anonymous Reviews—Are the Pros worth the Cons?* (Alexander McBimey, *GSA Today,* March 2003).

Most of us agree with the author that the review system has worked fairly well—especially so for writers who go along with prevailing views on the Big Bang/Accretion (BB/Ac) hypotheses. As stated by McBimey, there is a tendency for an advocate reviewer to have "an emotional reaction to new ideas the reviewer finds disturbing [and] ...to sidetrack work that may render our own obsolete."

These are indeed powerful incentives for rejection of new, definitive ideas, especially those that argue strongly against prevailing beliefs. History reveals how difficult it is for human nature to accept such revolutionary changes. Two examples are the heliocentric idea of Copernicus and the cleanliness-at-birth concept of Semmelweiss. Such reluctance seriously impedes the progress of science; it may well be the most significant negative aspect of the review system.

An excellent modern example of this point is the discovery of the Fourth Law of Planetary Motion (1980-1995) detailing how the planets attained their orbital spacing around our Sun—a mystery that first eluded Kepler in 1595. Although a plethora of definitive evidence supports this finding, it has often been rejected by peer reviewers. If permitted publication in journals with proper vetting, it will open doors to ever-more definitive solutions to planetary and universal anomalies. Three examples are the five-stage evolution of planets, the abiogenic origin of hydrocarbon fuels, and the accelerating expansion of the Universe.

I now consider the revolutionary Fourth Law and its ramifications incontrovertible and the most significant discoveries in cosmology since the 1905 work of Einstein. I agree with McBimey: "Anyone who has devoted months, if not years, to a piece of work deserves nothing less [than a proper review]." In my case, in addition to the 15 years spent solving the Fourth Law, more than seven years have passed without a successful review of the completed work.

Thank you again for your astute, timely choice for publication of this most accurate perspective of the current peer-review process. In the best interest of science, let us hope that McBimey's insightful article will spur scientists to alleviate the problems now impeding rapid scientific progress.

To: Dr. Cynthia K. Whitney, Editor, Galilean Electrodynamics, Arlington, MA 02476. June 28, 2003

It was a real pleasure meeting and talking with you at the NPA meeting in Storrs, June 9-13, 2003. Thanks a million for your interest and for the valuable suggestions made in your letter of 23 June; they are precisely the type I have sought these 30 years of putting all the pieces together. In response to your four appreciated suggestions, I would like to make these comments:

1) I like the change to "Principles of Planetary Origin & Evolution"; it's more definitive. And it does disentangle this work from biological evolution, which will help readers a great deal. My reference to Descartes' belief in Earth's "Sun-like core" (1644) is intended to add credibility to this fact; however, I learned of it some time after the LB/FLINE model came into being, and was very happy to have confirmed his great

insight.

2) I like the idea of beginning with the discussion about Descartes and solar system origin and nucleosynthesis, rather than beginning with the Fourth Law. But I should not have stated that this law might not apply in another solar system. Certainly, if other systems like ours do exist, it would apply. This law first eluded Kepler in 1595, more than a decade before discovering his first three laws. I'm confident now that certain aspects of the geometric solution to the law will be found in all other planetary systems different from ours; i.e., aspects pertaining to the relative masses of the interacting bodies of energy, the distances and the relative velocities of these masses during the dynamic origin of the SS in question: the mass-distance-velocity-time relationships that determine the number of planets in any solar system. Eventually, there should be volumes of information derived from these relationships.
3) I agree; nuclear synthesis should precede fuel synthesis.
4) I will send you a copy of the paper done in the order and continuity you suggested. I will work on bettering the labeling of the figures for clearer understanding (but this is somewhat difficult because the complicated visualization in my mind gets in the way of continuity of explanation). Writing is not my forte. I use "orbital velocity" and "speed" interchangeably simply for variety; no distinction is intended. The Fourth Law is so important because its geometric solution details the dynamic manner in which the orbital spacing of the planets occurred in accord with the Golden Ratio that pervades throughout Nature. (See enclosures for more reasons.) And to the original spacing and orbital velocities, we can compare the current positions and velocities, from which we can derive a wealth of information on the changes that have occurred since the dynamic birth of our SS; then we can calculate and accurately predict future changes in the system. This also verifies past events, and permits an understanding of how all dynamic solar systems came (and continue to come) into being. It dispels the Big Bang/Accretion hypotheses that Sagan called myths and Hawking admitted are conjectural—all after my one-way messages to them.

You are the only person to offer the constructive (or any type of valid) criticism sought during the past 30 years of development and presentations of the many facets of this revolutionary LB/FLINE model of origins and evolution. For that, I am very grateful. Certainly, I will follow your valuable suggestions, and will send a copy of the revised product for additional suggestions, changes and approval.

Memos
Do Planets Accrete or Evolve?

Theorists don't believe that Uranus and Neptune could have formed so far out; there, gas and dust were too sparse to coalesce into planets. Now a new computer model suggests that "sibling rivalry might be to blame for their banishment. Runty Uranus and Neptune may have grown up in tight quarters much closer to the sun, only to have the big bruisers, Jupiter and Saturn, fling them into the outer reaches of the solar system."

The reason for this logic? ".. .out on the nebula's fringes, matter was spread too thinly for anything like planets to form. In the best simulations of the process, cores for Uranus and Neptune fail to form at their present positions in even 4.5 billion years, the lifetime of the solar system." To quote one astronomer: "Things just grow too slowly in the outermost solar system. We've tried to form Uranus and Neptune at their present locations and failed miserably." So the astronomers theorize that the two planets did form closer in where the nebula was far denser, and then were flung into the outer reaches of the solar system.

A number of years ago, other computer simulations failed miserably in efforts to create planets properly spaced in their current orbital positions. The source of the problems leading to these (and future) failings resides in the Accretion hypothesis itself; planets did not accrete from gas, dust, planetesimals, comets, etc. There is a much simpler and more viable alternative that poses none of these problems with our Solar System or with an extra-solar system—a revolutionary definitive concept structured with five basic and irrefutable principles of Nature.

The solution (1980-1995) to the enigmatic Fourth Law of Planetary Motion clearly reveals how the planets were placed in their orbital positions around the Sun during the layout of our solar system some five billion years ago. The Five Laws of Planetary Motion (FL) clearly support a dynamic origin of our Solar System in which the nebulous planetary energy masses (first identified by Descartes in 1644) have evolved via Internal Nucleosynthesis (IN) (in compliance with Einstein's formula $E=mc^2$) into their current stages of a five-stage Evolution (E) common to all such spheres throughout the Universe. We have only to look to the Sun and planets of our own system to observe these five stages of planetary evolution. The three principles embodied in the FLINE model are inseparable and ongoing until each sphere's energy core is depleted and the planet (or moon) has entered its inactive fifth stage of evolution.

The fact that Pierre Simon Laplace's Accretion hypothesis of 1796 (with some modification) remains the dominant concept of planetary origins in spite of its highly speculative structure and much contradictory evidence, and in the face of even stronger evidence favoring the FLINE model, speaks volumes about the status of our planetary sciences. With two notable series of computer failures of recent times, one must wonder how much longer this antiquated hypothesis can survive the onslaught of opposing facts, and how much longer it will serve as a costly roadblock to understanding the full significance of the brilliant past, present and future discoveries in outer space.

In the absence of the FLINE model, the web being woven around the Accretion hypothesis becomes more and more tangled. In the new perspective, the solution to every anomaly of every planet and moon has its taproot deeply embedded in the FLINE model—a statement already proven true by its splendid record of accurate predictions.

A Field Guide to the New Planets

A Field Guide to the New Planets (Discover, March 2000) is an interesting article on the 29 new planets discovered orbiting stars like our own with pictorial data concerning the 29 gaseous giants. The article states: "the orbits in which [our planets] were born some 4.6 billion years ago have remained the same ever since. Until recently that was the accepted scenario. But now the detection of extra-solar planets has forced astronomers to re-examine such notions, because they present us with a paradox. Many are so monstrous in size, and hug their stars so closely, that they could not have formed in their present position. The searingly hot stars around which they circle would have melted their rocky [?] cores before they got started. Instead, it's assumed that they coalesced some distance away, then barreled inward over millions of years. And if such chaos characterizes the birth of extra-solar planets, could not similar disorder have reigned closer to home?"—A proposed scenario that presents more problems than it solves.

However, there is an alternative that warrants consideration—one in which the data does fit precisely and without need of the assumptions of the proposed scenario. Such extra-solar planets were predicted and explained by this revolutionary concept, even before these amazing discoveries were made. In it, there is no need for such disorder in solar systems; they are precise systems that adhere to the Four Laws of Planetary Motion (FL) and the processes of Internal Nucleosynthesis (IN) that drive the Evolution (E) of all active celestial spheres. These gaseous giants are simply in the early part of the second stage of the five-stage evolution common to all planets—all in full accord with size and the laws of thermodynamics. The revolutionary FLINE concept eliminates both the need for gaseous giants to form at much greater distances from their central stars and the need to sling planets out of their orbits in a chaotic manner.

The new Fourth Law of Planetary Motion that brought the FLINE model full circle explains how the planets attained their orbital positions around our Sun—an enigma that first eluded Kepler in 1595. And we have only to look to the skies to observe the five common stages of planetary evolution.

The Source of Earth's Water

On Tuesday evening, June 21, 2005, TV Station PBS in Georgia showed a one-hour video entitled *Origins*. The valiant efforts of featured scientists were aimed at "proving the Big Bang to be the true manner in which our Universe came into being." Their observational evidence seemed scientifically valid; however, their interpretations, logic, and conclusions—all based on the BB myth—sounded force-fitted, even ludicrous.

But when the same evidence is applied to the LB/FLINE, it fits easily, taking its place as substantiated evidence favoring the new model.

One excellent example is the source of Earth's vast amounts of water. The BB version attributes the source to infalling comets. The LB/FLINE recognizes the source as Internal Nucleosynthesis (IN) that drives ongoing Evolution (E) of Earth's water, land, atmosphere, and all things inanimate and animate. Every planet is a self-sustaining entity that creates its own substances via IN/E processes. There is no need to look to outer space for definitive answers that are right under our feet.

I am appalled at the conclusion that speculative misinterpretations of observations and subsequent data were cited as proof of the "BB myth" (as Sagan called it. Myth begets myth after myth.)

The NPA may be the only organization capable of ending this cycle of ludicrous speculations. Unfortunately, we cannot expect financial aid from Foundations or Government organizations committed to entrenched BB beliefs—no matter how speculative they remain.

To: David W. Grogan, Articles Editor, Discover Magazine, N.Y. 10011. September 15, 2005
Re: Frontiers of Science, (1) Planets, by Won, and (2) Geophysics, Earth's Inner Core, by Wood (Discover, Oct 2005)

During the past three decades, many scientists have openly recognized that mysteries of planets and planetary systems viewed via the Big Bang (BB) hypothesis are destined to remain "conjectural" (Hawking terminology, 1999), with no possibility of valid scientific solutions to problems such as those encountered in the subject articles. In reality, the BB gives no choice to scientists other than forcing them into awkward fairy-tale reasoning about their brilliant discoveries.

However, definitive solutions now are being found via the new well-substantiated alternative to the Big Bang hypothesis: the LB/FLINE model outlined in the enclosed article embracing 12 giant steps for mankind (some little known) during the past 2500 years.

Initially, around 500 B. C. the Pythagoreans gave us the first crucial tool for learning Nature's deepest secrets of its beauty, power and precision. Since Copernicus in 1543 pointed science in the right direction, much brilliant work by a number of scientists has brought the Copernican Revolution nearly full circle.

These findings obviate the need for conjecture, and open the floodgates to comprehending definitive origins and evolution of planetary systems, solar systems, galaxies, and ultimately, the functioning of all systems of our Universe in which everything is interwoven with everything else. Rather than conjectural science, substantiated facts should prove far more exciting to your readers, and much more effective in advancing the cause of science. They also prove Einstein right when he stated (paraphrasing): "The right answers will be simple and beautiful."

I urge you to give this revolutionary literature your most serious consideration for publication. If you would like more details, please let me know, and I will send pertinent papers and books written about these well-researched subjects during the past 32 years, but yet remain ahead of their inevitable time.

P. S. - The new Fourth Law of Planetary Motion (1980-1995) can be found in the *Journal of New Energy: An International Journal of New Energy Systems* (Fall 2003, pp154-160).

To: Dr. Domina E. Spencer, NPA , University of Connecticut, Storrs, CT February 16, 2006

Thank you for your kind invitation to give the John E. Chappell Memorial Lecture at the NPA Meeting in April, 2006. I am honored to present it on what might well become a memorable occasion in the history of science.

I feel that a maximum of one hour would be sufficient to cover both the subject and the questions that might ensue. Because the new Fourth Law of Planetary Motion detailing the spacing of planets around our Sun seems to interlock everything together, I believe that making it understandable is paramount to the future direction of scientific thought.

By now, Cynthia and Nash (AAAS) will have received a copy of the proposed presentation, along with a revision of the original Introduction, page 1 (minus the transparency graphics).

As usual, my wife and Kay will be with me. We look forward to seeing you again.

c: Dr. Cynthia K. Whitney, Ed., Galilean Electrodynamics.

To: Joel Martin, LDN. Subject: A BIG News item February 22, 2006

A retired LaGrange scientist, Alexander Scarborough, has been selected by an international scientific group to give the 2006 John E. Chappell Memorial Lecture at a joint meeting of the AAAS-SWARM-NPA. The event will be held at Tulsa University in April.

The host AAAS-SWARM is the southwest regional section of the American Association for the Advancement of Science, the largest and one of the most influential science organizations in the world.

The Natural Philosophy Alliance is an organization with a current membership of 215 international scientists dedicated to substantiated truth in science. Some 21 countries are represented in the NPA, whose creed, as written by the founder, Dr. Chappell, states: "The NPA is devoted to broad-ranging, fully open-minded criticism, at the most fundamental levels, of the often irrational and unrealistic doctrines of modern physics and cosmology, and to the ultimate replacement of these doctrines by much sounder ideas developed with full respect for evidence, logic, and objectivity."

Some seven or eight years ago, while presenting his new Fourth Law of Planetary Motion in the Poster Session at a national meeting of the AAAS, Scarborough accepted Chappell's invitation to join his fledgling group, which now has more than tripled its membership. Scarborough's Memorial Lecture is entitled "Origins of Universal Systems via the LB/FLINE Model: From Myth to Reality in 12 Giant Steps and 2500 Years". The paper stems from his latest book, *New Principles of Origins and Evolution* (Ninth Edition).

The foundation of his new model is structured on 12 critical discoveries during the past 2500 years. Included are the Four Laws of Planetary Motion that reveal the dynamic origin of our Solar System some five billion years ago. The First Three Laws governing our Solar System were discovered by Johannes Kepler in the early 17th century. However, Kepler failed in attempts to discover the Fourth Law governing the spacing of planets around our Sun.

Scarborough's discovery of the Fourth Law of Planetary Motion (1980-1995) melds together the first 11 discoveries, beginning with the Pythagorean geometry of 500 B. C. and progressing through the works of Aristotle, Copernicus, Kepler, Newton, Descartes, Olbers, Einstein, Hubble, Chladni, and others to bring the Copernican Revolution full circle.

"These findings should nudge science away from the Poe/Laplace, Big Bang/Accretion hypotheses, and toward a new direction of revolutionary thought: the Little Bangs (LB)/FLINE model of universal origins and evolution," Scarborough stated. He believes the Memorial Lecture could be a vital step toward establishing a firmer foundation for the new direction of scientific thought within the scientific community. "Especially since it does supply long-sought answers to many questions that remain unanswerable in the Big Bang version," he added. "For example, the geometric spacing of planets around our Sun."

c: File, Distribution (after 2-26-05)

Memo to: Joel Martin, LDN. Subject: News on Tulsa Conference **April 14, 2006**

ORIGINS of UNIVERSAL SYSTEMS:
From Myth to Reality in Twelve Giant Steps and 2500 Years
The 2006 John Edgar Chappell Memorial Lecture
Alexander A. Scarborough AAAS-SWARM-NPA Meeting
Tulsa, OK, April 3-7, 2006

The Conference held at Tulsa University was a huge success, with a few participants presenting more than one paper. Dr. Robert Heaston presented five significant papers, with one being a major breakthrough that locked in well with the four presentations by Alexander Scarborough, assisted by his daughter, Kay Scarborough, a resident of Marietta.

Heaston showed the derivation of a critical mathematical formula that provides the missing link in Scarborough's Little Bangs concept of universal origins. The formula reveals that compression is Nature's key to initial nuclear fusion in spheres of all sizes: stars, suns, planets, comets. Such compression, according to the new concept, initiates in the super dense black spheres continuously forming at the spherical perimeter of a Universe expanding at the speed of light into the absolute-zero temperature of infinite Space where all atomic motion ceases at the interface with our Universe.

Eventually, the warming compression within the dense spheres becomes sufficient to trigger eruptions of billions of brilliant nuclear masses of all sizes, embedded in gaseous dust-clouds, from the two poles of each spinning sphere. The results are billions of beautiful pinwheel galaxies, each consisting of billions of stars; the smaller masses eventually evolve into planets, usually locked in orbit around larger masses—all in accord with the Four Laws of Planetary Motion. A small black fraction (0.5% of ejected mass) remains at the center to hold each galaxy together.

This Little Bangs (LB) concept explains the large varieties of strange solar systems scientists have discovered in galaxies—and why many erroneously believe that stars accrete from gaseous dust-clouds. It also explains the accelerating rate of expansion of the Universe, no matter in which direction astronomers look.

Scarborough's solution to the enigmatic Fourth Law of Planetary Motion proved to be the fundamental key to providing answers about origins of solar systems, which in turn, provided clues to the fiery origins of galaxies.

"This Little Bangs (LB)/FLINE concept is the product of a lifetime of study that began at age12, and became more concentrated during the past 33 years. The first breakthrough occurred in 1973 when I realized the close relationship among the hydrocarbon fuels (gas, petroleum, coal), and how they were, and still are, made via universal laws. Our organization's ultimate goal is to ease science away from the Big Bang myth, and more toward substantial evidence," he added.

Scarborough, a native of Macon, Georgia, resides in LaGrange, Georgia, and is married to Mary Frances Akins, a native of Statesboro, Georgia. He holds degrees from the University of Georgia and Georgia Tech, is a veteran of WWII, and is retired from a career in R&D.

To: Dr. Robert Heaston, 220 Arlington Avenue, Naperville, IL 60565. April 20, 2006

As you suggested at the Tulsa meeting, I have attempted to rewrite the Fourth Law of Planetary Motion. To do so, it was necessary to write a Fifth Law concerning the gradual changes in orbits since the origin of the SS some five billion years ago. The original law attempts to combine the two.

April 18, 2006

Rewrite of Fourth Law of Planetary Motion, plus Fifth Law:

4. For each planet, its distance from the Sun during the original layout of the Solar System equals 1.618034 (the GR) times greater than the preceding planet. In reverse from Pluto, the distances between any two planets equal 0.618034 times the previous distance between two planets.
5. Over time, gradual changes in planetary orbits are functions of the distance-mass-velocity-time relationship.

I remain indecisive concerning my preference between the original Fourth Law that encompasses the changes, and the two Laws that separate the events. Obviously, I need your help. Please let me have your input regarding this critical decision; your suggestions, even to the point of another rewrite, would be deeply appreciated.

To me, the LB/FLINE model seems in synch with Einstein's relativity theories, except perhaps where gravity is concerned. Here, it seems to be more in synch with Newton's idea of gravity, especially if one assumes that gravity is inherent in the most fundamental energy strings—else how could strings attract and initiate the very beginning of universal origins, as they do in the very dense black spheres (a.k.a. holes) at the speed-of-light expanding perimeter of our Universe at the interface with infinite Space at absolute zero where all atomic motion ceases?

A TV video just this week on gamma rays proved interesting and exciting. The data shown, especially the distribution pattern of the observed intermittent gamma ray sources, fit precisely into the LB/FLINE model in which these sources coincide with the pattern of new galaxies continually born from these ongoing countless black spheres at, or very near, the perimeter of our expanding Universe.

In this scenario your new formula for compression as the source and/or cause of nuclear energy cores fits precisely into the black-energy-spheres-to-galaxies concept explained as the LB/FLINE model of universal origins. It is the missing link that completes the big picture.

Herewith is the news item submitted to the local paper. Unfortunately, LaGrange is one of those little cities that does not delve in science unless it's another spectacular myth from wire services, and based on the BB, so I'm not sure it will be printed. Looking forward to your informative response.

ROBERT J. HEASTON, Ph.D., P.E.
220 Arlington Avenue Naperville, IL 60565
Phone (630) 416-8338 FAX (630) 416-9203 Email robert@drheaston.com

To: Alexander Scarborough, 202 View Pointe Lane, LaGrange, Georgia 30241. 5 May 2006

My comment about the definition of your "Fourth Law of Motion of the Planets" was for you to define your law in a format similar to Kepler (George Gamow, *The Great Physicists from Galileo to Einstein,* pp. 28-33). For example:

I. The planets travel in elliptical orbits with the Sun at one focus. (Kepler, 1609)

II. For every planet, in equal intervals of time, equal areas are swept out by a line drawn from the Sun to the planet. (Kepler, 1609)

III. Comparing the orbits of the several planets, it is found that the times of revolution are related to the distances from the Sun by a proportionality: the squares of the periods of revolution are proportional to the cubes of their mean distances from the Sun. (Kepler, 1618)

IV. In another comparison of the orbits of the several planets, it is found that the distances between the orbits of any pair of planets have a common proportionality: the orbit of a planet relative to the one on either side of it is defined by the classic Greek golden ratio. (Scarborough, 19xx)

I re-read your letter, and I am not sure that my definition above of your Fourth Law is consistent with what you wrote. However, whatever source you use for the wording of Kepler's three laws of motion, adapt your definition of the Fourth Law (or even Fifth Law) to have a parallel structure with Kepler's definitions. Always spell out golden ratio, not state as *gr.*

Then, give an example to explain what you mean for inward and outward relationships and the golden ratio. Also, anytime you exactly state your Fourth Law, always state it with Kepler's three laws. Adopt a standard definition of your Fourth Law and the golden ratio and repeat it over and over again so that people remember exactly what you meant. That will be your heritage. You don't have to give every little detail in your formal definition, only enough to be consistent with Kepler and to guide others in doing the details. You might even give a paper at NPA 2007 on *The Laws of Planetary Motion* with a summary of Kepler's history (association with Tycho Brahe) and how you made your own discovery. Only briefly discuss LB-FLINE and emphasize that the FL stands for the Fourth Law.

Here is an interesting quote from the 2005 CD on the *Encyclopaedia Britannica* under "Kepler".

It may be noted that Kepler's laws apply not only to gravitational but to all other inverse-square-law forces and, if due allowance is made for relativistic and quantum effects, to the electromagnetic forces within the atom.

You may have discovered more than you realize.

With regard to my prediction of a third mechanism of generating a nuclear reaction by gravitational collapse, I am reworking that paper to include as much supporting evidence as possible. I still believe it, but right now I have put that paper aside because I am working on summaries of all 80 NPA 2006 papers (49 presentations and 31 *in absentia)* for the *NPA Newsletter.* I gained a greater appreciation of your efforts when I read your three papers and summarized them.

Thank you for mentioning me in your press release.

To: Robert J. Heaston, Ph.D., P. E., 220 Arlington Avenue, Naperville, IL 60565. May 10, 2006

I have slightly modified your suggested rewrite of the new Fourth Law, and slightly modified the Fifth Law to complete the intended messages. These changes seem essential to completion of the intended message and the full picture of the origin and functioning of our SS:

I. The planets travel in elliptical orbits with the Sun at one focus. (Kepler, 1609)
II. For every planet, in equal intervals of time, equal areas are swept out by a line drawn from the Sun to the planet (Kepler, 1609)
III. Comparing the orbits of the several planets, it is found that the times of revolution are related to the distances from the Sun by a proportionality: the squares of the periods of revolution are proportional to the cubes of their mean distances from the Sun. (Kepler, 1619)
IV. Comparing the orbits of the several planets, it is found that the distances between the orbits of any pair of planets have a common proportionality: the original orbit of a planet relative to the one on either side of it is defined by the classic Greek golden ratio. (Scarborough, 1995)
V. In additional comparisons of the orbits of the several planets, it is found that gradual changes in their distances and velocities have a common proportionality: all are functions of a distance-mass-velocity-time relationship. (Scarborough, 1995)

You gave an interesting quote from the 2005 CD on the *Encyclopaedia Britannica* under "Kepler": "It may be noted that Kepler's laws apply not only to gravitational but to all other inverse-square-law forces and, if due allowance is made for relativistic and quantum effects, to the electromagnetic forces within the atom." Your comment followed: "You may have discovered more than you realize."

Bob, for many years, I have recognized the ubiquity of electromagnetism as inherent in all forms of energy and matter, and the most fundamental substance in the Universe. I have a problem separating it from gravity; until proven wrong, I must believe they are one and the same. They are what hold together all systems universal, from start to finish, beginning with inert electromagnetic energies frozen throughout infinite Space: the source of black holes that spawn galaxies loaded with stars of all sizes embedded in gaseous dust-clouds, with many solar systems, mostly of the binary type. And both are vital to the foundation of the LB/FLINE model in which the FL must now stand for the five laws, rather than only four laws.

The two new laws, along with the works of Kepler, Newton, Einstein *et al.*, open the floodgates to unlimited knowledge of solar systems and the galaxies that continually spawn their births. Your formula for nuclear energy via compression fits precisely into this version when applied to black holes (a.k.a. black spheres) that grow ever denser before releasing their masses of nuclear energies in the form of brilliant galaxies. But I'm sure you are right in that I may have discovered more than is now realized.

At the last meeting, Cynthia and I briefly discussed how the new law(s) can be made easier to understand by using skeleton graphs to illustrate the principles, rather than the full diagrams. This should work well at the next meeting, or before, if I can perhaps have it published before then. Many thanks for your help.

To: Editor, Journal of the Royal Astronomical Society, Burlington House, Piccadilly, London, UK. Sept., 2006

Re: A Really Bright Future for Science!

My manuscript briefly presents the new Fourth and Fifth Laws of Planetary Motion; fuller details are found in the References. The two laws underpin the continuity of interrelationships comprising the revolutionary LB/FLINE model (1973-2006) of universal origins and evolution, and are the primary objectives of this paper submitted for your journal.

The revised 1995 Fourth Law presents a definitive understanding of the dynamic origin of our Solar System via the geometric spacing of planets around our Sun, all in full accord with the Golden Ratio.

The new Fifth Law defines and explains the gradual changes in orbits and velocities of all planets since the dynamic origin of our SS some five billion years ago; e.g., why Neptune and Uranus are the only two planets slowly drifting away from the Sun, while all others are drifting slowly toward the Sun.

Both laws are in sync with the First Three Laws discovered by Kepler early in the 17th century. Surprisingly, the Five Laws (FL), in conjunction with $E = mc^2$, open the floodgates to definitive understanding of the intimate relationship among SSs, galaxies, black holes (spheres), electromagnetism, gravity and space-time, all in accord with the LB/FLINE model.

The new model offers valid scientific solutions to a number of the greatest mysteries of the Universe—e. g., the mathematical spacing of planets in our SS, and the accelerating expansion of the Universe (inherent in the model without the need for a lambda-type cosmological constant or a dark substance)—validating its supportive evidence to the point of making it seem incontrovertible.

Presentations at national and international science meetings during the past decade have added much assurance of its scientific validity. Although it remains ahead of its time, the model has gained a firm beachhead via substantiated evidence. Fortunately, every relevant discovery further corroborates its scientific validity.

Originally too broadly encompassing, the Fourth Law (1995)—with the help of Dr. Robert Heaston—has been revised into the Fourth and Fifth Laws in compliance with Kepler's style and for improved clarity. Detailed graphs are included.

To: Letters Editor, P.O. Box 37012, MRC 951, Washington, D.C. 20013. September 30, 2006

In stating that planetary orbits are concentric, *The Planet Hunters* by Robert Irion (*Smithsonian*, Oct 2006) commits a grievous error in direct opposition to Kepler's first law of planetary motion (while ignoring his other two laws):

1. "The planets travel in elliptical orbits, with the Sun at one focus." (1609)

The differences in distances between the Apogee and Perigee of elliptical orbits (shown relative to average distances from the Sun) of each of three examples illustrate the elliptical nature of planets:

Difference: A minus P: Mercury: 14.88 MM, Earth: 2.79 MM, Jupiter: 46.5 MM
Avg. distance from Sun: " 35.99 MM, " 93 MM, " 483.6 MM

Elliptical orbits are definitive (but not the only) clues to the dynamic origin of the SS some five billion years ago. A quick review of the new fourth and fifth laws should be helpful. All orbits except Earth's (0°) are inclined by various degrees to the ecliptic. Mercury's ecliptic inclination of 7° 00 15" is almost identical to the Sun's axial inclination of 7°- 15°. Under the new LB/FLINE model this would be expected, since the smaller cometary mass passed the Sun at that angle from a distance of only O.618 AU, accelerating the Sun's equatorial mass accordingly.

Mercury, the closest and the first planetary mass to break from the mother mass, naturally retained the same angle during the dynamic birth of our SS some five billion years ago. These results make it obvious that distance and size are critical to the final shape and relative size of each ellipse.

All planetary movements, spacing, and angles were, and still are, governed by the five laws of planetary motion. The new fourth law governs the geometric spacing of the ten orbits. The new fifth law governs the small changes (past, present, and future) in the ten orbits; the maximum changes have occurred in the orbit of Pluto, slowly moving closer in toward the Sun, while being forced into the highest degree of inclination during the past five billion years.

The basic problem here resides squarely in the BB hypothesis being diametrically opposed to the five laws of planetary motion it cannot explain: laws that provide a solid foundation of powerful supportive evidence for the revolutionary LB/FLINE model of universal origins.

From: SMITHSONIAN INSTITUTION, Victor Building, Washington, D.C. 20001.

To: Mr. Alexander Scarborough, Ander Publications, LaGrange, GA 30241. **October 20, 2006**

Thank you for your recent letter regarding our article, "The Planet Hunters" by Robert Irion, which appeared in the October 2006 issue of SMITHSONIAN magazine.

We greatly appreciate your taking the time to send us your comments and are forwarding them to Mr. Irion and his editors.

Karla A. Henry, Reader Services, SMITHSONIAN

At this writing, observational evidence indicates that the new fourth law apparently might not apply to known extra-solar systems comprised of a low number (1-3) of planets. With a high number of 10 to 12 or more original planets, our Solar System is the only one known to consist of geometrically-spaced planets in accord with the fourth law.

This evidence confirms that a solar system like ours must be very rare. As shown in another section, the prerequisites to form one are very strict and obviously unyielding.

Estimates of zero, one or two in each galaxy signify that other Earth-like planets are indeed very rare throughout the Universe. At this time, the fourth law may be considered a special case in solar systems.

In contrast, the fifth law applies to all solar systems in all galaxies. The gradual displacement of planets from their original orbits occurs as a function of time, and in accord with the mass-distance-velocity-time relationships.

Earth's "Sun-Like" Core—What You Need to Know about It

"Journey to the Center of the Earth" (*Discover*, July, 2007) by Susan Kruglinski poses an extremely dangerous, prohibitively expensive, and totally impractical mission to the center of Earth. The thousands of tons of liquid iron poured into a hole would never make it beyond an outer layer of the dense molten lava already under extreme pressure and temperature—even if the bored hole did not serve only as another volcanic-type outlet, which it certainly would do.

In reality, all that's necessary to understand Earth's core is already clearly spelled out in the scientific literature: a plethora of definitive evidence now confirms Descartes' belief, recorded in 1644, that Earth's interior is "Sun-like". His common-sense deduction can be definitively understood simply via observations and comprehension of causes of volcanic actions pouring out hot lava that bring forth all the land, atmosphere and water comprising and shaping Earth's ever-growing surfaces: consequences underpinned by the five laws of planetary motion and $E=mc^2$.

These ongoing processes would not be possible without Earth's core of quark-gluon nuclear energy from whence all atomic and subsequently, all molecular matter comprising our planet was, and is, formed continually—inclusive of all common elements—from hydrogen to uranium. Uranium is not the source of Earth's energy; it is simply the heaviest element produced via the nuclear energy processes that are common in all stars and in all planets. Much substantiated evidence clearly reveals that planets are simply small stars that evolve through the five stages of planetary Evolution via Internal Nucleosynthesis (the transformation of nuclear energy into atomic matter) so readily realized by simple observations.

All the forever-ongoing processes of universal systems, inclusive of the revolutionary LB/FLINE model of origins and evolution, are well-grounded in mathematics, physics, and chemistry; all are centered around $E=mc^2$, the energy-matter relationship. An understanding of Universal origins and evolution via a unified field theory will be accomplished only when the proper relationship among these disciplines is finally accomplished.

In conclusion, science has only to follow the pathway set by Descartes to find all the right answers without a costly, dangerous and foolish mission to the nuclear energy center of Earth. The LB/FLINE model of universal origins, offering the right answers, seems a far better route to pursue if we wish to confirm the true nature of Earth's "Sun-like" core.

Asteroids Vesta and Ceres: Predictions and Right Answers

The article, *Spacecraft to Probe Asteroids Vesta, Ceres* (*ajc.com*, Jul 6 '07), tells that NASA is set to launch the spacecraft, Dawn, on a mission to learn more about asteroids, Vesta and Ceres. Dawn will encounter Vesta in 2011 and will meet up with Ceres in 2015 in the search for clues about the birth of our Solar System.

Some of the questions scientists hope to resolve via the $344 million ($446 million, according to *Time* magazine) mission: Why are Vesta and Ceres so different geologically? How do size and water affect planet formation? What does the evolution of the asteroids say about Earth's formation?

Dawn will orbit each body, photographing the surface, and studying the asteroid's interior makeup, density and magnetism. When the data arrive back at Earth, scientists will discover that they fit precisely with the predictions and principles of the LB/FLINE model of universal origins, while presenting additional powerful arguments against the prevailing BB/Accretion hypotheses. Posterity will ponder why so much money is funded for projects in search of answers that already pervade in the scientific literature. (1)

Perhaps the answers to the three questions posed above—in accord with the LB/FLINE model—would be of interest to the reader.

Heinrich Olbers was the first person to recognize and record in 1802 the fact that asteroids are remnants of a disintegrated planet. The new model (1973-1995) attributes the cause to its unfortunate position balanced between the powerful gravities of Jupiter and the Sun. Contrary to BB beliefs, the remnants—or gaseous dust-clouds—can never coalesce into a planet. The five laws of planetary motion reveal the fiery dynamic birth of the SS some five billion years ago. Once in geometrically-spaced orbits the small fiery masses are continuing their ongoing planetary evolution through four additional stages of evolution at rates in accord with mass (size)—a fact that is readily confirmed by simple observations and a plethora of authenticated data in the scientific literature.

Vesta is a typical asteroidal remnant, a cratered shell of its former comet-like self, one that may or may not yet have totally evolved from its fiery birth. Its cratered and sunken surface features will attest to its out-gassing from the encased nuclear energy core, evolving in the same manner as the two moons of Mars, Phobos and Deimos, evolved.

Ceres remains the largest mass from the disintegration event, a mass large enough to evolve as a spherical planet, however small, that remains in its original fifth orbit, properly spaced between Jupiter and Mars in full accord with the new fourth and fifth laws of planetary motion—laws that the BB is incapable of explaining.

These principles of the LB/FLINE model furnish the right answers to the questions being sought by the Dawn mission: their huge differences in size make it easy to understand the why and how of all their other differences. As with planets, mass is the determining factor in rates and results of their evolution; Earth is no exception to these principles as it revolves and evolves in its third orbit around the Sun.

All clues waiting to be discovered by the Dawn spacecraft will prove to be definitive evidence of the LB/FLINE model of universal origins, while casting serious doubts about the BB/Accretion hypotheses. Thus, the costly mission will not have been a total waste of time and money.

Addendum: The plethora of definitive evidence offered by the Five Laws of Planetary Motion and the LB/FLINE model of universal origins they underpin leave little, if any, leeway to doubt the scientific validity of the intimate relationship among these discoveries.

Any concept (e.g., BB/Accretion) failing to account for the definitive Five Laws of Planetary Motion and their profound significance is destined to miss pivotal insights: crucial keys to origins and the interconnected functioning of all universal systems; i.e., the theory of everything.

20 Things You Didn't Know about our Solar System

July 26, 2007

1. Kepler's Three Laws of Planetary Motion:
   (a) The orbit of a planet is an ellipse, of which one focus is the Sun. (1609)
   (b) For each planet a straight line joining it to the Sun sweeps over an equal area in equal time. (1609)
   (c) For each planet, the cube of its distance from the Sun is equal to the square of its time of revolution. (1619) (i.e., the greater the distance, the slower the velocity.)

2. Scarborough's new Fourth and Fifth Laws of Planetary Motion:
   (d) The original orbit of a planet relative to the one on either side of it is defined by the classic Greek Golden Ratio. (1995, modified 2006)
   (e) Gradual changes in distances and velocities of planets are functions of a distance-mass-velocity-time relationship. (1995, modified 2006)

3. The Pythagorean Theorem (ca. 500 B.C.) and the GR found within it are crucial keys to the solution to the 400-year mystery of the roughly geometric spacing of planetary orbits (known as Bode's Law). They made the discovery possible.

4. The Fifth Law explains why Uranus and Neptune are the only two planets drifting away from the Sun, while all others are drifting towards the Sun. The orbital changes are small for the five-billion-year time frame.

5. In 1644, Descartes, the inventor of analytic geometry, became the first person to record Earth's interior

as "Sun-like", even though he had no knowledge of nuclear energy or $E=mc^2$, the crucial keys to the evolutionary nature of all things universal, including stars, planets, moons, comets, etc.

6. The Five Laws reveal the fiery, dynamic origin of the geometric beauty of our SS, thus enabling scientists to decipher its history definitively.

7. The interchangeable relationship between energy and matter is best expressed by the famous formula, $E=mc^2$, that led to nuclear bombs that brought an end to WWII, (and saved this author's life by making it unnecessary to continue the mission to invade Japan. Our troop ship was diverted into Boston Harbor).

8. The energy-matter relationship is the crucial key to understanding the five stages of planetary evolution at rates in accord with mass. Simple observations of the planets confirm these stages and their relationship to mass and their rates of evolution.

9. Astronomy is a science of observations and logic; the two are inseparable.

10. In 1802, Heinrich Olbers correctly identified the asteroids as remnants of a disintegrated planet, a fact now confirmed by observations, logic and overwhelmingly persuasive evidence.

11. In 1794 Ernst Chladni published the first work arguing for the cosmic origin of cometary fireballs and the generic connection between fireballs, meteorites and meteors. Substantiated evidence adds strong support to his findings; e.g., the burned-out shells of Phobos and Deimos (the two inactive moons of Mars) are former comets. Neither asteroids nor comets can coalesce.

12. In a write-in contest predicting the results of the pending collisions of 21 fragments of Comet SL9 on Jupiter in 1994, I was fortunate to be the only person to predict the sheer power of the 21 explosions at the thin cloud-tops of the giant planet. My predictions were based on Chladni's findings: Comets are fireballs of nuclear energy, ejecta from larger masses of energy.

13. The relatively large number of geometrically spaced planets (10+) in our SS met the rigid prerequisites of two energy masses of proper relative sizes interacting as the smaller mass passed the larger Sun at a precise distance and velocity that enabled each nebulous planetary mass to break into orbit at every GR spacing—a very rare occurrence throughout the Universe.

14. All the above evidence interlocks into a feasible concept of the origin and evolution of our SS: The FL-IN-E model—(FL) the Five Laws of Planetary Motion, (IN) Internal Nucleosynthesis that drives all (E) Evolution, proving Descartes right.

15. The FLINE model definitively explains the abiogenic origin, evolution, and intimate relationship of the hydrocarbon fuels (gas, oil, coal): ongoing nuclear and chemical processes of Earth's interior.

16. The expanding Earth and its ongoing increases in sea level and land via volcanism are easy to explain via the new model.

17. The ongoing processes of IN and E provide definitive reasons species come and go; e.g., the age of dinosaurs and their extinction.

18. Earth and all other planets are self-sustaining entities that need no miracles from outer space. The FL-IN-E model seems destined to displace the ancient accretion hypothesis.

19. The new model definitively explains the origin and fate of planetary rings as ejecta that eventually settles back onto their planet of origin.

20. The new FLINE model's two fundamental principles—the Five Laws of Planetary Motion and the energy/matter relationship expressed by $E=mc^2$—offer definitive insights into origins and evolution of planets, moons, comets, asteroids, solar systems and the galaxy that spawned them. The same principles underpin all universal systems, bringing science ever closer to a theory of everything.

Modifications to Kepler's Laws of Planetary Motion

To understand the revolutionary LB/FLINE model, one must first understand the Five Laws of Planetary Motion:

I. The planets travel in elliptical orbits, with the Sun at one focus. (Kepler, 1609)
II. For every planet, in equal intervals of time, equal areas are swept out by a line drawn from the Sun to the planet. (Kepler, 1609)
III. Comparing the orbits of the several planets, it is found that the times of revolution are related to the distances from the Sun by a proportionality: the squares of the periods of revolution are proportional to the cubes of their mean distances from the Sun. (Kepler, 1619)
IV. Comparing the orbits of the several planets, it is found that the distances between the orbits of any pair of planets have a common proportionality: the original orbit of a planet relative to the one on either side of it is defined by the classic Greek golden ratio. (Scarborough, 1995, modified 2006)
V. In additional comparisons of the orbits of the several planets, it is found that gradual changes in their distances and velocities have a common proportionality: all are functions of a distance-mass-velocity-time relationship. (Scarborough, 1995, modified 2006)

Because of the disintegration of Planet Asteroids (Heinrich Olbers, 1802), the five laws definitively reveal why Pluto now is the 9th planet, but remains in the 10th planetary orbit.

CHAPTER IV

MOONS, PLANETARY RINGS, COMETS AND ASTEROIDS

We must forsake such beliefs that Earth is flat, that its core is iron or rock, that fuels were made from fossils, that energy is finite, that planets are created from dust and gases, and that continents still drift. Such concepts have too long misled scientists down blind alleys. Author (1980)

Origin of Our Moon

While it appears to be pale silver in the daytime, our Moon is actually a dark brown color. Brown is the color of cooled lava, pumice (volcanic glass) and igneous rocks that comprise the surface. Jagged, rocky mountains stretch across part of the Moon.

The silver sphere is 2,160 miles in diameter and one-eightieth the mass of Earth. Its surface gravity is only one-sixth as strong as Earth's surface gravity—too weak to retain an atmosphere. Any water or light gas vapors formed in the past would have evaporated immediately into space. With its mountains, extinct volcanoes and moonquakes, it is easy to visualize the Moon as simply a small Earth without atmosphere or water. One can expect to find most of the same elements comprising both spheres.

However, due to different internal conditions affected by size difference, the ratios of created and of retained elements should, and do, vary considerably. Between the first landing on the Moon in 1969 and the last one in 1972, some 850 pounds of rock samples were brought back for analyses. The abundance of some of the main elements—silicon, magnesium, iron, manganese—in the two spheres matched. Refractory substances such as aluminum, calcium oxide, chromium oxide and titanium that are difficult to vaporize were quite different, as predicted by the FLINE concept. The Moon samples showed twice as much as Earth's contain. The biggest differences showed up in the more volatile substances, such as potassium and sodium: the Moon has much less than Earth. No trace of water was found on the Moon.

In 1982, Kirk Hansen of the University of Chicago suggested that the changing rotation rate of the Earth determines the rate of tidal dissipation over geological time. His calculations argue favorably for the simultaneous creation of Earth and Moon, thereby adding credibility to the contemporaneous geometric birth of the "double-planet" expounded in the FLINE model as early as 1980. According to the FLINE (formally the GBSST) concept, Earth and Moon were formed contemporaneously from the SEM as it sped beyond the Sun.

Coupled into a giant dumbbell formation since then, the two masses trace an intertwined pattern in their revolutions around the Sun. The center point of the dumbbell's mass, rather than the center of the Earth, moves in a smooth ellipse around our Sun causing each of the two masses to follow a serpentine path. The results from the Clementine survey of the Moon yielded the first complete global portrait of Earth's orbiting partner. Among the findings in 1994 are: 1) volcanic activity as recent as a billion years ago; 2) a wildly variable crust;

3) the possibility of ice in the shadow of the South Pole; 4) a crater large enough to span the USA from the East Coast to the Rocky Mountains; and 5) a fresh-looking crater that may have been made in the 12th century and recorded by monks. Scientists had thought that nothing much had happened on the Moon in the last three billion years.

The discoveries brought the realization that scientists do not understand the Moon as well as they thought they did. Clementine's 71-day rendezvous with the Moon revealed a topography marked by steep peaks rising to ten miles higher than the lowest valleys. The deepest valley is rimmed by the highest peaks: a vital clue to the early stages of formation when forces of isostacy and ejection (internal forces) worked together to push up the huge rim of pliable, amorphous material that solidified as mountains, while the remaining ejecta sank, filling the void below and forming the deep valley. Evidence of recent lava flows in the Schroedinger basin suggests that the Moon was erupting with volcanoes perhaps two billion years after it was believed to have settled down. Such volcanoes represent the second stage of mountain-building in which ejecta must find a way out from beneath the now-solidified crust. Just as Earth and other planets, virgin ejecta push through crystal places of least resistance to its pressurized flow to build the tall volcanic outlets. This new evidence fits precisely into the FLINE concept in which these surface features can be traced to their source: Internal Nucleosynthesis.

Additional evidence came in 1995 from McDonald Observatory in Fort Davis, Texas. Observations made during a total lunar eclipse in 1993 (released in 1995) show the faint flow of sodium gas some 9,000 miles in the atmosphere surrounding the Moon. Here the IN process appears to be nearing its final stages, as attested by occasional small moonquakes and the weak out-gassing from craters.

The one big question left unanswered in many minds by the space probes is the Moon's origin. A growing consensus among astronomers favors the "giant impact" hypothesis in which the Moon may have gotten its start 4.5 billion years ago when a planetary projectile about one-seventh Earth's mass collided with our mother planet. The energy of the collision crushed and vaporized major parts of the two masses, sending out a high-velocity jet material at temperatures as high as 12,000 degrees F. Within a few hours, some of it came back together far enough away from Earth to remain in orbit as our moon. Proponents of this concept claim that it appears to explain the chemical findings from the Apollo mission; e.g., the moon rock brought back lack of water, sodium and other volatile materials—precisely the materials that would boil away in the rapid vaporization after impact. And some scientists believe the concept explains why veins of gold and platinum lay shallow enough in Earth's crust to be mined!

But not all scientists are satisfied with such a scenario. First, two of the above statements concerning sodium are contradictory. The 1993 observations of the faint glow of sodium gas surrounding the Moon up to an altitude of 9,000 miles directly contradict the 1980s impact scenario in which all sodium boiled away 4.5 billion years ago. Further, the potassium and sodium reported in the early analyses of the Moon rocks could not have survived such an impact. And why would gold, platinum and other heavy metals be found only in a projectile and not in the larger mass of Earth? How were these materials made in the projectile? Would the missile have delivered all of Earth's other mined materials? Does this impact scenario apply to all other moons of all other planets? Aren't spherical moons all made in the same manner? Did the moons of Jupiter and Saturn result from projectiles bouncing off their clouds? Why didn't the exploding "asteroids" of Comet SL9 bounce off Jupiter to form more moons? How dependable were the computer simulations that seemed to confirm the hypothesis? Were their results any more dependable than the erroneously predicted results of the powerful computer simulations of the collisions of Comet SL9 with Jupiter, as detailed later in this chapter? Wouldn't the same type of simulation erroneously show that all moons of all planets were formed in similar impact scenarios? The fallacy of the impact hypothesis becomes obvious under the pointed finger of questions. As one scientist put it, "Books and articles supporting this giant-impact hypothesis of lunar origin are more a testament of our ignorance than a statement of our knowledge." As with the fallacious dinosaur theory, why is it necessary to look to outer space when better answers can be found more readily by looking to the nucleosynthesis processes within planets? In reality, all anomalies relevant to all moons and planets are explainable within the realm of the FLINE concept.

For example, the Moon once generated a magnetic field, which may have been nearly twice as strong as the present-day magnetic field of Earth, as shown by Runcorn *et al.*, who used magnetized lava rock from the Moon as evidence. The proof of the declining strength of the Moon's magnetic field is powerful evidence that such magnetism is (or was) of nuclear energy origin rather than of iron or rocky core origin. The large decline is attributed to the dwindling size of the core as its energy transforms into matter. Other observations and studies of moons and planets have revealed magnetized lava rocks, the increasing of interior temperatures as a function of depth, the size and density of cores, the equatorial bulges, the convection of materials, the volcanic layers, the chemical composition of crusts, the ejecta and impact craters, the advanced stages of evolution, etc. All are symptomatic of nuclear cores. In contrast to the impact concept, evidence supporting the FLINE version clearly shows that Earth and Moon formed contemporaneously, and both have evolved via Internal Nucleosynthesis into the rocky fourth stage of evolution.

The Moon is nearing the fifth and final stage: an inactive sphere. In the final analysis, the Moon is nothing but a small and shining planet that evolved alongside Earth and in the same manner as Earth, partners from the beginning. Shelly expressed his feelings beautifully when he wrote: "....that orbed maiden, with white fire laden, whom mortals call the moon."

To: David H. Levy, c/o Parade Magazine, New York, N.Y. 11-06-00

While very interesting reading, your article, *Why We Have a Moon* (*Parade Magazine*, Nov. 5, 2000), presented the prevailing hypothesis of the manner in which our Moon was created: the Hartmann-Davis Giant Impact concept of its origin. In this incredulous version, "a small planet, with unimaginable energy and deafening noise, sideswiped Earth about 4.5 billion years ago, bounced off and, seconds later, tore right back into our planet with a super colossal force. That Mars-size world broke apart, and huge chunks of Earth's crust flew off into space. Two rings of debris, their particles much larger than those in the fine-grained rings of Saturn, grew and circled Earth...pieces of the outer ring slowly gathered together around its largest chunks. In just a year, those pieces formed a large new world, a world we can still see. That world we call the Moon."

Any concept based on the Accretion hypothesis is already fatally flawed, with no chance whatsoever of being blessed with sound logic or factual evidence. This concept fails to meet any of the prerequisites for a valid theory as specified by true science. It is pure speculation, with no basis in fact; known facts about our Moon's origin present a powerful argument against this version. The facts, explained elsewhere in my books (1975-1996) and in other literature, are a vital part of the FLINE model of planetary origins and evolution. In brief, moons originate and evolve through the same five cycles of evolution as planets do; they are, in fact, simply smaller planets.

Planets come in all sizes, from the smallest sphere to the largest giant exo-planet. The too-close proximity of each giant gaseous exo-planet to its central star is a powerful argument against the Accretion concept, but is strongly supportive of the FLINE model. Many questions always will remain unanswered via the Giant Impact concept, but one question alone can illustrate the fallacy of this hypothesis: Considering all the moons of Jupiter, Saturn, and other planets, and realizing that all spherical moons were made in identical manner, did all these moons bounce off the thin cloud tops of their respective planet to become what they are today? If this sounds highly unlikely, then we must conclude that our Moon, or any other large moon, could not have been created via the giant impact-accretion hypothesis.

Meanwhile, the FLINE model of creation of moons presents a valid scientific alternative soundly structured with irrefutable facts that leave no questions unanswerable or non-provable. An in-depth comparison of the two concepts readily exposes the ludicrous logic of the giant impact concept, which already is becoming a genuine embarrassment to science. Herewith is a flyer describing my current book, *The Spacing of Planets: The Solution to a 400-Year Mystery*; it is on-line at iUniverse.com.

My newest book is due on-line in January. The books contain answers to the many questions left unanswered by the Accretion hypothesis. Additionally, they detail the FLINE model's accurate predictions and interpretations of the results of the crashes (actually powerful explosions) of Comet SL9 on Jupiter. That event alone is conclusive evidence of the true nature of comets, which, in turn, is irrefutable evidence against the Giant Impact/Accretion hypothesis.

Recent Developments Concerning the Lunar Cataclysm Hypothesis

In regard to *Support for the Lunar Cataclysm Hypothesis from Lunar Meteorite Impact Melt Ages (Science*, 1 December 2000), suffice it to say that the cogent evidence presented by B. Cohen *et al*., becomes exciting when more logically interpreted within the realm of the FLINE model. In reality, the data serve as powerful supportive evidence for Allan Mills' famous experiment that duplicated the surface of the Moon by forcing gases up through a layer of thick mud, thereby revealing the manner in which the crater-covered surface was formed via the ejecta concept of the FLINE model.

The event of a short, intense period of bombardment in the Earth-Moon system to account for these craters is a highly unlikely occurrence; this hypothesis isn't needed to explain the multitude of craters that can cover the entire surfaces of moons and planets. All that is needed is an understanding of Nature's principle of Internal Nucleosynthesis, the creative force without which evolution and ejecta craters would not be possible. Of course, this does not preclude the probability of occasional impact craters being formed when heavy ejecta matter fell back onto the surface, or when, on very rare occasions, an outer space object could have added to the sum total. In the same issue of *Science* magazine, *Beating Up on a Young Earth, and Possibly Life* expounded on the lunar cataclysm. To quote: "Astronomers still don't have any good idea of the alleged lunar cataclysm's source. Simulations show that the gravity of earth and other terrestrial planets would have cleared the inner solar system of threatening debris within a few hundred million years." According to the researchers, such "a cataclysm would require the breakup of a sphere larger than 945-kilometer Ceres, the largest Asteroid; and the chance of that happening any time in the past 4.5 billion years is nearly nil."

As a last resort, astronomers look to the outer reaches of the solar system, and speculations abound. For example, "Neptune and Uranus could have tossed icy debris, along with some asteroids, inward in sufficient quantities to resurface the Moon, give Mars a warm and wet early atmosphere, and sterilize Earth's surface with the heat of bombardment." One astronomer is toying with the idea that Uranus and Neptune started out between Jupiter and Saturn, where his simulations suggest they could have orbited for millions of years before flying out into the lingering debris beyond Saturn and triggering a late heavy bombardment! "That's my fairy tale," he correctly stated—a statement with which the new Fourth Law of Planetary Motion and the FLINE model fully agree.

The most amazing thing about this situation is that after 27 years of the evolving FLINE model's irrefutable facts and impeccable record of accurate predictions and years after the discovery of the Fourth Law, many astronomers still prefer fairy-tale speculations over well-substantiated, scientifically valid arguments. Contrary to these *Science* articles, the origins and evolution of moons have been proven beyond any doubt by valid scientific methods. Even though the new model's powerful supportive evidence shouts loudly against the establishment's favored, but erroneous, hypothesis, it continues to fall on deaf ears. The blame must be placed squarely on the shoulders of advocates of the prevailing Big Bang/Accretion hypothesis.

New Evidence Substantiating the FLINE Model of Creation of the Rings and Moons of Jupiter

(Ref: *The Spacing of Planets: The Solution to a 400-Year Mystery*, Scarborough, 1996). On September 15, 1998, scientists announced to the news media that "the faint rings around Jupiter come from clouds of dust that are the result of cosmic debris battering Jupiter's small moons, according to data from the Galileo spacecraft." "The rings - which are nearly invisible to even the best telescopes - clearly show their relation to the orbits of four small inner moons," the scientists said. According to Joseph Burns of Cornell University, "Pictures are the smoking gun that allows us to say this theory works."

Some questions become obvious: If it is true that planets accreted from dust, gases and planetesimals, one might question why the matter comprising the planetary rings remained in orbit rather than accreted into the greater mass. Why would this "cosmic debris" be battering Jupiter's small moons to create even more dust rather than accreting as larger bodies? Or is the condensation/accretion hypothesis of planetary creation erroneous? Can the relation of the rings to the orbits of the four small moons be interpreted as something more significant than a battering action that makes dust? What can be learned by comparing the rings of Jupiter to the rings of Saturn?

A new perspective on the subject offers some answers. Three fundamental and inseparable realities of nature comprise a revolutionary paradigm of planetary origins and evolution: the Four Laws of Planetary Motion (FL), Internal Nucleosynthesis (IN), and Evolution (E). In any sphere, neither E nor IN can exist without the other. A critical connection exists between planetary evolution and the essential source of energy that drives planets through the five stages of evolution common to all such spheres - until each core is depleted of energy, leaving only an inactive (dead) sphere (e.g., Mercury) in orbit. This paradigm provides an understanding of the role of the inner core in various Earth processes, including the generation of the geomagnetic field and the thermal evolution of Earth. The IN/E concept leads inevitably to the mathematical solution to the pending Fourth Law of Planetary Motion (1980-1995) that first eluded Kepler in 1595.

The Four Laws (the third fundamental reality) bring the FLINE paradigm full circle by explicitly revealing how nebulous planetary masses, including all exo-planets, attained their orbital spacing around a Sun. Viewed in the perspective of the FLINE paradigm of origins of solar systems and the orbital spacing and subsequent evolution of planets, the data evolves into a much clearer picture of how planetary rings were, and are, created. First, we must recognize that all planets began as fiery masses of energy (small stars, the first stage of evolution) placed in orbit in full accord with the FL. During the first two stages of planetary evolution, these violent Sun-like masses ejected plasma of ionized gases, dust and solid matter that were rocketed into orbits as dust and small moons. In reality, the pictures identified as the "smoking gun" in support of the prevailing ring theory clearly reveal a pattern that fits precisely into the FLINE scenario.

To corroborate this viewpoint, we can look to Saturn's older rings (now at, or near, peak accumulation) and observe their beautiful multi-patterned striations attained by the multiple ejection of selective matter created via Internal Nucleosynthesis and jettisoned into orbits at various times and at various distances around the mother planet. All planets go through the early ejecta ring phase of evolution. The rings always end up circling the equator - just as the young and growing rings of Jupiter now appear to be in the process of accomplishing. Planetary size governs the rate of evolution through the five stages: the smaller the size, the sooner the evolution through each of the five stages. This tells us that Saturn is ahead of Jupiter in the forming of rings;

thus, its orbiting masses are a preview of the many more rings that will form eventually around Jupiter in similar manner.

Of course, the early rings of our smaller Planet Earth and other small planets have long since disappeared onto planetary surfaces and into space. At the latest count of astronomers in mid 2001, Saturn has edged into first place as the planet with the most known moons: 30, to Jupiter's 28, Uranus' 21, and Neptune's 8. The new survey turned up 12 new moons around the ringed planet to surpass Jupiter, whose number of moons has not been updated. But not to worry; in the long run, Jupiter, due to its larger size and slower pace of evolution, will regain the lead in this interplanetary battle of numbers of ejected moons.

The Far Side of the Moon

USA and Soviet lunar-orbiting craft have photographed portions of the far side of the Moon. Painstaking analysis of computer tapes indicates that some of the images provide an accurate, multi-wavelength portrait of parts of the far side. The images from Mariner 10 and Galileo, each depicting a different part of the far side, are helping astronomers decipher, in detail, the composition of the Moon's hidden half. Surface composition provides crucial clues to the nature of the volcanic eruptions and other upheavals that shape the surfaces of spheres.

Scientists have already mapped much of the chemical composition of the Moon's near side. Analyses of the lunar rock samples brought back by U.S. astronauts and unmanned Russian craft have proven extremely helpful. Two types of terrain form the Moon's outer surface. The light-colored highlands represent the brighter, lower-density minerals. In contrast, the dark plains are regions identified by Galileo and his Renaissance colleagues as maria, a Latin word meaning oceans. To them, the dark plains appeared to be bodies of water.

Researchers gradually agree that maria are the results of volcanic eruptions: lava flows from 4 billion to 2.5 billion years ago. Images from the spacecraft confirm that the maria are much less abundant on the far side than on the near side of the moon. This suggests earlier and then-milder volcanic activity on the far side—the latter due to the earlier, thicker crust there. But why would the crust be thicker on the far side? In addition, when did the maria form? Evidence suggests that early volcanic activity that formed maria was extensive; some maria existed more than 4 billion years ago.

According to the IN concept, the Moon's crust formed in the same manner as all other spheres of the SS, passing through the common stages of evolution while cooling from hot energy to cold matter. During the transition stage, the Moon's thin crust struggled to survive and grow, eventually encapsulating the sphere with a rough, flexible blanket. Ever thickening, the soft crust became a thick, bubbling, mud-like caldron, stirred by escaping gases, matter and fireballs, all newly created within. These escape outlets hardened into circular patterns now erroneously identified as impact craters.

Meanwhile, the heavy atmosphere thinned and eventually vanished. Volcanic flows became common, covering many of the craters, even as others were being formed. Volcanic mountains and lava plains formed from huge outpourings of virgin materials that found the paths of least resistance to their upward pressures. Fireballs shot violently from within and fell back to the surface, sculpting the landscape with a multitude of ejecta and impact craters into the circular patterns and dark maria visible today. Instruments that view the surface at several wavelengths can detect the dark, underlying maria, called crytomaria, mixed with the lighter-colored shroud of highland crust.

Patrick Moore, in his book, *New Guide to the Moon* (1976 edition), presents a strong case for the formation of the craters of the Moon by internal forces rather than by impacting meteorites. Included is convincing evidence by Allan Mills, who successfully produced model craters strikingly similar to those of the Moon. He accomplished this proof by introducing a pressurized stream of gas below a particular material, which behaved as if liquid while the bed expanded. Bubbles appeared, venting through the muddy material to produce the crater-like appearance of the Moon.

To stunned astronomers, the two mismatched sides of the Moon "seemed like different worlds," and [erroneously] "nobody really knows why." The explanation appears simple enough in the definitive LB/FLINE model, as detailed in *The Spacing of Planets: The Solution to a 400-Year Mystery* (A. A. Scarborough, 1996, pp 91-94). In this well-substantiated model, the mismatched sides are a natural consequence of the detailed, one-rotational conditions prevailing during the early evolutionary stages of the Moon as it revolved around a former central star: Earth. Analyses of views of the far side show that the Moon's hidden half contains far fewer maria than the near side, and far lower concentrations of titanium oxide than do many maria on the Moon's near side. This evidence and the analysis of moonquakes and human-generated disturbances on the Moon have revealed that the lunar far side has a much thicker crust than on the near side. To the obvious question of why, the IN concept offers a reasonable answer. As contemporary partners for some five billion years, Earth and Moon evolved with the near side of the Moon always facing Earth. During their evolution from energy masses to rocky spheres, the lunar far side remained exposed to the severe cold of space, while the near side basked in the heat from Earth's nuclear mass, a small secondary sun. These extreme conditions, along with the slightly larger centrifugal force acting on the far side, caused the crust there to evolve quicker and thicker than the much warmer near side evolved.

Under these extremely different conditions, the two sides naturally evolved differently. Noticeably more craters and titanium oxide on the near side are indicative of a later, slower, longer stage of ejecta-cratering. The difference in the amounts of titanium oxide is attributed to the fact that larger volumes of heavier elements such as titanium are produced in the later stages of evolution of any sphere. The steady increases in internal temperature and pressure - caused by the huge energy-to-matter expansion within—as the sphere becomes more tightly encapsulated accounts for the larger amounts of titanium oxide produced at a later time and over a longer period of time on the near side than on the far side of the Moon. The thinner crust adds another advantage to the near side by providing an easier upward route for the chemical ejecta. One interesting observation here: The titanium produced on the Moon is analogous to the tellurium produced on Venus, the iridium produced on Earth, the magnesium produced on Neptune, and the abundant metals observed at the impact site of the G fragment of Comet SL9 on Jupiter—all are in situ ejecta products of nucleosynthesis. In summary, the difference in the number of craters, the thickness of the two crusts and the amounts of titanium oxide are indicative of the differences in the rates of cooldown and solidification of the crust on each side of the Moon. These different rates, in turn, are functions of the outside temperatures to which the two sides were exposed and to the steady increases in internal temperature and pressure as encapsulation progresses as a function of time.

Other questions challenge the impact craters hypothesis. If by impact, why such a noticeable difference in the number of craters on the two sides of the Moon? In addition, why so many craters altogether on such a tiny target? Why are all craters lined up precisely with the center point of the spherical moon? Why don't any of the craters show a skewed configuration typical of a near-miss projectile hitting the surface at a sharp angle? In conclusion, the evidence supporting the FLINE concept argues strongly against the giant impact hypothesis of the Moon's origin, while presenting a valid case for the origin of Earth and Moon as contemporary partners during their simultaneous evolution via nucleosynthesis from energy masses to rocky spheres.

Although beginning their evolution simultaneously, Earth and Moon have evolved at different rates because of the size factor. The Moon's energy core is nearing depletion, as attested by very weak signals from within.

Approximately two-thirds of Earth's core has been depleted in nearly five billion years, leaving an estimated 2.5 billion years or more until it becomes inactive. Meanwhile, its orbit will have moved approximately 20% closer to the Sun. During this long interval, Earth's climate will cycle through many changes while growing ever more barren. Our planet will follow Mercury, Moon, Pluto and Mars in becoming inactive spheres. Expectations that humankind can survive the gradual drift into these slow, drastic changes seem unreasonable.

How Planetary Rings Were Formed

In 1610, Galileo first observed Saturn's peculiar form - recognized several years later by Christian Huygens as rings. The next set of rings was detected in 1977 by James Elliot and his colleagues at Cornell University when they noticed that the light from a star blinked several times just before Uranus occulted it. Voyager 1 sailed past Jupiter in 1979 and Saturn in 1980. Voyager 2 visited Jupiter, Saturn, and Uranus, finally sailing past Neptune in 1989. The spacecraft made the first detailed images to confirm the rings, and discovered rings around Jupiter and Neptune during the extensive SS tour.

How can the formation of these rings be explained? By observation, the rings in each case are located at the equator, the circumference of maximum centrifugal force. In the perspective of the IN concept, substances (gas, liquid, solid) were forcefully ejected from these spheres and collected at the circumferences of maximum force, and remained suspended in orbits around each active planet. The greater the escape velocity/mass ratio, the higher the orbit is of each substance. Regardless of any change in the tilt of the planetary axis, the rings will remain firmly in equatorial orbits while their velocities decrease over billions of years, and eventually will drop onto the planets to become small parts of the crust.

The same principles can be applied to other satellites of planets, whether Nature-made moons or man-made craft, provided the orbiting masses remain below a critical mass/velocity/gravity/distance relationship. Any mass exceeding this yet-to-be calculated relationship will gradually drift away from its central body. Uranus, Neptune and our Moon are examples of orbiting bodies known to be undergoing such drifting. To some degree, when blasting satellites into orbit, mankind replicates Nature's procedure for orbiting the materials that formed planetary rings and some of the smaller moons. In this perspective, the discoveries of the rings of Jupiter, Uranus and Neptune were predicted by the IN concept before the author learned of their existence. By the same logic, it can be predicted that Pluto, because of its small size, is well beyond this ring stage of evolution. Why rings on the larger outer planets and none on the smaller inner planets?

The reasons can be found in their sizes, which control the relative rates of planetary evolution. In the gaseous stages of their transformations, all planets had rings of ejecta: gases, solids, and liquids that quickly froze solid. Like a flower in full bloom, Saturn is at the zenith of planetary rings, the goal that Jupiter will attain some time in the distant future—the stage through which all other planets of our SS already have progressed. The smaller the planet, the earlier it passed through the ringed stage of its evolution, which ended when all materials had fallen out of their orbits in full accord with the natural laws of gravity. Nothing can stay in orbit forever. Uranus and Neptune are in the twilight time of passing out of their ringed stage, while Pluto should have passed out of it a long time ago. Ejecta created via Internal Nucleosynthesis is the key phrase that firmly ties together all things: planetary rings, craters, volcanoes, planet quakes, moonquakes, fireballs, comets, meteorites, the moons of Mars (Phobos and Deimos), lightning, electromagnetism and all other phenomena involved in the evolution of physical worlds. All are interconnected; all can be traced to a common origin: a nuclear energy core in each sphere. In this perspective, all past and future discoveries of the space probes will be understood.

Jupiter's Rings: A Giant Leap Forward

Until early 1992, the prevailing Accretion Disk theory (the condensation/planetesimal/accretion hypothesis) taught that planetary rings consist of leftover debris that did not accrete. In February of that year, the news media announced that the Ulysses spacecraft boomeranged past Jupiter on Saturday, flying through *intense radiation and an orbiting ring of volcanic debris* on its way to study the Sun. The underlined words above represent a giant leap forward in the right direction of thought in the scientific community concerning the origin of planetary rings. The nuclear energy core of Jupiter accounts for both the volcanic debris and the intense radiation experienced by Ulysses. Both findings bring current beliefs about planetary origins sharply into question. Since both findings are products of Internal Nucleosynthesis and were predicted by the IN concept, they are powerful evidence supporting the FLINE model that foretold these discoveries as far back as the mid-70s.

Neptune's Rings are Fading Away

The first complete images of Neptune's outer rings to be taken in over a decade show that some parts of them have dramatically deteriorated and one section is close to disappearing altogether. Whatever is causing the arcs to deteriorate is acting faster than any mechanism regenerating them. "The system is not in equilibrium," says Eugene Chiang. "Our whole understanding is up in the air." By observation, the rings in each case are located at the equator, the circumference of maximum centrifugal force.

In the perspective of the ITEM concept, substances (gas, vapor or solid) are forcibly ejected and suspended in balanced orbits around the planet. The greater the escape of velocity-to-mass ratio, the higher the orbit of each substance. Regardless of any change in tilt of the planetary axis, the rings remain firmly in their equatorial orbits. One might question whether primordial disks would tilt with the planet or remain in their original plane of formation, if remnants of a gaseous dust-cloud. The same principles and questions can be applied to the other satellites, or planetary moons, of all planets. As a general rule, the larger the moon, the greater the distance from its planet—until the geometric mean is reached—mimicking the Solar System. Beyond that point, the moon sizes are progressively smaller. Like the rings, moons follow the tilt of their planetary sphere.

To some degree, when sending satellites into orbits, mankind replicates Nature's process of orbiting the materials that formed its planetary rings and some of its moons. The rings of Jupiter, Uranus and Neptune were predicted by the ITEM concept before the author learned of their existence. By the same logic, Pluto, because of its small size, can be predicted to be beyond this stage; however, its great distance from the Sun might act as a preservative of rings similar to Neptune's. But why rings on outer planets and not on the inner ones? The reasons lie in size and distance, which, in turn, control the rate of planetary evolution. In the early gaseous stages of their transformations, all planets had rings of ejecta materials. Like a flower in full bloom, Saturn is at the zenith of planetary rings, the goal that Jupiter will attain someday, and the stage through which Uranus and all smaller planets have progressed. The smaller the planet, the earlier it moved through the ringed stage of its evolution, which ended with the materials falling out of their orbits in accordance with natural laws governing decreasing rotational speeds and the gradual decreasing of centrifugal forces.

Neptune and, perhaps, Pluto are planets in the twilight zone of change into spheres that are losing the last remnants of their rings. Ejecta is the key word that ties everything together: planetary rings and craters, moon craters, comets, volcanoes, earthquakes, planet quakes, moonquakes, fireballs, meteorites, the moons of Mars (Phobos and Deimos), lightning, electromagnetism and all other phenomena involved in the evolution of physical worlds. All are interconnected and can be traced to their common origin: the internal transition of energy to matter in each sphere. In this ITEM perspective, all the exciting recent and future discoveries by space probes will be explainable. By confirming the 1991 prediction and explanation of the LB/FLINE model of universal systems, this critical discovery by Pater, Chiang *et al.*, is powerful evidence for the new model.

Comet Halley and Its Last Farewell

In March of 1986, five space probes encountered Halley's Comet to photograph its nucleus and to analyze its composition during its most recent periodic visit to the Sun. The photographs made by Giotto, the closest probe, revealed some real surprises. The picture and data beamed to Earth showed a velvet black, peanut-shaped nucleus with a surface full of pits. According to the IN concept, the surface is the carbonized crust encapsulating its nuclear energy mass, and the surface pits are ejecta holes or friction marks formed in the past by powerful jets of forcibly ejected materials.

The Vega probe reported evidence of water, carbon dioxide and hydrocarbon molecules among its findings, all products of combustion. Temperatures within the comet's thin plasma coma were recorded to be well above the boiling point of water. Computer-enhanced color images of the comet were identical to those of the background stars, indicating similar brightness and high temperature gradients throughout each of the photographed bodies. These likenesses received little notice and less publicity, but remain a significant factor in deciphering the true nature of comets. In spite of this contradictory evidence, many scientists still cling to the belief that fiery comets are "rocky snowballs"!

Violent, bright jets of gas and fine matter spew profusely from vents (or craters) in the nucleus. The jets appear to be firing out through two or three vents comprising only a small fraction—about 10 percent—of the total surface, leaving some 90 percent of the surface inactive. This fact of no losses from the black surface is highly significant: It means that materials forming the tails and the coma (with a radius of 600,000 miles) could have come only from within the nucleus, the source of their creation.

Dimensions of the nucleus were estimated at 16 x 8 x 8 kilometers. Its surface area is approximately four times larger than expected, and thus, the low albedo of two-to-four percent means that Halley's Comet is the darkest of all known bodies in the SS.

Scientists have difficulty in reaching firm conclusions about the nature of the observed jet-like features, and in interpreting new information to fit the concept of a dirty snowball of pristine materials that supposedly formed the SS. The basic reason for these difficulties lies in the fallacy of the antiquated accretion disk hypothesis that evolved from the Kant-Laplace hypothesis of a gaseous dust-cloud origin of the SS—a tragic myth that continues to lead many scientists down blind alleys in which nothing seems to fit together.

The Giotto pictures confirm that inert shells do develop on cometary surfaces, much like crust forms on planets—an exciting quantum leap. The next step is for skeptics to realize this startling fact. From there, it will be much simpler to understand how moons, planets and their systems were made in compliance with the laws of thermodynamics.

Both Giotto and Vega spacecrafts carried aboard them sensitive instruments to identify gaseous molecules in Comet Halley's coma. They found a veritable stew of molecular fragments known as free radicals, confirming conclusions made previously from telescopic observations. Some 75-80 percent of the radicals pertained to water, a ratio common to Earth's water-to-land ratio produced via nucleosynthesis!

The remaining free radicals suggest that carbon dioxide, ammonia, methane and hydrogen cyanide would be the end products if these highly reactive radicals could get closely enough together to interact and combine. However, most remain too far apart to combine into these compounds. It is no coincidence that the same compositional radicals are found in stars, and the same end products they form are found on all planets and moons of the SS. All were, and are, produced via nucleosynthesis within the fiery core of each evolving mass.

Other than these gases, and not surprisingly, a major discovery was the existence of tiny, solid CHON particles. CHON is an acronym for the chemical symbols of the compositional elements: carbon, hydrogen, oxygen and nitrogen. While these compounds are not biological in origin they are indicative of compounds that might appear as black soot made via nucleosynthesis within the fiery comet. (Note the similarity with the nucleosynthesis of hydrocarbon fuels in Chapter I.) This accounts for the velvet-black surface of the nucleus, perhaps a carbonized insulator against the tremendous heat of its core, and the near-absolute cold of space—analogous to the formation of crust on all planets and moons.

Large quantities of very fine particles were gathered by the dust collectors aboard Giotto and Vega. Chemically, these motes consist of many of the same elements comprising terrestrial rocks: iron, magnesium, silicon, oxygen, etc. The presence of these virgin elements, created via nucleosynthesis within the comet, further validates the FLINE concept and its predictions.

Since 1987, observers have detected methanol in seven comets. It now appears that this compound ranks third in abundance among cometary ejecta, behind water and carbon monoxide. The methanol can amount to five percent or more (relative to water) of a comet's volatile matter.

The results from Halley's Comet reveal its true nature. Rather than "dirty snowballs", comets are what they seem: nuclear fireballs ejected from larger nuclear masses (much like flares ejected from the Sun), and powered in their orbits like nuclear-fueled jet engines. This mental picture was brought sharply into focus on the TV program, *Science Frontiers,* on June 7, 1994. Some eight years after the big event, the public was privileged to watch video close-ups of Halley's Comet zipping through the sky, looking precisely like a jet-propelled rocket in stabilized flight (see illustration).

Records show that the first recorded sighting of Halley's Comet was 240 B.C. It has returned periodically every 76+ years, each time diminished in size as energy transformed continually into ejected ionized matter. The original gigantic comet must have been a terrifying sight in its early years. In 1456, it frightened everyone in Europe so much that Christian churches added a prayer to be saved from "the Devil, the Turk and the Comet". As late as 1910, it still furnished a good, but smaller show. But in 1986, it was a disappointment to observers, who had to look closely to find it.

Then in February, 1991, when Halley's comet was 1.3 billion miles from the Sun, about midway between Saturn and Uranus, it suddenly expanded to 180,000 miles across and shone more than 1,000 times brighter than normal. The event was startling and unique, totally unexpected so far from the Sun, and remains unexplainable in the perspective of the "dirty snowball" concept.

However, viewed in the FLINE concept, the spectacular event is understandable: the comet simply exploded in its death throes. It was the last sighting of the famous comet—the last farewell—never to return on another mission to brighten Earth's skies. Other known comets have had catastrophic endings; some have split into two or more pieces, others have crashed into larger spheres, including the Sun. The Biela Comet broke in two in 1846 and has, since, disappeared. In reality, each fiery comet exists for a relatively short time, usually measured in centuries, before becoming extinct. Rather than primordial remnants (dirty snowballs) left over from the formation of planets, they are more like a fiery species that refuses to become extinct.

Comets and Asteroids: Keys to Planetary Origins?

The ever-present dangers of comets and asteroids crashing into Planet Earth and wreaking tremendous havoc worldwide have caught the attention of the public. The extinction of the dinosaurs is often used as an example of what can happen when a large object from outer space creates a nuclear winter—although this scenario remains in dispute. Many scientists believe that volcanism was the culprit—and there is much corroborative evidence favoring this concept.

To understand Earth's mysteries—past, present and future—we must understand the origins of solar systems and the evolution of planets. But to fully comprehend planetary origins, we must know, beyond doubt, the true nature of comets and asteroids and their sources. In spite of much evidence to the contrary, many scientists cling to the belief that comets are gigantic masses of ice, snow and dirt from outer space far beyond our Solar System. This icy-conglomerate hypothesis of comets was brought forward in 1950 by Fred Whipple, and modified somewhat by Gerald Kuiper in 1951. In 1987, William Hartmann and his colleagues defined a comet as "a body formed in the outer Solar System containing volatiles in the form of ices and capable of developing a coma if its orbit brings it closely enough to the Sun".

The accretion concept of the origin of our Solar System and the evolution of planets is based on the belief that comets, asteroids, dust and gases are pristine materials that came together to form these systems. Keep in mind that this hypothesis is structured with speculation, with no basis in fact. But it soon may be facing a serious challenge to its validity. Let's see why. When recent close-up photos of comets are examined carefully, they present some startling revelations. Always conspicuous is a brilliant white mass that looks precisely like a white-hot fireball blasting from the rear of the relatively tiny, pitch-black nucleus (Figure 7), followed by a number of jet streams formed into a fan-shaped pattern of material (Figures 9,10). Ancient drawings of comets and recent photos reveal separate jet streams usually numbering between two and ten or more. This pattern of distribution gives an indelible impression of jet streams forcibly ejected at various angles (against the solar wind) from several portholes, or craters, in the nucleus.

These comet tails, millions of miles long, consist of ionized gases and dust comprised of a variety of chemical elements like those observed on the surfaces of the Sun and Earth. An ultraviolet image reveals that the hydrogen coma of comet Kohoutek had a diameter of more than 3,000,000 miles, nearly four times the diameter of our Sun. Such a larger spherical coma is typical of all comets; for equal-sized comets, the younger the mass, the larger the coma. The fact that this thin veil of hydrogen extends so far in front of the tiny nucleus signifies a powerful and constant driving force capable of putting, and sustaining, it there against the resistance of the solar wind.

These exciting photos of Halley's Comet clearly reveal its fiery nature. Note the brilliant white jets firing from the rear of the fireball's irregular crust, locking the nucleus in a precise flight position in each picture—much like a jet-powered rocket. When the core's nuclear energy finally is exhausted via conversion into matter, the nucleus may remain as a pitted, ejecta-cratered chunk resembling the two moons of Mars: Phobos and Deimos. Note: Time elapse for photos was approximately one hour.

The article entitled *Rocky Relics* (*Science News*, 5 Feb 1994) discusses near-Earth asteroids (NEAs) and the possible relationship of comets, asteroids and meteorites. It states that in 1992, the asteroid known as 1979VA was identified as the same object that had been identified in 1949 as a comet (Wilson-Harrington) with a "faint, but definite tail....Researchers now say that the comet and the asteroid are one and the same." This finding adds much confirmatory evidence and credibility to the same conclusion reached by the IN concept in the 1970s. It represents another giant step in the right direction.

Another quote from the same article states that "several bodies classified as asteroids may once have been comets. One candidate is the asteroid 3200 Phaethon. It follows the path of small bodies that produce the Geminid meteor shower, the flashes of light visible each December. Meteor showers are typically associated with comets. Tracks of dust expelled by active comets as they pass near Earth's orbit produce the flashes as the dust burns up in the atmosphere. Following this line of reasoning, Phaethon may be a comet masquerading in its old age as an asteroid."

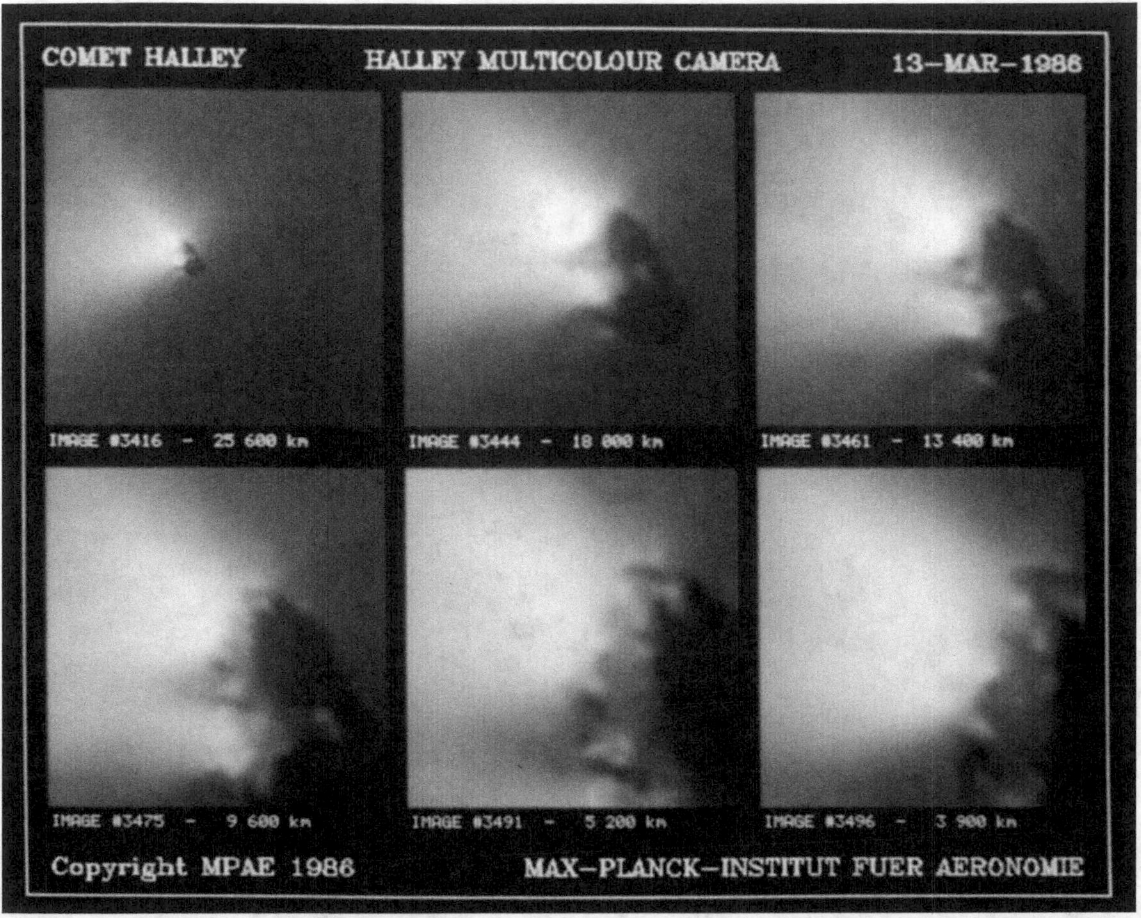

Photos: Max-Planck-Institute Fuer Aeronomie, Lindau/Harz, FRG; taken by the Halley Multicolour Camera on board ESA's Giotto spacecraft. Interpretation is by the author.

These exciting photos of Halley's Comet clearly reveal its fiery nature. Note the brilliant white jets firing from the rear of the fireball's irregular crust, locking the nucleus in a precise flight position in each picture -- much like a jet-powered rocket. When the core's nuclear energy finally is exhausted via conversion into matter, the nucleus may remain as a pitted, ejecta-cratered chunk resembling the two moons of Mars: Phobos and Deimos.
Note: Time elapse for photos was approximately one hour.

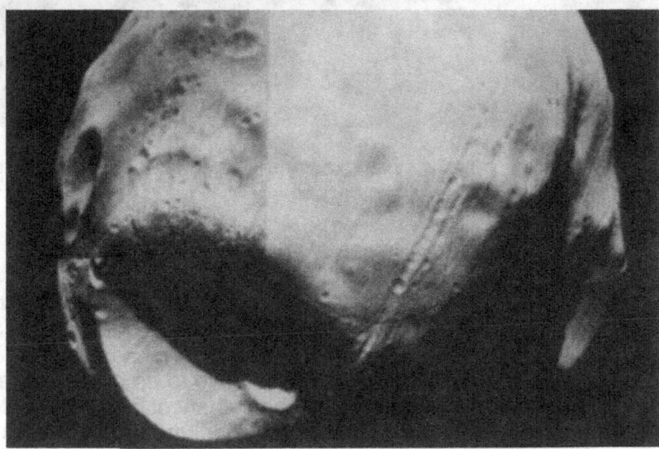

IMAGE; JPL/NASA

Phobos, a burned-out cometary nucleus.

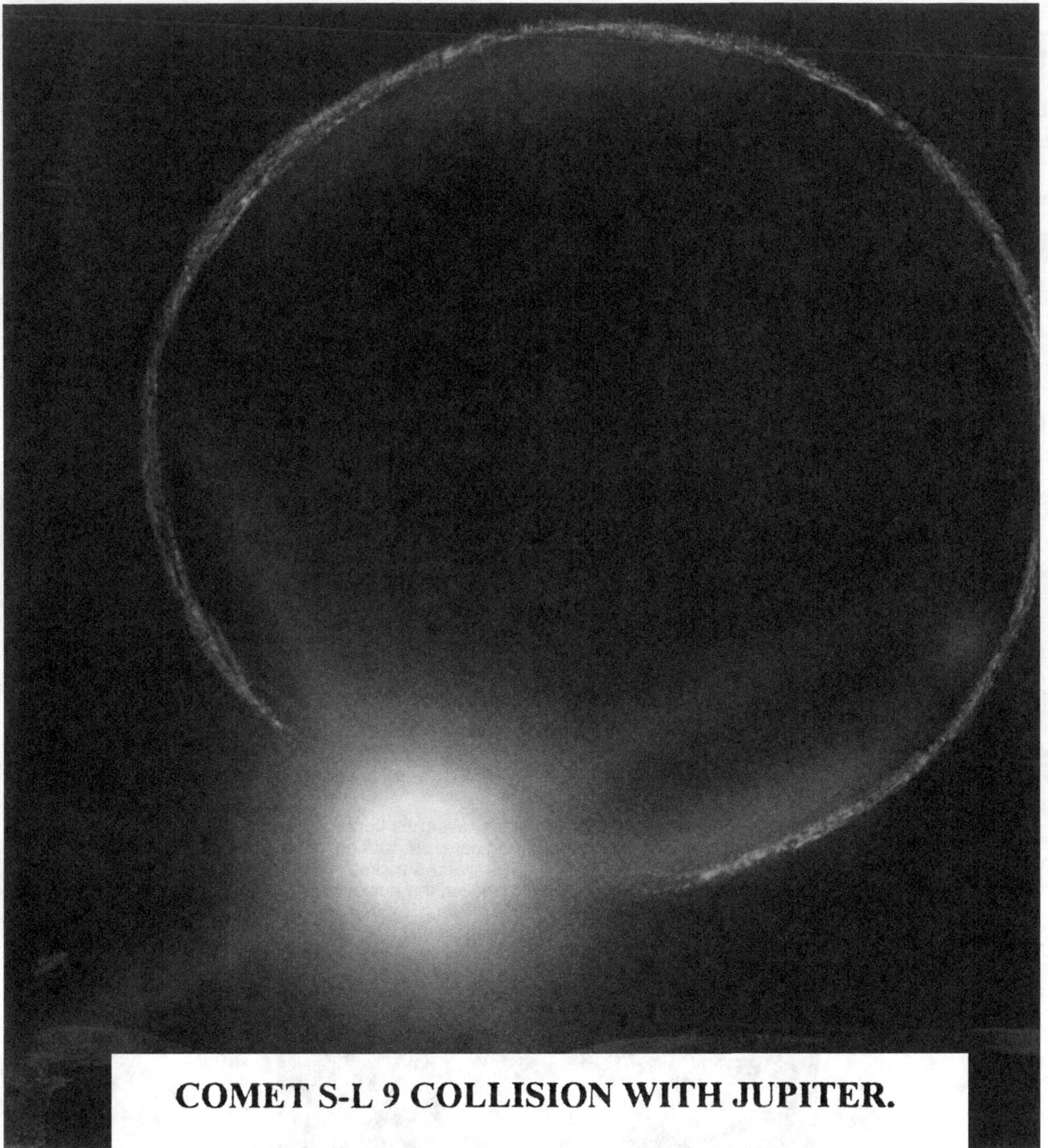

COMET S-L 9 COLLISION WITH JUPITER.

A stunningly spectacular explosion of one fragment of Comet SL-9 on Jupiter in July 1994 that far exceeded all predictions. Fragment G alone propelled a fireball thousands of kilometers above Jupiter's stratosphere and is thought to have yielded at least 6 million megatons of energy. A megaton is the equivalent of a million tons of TNT. To expend an equal amount of energy, a Hiroshima-type bomb would have to be detonated every second for ten years. One collision was estimated to be the equivalent of 500 million atomic bombs. The fragments exploded at the cloud tops, another crucial clue to their true nature: nuclear energy fireballs. The picture shows the actual explosion of a nuclear fireball rather than of a dirty rocky snowball. This information is devastating to the Big Bang theory.

In his *Principles of Philosophy*, René Descartes (1596-1650), the inventor of the analytic geometry used in the recent solution to the Fourth Law of Planetary Motion, defined Earth's interior as being "Sun-like"—a brilliant insight, especially since Internal Nucleosynthesis was unknown in his time. Without any evidence, he accurately defined the true nature of all active planetary cores. Planets and moons can still show clear evidence of being cometary masses, and recently have been recognized as such by scientists—although they stop short of the Sun-like version. In 1988 scientists detected traces of sodium and potassium as components of the Moon's extremely thin atmosphere. Three years later, researchers discovered that the atmosphere stretches out into a long tail about 21,000 kilometers in length and pointing away from the Sun. A corona of atmospheric atoms was found to extend 7,000 kilometers above the lunar surface—much like the coma of a comet. Their observations suggest that the Moon's atmosphere resembles that of a comet with an extremely tenuous tail; they also serve as powerful corroborative evidence of the evolutionary concept of planetary spheres from their original energy masses.

Some astronomers concede that the Moon does have the appearance of a comet. This is a giant step toward confirming my 1973 realization of the manner in which all solar system spheres were, and are, made via Internal Nucleosynthesis within their energy cores. Each sphere is a self-sustaining entity that does not require help from outer space to account for its ever-changing surface features and species. These facts led to the solution to the proposed Fourth Law of Planetary Motion, defined Earth's interior as being "Sun-like".

Planets and moons can still show clear evidence of being cometary masses, and recently have been recognized as such by scientists—although they stop short of the Sun-like version. Their observations suggest that the Moon's atmosphere resembles that of a comet with an extremely tenuous tail; they also serve as powerful corroborative evidence of the evolutionary concept of planetary spheres from their original energy masses. Some astronomers concede that the Moon does have the appearance of a comet.

This is a giant step toward confirming my 1975 realization of the manner in which all solar system spheres were, and are, made via Internal Nucleosynthesis within their energy cores. Each sphere is a self-sustaining entity that does not require help from outer space to account for its ever-changing surface features and species. These facts led to the solution to the proposed Fourth Law of Planetary Motion (1980-1995) that first eluded Kepler in 1595, and now explains how the nebulous planetary masses attained their orbital spacing around the Sun.

In like manner, the Four Laws (FL), along with the corroborative evidence presented herein, also explains the origins and evolution of the exo-planets. All active planets continue as self-sustaining entities until their energy cores are depleted (all energy transformed into atomic matter), at which point they enter the fifth and final stage as inactive, or dead, spheres. Valid arguments can be deduced from the evidence above to identify and confirm the sources of comets.

Their ongoing birth-and-death cycle points clearly to the connection between the craters (primarily ejecta type, but some impact) of planets and moons as the by-products of periodic ejections of these smaller cometary masses from larger energy masses. But our Sun will always be the main source of the fireballs of energy we call comets. Many comets survive their cycles without exploding or falling to fiery deaths into larger masses. What's left after the cometary burn-outs are recognized as asteroids of a volcanic nature and appearance. Close examinations of the ejecta craters of the two moons of Mars (Phobos and Deimos, Figure 8) confirm their volcanic and ejecta nature beyond doubt.

But what is, or was, the source of other types of asteroids? In 1802, Heinrich Olbers, Germany's leading expert on comets, recognized the asteroids orbiting in the fifth orbit from the Sun (between Mars and Jupiter) as fragments of a planet that has disintegrated, exploded. This accurate interpretation accounts for the more than 30,000 asteroid fragments, most of which would have begun their existence as small, fiery comets and smaller, shattered fragments we identify as meteorites: lumps of stone or metal that survived fiery passage through Earth's atmosphere. In all probability, the surviving asteroid fragments represent only a fraction of the original number of shattered pieces in orbit.

Evidence of current orbital patterns indicates the probability that three small planets existed in three distinct orbits before their disintegration into fragments. The Phi geometry in Chapter III shows that the Asteroids planet might have gone into orbit as a single mass at 4.24 AU. Later, due to its vulnerable position between the Sun and the giant Jupiter, it broke free into three fiery masses that settled near their current orbits (2.2 AU, 3.2 AU, 3.9 AU) before the explosive disintegration of each of the three potential planets resulted in the three wide belts of orbiting asteroids.

In 1997, some 500 pictures of the heavily cratered asteroid, 431 Mathilde, were recorded by the NEAR spacecraft as it passed within 1200 kilometers (km) of the 50-km rock. Of the numerous craters on its surface, five measured more than 20 km in diameter. Casual examinations of the photos and these figures raise some key questions: How could this asteroid exist after its alleged collisions with so many huge projectiles, especially if comprised of accreted "rubble loosely bound by gravity"? Why the low density and dark color of the rock? Why does it rotate so slowly (once every seven days) after such alleged battering by the projectiles?

In reality, no logical answers seem possible in the current perspective of the accretion or leftover debris hypotheses of the origin and nature of asteroids. Research provides data, but data alone don't provide answers; it remains useless (and often harmful until accurately interpreted). When these findings are analyzed in the perspective given to science by Olbers in 1802, the answers to these questions begin to make sense. As is well-known, hot masses (e.g., lava balls) have quick-cooling surfaces: a dark, encapsulating coat that grows ever thicker as a function of time. During progressive stages of cool-down of asteroids like Mathilde in the super-cold vacuum of space, out-gassing and dramatic shrinkage of solid matter undergoing entrapment result in the collapse of huge areas looking much like smooth, shallow bowls. The bowl-shaped areas usually are accompanied by smaller porthole-like craters scattered about, small craters that had once served as outlets for the hot, pressurized gases and other material. Mathilde-type asteroids could have evolved only from the scattered magma from their exploded planets - which accounts for their lower densities and the darker color and surface features revealed so clearly in the photographs. By similar logic, we can reason that stony asteroids are remnants of the original surfaces of the planets, while the nickel-iron masses came from within, or near, the planetary energy cores, either before or during the dramatic breakup of the three planets. Thus, the creation of different types of asteroids can be accounted for by the remnant masses undergoing rapid cooling in the super-cold vacuum and weightless conditions of space. Whoever determines the correct age of any stony or metal asteroid should establish precisely how long ago the three planets exploded into the tens of thousands, or more, asteroids. While arguing strongly against the current beliefs of many scientists that planets formed from accreted rocky snowballs or icy mud balls, these findings clearly show the close relationship among the fiery comets, cratered asteroids and moons, planets, and our Sun, thereby adding more credibility to the FLINE concept of our planetary origins. What happens to comets that do not meet catastrophic deaths before all their internal energies become depleted?

We can look to Mars for two good examples of burned-out cometary masses: its two moons, Phobos and Deimos. Both of the small, irregular bodies exhibit typical characteristics of burned-out comets: ejecta craters and parallel striations or grooves. The densities of both moons, like the volcanic rocks of earth, are much lower than that of Mars. From certain angles, Phobos even bears a striking resemblance to the rear end of a jet engine. Structured with the brilliant works of Chladni, Kepler, Olbers, Descartes, Mills, *et al.*, and corroborated by recent space discoveries, the FLINE paradigm offers a valid alternative to current beliefs about planetary origins and evolution - a paradigm made possible via evolving comprehension of the true nature, history, and relationship of comets and asteroids. These often terrifying missiles are indeed crucial keys to planetary origins and evolution.

Hale-Bopp: A Great Fiery Comet

On July 23, 1995, two amateur astronomers independently discovered their namesake comet, Hale-Bopp, blazing away beyond the orbit of Jupiter 25,000 times brighter than Comet Halley did when it was at the same distance from the Sun. As it swung ever closer, the new Hale-Bopp mass put on a spectacular show in March and April 1997, a spectacle reminiscent of the early appearances of Comet Halley between 12 B.C. and 1682 (and to an even lesser degree until 1910) when its brilliance caused a sensation in the sky every 76+ years.

How Comets Stay Active in Near-Absolute Zero Cold

Assuming comets are ice-rock masses, temperatures in the far reaches of the Solar System are too low for compounds to sublimate from their hard-frozen surfaces. However, astronomers have discovered that some comets, including Halley, experience brief flare-ups of activity, and others such as Schwassmann-Wachmann 1 (SW1) retain their persistent coma. The claim of some astronomers that sublimation of specific gases accounts for these phenomena in near-absolute zero conditions seems unrealistic. In the cold, dim environment of the outer SS, specific radio emissions characteristic of any sublimated gas molecules would be so faint as to be undetectable. In addition, a coma would be non-existent, leaving the comet's small nucleus invisible even through powerful telescopes.

But the nucleosynthesis (fireball) version circumvents these problems. In it, the comet continuously produces trails of gas and ultra-fine dust; however, with increasing distances from the observer, these tails soon become invisible while still sending out detectable radio emissions. For example, faint detection of CO has enabled astronomers to estimate that the very distant SW1 is emitting CO at the rate of about 2 tons per second. "This is remarkable," stated one researcher, "because it is comparable to the out-gassing rates of comets" [in the inner SS]. The style of out-gassing is capable, too. A slight Doppler shift in the recorded line implies that the gas is spewing out of SW1 in a jet, supposedly like that produced by water vapor in comets closer to the Sun. The expulsion of other molecules (gases) such as nitrogen cannot be ruled out: astronomers have no way to detect them at such distances at the present time. Without the continuous activity of nucleosynthesis in these very distant comets, the existence of coma and production of gases and dust would not be possible. Perhaps some day soon, improved detection methods will enable astronomers to detect and verify that the other gases and matter do continue spewing in a jet-like fashion from comets in the deep cold dimness of the outer SS—just as predicted in the 1970s by the IN theory. This will be another giant step favoring the FLINE model of planetary origins and evolution over the antiquated accretion concept.

More on Comet SL9's Collision with Jupiter: Some Stunning Results

In March of 1993, a fragmented comet was discovered moving away from Jupiter. Because of their bright, in-line formation, the Shoemaker-Levy 9 (SL9) masses were immediately dubbed "a string of pearls". Scientists designated the glowing fragments of SL9 by the letters A through W. The following anomalies will be viewed first in the perspective of current beliefs about our planetary origins and then in the perspective of the FLINE concept. The reader will note a major difference in the understanding of what really happened in each of the mysterious events.

On July 16, 1993, the segmented comet reached its greatest distance from Jupiter before heading back toward the giant planet. Inexorably moving in a lengthening train toward their target, some of the fragments appeared to be breaking up, some growing dimmer. Exactly one year later the bombardment of Jupiter began. Jupiter's radio noise sounds like regular background white noise, except that it ebbs and flows in what has been described as the sound of the surf along a shore. Forty-three minutes after the first impact occurred (the time it takes for radio signals to travel from Jupiter), the Jovian sound changed to a "rat-a-tat" static that lasted for nearly seven minutes.

A quote from the *Astronomy* magazine reads: "Radio astronomers expected radio emissions at high frequencies to drop when the SL9 comet crashed into Jupiter in July, 1994. Instead ... the JPL in Pasadena, California, found emissions increased 20 to 30 percent at a frequency of 2.3 gigahertz. Astronomers thought dust kicked off the comet fragments would absorb the electrons spiraling around Jupiter's magnetic field that causes the radio waves. But apparently the dust somehow created extra electrons that intensified the radiation through an, as yet, unknown mechanism."

In the new perspective, the "string of pearls" glowed so brightly because they were exactly what they seemed: freshly exposed nuclear fragments of a disintegrated comet. Any other masses of their small sizes would not have glowed like a string of pearls for such easy visibility. The two smaller masses, J and M, disappeared, due perhaps to either dissipation, or an over-coating of carbon. The unexpected increase, rather than the expected decrease in electrons, is precisely what happens when a huge amount of energy is instantaneously added to a system. The "extra electrons" came from the exploding nuclear fragment. Any dust, if present, would have played an insignificant role. Only a nuclear explosion could have caused the seven-minute "rat-a-tat" static.

Beginning July 16, astronomers were stunned by the tremendous explosion caused by the fragments of Comet SL9 crashing against the thin cloud-tops of Jupiter. Each powerful collision far exceeded supercomputer calculations. Comparisons of expectations with actual results were startling, even shocking to scientists. The *Astronomy* magazine (Nov 1994) published a well written article evoking full agreement that pre-crash expectations were among the casualties. The spectacular explosions were indeed vastly more brilliant and powerful than predicted by supercomputers.

The information in this paragraph is from the record of predictions made by researchers (*Eos*, 5 Jul 1994). "Prior to the big event, three-dimensional simulations of the impacts on Jupiter were performed on the 1840-node, Ontel Paragon, the world's most powerful parallel computer, located at Sandia National Laboratories, Albuquerque, NM. The key results of the fireball simulation for a <u>three-kilometer</u> fragment impact were moderate, but seemed sufficient for the event. At 70 seconds after impact, a spherical shock wave has reached a diameter of 700 km and an altitude of 900 km above the clouds. Hidden within the spherical shock wave is the fireball itself, which is a rapidly rising cloud of cometary debris and Jovian atmosphere at a very high temperature (1700 K). Its apparent magnitude of 2 is about one-fiftieth as bright as Jupiter itself. The visible fireballs will be orange, fading to red, and most of the energy will be radiated in the near infrared." Adding to the predictions, an article in *Science News* (9 Jul 1994), headlined *The 200,000-Megaton Meeting*, stated that each chunk is packing a punch that may exceed 200,000 megatons of TNT, and the expanding fireball may be as wide as 100 km when it emerges through the cloud tops within a minute after collision. These recorded predictions later could be classified as puny and far off target when compared with the actual show put on by the impacts.

The collisions were stunningly spectacular, the greatest ever witnessed in recorded history. Instead of small, orange flashes one-fiftieth as bright as Jupiter, the huge, white-hot explosive flashes were up to fifty times brighter than Jupiter. Expended energy estimates shot upwards to 250 million megatons of TNT, then to <u>600 million megatons, creating temperatures of more than 30,000 degrees C</u>. Fragment G, alone, propelled a fireball thousands of kilometers above Jupiter's stratosphere, and is thought to have yielded at least six million megatons of energy. (A megaton is the equivalent of a million tons of TNT.) To expend an equal amount of energy, a Hiroshima-type bomb would have to be detonated every second for ten years.

Infrared radiation (heat) from the explosion was so great that detectors at Keck Observatory were overwhelmed, or saturated. One collision was estimated to be the equivalent of 500 million atomic bombs. Astronomers reported seeing fireballs (from two of the smaller fragments) that erupted up to 1000 km above the cloud-tops. After fragment C hit, astronomers at Keck Telescope in Hawaii took infrared photos. The views show two glowing ovals, each about the diameter of Earth, left by fragments A and C. Other computers proved equally wrong. First, the fragments exploded at the cloud-tops (a crucial clue to their true nature) rather than penetrating deep enough to dredge up Jovian material, disappointing many who hoped the impacts would reach Jupiter's water-rich lower atmosphere. Researchers reported that infrared spectra from Galileo suggest that comet fragment G exploded high in Jupiter's atmosphere, never penetrating the uppermost cloud layer. Analysis of much weaker spectra from the R fragment supports this conclusion. Another team of researchers deduced, from absorption lines in their spectra, that relatively little atmosphere resided above the G fireball, indicating that the fragment had not plowed into Jupiter's thin cloud-tops. The anticipated problem of separately identifying Jupiter's water and the supposedly icy water of the comet never materialized. After the largest collision by fragment G, astronomers were puzzled by their failure to find the chemical signature of water in the clouds created by the comet's impacts on Jupiter. "It's puzzling, but we will continue to look for water," stated one astronomer. Scientists studying one of the early impacts found evidence of two chemical species never seen before on Jupiter: methylene (a carbon compound) and the hydroxyl ion. Both species would seem to require high-heat to be formed.

Together, these findings raise serious doubts about prevailing beliefs concerning the nature of these brilliant streamers across the sky. If they really are "dirty snowballs", there should have been obvious signs of copious amounts of water released in the collisions. And how will scientists explain the gigantic discrepancies between the supercomputers' puny predictions and the unexpected stunning immensities of the explosive collisions? There can be only one reason for these huge discrepancies. Computers will put out correct information only when correct information is fed into them. Had they been fed information based on the correct assumption that the fragments were of a nuclear energy nature rather than an ice-rock composition, the computer would have been precisely in line with the stunning results. The use of this correct assumption in a science contest of write-in predictions enabled me to correctly predict that mankind would witness the greatest explosions ever recorded within our Solar System. That prediction became a laughing matter among the experts, who, even today, still cling to the concept of rocky-snowball comets simply because it plays a crucial supporting role in the current Big Bang/Accretion theories about our planetary origins.

Further, in 1794, the true nature of comets was identified and published by the father of acoustic science, Ernst Chladni, who correctly argued for the cosmic origin of cometary fireballs and the generic connection between fireballs, meteorites, and meteors. The first proof came in January, 1866 via spectroscopic observations of Comet Tempel-Tuttle that clearly showed it was shining by both reflected sunlight and glowing gas, and by its own light from the bright nucleus region that was too small to be seen through the telescope of the observer, William Huggins. And thermal radiation from the region of a nucleus was observed first by Cal Lampland as early as December, 1927. Just recently, comets were found to have a small magnetic field and to give off x-rays: all these properties are inherent in nuclear masses; all were anticipated by the fireball concept. The obvious answers to all questions concerning these memorable collisions lead inexorably to the conclusion that the spectacular explosions were indeed of a nuclear energy nature. This rare event should be the final nail in the coffin of the "dirty snowball" hypothesis, a false speculation that too long has misled scientists down blind alleys. In reality, comets flashing through outer space are precisely what they appear to be: masses of hot nuclear energy that should not be confused with inactive asteroids composed of rocky matter that does not burn brightly in their flights through the vacuum of space, and thus remain invisible to the naked eye. The evidence favoring the ancient belief that comets are of a fiery nature is overwhelmingly persuasive. In spite of this powerful evidence, astronomers are not ready to change their minds.

But there is hope: NASA's Deep Impact Mission is scheduled to launch a robotic spacecraft in January, 2004 to arrive at Comet Tempel 1 in July, 2005. Once there, it will launch a heavy projectile, allegedly, to blast a hole in the celestial body some seven stories deep and about the size of a football field in an attempt to peer beneath its surface. Researchers hope the impact will yield dramatic scientific breakthroughs - and it will!

The 21st century will be an exciting time of change in the direction of scientific thought about our planetary origins, thanks to Einstein's ubiquitous $E=mc^2$ that accounts for all the majestic beauty and awesome power of energy masses: from the smallest comet to the largest star.

This new knowledge about the true nature of comets should dispel the notion of "dirty snowballs" from which our planets supposedly came into being. If not "dirty snowballs", then the building materials for planets, moons and systems universal must be the one thing that is distributed throughout the Universe, the one thing that comprises all atoms and molecules of matter, the one thing that furnishes a solid basis for understanding all anomalies of our SS, the one thing capable of explaining all the stunning results of SL9 impacts on Jupiter: nuclear energy. The picture below clearly shows the white-hot nuclear power of nine Comet SL9 fragments on their way to colliding with Jupiter.

Belief vs. Science: Which is more important? (April 2001)

On the evening of March 24, 1999, Stephen Hawking, the author of the highly successful book, *A Brief History of Time*, was to receive the American Physical Society's Lilienthal Award in Atlanta, and to give an address titled, *The Universe in a Nutshell.* According to the follow-up news article, "the ideas discussed were no longer new and are not universally accepted. But his talk was less an outline of new ideas than a public event featuring the famous scientist." According to the report in *The Atlanta Journal-Constitution*, "many physicists question whether Hawking's highly mathematical, highly speculative work will produce the unified field theory he and others have sought, combining Einstein's General Theory of Relativity (which deals with the largest physical phenomena in the universe) and quantum mechanics (the behavior of the most elementary particles). But they all agree that no one has looked further into the nature of space and time. ....*A Brief History of Time*, his first book for lay readers, made him a celebrity beyond the confines of theoretical physics." The news report continued: "The question, among some physicists, is whether cosmology is actually physics, whether it is science or science fiction. Its hypotheses are conjectural when pressed to the limit, and its conclusions can be neither proved nor disproved." Even Hawking has acknowledged the dilemma. "Cosmology isn't science," he told *Discover* magazine, "if it doesn't predict what we can observe." [Viewed in the LB/FLINE model's non-conjectural version, cosmology does qualify as science.] To pursue the issue further, a quote from *The World and I* magazine (Mar 2001, p 251) seems in order: "A point that keeps appearing throughout Ingram's essays is the subjectivity of science. 'We scientists sometimes fool ourselves into thinking that our work is always objective. It is instilled in us that our experiments must be totally unbiased. They must be double-blind, randomized, and controlled. There is an object lesson here for anyone out there who still believes that science is an unbiased route to the truth,' says Ingram. 'It's not that scientists are dishonest; it's just that they start with different preconceptions. After all', he says, 'the answers you get are constrained by the questions you ask. Different scientific approaches can lead to different versions of the truth. Sometimes it's belief, not science, that is most important, until the data are overwhelmingly persuasive.'"

One of the best examples in which the data are overwhelmingly persuasive is the issue of the true origins and evolution of our hydrocarbon fuels: gas, oil, and coal via Internal Nucleosynthesis and polymerization; its many facts are irrefutable. Ingram hits on another basic truth about science and society: that it is difficult to interest people in learning about science unless you give them a practical reason for knowing it. At this writing (August 2001), the new "energy crisis" now threatening our economy stems directly from the antiquated Fossil Fuels hypotheses of the 1830s and 1920s: that gas, petroleum and coal are products of decay, marine sediments and plants, respectively, and thus, abound in very limited quantities. Nothing could be further from the truth.

The FLINE model's many provable facts about the origins, evolution and close relationship of these fuels should, and would, excite the public if permitted equal treatment by the news media and scientists, with financial backing equal to that of the antiquated Fossil Fuels hypotheses. Such fair treatment would go a long way in keeping any potential energy crisis at bay - and would give people a practical reason for learning about science. In any era of science, the greatest joy is discovery of a revolutionary truth, but tragedy too often resides in the reasons for, and the consequences of, its suppression.

Understanding the origins and evolution of hydrocarbon fuels can open many doors to understanding a cosmological science that is capable of predicting what we can observe. But first, one must be willing to let go of all antiquated beliefs and preconceptions about origins, while investigating the newer concepts. The results of all discoveries must be interpreted in perspectives newer and more factual than Poe's antiquated Big Bang. At this time, belief still reigns over science - but who knows for how much longer?

Letters and Memos

To: Glenn Strait, The World & I, Washington, D.C. **Jan 8, 1998**

Thanks for the press release dated 98-01-05 concerning the Lunar Prospector mapping the magnetic patchwork of the Moon, and for the statement that you are beginning to understand the FL/IN theory. The revelations in the press release, as well as other space findings (on Mars, etc.), continue to corroborate this new paradigm of planetary origins. *The patchwork of weak magnetic fields on the Moon's surface is a remnant of the strong magnetic field inherent in the energy mass at the birth of the Moon. As you may realize, all planets and moons began as fiery masses of nuclear energy with powerful electromagnetism inherent in each mass. This has been confirmed by the proposed Fourth Law of Planetary Motion, the principles of planetary evolution via natural laws (including the Laws of Thermodynamics), the recent findings on Mars (similar surface magnetic fields, rocks, dry river beds, volcanoes, etc.), and other space discoveries.*

This Moon project will simply add more confirmatory evidence of the prior existence of the sphere's original strong magnetic field that grew ever weaker as its core energy transformed into atomic matter that initially encapsulated it via atmospheric gases (e.g., see large planets), followed by crustal formations. This is the evolutionary pattern of all planets, including exo-planets, moons, comets and asteroids (a special case). All the smaller planets of our Solar System should now show similar patterns of remnant magnetism. And perhaps most of the irregularities of the remnant magnetism can be attributed to the last effects imposed on the surface by the dying energy core—via its final-stage gasps, so to speak. The discovery that "electrons were coming from the [Moon's] surface" raises some questions about their source. Do they come from outer space or from inside the sphere? My guess is the latter source rather than the solar wind source suggested in the article. Here we have two tentative conclusions, each based on a different perspective of the Moon's origin and evolution.

Can we look at Earth's field of electrons (in both perspectives) and find the correct answer? Perhaps they stem from both sources.

I agree with the statement that: "Magnetic field information also could provide constraints on the physical processes undergone by the Moon in its evolution, such as how the core formed, the thermal evolution of the crust, tectonic processes and erosion." This line of reasoning, if followed, inevitably will bring scientists ever closer to the FL/IN paradigm. Once this goal is reached early in the 21st century, the answers to the remaining anomalies of solar systems will fall into place, and science will benefit immensely from this breakthrough.

Two crucial keys will be the discovery and recognition of the true nature of comets and asteroids. I'm convinced that the FL is the most monumentally significant discovery about our Solar System since Kepler's discovery of the First Three Laws of Planetary Motion. Glenn, at what point along the way would it be feasible to publish the FL/IN paradigm as a serial article in your magazine? It does seem to me that with a scoop of this magnitude, at this stage of development, and with the persistent encroachment of confirmatory discoveries, time is of the essence.

P.S. A paper on the solution to the Fourth Law is being submitted next week to a leading journal.

To: Glenn Strait, The World & I Magazine, Washington, D.C. 20002. 01-26-98

One night last week, I watched with trepidation as a science program about comets unfolded. The plan is to land a small drilling mechanism on a comet and bring back samples for analysis. In the film, scientists already were designing the sampling mechanism. I'm not sure of the target date for the landing - it could be around 2007 or later. The plans are based on the assumption that comets are "dirty snowballs"—frozen leftovers from the creation of our Solar System. However, the evidence obtained during my research of the past 25 years on the subject of planetary origins clearly indicates that comets are fiery masses of nuclear energy. If true, the consequences of landing on such a mass will be monumentally significant, to say the least. If triggered, the explosion would be similar in force to those of fragments of Comet SL9 on Jupiter in July, 1994. But better a mechanism than a person.

However, in June, 2000, the Deep Space 1 probe is scheduled to fly past Comet West-Kohouteklkemura, taking close-up pictures and studying the coma surrounding the comet's dark nucleus. It may also view the two Martian moons, Phobos and Deimos, which I believe are actually shells of burned-out comets. In view of this information, it seems wiser to sample one of these moons rather than the comet. Certainly, it would be safer and easier to obtain samples representative of matter produced by a comet to form its dark shell. But, of course, this would not settle the issue of the true nature of comets; it is perhaps mandatory that the landing take place as planned to verify their nature. I only wish it could be done sooner. A more elaborate explanation of comets can be found in my book, *The Spacing of Planets*. Herewith are copies of pages 99-108.

To: Walter Anderson, Editor, Parade Magazine, New York, N.Y. May 11, 1998

I would like to comment on David Levy's article, *Was Chicken Little Right?*, concerning comets and asteroids. Substantial evidence reveals that a comet consists of a small shell containing an extremely powerful mass of energy capable of exploding instantaneously upon contact with anything as thin as the outer atmosphere of a sphere. Examples are: (1) the 21 explosions of fragments of Comet SL9 at Jupiter's cloud-tops in July, 1994; and (2) the upper-atmosphere explosion of the 1908 Siberian comet.

Asteroids, rather than comets, create the type of huge 100-mile diameter craters described by Levy. But such craters are far more often created by ejecta forces from within a sphere—which accounts for the multitude of craters on Mercury, Callisto, our Moon, and on surfaces of other spheres. Cogent evidence of the true nature of comets can be found in *Comets* by Donald K. Yeomans (1991), and in my books: *Undermining the Energy Crisis* (1979); *New Concepts of Origins: With White Fire Laden* (1986); *The I-T-E-M Connection* (1991); and *The Spacing of Planets: The Solution to a 400-Year Mystery*. My books also differentiate between comets and asteroids: comets are fireballs of "Sun-like" matter (Descartes' description of planetary cores)—actually Sun-like nuclear energy/matter. Cogent evidence clearly reveals that Asteroids are fragments of a disintegrated (actually 3) planet(s), just as Olbers stated in 1802. They can, and do, create large craters by crashing into larger masses. However, these are very rare events and, contrary to popular dogma, such collisions are not the cause of the vast numbers of craters on Mars, our Moon, Callisto, etc. As proven by Allan Mills in the 1980s, the vast majority of such craters caused by ejecta forces are detailed in my aforementioned books. Since each planet is a self-sustaining entity that creates its own atmosphere, crust, water, life, etc., via Internal Nucleosynthesis (IN), there is no need to look to comets or other outer space objects for sources of such matter. The key factor in the evolution of planets is that Internal Nucleosynthesis (IN) and Evolution (E) are inseparable: one cannot exist without the other. This is the basic underpinning of the proposed FL/IN/E paradigm of origins that is now strongly supported by my proposed Fourth Law of Planetary Motion (FL), detailing how the fiery masses that evolved into planets attained their geometrically spaced orbits around the Sun and why the last two planetary orbits are unorthodox.

The comet crashes on Jupiter were indeed "the most incredible explosions ever seen on another world". Each powerful explosion far exceeded all prior supercomputer calculations that were based on the "snowball" hypothesis. Instead of the predicted small, orange flashes one-fiftieth as bright as Jupiter, the huge, white-hot flashes were up to fifty times brighter than Jupiter. Fragment G alone yielded at least six million megatons of energy. To expend an equal amount of energy, a Hiroshima-type bomb would have to be detonated every second for ten years! These powerful explosions cast serious doubt on the "snowball" comets hypothesis while conclusively confirming my 1970s prediction of their true composition: nuclear energy. Supportive evidence of the latter removes the remnants of any doubt.

The several iridium layers (and all other layered matter throughout geological history), including the layer marking the demise of the dinosaurs, can be explained far more logically by the FLINE paradigm than by the alleged collisions of missiles from outer space. Substantive evidence proves that the dinosaurs actually died out over a five-hundred-million-year time frame as a result of atmospheric changes attributed to a dynamic, evolutionary Earth driven by the energy of its internal engine.

While the prevailing hypotheses make exciting reading and movies, they do grave disservice to true science, to present-day science students and to posterity. Ironically, the more factual truth about our planetary origins and evolution creates even more exciting scenarios. True science should remain open to presentations of both sides of any controversy. It is indeed unfortunate that a naive public hears and sees only the less factual side of these issues rather than the full story.

To: Glenn Strait, The World & I, Washington, D.C. 20002. 07-03-98

Thanks for the news release from Seattle's University of Washington's Professor Donald Brownlee, forwarded to you by Steve Maran, American Astronomical Society (AAS). I noted with interest that Maran disclaimed endorsement by the AAS; unfortunately, the reason was not specified. I am disappointed that the Stardust craft will pass only 75 miles from the main body of the comet, Wild 2, rather than attempting to land on it. It is not likely that capturing samples of the dust particles and ionized gas from the comet will reveal anything beyond what is already known about the composition of cometary ejecta. The problem is, and always has been, in interpreting the data obtained from previous analyses of such ejected matter. When correctly interpreted, the findings argue strongly against the "dirty snowball" belief, while offering even stronger evidence favoring the fireball concept.

The most effective way to gain any real insight into the true nature of the comet is to carefully measure the temperatures in and around the nucleus and the coma. However, the mission is being run on the false assumptions that "comets are about equal parts ice and dust", and thus, "the particles will be cryogenically preserved interstellar dust left from the birth of the solar system. ......such grains can be found only in the outer solar system because heat has destroyed them nearer the Sun." How does Brownlee compromise the contradictory fact that many comets have the Sun as one focus of their orbits? Dust particles gathered from outer space and from different levels in our atmosphere naturally coalesce into larger particles, which is the nature of such lightweight matter. However, this reasoning cannot be carried over to apply to massive boulders (asteroids, etc.) and comets coalescing to form larger masses.

In real life, such larger spheres are formed in full accord with the FLINE paradigm's three chronological, inseparable and ongoing events: The Four Laws of Planetary Motion (FL), Internal Nucleosynthesis (IN) and Evolution (E). In any active sphere, IN and E are inseparable; one cannot exist without the other. In our Solar System, the Four Laws of Planetary Motion must be added as the initial (and ongoing) factor that accounts for all motions of the planets, including their orbital spacing, thereby adding substantive evidence to this revolutionary paradigm (see enclosed flyer for more details). At least Brownlee got the name right: Stardust—truly the ejecta spewed from every fiery star, sun, early-stage planet and active comet throughout the Universe can be called stardust.

However, its analyses are misinterpreted continually as force-fitted evidence that comets are pristine snowballs rather than the fireballs in which the virgin dust particles and ionized gases are created via Internal Nucleosynthesis. As revealed by the extremely powerful explosions of Comet SL9 at Jupiter's very thin cloud-tops in 1994, newly-formed comets are easily triggered into nuclear explosions. Rather than "the preserved building blocks of the outer planets", comets are ejecta energy from larger masses of nuclear energy comprising all stars. Perhaps Brownlee will "find clues about the formation of the Solar System and perhaps of the Universe itself", but they must be interpreted accurately. Monitoring the temperatures and, perhaps, triggering an explosion of the comet could add immeasurably to these interpretations. In time, why the AAS withheld endorsement of the project will become clearer. The results of the $200 million mission could alter the direction of scientific thought on the origins and evolution of planets.

To: Glenn Strait, The World & I Magazine, Washington, D.C. 20002. **07-08-98**

Thanks for the information received today on our Solar System's hottest planetary surface: Moon Io—truly exciting news. In answer to your questions concerning the fiery nature of Io versus the quiescent nature of our Moon (and Mercury) (three spheres of approximately equal size), the FLINE paradigm offers this reasoning: Planets evolve through five common stages of evolution in accordance with size. However, our Moon and Mercury are entering the fifth and final stage (inactive), while Io is only in the fourth stage (rocky), but is not as far along in this stage as Earth's evolution. Io's appearance, activity and high temperature now are similar to those of Earth during that specific time frame for the early part of our planet's fourth stage of evolution. With this information in mind, logic dictates that some factor(s) in addition to size played a crucial role that accounts for these differences in the stages of evolution of these three spheres. The options: (1) Jupiter is playing an active role in creating tidal friction heat in Io; or (2) Jupiter played some other crucial role in the past. The first option represents the prevailing belief, *which can be readily nullified by a simple analogous comparison of the Jupiter-Io system to the Sun-Mercury system in which no tidal friction heat is created in Mercury.*

The second option fits neatly into the FLINE paradigm: Io is much younger than both our Moon and Mercury; in reality, it is the youngest moon of Jupiter and is in the smallest orbit of Jupiter's four large moons. As such, we can reason that a later birth of Io accounts for its later start in the evolutionary time frames. But how and why was Io born at a later time than our Moon (and Mercury)? And is there evidence to support this theory? Based on substantive evidence, the FLINE paradigm teaches that Earth and Moon are contemporary spheres, while Jupiter and Io are akin to a mother-and-child relationship. Some time after the original layout of the Solar System (in full accord with the Four Laws of Planetary Motion), and after Jupiter had entered its second (gaseous) stage of evolution, the original energy mass we now call Io was ejected from Jupiter's gigantic energy core and sent into orbit around, and close to, Jupiter. Then came nucleosynthesis in the fiery masses: stars, suns, planets, brown dwarfs, etc., throughout the Universe.

As you know, nucleosynthesis is the transformation of energy into matter ($E=mc^2$) under conditions of high temperatures and pressures within a nuclear energy mass. Such atomic matter can be observed on all fiery masses, including our Sun, stars, planets, moons, comets, etc. In all these cases, active E is possible only via active IN.

I hope this answers your request of how I explain the data presented in the news release. In summary the FLINE model's Four Laws of Planetary Motion (FL) placed the fiery planetary masses in GR-spaced orbits during their initial stage in which Internal Nucleosynthesis (IN) begins driving their five-cycle Evolution (E). Comets are smaller versions of these processes; CO is an IN product common to both.

To: Glenn Strait, The World & I Magazine, Washington, D.C. 20002. 07-07-99

Thanks for the AAS News Release received yesterday concerning Comets Wirtanen, Hale-Bopp and Halley. The two common threads throughout are: (1) the relationship between the comet's distance from the Sun versus coma size; and (2) the jet-black color of the nucleus. In the perspective of the snowball belief, the standard explanation of the increase in coma size as the comet approaches the Sun is well-known. But several questions persist: How can a coma consisting of hydrogen up to one million miles in diameter form a spherical envelope that continues to grow around such a tiny nucleus when approaching against the Sun's solar wind? Why does the coma remain so large at such great distances from the Sun - well beyond its warming effect?

For example, Comet Halley's coma was almost 500,000 km at a distance of 8.5 AU from the Sun (almost to Saturn), and Comet Hale-Bopp's coma at 8.66 AU was more than a million km across, or nearly ten times larger than Saturn! Why do ejecta forming the hydrogen coma continue at these great distances? If a snowball, why is the nucleus always jet-black? Will Comet Wirtanen's nucleus at its aphelion show through the telescopes as a black dot or a brilliant dot? When comets are viewed as energy masses, or fireballs, as depicted in the FLINE model of planetary origins, logical answers to these questions simply fall into place. Too long for this letter, they can be found in my book, *The Spacing of Planets: The Solution to a 400-Year Mystery* (Chapter IV: *Moons, Planetary Rings and Comets*, beginning on page 87). One last comment: I was relieved to learn that the missions to comets do not include a plan to try to land a man on Comet Wirtanen. From the viewpoint of safety and discovery, the instrument landing, per se, should be exciting enough.

Lunar Prospector Findings on the Moon's Metal Core

In the AAS press release of August 10, 1999, the authors are in agreement with the FLINE model in their conclusion that the Moon's core is smaller than the cores of the inner planets of the Solar System (SS). One principle of the new model is that the sizes of all cores of planets and moons created simultaneously as a solar system are proportional to their total mass: the smaller the mass, the smaller the core.

At the other end of the scale, the largest planet, Jupiter, at our SS's geometric mean, has the largest core. The initial reasons for the various sizes of the planetary masses (and their corresponding core sizes) are explained by the forces acting on the decelerating smaller mass as it acted in unison with the Sun during the dynamic layout of our Solar System in full accord with the Four Laws of Planetary Motion some five billion years ago. Size and distance from the Sun are the two principal factors explaining the compositional differences among the planets and moons as they evolve(d) through their five common stages of evolution. Meanwhile, the core of each planet dwindles as a function of time - a consequence of the slow transformation of its energy mass into the sphere's compositional matter.

The core-size evidence in the press release strongly supports the FLINE model. Unfortunately, the core size and composition of the Moon are the only conclusions in agreement with the FLINE model. The conclusion drawn from the results of the magnetic tests is premature: how can one differentiate between the actions of the magnetism of a small metal core and a small nuclear energy (or plasma) core solely on the basis of these findings? In the FLINE model perspective, whether this evidence is indicative of remnants of a small nuclear core or of a metal core remains debatable. If researchers would only consider possible scenarios for interpreting their findings!

On previous occasions, we have discussed the highly speculative hypothesis that a Mars-sized object collided with Earth and bounced away to become the Moon. Were all the moons of other planets made in similar manner? If so, how could so many moons have bounced off the thin, deep clouds of Jupiter? The FLINE model teaches that all spherical moons were created and evolve(d) in a manner like that of the planets. As one scientist put it, "Books and articles supporting this giant-impact hypothesis of lunar origin are more a testament of our ignorance than a statement of our knowledge." Full details can be found on pages 87-94 in my book, *The Spacing of Planets: The Solution to a 400-Year Mystery*; they present powerful arguments for a different and definitive origin of our Moon and other moons.

To: Glenn Strait, The World & I Magazine, Washington, D.C. 20002. 11-22-99

Thanks for the AAS Press Release (Nov 17), *Jupiter May Have Moved in Toward Sun, Probe Data Suggest*. My thoughts are that two responses would be helpful: one to you as well as the one to the researchers, as suggested, through you, and enclosed herewith. Please feel free to improve its composition where feasible. It was encouraging to read that "the Galileo spacecraft's suicide probe as it plunged into Jupiter's rolling atmosphere in December, 1995 has stamped a huge question mark over the prevailing models of how our solar system began." In my 1995-1996 book, *The Spacing of Planets* (p 67), I had written a similar message: "Just as Galileo himself drove the dagger into the heart of the Ptolemaic version by promoting the Copernican idea of a Sun-centered Solar System, the Galileo orbiter is plunging a dagger into the heart of the prevailing dogma about planetary origins."

It is gratifying to learn that scientists are now beginning to question prevailing models—the first critical step toward understanding and eventually recognizing the truths inherent in the FLINE model of planetary origins and evolution. The next step will be the realization that the Accretion hypothesis is fundamentally flawed and should be discarded from textbooks and replaced by the more logical and factual model. Your stated realization that the reported data fit in well with the FLINE model is good news. Indeed, the large quantities of heavy noble gases in Jupiter's atmosphere are precisely as predicted and explained by this revolutionary model for any planet in the second stage of planetary evolution via Internal Nucleosynthesis (IN). While the overwhelming evidence for IN is recorded throughout the book, it is concentrated more in Chapters II & III - some 42 pages of what appears to be irrefutable evidence of IN.

The Fourth Law of Planetary Motion, explained in Chapter I of the book, shows that Jupiter has moved only 1.46 AU, or 21.3%, closer to the Sun since its origin some five billion years ago—not enough to be a factor in the speculations of the researchers concerning its orbital changes or movements from far distant places. As based on the Accretion hypothesis, the illogical explanations offered in the report explaining Jupiter's origin and orbital changes call for more and more speculations. It reminds me of the tale about the three blind men, each feeling and describing a different part of an elephant, and each arriving at a much different - and entirely wrong - conclusion about it. One must ask how long this situation can go on before scientists do seriously "question.... prevailing models of how our Solar System began".

None of the wild speculation in the report would have been necessary if only they had been advocates of the viable alternative: the definitive FLINE model of planetary origins whereby all of Jupiter's anomalies can be explained easily, and proven conclusively, without the need for speculation. However, it does appear that more and more scientists are leaning in its direction, whether or not they have heard about my writings of the past 26 years. After 22 months, my manuscript on the solution to the proposed Fourth Law of Planetary Motion is still in the hands of referees at the Royal Astronomical Society, London. I believe that if they could find something fundamentally wrong with it, an early (and eager) response would have been received. One can only wait and hope.

To: Brig. Gen. Gordon C. Carson III, Livingston, TX. Feb 3, 2001

Thank you for your letter of January 28 concerning the readings about the old and new concepts of origins and evolution, and for your words of encouragement to keep up the good work. The responses you have received from colleagues reading *The Spacing of Planets* sound much as expected: "mostly good and some not-so-good". I imagine the good comments came from readers who read with open minds void of any strong bias favoring prevailing beliefs. The not-so-good comments probably came from strong believers in the popular Big Bang/Accretion version of universal origins. This, of course, is understandable when we look at the history of significant breakthroughs in opposition to beliefs of their time. But time and facts remain steadfast and powerful allies of truth.

I would be genuinely surprised if any of the unfavorable comments contained a valid rebuttal of any facet of the LB/FLINE model. If so, it would be the first one since conception of the initial facet of the concept in 1973: a theory solidly constructed with irrefutable evidence, against the erroneous concept of "fossil fuels" and in strong favor of the energy fuels concept of the origins and close relationship of these hydrocarbon fuels. In the past, the only rebuttals given were simply, "I don't believe it." - always said before showing a willingness to take the time to learn about its volume of supportive, intertwined facts. The concept needs a rebuttal or two if any weak points in the LB/FLINE are to be found, and strengthened or clarified. If you do receive a definitive one, please forward it to me for this essential purpose.

Once one understands the basic principles of how these fuels came into being, it is relatively simple to understand how all matter came into being via the same principles. Simply by applying Nature's laws of physics and chemistry and correctly interpreting observations and relevant data, one can arrive at pieces of the puzzle that always interlock precisely with the previously fitted pieces. Unfortunately, proponents of the Big Bang/ Accretion beliefs remain firmly in control of the mind-set of most scientists and the general public. That hold isn't likely to diminish in my time.

To paraphrase the astute observation of a great physicist, Max Planck, science must advance slowly, one funeral at a time. Not having a Ph.D. with a mind trained in prevailing beliefs did offer a big advantage in being able to see things in new perspectives. Chemistry and chemical engineering degrees, along with my career in industrial R&D (applied), problem solving, etc., boosted and broadened that tremendous advantage, without which I would not have been capable of fitting all the pieces together - even with a Ph.D. Fate did play a major role.

To: Glenn Strait, The World & I Magazine, Washington, D.C. 20002. 03-12-01

Thanks for the four e-mails concerning Comet Hale-Bopp, Ganymede, Io and "bucky" balls. The printing from this end was faulty, but I will try to decipher it well enough to supply answers in accord with the FLINE model. Meanwhile, Kay is installing a new printer that better matches her new computer.

The news release that Comet Hale-Bopp, now at a distance halfway between Saturn and Uranus, retains its huge coma and is still spewing trails of dust, gas and chemicals—precisely as predicted and explained by the FLINE model—is truly exciting (Ref: *The Spacing of Planets,* p 103,1996). Quoting the primary reason: "Without the continuous activity of nucleosynthesis in these distant comets, the existence of comae and the production of gases and dust would not be possible [at great distances from the Sun]. Perhaps some day soon, improved detection methods will enable astronomers to detect and verify that...gases and matter do continue spewing jet-like from comets in the deep, cold dimness of the outer SS - just as predicted in 1973 by the IN theory. This will be another giant step toward proving the FLPM/IN concept of our origins."

Comets are fireballs, not dirty snowballs; they played no vital role in the formation of planets. The interpretations researchers give for the observations of Ganymede's surface features are bringing them ever closer to the FLINE model. But they need to go a giant step further by identifying the driving force behind these ongoing changes in its surface features: the fiery energy core, the nucleosynthesis engine that basically drives all planetary changes, volcanism, evolution, life, etc.

The same principles quite naturally apply to younger Io, whose dramatic differences are explained by the new model. Becker *et al.* should consider how "bucky" balls are made in the carbon-black industries that furnish carbon-black to tire manufacturers and other industries - and we live in a carbon-dominated world. Earth's internal manufacturing system is quite capable of making "bucky" balls containing trapped noble gases; there is no need to claim that they came from outer space. To do so, researchers are making the same mistake made by proponents of the belief that our iridium layers came from outer space, or that the Moon's titanium came from outer space. Quoting again from *The Spacing of Planets,* p 93, "The titanium produced on the Moon is analogous to the tellurium produced on Venus, the iridium produced on Earth, the magnesium produced on Neptune, and the abundant metals observed at the impact site of the G fragment of Comet SL9 on Jupiter; all are in situ ejecta products of nucleosynthesis."

If from outer space, these metals would not have been so selective of landing sites; each is the result of internal conditions that prevailed during each sphere's time-frame for making and ejecting a specific chemical. In addition to the manuscript on the solution to the Fourth Law of Planetary Motion and its FLINE model, which was mailed to you on March 12, I have a short manuscript, *Comets and Asteroids: Keys to Our Origins?*, which details the complete story about their true nature, origins, etc. If you would like a copy for consideration to publish, just let me know.

From: Amanda McCaig, Editorial Assistant, Astronomy & Geophysics, The Journal of the Royal Astronomical Society, London. 2001-03-28
Re: On the spacing of planets: The proposed fourth law of planetary motion - our ref. qy 016

We apologize for all the delays in handling your paper, which we accept fell below the standard expected of any academic publication. However, we have had enormous difficulty in finding anyone willing to referee your paper. Both the eventual referees agree that your paper is not suitable for publication in A&G. To be specific, we must reject your paper on the following grounds: 1.The author does not cite any modern scientific work other than his own ideas; 2. It is not clear what deficiencies in currently accepted theories the author wishes to address; 3. The author's ideas are presented as assertions of fact without logical proof or any supportive observations. Any paper that does not meet these academic conventions must be regarded as fundamentally flawed.

We cannot, therefore, accept this paper for publication in A&G. Please note the Editor's new address is: Dr. Sue Bowler, Editor, Department of Physics & Astronomy, University of Leeds, Leeds LS2 9JT Department of Earth Sciences, The University of Leeds, Leeds LS2 9JT.

To: Amanda McCaig, Department of Earth Sciences, University of Leeds, UK. April 10, 2001
Re: qy016 On the Spacing of Planets: The Proposed Fourth Law of Planetary Motion.

Thank you for the valuable feedback on my paper submitted in January, 1998 to the Royal Astronomical Society for peer review and publication. I had predicted openly that you would have the enormous difficulty, as you stated, in finding anyone willing to referee the paper; the reason may well be that it is too revolutionary, somewhat analogous to the situation when Copernicus first placed the Sun at the center of the Solar System. The Fourth Law that first eluded Kepler in 1595 will have an equal impact on changing the way scientists view the origins of solar systems and the evolution of planets and moons.

I especially appreciate your giving me the three specific reasons for rejection of the paper. Please allow me to offer some thoughts on each one in numerical order:

1. During the 23 years of carefully separating myths from facts and finding the solution to the elusive Fourth Law, I was equally dutiful in rendering proper credits in the body text to the person(s) responsible for each discovery utilized in the paper. All of the main ideas were mine, which, again, like the heliocentric idea of Copernicus, needed no contributions from the modern beliefs it opposes and exposes. Keep in mind that this is a revolutionary alternative to accepted theories, and like the Copernican idea, it is a definitive concept that, on its merits alone, will stand the test of time. It is the only known concept capable of furnishing provable solutions to all planetary anomalies.
2. The paper was already too long to leave room to discuss the many implied deficiencies of current beliefs: e.g., Laplace's Accretion concept and any interpretation derived via Poe's 1848 Big Bang. Both concepts are widely known and needed no comments other than those implied by the LB/FLINE model. An honest effort was made to stay within the bounds of the Four Laws of Planetary Motion proving the fiery dynamic origins of solar systems and, subsequently, the five-stage evolution of planets (including exo-planets) and moons.
3. Quite to the contrary, my ideas were not assertions of fact without logical proof or any supportive observations. Logical proof resides in the Seven Diagrams clearly revealing how each nebulous planetary mass attained its original and current orbital spacing—no other concept can offer a definitive solution to this mystery.

These Diagrams required 15 years and much logic based on a research background and the ability to separate myth from fact. Once in orbit, fiery spheres follow the second law of thermodynamics as they cool through their five stages of evolution. Mankind has only to look to the skies to observe each stage of planetary evolution. Convincing proof of the Four Laws and the subsequent planetary evolution resides in the absolute fact that Internal Nucleosynthesis is essential to Evolution; one cannot exist without the other. The supportive observations of planetary evolution via internal and external discoveries of all planets are voluminous. Everything evolves continually and in full accord with natural laws; there are no known exceptions. Logic and voluminous supportive evidence in my book, *The Spacing of Planets: The Solution to a 400-Year Mystery,* provide sound assurance that the FLINE model contains no fundamental flaw.

The proven origins, evolution and close relationship of hydrocarbon fuels (1973-1980) are facets of the FLINE model—an example of a precise and logical concept with voluminous supportive evidence. It is unfortunate that this definitive LB/FLINE model, again as with Copernicus, offers such powerful evidence for a viable alternative to prevailing beliefs. In any era of science, the greatest joy is discovery of a revolutionary truth; however, tragedy too often resides in the reasons for, and the consequences of, its suppression. If anyone in the RAS can find a fundamental flaw in this new FLINE concept, please make me aware of it.

Memos
A Question of Ethics in Science

The third part of a series of four videos on *Savage Planet*, entitled *Deadly Skies*, was shown on GPTV, PBS on November 19, 2000, from 7:00-8:00 p.m. The featured astronomers included David Levy and Carolyn Shoemaker as experts on asteroids and comets. For the first time ever, the presentation actually attributed the source of asteroids to an exploded planet that once existed between Mars and Jupiter. Since that event - the most powerful explosion in the history of mankind - the resulting fragments have been falling periodically onto Planet Earth for an unknown number of years.

This sudden about-face conclusion of scientists raises questions: Are the asteroids a product of the Big Bang (BB) (as has been the standard claimed for some five decades), or are they actually remnants of a disintegrated planet, as first stated by Olbers in 1802—and a fact that has been a vital principle of the evolving LB/FLINE model since 1980? What do the asteroids tell us about the origin of our Solar System? Do planetesimals (asteroids) really accrete into planets or do they break into smaller pieces over time (as stated in the video)? Why this sudden acceptance of asteroids as remnants of an exploded planet after so many decades of teaching and promoting their creation via the gaseous-dust accretion concept? Could such original gas and dust have evolved previously from the alleged Big Bang?

The answers to these questions, as supplied by the LB/FLINE model of origins during the past two decades, speak loudly and clearly against Poe's BB hypothesis (1848) and its inherent, but antiquated, Accretion concept initiated by Laplace in 1796. Simultaneously, the model proves that Olbers was correct in his assertion that asteroids are fragments of an exploded planet (actually three spheres). Remember that the fragments are clearly in three definitive orbits between Mars and Jupiter.

The fact that scientists now accept another facet of the LB/FLINE model (the true nature and source of asteroids) is a giant step in the right direction. The next giant step will come when they eventually recognize and admit the true nature and sources of fiery comets.

The second surprise came when the video showed much more powerful explosions of Comet SL9 on Jupiter than had ever been depicted in previous programs on the subject. To their credit, they did not yet go so far as to admit that these powerful explosions were of a nuclear nature (as proven by the FLINE model), but the enhanced pictures gave that impression. Eventually, astronomers will be forced to admit the nuclear nature of comets, and to understand the sources of their ongoing origins. The acknowledgement of their nuclear nature, along with this long-overdue recognition of the true nature of asteroids, clearly will undermine both the BB and the Accretion hypotheses.

Of all the questions raised here, one issue stands out conspicuously: Why the sudden switch from the belief that asteroids are early accretion products of the BB to the belief that they are remnants of an exploded planet? Why does science continue to ignore and reject the LB/FLINE model in spite of its substantiated evidence and its impeccable record of accurate predictions during the past two decades? *Deadly Skies* and similar TV programs continually bring scientists ever closer to the LB/FLINE model of origins and evolution.

The question of ethics in science eventually must arise: At what point will it become obviously unethical to continue ignoring the LB/FLINE model described in this book and in previous publications of the past two decades, while continuing to "discover" its inseparable principles of creation via ongoing universal nucleosynthesis and its accurate interpretations of relevant data; e.g., the nuclear explosions of Comet SL9 on Jupiter, predictable neither by any other known concept nor by the world's most powerful supercomputers? Is this encroachment method ethical or unethical science?

Addendum:

The *greatest obstacle to advancement in the sciences of origins and evolution is the illusion of knowledge. If you had to identify, in one word, the reason scientists have not achieved their goal of understanding universal origins and evolution, the word would be "consensus". In science, the greatest joy is discovery of a revolutionary truth; the saddest tragedy resides within the reasons for, and the consequences of, its suppression.* Alexander Scarborough, Jan-Mar, 2001

NPA Talk, UConn, Storrs, Ct. June 9, 2003

New Principles of Origins and Evolution: The Imminent Scientific Revolution

Before his untimely death from cancer, Carl Sagan stated, "Science is an ever-changing set of myths." Stephen Hawking later added, "Cosmology's hypotheses are conjectural when pressed to the limit."

Many scientists recognize truth in these bold statements of Sagan and Hawking, but most remain unaware of viable alternatives capable of changing the way scientists think about our origins and evolution. Quoting from the NPA Newsletter, "If anything is clear in Thomas Kuhn's familiar book, *The Structure of Scientific Revolutions*, it is that no revolution has occurred nor will occur in the future unless there is a proposed alternative theory to replace the older one." Developed over a period of 30 years, my paper today offers a valid alternative theory: The LB/FLINE model of origins and evolution; i.e., the Little Bangs/Four Laws of Planetary Motion-Internal Nucleosynthesis-Evolution. These four inseparable, ongoing principles offer definitive knowledge of the origin and functions of our Universe.

The new Fourth Law of Planetary Motion is the crucial discovery that permits us to see clearly how and why everything is connected to everything else in the Universe. As we shall see, this new model is supported by a plethora of well substantiated, overwhelmingly persuasive evidence. Underpinning the laws of Nature is Huntley's book, *The Divine Proportion: A Study in Mathematical Beauty*. In it, for example, the successive chambers of the nautilus are built on a framework of the logarithmic spiral based on Phi, the Golden Ratio that pervades throughout Nature.

Scientists know that the Universe and its contents operate under specific Laws, which are being discovered at a reasonable rate. Understanding these Laws will enable scientists to reach a definitive understanding of the powerful forces that continuously create the beauty of our Universe. Keeping in mind that everything evolves, and nothing is as it once was, let's examine some of these Laws governing the forces that drive that ever changing beauty.

The LB concept envisions a spherical Universe expanding at the speed of light. In the state of absolute zero at its perimeter, energy particles, both old and newly formed, collapse into the densest form of energy: black holes. Eventually, each black hole is triggered into spewing from its two poles a pinwheel-shaped galaxy of white-hot energy masses of various sizes which we identify as fiery stars, gases and dust particles.

Thus, the observed acceleration of speed of universal galactic systems as astronomers look ever farther outward is a natural phenomenon of the LB concept; no lambda factor is necessary—Einstein correctly called it his greatest blunder. Additionally, the LB concept eliminates the need for the Accretion hypothesis. The LB/FLINE model also offers a definitive solution to the mystery of the 2.7 K background radiation. Quoting the *Encyclopedia of Physics* (1991): "Every object in the universe continuously emits and absorbs energy in the form of electromagnetic radiation...[which] originates from the internal energy associated with atomic and molecular motion and the accompanying accelerations of electrical charges within the object."

The LB/FLINE model attributes this internal action to the Internal Nucleosynthesis within closed systems (e.g., planets) as perfect emitters and absorbers that radiate into a vacuum; i.e., 2.7 K blackbody radiation. After the fiery birth of galaxies is visualized, we can concentrate on the subsequent FLINE model in which various sizes of the dynamic energy masses of a galaxy interact via gravity before evolving into solar systems with one or more planets, moons, and other gaseous and solid bodies revolving around a central star. Astronomers have discovered more than 100 extra-solar systems far from our Solar System. Too close to their central star to have accreted there, the planets of those extra-solar systems obviously could have formed only via the LB/FLINE concept—which also explains the huge numbers of binary-star systems in galaxies. The FLINE model consists of five monumentally significant, intimately interwoven discoveries about the dynamic origin of our SS and the evolution of its planets and moons:

1. Our heliocentric SS by Copernicus.
2. The dynamic First Three Laws of Planetary Motion* by Kepler.
3. Gravity by Newton.
4. Internal Nucleosynthesis/Evolution (IN/E)** by Descartes, Einstein, Scarborough, *et al.*
5. The new Fourth and Fifth Laws of Planetary Motion (FL)* tie together all five principles in ongoing, inseparable relationships of IN and E.

Notes: (*FL includes all Five Laws of Planetary Motion.) (**IN/E: Internal Nucleosynthesis/Evolution are inseparable in all active spheres; one cannot exist without the other.) A vast amount of substantiated evidence supporting these five intimately interwoven principles reveals a clear understanding of the dynamic origin of our SS and the ongoing five-stage evolution of its planets and moons.

The FLINE Model consists of Three Chronological, Inseparable and Ongoing Realities of Our SS:

1. The Five Laws of Planetary Motion (FL).
2. Internal Nucleosynthesis (IN).
3. Evolution (E).

Next, we shall see a good example of how all matter evolves via Internal Nucleosynthesis; i.e., the ongoing transformation of nuclear energy into atomic elements in accord with $E=mc^2$, and subsequently, into molecules that comprise all planetary matter. This example of the origin of Earth's hydrocarbon fuels—gas, petroleum, coal—makes it easy to understand the ongoing creation of all planetary matter. This picture shows the three-layer system of HC fuels in Earth's crust. This system was confirmed in a research project at Georgia Tech in 1985-1986, led by Dr. Karam, head of the Nuclear Engineering Department. In ascending order, gases first form deep within Earth's crust, while petroleum forms at intermediate depths, and coal is on, or within one mile of, Earth's surface. Let's see why this pattern is crucial to understanding our origins and evolution! After the creation of C & H atoms via Internal Nucleosynthesis (i.e., the transformation of nuclear energy into atoms), four H atoms, with a natural affinity for joining with one C atom, form one molecule of the countless members of CH_4 molecules comprising the tremendous supplies of methane gas found worldwide in Earth's crust. Through polymerization, these methane molecules can link into chains of ever heavier gases, starting with ethane and moving into propane and butane. The next stage of polymerization is pentane, a lightweight liquid always found in lightweight petroleum. Additional polymerization forms ever longer carbon chains of heavier-weight petroleum.

Scientists know that every drop of crude oil contains tiny particles of coal. Eventually, much of the crude oil erupts onto Earth's surface, inundating low-lying swamplands, encapsulating live plants, and filling tunnels in which it subsequently polymerizes further, cross-links and solidifies as beds and veins of coal. We can summarize in this manner: The Theory of Internal Transition of Energy to Fuels. Energy--Atoms--Molecules--Gas--Oil--Coal. (nuclear), (C, H, etc.), (CH_4, etc.), (methane), (oil), (coal). We can then broaden the concept to include all universal matter: The Universal Law of Creation of Matter. Energy--Atoms--Molecules--Gas--Liquid--Solid. Note: In both Laws, all reactions are reversible by mankind. From these findings, we can conclude that HC fuels were not made from fossils, nor did they come from outer space (as claimed in the article, *Mining for Cosmic Coal* (*Astronomy,* June 2002). This article, stemming from BB beliefs, is an excellent example of the sophistry in many publications supporting prevailing hypotheses—exactly as predicted by Kuhn in his 20th century book, *The Structure of Scientific Revolutions*. Along with an impeccable record of accurate predictions, *The Spacing of Planets* and *New Principles of Origins and Evolution* present much greater detail and far more substantiated evidence for the inseparable connection between Earth's beauty and its Internal Nucleosynthesis. Upon depletion of Earth's nuclear energy core, the beauty of Earth will fade away; our planet will become as barren as Planet Mars. (Record: methane hydrates, hydrocarbon fuels, extra-solar systems, binary-star systems, accelerating universe, the tremendous power of Comet SL9). As science moves ever deeper into the inevitable Scientific Revolution initiated by Copernicus, and brought full circle by the LB/FLINE model, these revolutionary principles of origins and evolution will be without the need for conjecture or myths. Filling in the vast amount of remaining details will present many golden opportunities for all scientists to participate in the imminent scientific revolution.

Comets and Asteroids: The Differences

June 8, 2004

The movie drama, *Deep Impact*, shown on CBS, June 5, 2004 from 8:30-11:00 p.m., made it crystal clear that a vast majority of the public, along with many scientists, remain confused over the differences between a comet and an asteroid. The plot was to send a manned spacecraft to an incoming comet, land on it, bore holes in the nucleus, and plant explosives that would blow the comet off its collision course with Earth. The serious problem here? As was seen in 1994, 21 fragments of Comet SL9 crashed one by one onto the thin cloud-tops of Planet Jupiter, each resulting in a tremendous explosion; comets are nuclear fireballs easily triggered by the slightest contact with clouds.

My accurate predictions stood alone because they were based on the nuclear nature of comets. All other predictions, even those made by the best computers, failed because they were based on the "dirty snowball" myth that yet prevails at this writing. As correctly identified by Heinrich Olbers in 1802, asteroids are the solid remnants of a disintegrated planet (actually three smaller spheres). The chances are nil that any of them contain nuclear energy. Those that might have once contained it should now be inactive, with tell-tail signs that identify them: holes, long, smooth troughs, craters, etc. Two prime examples are the moons of Mars: Phobos and Deimos. Asteroids should pose no explosive threat to anything landing on them.

On January 8, 1866, Angelo Seechi made spectroscopic observations of Comet Tempel 1, and reported the presence of the three bands that Donati had seen in Comet 1864 II. Their spectroscopic observations of Comet Tempel I in 1866 clearly showed it was shining by both reflected light and glowing gas. Seechi concluded that at least some of its light was intrinsic, not reflected sunlight (Ref: Yeomans, Donald K., *Comets: A Chronological History of Observation, Science, Myth, and Folklore*, 1991). These observations by Donati and Seechi clearly revealed for the first time that Comet Tempel I was in its final burnout stage of evolution from a nuclear energy mass into a low-density asteroid much like Phobos and Deimos, the two moons of Mars. This explains why the delayed nuclear explosion on Comet Tempel I, although huge, was relatively small when compared to those of Comets SL9 (1994) and Halley (Feb 1991).

The Deep Impact Spacecraft to Comet Temple 1: Prediction and Reasons

June 27, 2005

On July 4th, Americans should witness some cosmic fireworks when the Deep Impact spacecraft will fire a 370-kilogram copper missile at Comet Tempel 1. The results will depend on two things: (1) hitting the target; and (2) the age and evolutionary stage of the comet. Assuming a direct hit, one of three things should happen: (1) a very powerful explosion will occur if the comet is still in its active energy stage; (2) if now an older, inactive and loosely bound object, the comet might shatter into pieces; or (3) if solid enough, the comet should survive a deep hole, perhaps 10-15 feet in depth, blasted by the missile.

The three choices are listed in their order of probability; each is based on the two conditions mentioned in the first paragraph. The best history of comets can be found in the book, *Comets* (1991), by Donald K. Yeomans. In it, he credits the 1794 publication of Ernst Chladni as perhaps the first to recognize the cosmic origin of comets as fireballs and the generic connection between fireballs, meteorites, and meteors. Chladni also proposed that certain metal-rich masses, such as iron meteorites, could not have been produced by processes on Earth. In 1866 William Huggins observed that Comet Temple 1 was shining by its own light [glowing white-hot gas] as well as by reflected sunlight. Periodic Comet Temple-Tuttle had been observed in 1366 by astronomers in China, Japan and Korea. Upon its return in 1822, Comet Encke's motion (its orbital size and period) appeared to be decreasing with time. In 1835, Friedrich Bessel observed sunward jets, or emanations, arising from the head of Comet Halley in 1835. He reasoned that these rocket-like emanations, whatever their cause, must affect the motion of the comet itself. The cause is now recognized in the new LB/FLINE model as the fiery nature of active comets that are simply ejecta from larger energy masses. Comets are relatively short-lived, with active life cycles up to several thousand years. Each burned-out nucleus can survive for an indefinite time—Phobos and Deimos, the two moons of Mars, are excellent examples of former comets.

As it rounded Jupiter in 1993, young Comet SL9 separated into 21 fiery fragments that, in 1994, exploded one by one at the thin cloud-tops of the giant planet. The extremely powerful explosions could have been caused only by masses of nuclear energy. In a write-in scientific contest to predict results of these collisions, my prediction of the most powerful explosions ever witnessed in our Solar System was the only one that proved accurate. Predictions of the best supercomputers were puny by comparison. These results clearly defined the true nature of comets: fireballs of nuclear energy. They are not "dirty snowballs" (leftovers from a Big Bang) that

BB advocates believe coalesced into planets.

For a definitive understanding of how planets came into being, one must first understand the Four Laws of Planetary Motion (FL), Internal Nucleosynthesis (IN) and Evolution (E). To comprehend our planetary evolution, it is essential to understand how and why FL, IN and E are inseparable, why one cannot exist without the others. So if we witness a powerful explosion in the sky on July 4th, it will be confirmation that Temple 1 was still an active comet.

Deep Impact Exposes Comet Secrets

In July, 2005, NASA's spacecraft smashed a high-speed 800-pound probe into Comet Temple 1, creating a spectacular explosion whose appearance and subsequent data cast serious doubt on the belief that comets are icy snowballs. Detailed pictures of the comet's surface reveal large regions of flat and smooth, solid rock, while other areas were speckled with ejecta craters. The surface characteristics and the larger-than-expected quantity of organic material emanating from the powerful explosion are typical of nearly exhausted, last-stage comets in which their source of nuclear energy finally becomes exhausted in one last gasp, converting them into burned-out asteroids like the two moons of Mars: Phobos and Deimos.

Had this old-age, low-density, 165-mile comet been other than a hard rock, its final powerful explosion would have shredded Temple 1 into oblivion in the same manner in which Comets Halley and SL9 disappeared prematurely via extremely powerful explosions while researchers watched in amazement. Additionally, the general public was privileged to view on TV the 21 successive nuclear explosions of the fragmented younger Comet SL9 at Jupiter's cloud-tops in 1994. The reason the author was the only scientist to correctly predict these powerful explosions in a write-in contest is because his prediction was based on Comet SL9 being a nuclear-energy fireball, while the others were based on icy snowballs. This evidence clearly undermines Big Bang beliefs, while simultaneously adding much scientific validity to the LB/FLINE model of the universal origins that views comets as ejecta from larger masses of nuclear energy; e.g., our Sun. Generally, comets remain active for a few thousand years before their nuclear cores are exhausted.

Comet Tempel 1 and other Comet Examples
(2005)

Material ejected by the collision spewed thousands of miles into space. The initial bright explosion was followed by secondary, more diffuse, explosions. The two explosions are indicative of a last-stage comet, just as predicted by the LB/FLINE model in the author's computer message to the NPA organization on June 27, 2005. Comet Tempel 1 might now be called the typical Asteroid Tempel 1—if it can be correctly assumed to have no remaining pockets of nuclear energy.

"After noting the decreasing intrinsic brightness of several short-period comets as successive apparitions, Vsekhjsvyatskij estimated cometary lifetimes to be no more than a few thousand years." Yet short-period comets are seen in abundance in the inner Solar System. Short lifetimes, coupled with the inefficiency of capturing new interstellar comets with planetary perturbations, led Vsekhsvyatskij to suggest the planets as a more fertile source of the short-period comets. In 1930 and 1931, he argued that volcanic eruptions from the major planets were the source of comets.

Twenty years later he advocated similar eruptions from the satellite systems of the planets—mainly Jupiter—as the source, because the escape velocity from these small satellites was only a fraction of that required to escape Jupiter: 60 kilometers per second. In Vsekhjsvyatskij's view, short-period comets continually issued forth from satellite eruptions. The LB/FLINE model teaches that comets are fireballs ejected periodically from larger masses of fiery nuclear energy: i.e., our Sun and its still active planets.

When Spain's Instituto de Astrofisica de Canarias reported on 28 July 2000 that an ordinary-looking comet was breaking up, some of the world's top telescopes watched its subsequent disintegration until nothing was left. The French-Finnish SWAN instrument on the SOHO spacecraft had already been observing Comet LINEAR by ultraviolet light for two months, and continued to watch it until the remnants faded from view in mid-August. Today the SWAN team reports in the journal, *Science*, that their observations showed four major outbursts in June and July. The fragmentation seen by SWAN began on 21 July, almost a week before observers on the ground noticed it.

Validating the LB/FLINE Concept

Turkevich's 1964 findings that the Moon's surface is volcanic rock basalt with a high degree of titanium is powerful confirmation of the five-stage evolution of planetary spheres, all in full accord with the LB/FLINE concept. This volcanic method of forming spherical crust is common to all planetary/moon spheres, although the composition of each one's crust and atmosphere will vary in accordance with the variable conditions within each sphere. Such conditions include the nuclear energy core's variables of temperature and pressure that dictate the types and quantities of each element comprising the composition of the surface of each sphere. In turn, these variable conditions are functions of the size (mass) of each sphere, its distance from a central sun, and time. Each sphere is a self-sustaining entity that creates its own compositional matter as it evolves through the five common stages of planetary/moon evolution. Any conclusion predicated on the mythical BB/Accretion hypotheses will itself be mythical and inaccurate. Universal elements are products of ongoing nucleosynthesis within all active spheres of the Universe.

Worlds Out of Balance (Discover, Dec. 2003)

To stunned astronomers, the two mismatched sides of the Moon "seemed like different worlds," and [erroneously], "nobody really knows why." In the definitive FLINE model, the explanation appears simple enough, as detailed in *The Spacing of Planets: The Solution to a 400-Year Mystery* (A.A. Scarborough, 1996, pp 91-94). In this model, the mismatched sides are a natural result of the one-rotational conditions prevailing during the early evolutionary stages of the Moon as it revolved around a former central star: Earth. The model's definitive Five Laws of Planetary Motions detail the dynamic origin of our Solar System, while explaining planetary ellipses, ecliptic angles, the geometric spacing of its planets, and many other anomalies of planetary systems, including exo-planets.

References:

(1) A.A. Scarborough, *The Spacing of Planets: The Solution to a 400-Year Mystery,* (1996). (Ander Publications, LaGrange, GA 30241).

(2) A.A. Scarborough, *New Principles of Origins and Evolution: Revolutionary Paradigms of Beauty, Power & Precision, (2002).* (Ander Publications, LaGrange, GA 30241).

Spacecraft aims to 'catch' comet's tail

PASADENA, Calif.(AP)—A spacecraft is on track to catch a comet by its tail later this week, capturing hundreds of specks of dust from the shimmering cloud that envelops the dirty ball of ice and rock, according to NASA. "The Stardust spacecraft is expected to fly within 186 miles of Comet Wild 2 on Friday, collecting samples and snapping photographs. At the time, the comet and probe will be 242 million miles from Earth. During the flyby, Stardust should capture roughly 1,000 particles of dust ripped from Wild 2 (pronounced vilt-two) by streams of gases boiled from the comet's surface by the warming rays of the Sun. Scientists are eager to study the dust since it represents pristine examples of the building blocks of our Solar System, preserved for billions of years by the cold of space."

It seems unbelievable that science still clings to the myth of "dirty snowballs" almost 18 years after publication of *New Concepts of Origins: With White Fire Laden* (Ander Publications, A.A. Scarborough, 1986) in which the origin and evolution of the fiery nature of comets were detailed. Viewers in 1986 witnessed what may be the last farewell to the diminished brilliance of Comet Halley. Quoting from the *I-T-E-M Connection: How Planet Earth and Its Systems Were Made by Means of Natural Laws* (Ander Publications, A.A. Scarborough, 1991): "Then in February, 1991, when the comet was 1.3 billion miles from the Sun, about midway between Saturn and Uranus, it suddenly expanded to 180,000 miles across and shone to more than 1,000 times brighter than normal. The event was startling and unique, totally unexpected so far from the Sun, and unexplainable in the perspective of the 'dirty snowball' concept."

"However, when viewed in the ITEM perspective, the event was predictable and understandable: the comet (nuclear-energy mass) simply exploded. It was the last sighting of the famous Halley's Comet—the last farewell—never to return on another mission to brighten Earth's skies. Other known comets have had catastrophic endings; some have split in two, while others have crashed into larger spheres, including the Sun. Comet Biela broke in two in 1846 and, has since, disappeared. Comets have short lives, measured in centuries."

Comet Halley's explosion was of the same nature as the powerful explosions of the 21 fragments of Comet SL9 at Jupiter's thin cloud-tops in 1994; these accurate predictions were possible only by the realization of the fiery nature of comets. Computer predictions based on "dirty snowballs" were extremely puny by comparison. "What happens to comets that do not meet catastrophic deaths, and all their internal energies are depleted? We can look to Mars for the best examples of burned-out cometary masses: its two moons, Phobos and Deimos."

A short article, entitled *European satellite takes off for comet*, appeared in *The Atlanta Journal-Constitution* on March 3, 2004. It read, "A European Space Agency spacecraft sped away from Earth's gravity Tuesday after a flawless launch from Kourou, French Guiana, beginning a ten-year journey to rendezvous with an icy comet in a search for answers about the birth of the Solar System and the origins of life on Earth."

This is another example of problems in science. By the time the satellite reaches the "icy comet", scientists will have learned that active comets are fireballs of nuclear energy ejected periodically from larger masses of nuclear energy, rather than icy remnants left over from the mythical Big Bang (BB) that never happened. They will have confirmed that the moons of Mars (Phobos and Deimos) are shells of inactive, burned-out comets that had been ejected from Mars millions or perhaps billions of years ago. Unfortunately, the "icy comet" holds no answers about the birth of the Solar System or the origins of life on Earth; those answers are found here on Planet Earth. But it should reveal to scientists the true nature of comets, while confirming the mythical nature of the BB hypothesis and the factual nature of the LB/FLINE model of origins and evolution.

The article, *Scientists to Earth: Asteroids will hit someday* (*Atlanta Journal-Constitution*, April 8, 2004), states, "If such a rock crashed into the atmosphere, it would explode at an altitude of 10,000 to 20,000 feet with the force of a 100-megaton nuclear weapon. Such a blast could destroy an entire metropolitan area. It is only a matter of time..."

In general, asteroids would not explode in Earth's atmosphere; they are known to crash onto the planet's surface and leave their mark as deep craters. The Great Meteor Crater of Arizona is a good example. Perhaps if the meteorite is a burned-out comet, it could break into pieces, but these bodies are rare. Phobos and Deimos, the two moons of Mars, are examples of this type asteroid. However, if a fiery comet hit Earth's thin cloud-tops, it would explode with the force of a 100-megaton nuclear weapon, which could destroy an entire metropolitan area. Such an explosion occurred in 1908 over the P. Tunguska River basin in Siberia, caused by an aberrant fragment from periodic Comet Encke. The key point here: the crucial differences between asteroids and comets.

The short article, *Cometary encounter* (*Science News*, July 3, 2004, p13), deals with the images of Comet Wild-2 (pronounced "vilt-two") taken during the flyby of the Stardust spacecraft, "images showing craters, spires, mesas, and jets of gas and dust peppering the comet's craggy surface." Scientists "were totally stunned" because they had expected to see "a subdued old surface like someone had dumped powdered charcoal on a body," because a previous cometary encounter (Halley's Comet) had revealed a charcoal-black surface.

"The 100-meter-high spires and other features suggest that the comet isn't a loose agglomeration of smaller pieces, but is a single porous mass..." Further, "The craft detected nitrogen-carbon bonds that are essential to DNA." These findings are indicative of asteroids (burned-out rather than active comets). Since Comet Wild-2 is not mentioned in the authoritative book, *COMETS: A Chronological History*, by D.K. Yeomans, one must wonder why. Further, the close-up craggy portrait of the mass, taken by the Stardust spacecraft, does not look like a comet; comets are fireballs, as illustrated by Comet SL9 whose 21 fragments crashed on Jupiter's thin cloud-tops in 1994 in the largest explosions ever witnessed in our Solar System.

One could speculate that it might be a burned-out shell of a former comet (a type of asteroid), as are the two moons of Mars: Phobos and Deimos. But, if true, perhaps its surface color should be several shades darker than the portrait shows. And it should have been recorded in Yeoman's thorough listing of observed comets, beginning in 1002 B.C. in China. Or could it simply be an asteroid? Perhaps its lighter color can be attributed to the final stage of cometary burnout under specific conditions that could account for its asteroidal appearances.

Phoebe: A Large Saturn Moon

The article, *Portrait of Phoebe: Cassini images a large Saturn moon* (*Science News*, June 19, 2004, p 387), reveals that the spacecraft found "a world of dramatic landforms, with craters everywhere, landslides, and linear structures such as grooves, ridges, and chains of pits." Speculation abounded: 1. "A passing comet or satellite gouged a 50-km hole in Phoebe. Fragments of this collision formed the other moons, which are typically about 20 km in diameter." 2. "Phoebe coalesced near its present location from the nebula of dust, gas and ice that swaddled the young sun...making it the only known moon that's survived at Saturn's location for the entire 4.5 billion-year-history of the solar system." These two speculations illustrate the sophistry being forced upon scientists via the lingering belief in the mythical BB/Ac hypotheses. The LB/FLINE model teaches that Phoebe and the other 30 moons of the ringed planet, and the rings themselves, are ejecta placed in orbit by the powerful nuclear forces of Saturn's nuclear energy core: the same explosive force that created the moon's "dramatic landforms, craters, landslides, and linear structures such as grooves, ridges, and chains of pits"—just as Earth's nuclear-energy core has created similar landscapes during the evolution of our Blue Planet and its Moon.

Such evolutionary changes would not be possible without planetary energy cores that drive the evolution of all planets and moons: Internal Nucleosynthesis (IN) and Evolution (E) are inseparable; one cannot exist without the other. In *The First Days in Saturn's Orbit* (*The Planetary Report*, July/Aug 2004), the Cassini-

Huygens spacecraft flew close by Saturn's outermost satellite, and sent back views of Phoebe that stunned scientists with the sheer number of craters dotting its surface. The article stated, "Most of the craters are undoubtedly very old, but some relatively fresh craters have a bright white material - almost certainly water-ice - cascading down the interior slopes." Scientists concluded, "The water-ice clearly indicates that Phoebe is not an asteroid, but is instead, a wanderer from the outer Solar System." This conclusion was drawn from belief in the mythical Big Bang, an event that evidence reveals never happened.

The new LB/FLINE model of origins and evolution clearly classifies Phoebe as a typical asteroid most likely captured after its formation in, and escape from, the Asteroids belt, an energy-mass ejecta from Saturn. Its larger number of craters were formed in the same manner as those on our Moon, Mercury, the two moons of Mars (Phobos and Deimos), and on all other heavily cratered surfaces of other spheres: either via ejecta from the high-pressured forces of Internal Nucleosynthesis (IN) or via sinkholes during cooldown of the hot surfaces created via IN. In like manner, water-ice is one of the most common IN products found on planets, moons, asteroids, comets, etc., simply because all such bodies undergo Evolution (E) via IN in the same manner. IN and E are inseparable; one cannot exist without the other.

Hot space rocks

(*New Scientist*, 18 June 2005) "Some asteroids that formed early in the Solar System's history were covered by oceans of molten rock, or magma. Richard Greenwood of the Open University in Milton Keynes, UK, and colleagues measured the ratio of two oxygen isotopes in meteorites known to have come from two different asteroids. They found that the ratio of isotopes was identical in every meteorite from the same asteroid, regardless of what part of the asteroid they came from. This suggests that at one time the asteroids must have been molten enough for these isotopes to be distributed evenly throughout." (*Nature*, Volume 435, p 916).

Asteroid or Comet: That is the Question!

July 7, 2005

The well publicized collision on July 4th, 2005 between a tiny spacecraft and the Manhattan-sized comet,Tempel 1, aroused much interest in the nature of comets, whose birth and death processes, via the Big Bang (BB), are still fraught with fundamental questions. Preliminary results of the Tempel 1 crash indicate the comet to be in an advanced stage of evolution, with only a small remnant of its nuclear core intact (but now gone)—the equivalent of (or just beyond) Earth's fourth (rocky) stage of evolution—which explains its dark, cratered appearance of an asteroid, and why the impact was more spectacular and much brighter than expected and "material ejected by the collision spewed thousands of miles into space."

Comet Tempel 1 has become Asteroid Tempel 1. When these results are interpreted in the BB perspective, they do raise more questions; when interpreted in the LB/FLINE perspective, the right answers fall easily into place.

A Strange Lump in Space

In this article, *A Strange Lump in Space*, new images from the Cassini space probe of Saturn's moon, Hyperion, show what astronomers are calling "a rubble-pile moon" pounded by a huge number of meteors over a long period of time. Hyperion averages 165 miles across, and seems to be made of ice partially covered with a dark coating. The origin of the moon and its completely cratered surface remains a mystery to the researchers, and an impossibility via the Big Bang model of universal origins. The pattern and texture of these closely packed craters clearly reveal why moons and many asteroids have cratered surfaces that could not have been created via meteors from space.

In reality, this photograph is powerful evidence of the manner in which moons and many asteroids (from a disintegrated planet) create their own cratered surfaces from within via Internal Nucleosynthesis. Rather than the pattern of irregularly spaced impact craters, they are tightly packed ejecta craters. And rather than "a rubble-pile moon", Hyperion was once a solid asteroid with a nuclear energy core: the driving force behind its heavily cratered surface and its much lower density. Notice also the large central area that collapsed when its internal support (the source of the ejecta) finally was exhausted.

Comet Sampler

Quoting from *Fire Meets Ice, Science News*, March 25, 2006, "The first study of comet dust brought to Earth by a spacecraft has revealed several minerals that could have formed only at the fiery temperatures close to the Sun or another star. The findings come as a surprise because comets, frozen relics of the early Solar System, were born beyond the orbit of Neptune and spend most of their time there."

Analytical tests on particles from Comets Wild 2 in 2004 and Tempel 1 in 2005 reveal the grains as high-temperature minerals: green-silicate-crystal olivine as well as minerals rich in titanium and aluminum. These results clearly confirm the fiery origin of active (and previously active comets), and further confirm them as fiery (or previously fiery) ejecta from larger energy masses, probably our Sun. Additional tests and time should confirm the latter source.

Asteroid or Comet: That is the Question!
(AAAS-SWARM-NPA Meeting, Tulsa, OK, April 2006)

The well publicized collision on July 4th, 2005 between a tiny spacecraft and the Manhattan-sized comet Tempel 1 aroused much interest in the nature of comets. The questionable nature and source of comets and asteroids warrant a closer examination of the BB version versus the LB/FLINE version via Seven Critical Questions:

1. What is the nature and source of comets?
2. What is the nature and source of asteroids?
3. Do comets originate within the Solar System or in outer space?
4. Do comets disintegrate, deplete, or burn out as functions of time; if so, by what processes are their numbers replenished?
5. Assuming they avoid catastrophic endings, do active comets disintegrate completely into meteor streams or do they evolve into inactive, low-density asteroids like the two moons of Mars?
6. Were the craters of our Moon, Earth and other planets formed by comets, asteroids, or ejecta material?
7. Were, and are, planetary rings, including small moons, formed with leftover BB materials or with virgin ejecta?

Fundamental Differences between Comets and Asteroids

1. Pictorial evidence reveals that comets are relatively small masses of nuclear energy, the same type found in stars, suns, and planets. Like these larger masses, comets are undergoing Internal Nucleosynthesis

(IN), the transformation of energy into atomic elements/matter in full accord with $E=mc^2$. In this manner, the core of a comet produces distinct tails of ions and virgin dust particles, along with a hydrogen coma, usually a million miles or more in diameter. While these processes may be exacerbated near the Sun, they continue functioning in outer space as far out as telescopes can detect, and do so until the energy core is depleted. When comets are triggered prematurely, they can release explosive forces beyond the wildest expectations. Comets are constantly replenished as ejecta from larger energy masses.

2. Pictorial evidence confirms that Asteroids are fragments of a disintegrated planet, as first recorded by Olbers in 1802. Large numbers of fragments from internal locations originally contained nuclear cores that may now be depleted. Evidence of this can be seen in the cratered and sunken surface areas formed during the depletion cycle of their former energy cores, and subsequent cooldown in the near-absolute-zero coldness of space—the same features as seen on surfaces of burned-out cometary shells like Phobos and Deimos, and on all planets beyond their second stage of evolution.
3. Neither comets nor asteroids can qualify as building blocks of planets and moons, or as leftovers from a BB. Both types are products made from larger masses.

References

(1) Donald K. Yeomans, COMETS*: A Chronological History of Observations, Science, Myth, and Folklore*, (1991). John Wiley & Sons, NY.

(2) Alexander A. Scarborough, *The Spacing of Planets: The Solution to a 400-Year Mystery*, (1996), 8th Edition, Energy Series. (Ander Publications, LaGrange, GA 30241).

(3) Alexander A. Scarborough, *New Principles of Origins and Evolution: Revolutionary Paradigms of Beauty, Power and Precision*, (2002), 9th Edition, Energy Series. (Ander Publications, LaGrange, GA 30241).

(4) Alexander A. Scarborough, *Bringing the Copernican Revolution Full Circle via a New Fourth Law of Planetary Motion, Journal of New Energy*, Fall 2003. Proceedings, NPA, International Conference, June 9-13, 2003.

Break-up of a Comet

As viewed by the Hubble Space telescope, Comet 73P/Schwassnan-Wachmann 3 is continually breaking up as it passes near the Sun every 5.4 years (*Big Breakup*, *Science News*, 6 May 2006). The breakup into several dozen pieces followed the same pattern as the breakup of Comet SL9 into 21 fragments as it approached and rounded Jupiter in 1993, followed by a powerful explosion of each fragment upon contact with the thin cloud-tops of the giant planet. The 21 extremely powerful explosions clearly confirmed the true nature of comets: fireballs of nuclear energy, formerly ejected from larger masses of nuclear energy: e.g., our Sun. The breakup of such energy masses, followed by extremely powerful explosions, is a fairly common occurrence among short-period comets. While the LB/FLINE model predicted and explains the nature of comets, such nature remains a "great mystery" via the BB concept: "How can a body as weak as meringue come together on a kilometer scale, then fall apart?" The simple answer, beyond question, resides in the true nature of comets.

Enceladus: Saturn's small moon visited by the Cassini spacecraft. — *June 2, 2006*

The Whole Enceladus (*Science News*, April 6, 2006) by Ron Cowen discusses the NASA Cassini spacecraft's visit to Saturn's small 500-kilometer-wide moon, Enceladus. Surprisingly, Cassini found a giant geyser of water vapor plume soaring 175 feet above its surface in the vicinity of 100-meter-wide linear cracks at the south pole. "An internal heat source probably drives the geyser, which looks like Yellowstone's Old

Faithful," according to the article. "The mix of inorganic compounds and hydrocarbons found in the plume, as well as organic compounds detected on nearby regions of the Moon, suggests that a rich, warm organic soup lies beneath the surface."

The basic findings about Enceladus were reported in nine articles in the March 10 *Science* magazine. All the findings are powerful confirmation of the FLINE model of planetary origins that both predicts and explains them. This first became possible in 1975 when publication of *Fuels: A New Theory* opened the floodgates to understanding the true source of internal heat that drives all planetary evolution: nuclear-energy cores.

The Five Laws of Planetary Motion explain how the Solar System formed dynamically in geometric order some five billion years ago, followed by a five-stage evolution of the planets into their current stages at rates in accord with size.

The five stages are readily observed and identified while the planets revolve around our Sun in full accord with these laws; e.g., Earth is in the fourth (rocky) stage, while Jupiter remains in the second (gaseous) stage. The five laws make it possible for scientists to measure the gradual changes that have occurred in the orbits and velocities of the planets since the dynamic origin of the Solar System some five billion years ago; e.g., Neptune and Uranus have drifted outward, away from the Sun. Additionally, they make it possible to predict future changes as functions of time.

One of the most critical confirmations of the FLINE model was not mentioned in the reports. The tiny size of Enceladus and its still-active core are definitive clues that attest to its young age: powerful evidence that the young moon began as a small mass of energy ejected from Saturn's nuclear core in relatively recent times.

Jupiter's Io is another example of a still-active small moon ejected relatively recently as an energy mass from its mother planet's nuclear core. Because of their small sizes, both moons already have reached the fourth (rocky) stage of planetary evolution - according to the LB/FLINE model - while their giant mother planets yet remain in the second stage (gaseous).

The new model also explains why "more heavily cratered, older terrain lies next to smooth, younger terrain. The smooth patches indicate that the south-polar area has undergone recent fracturing or upheavals" and outpourings, all common characteristics of the five-stage evolution of planets and spherical moons. In summary, every discovery about Enceladus by the Cassini spacecraft follows the pattern of previous space discoveries in validating the LB/FLINE model of universal origins.

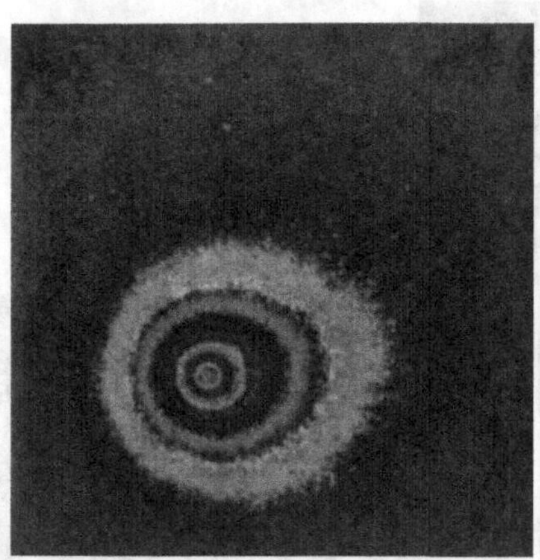

Color-enhanced Halley's Comet showing the concentric spherical layers of enveloping matter indicative of temperature differentials. Photos of nuclear stars in the background showed identically-colored layers and patterns.

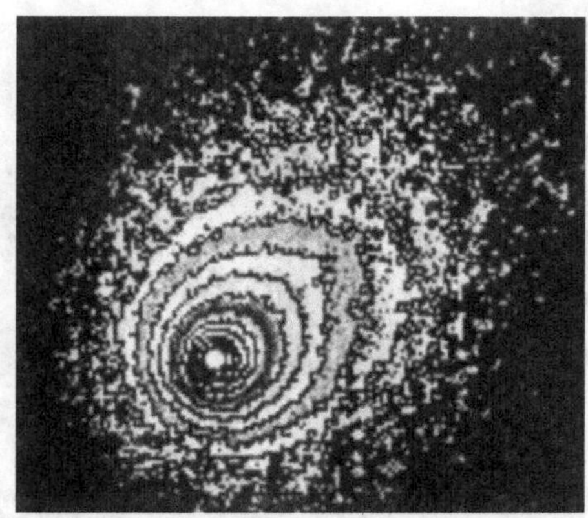

A typical false-color image that is always identical for comets and background stars.

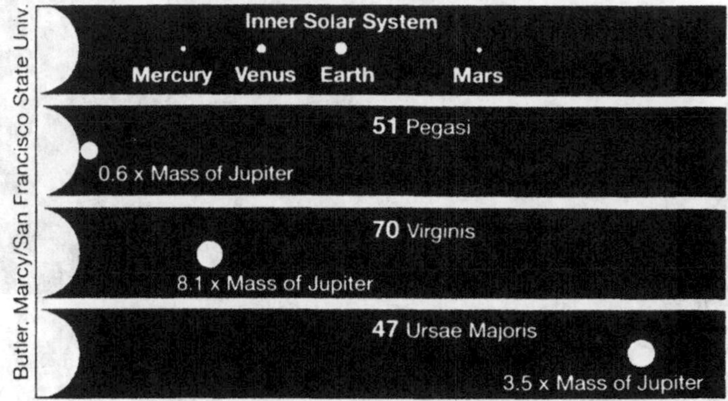

The relative locations of planets in our Solar System and the newly discovered planets orbiting 51Pegasi, 70 Virginis, and 47 Ursae Majoris.

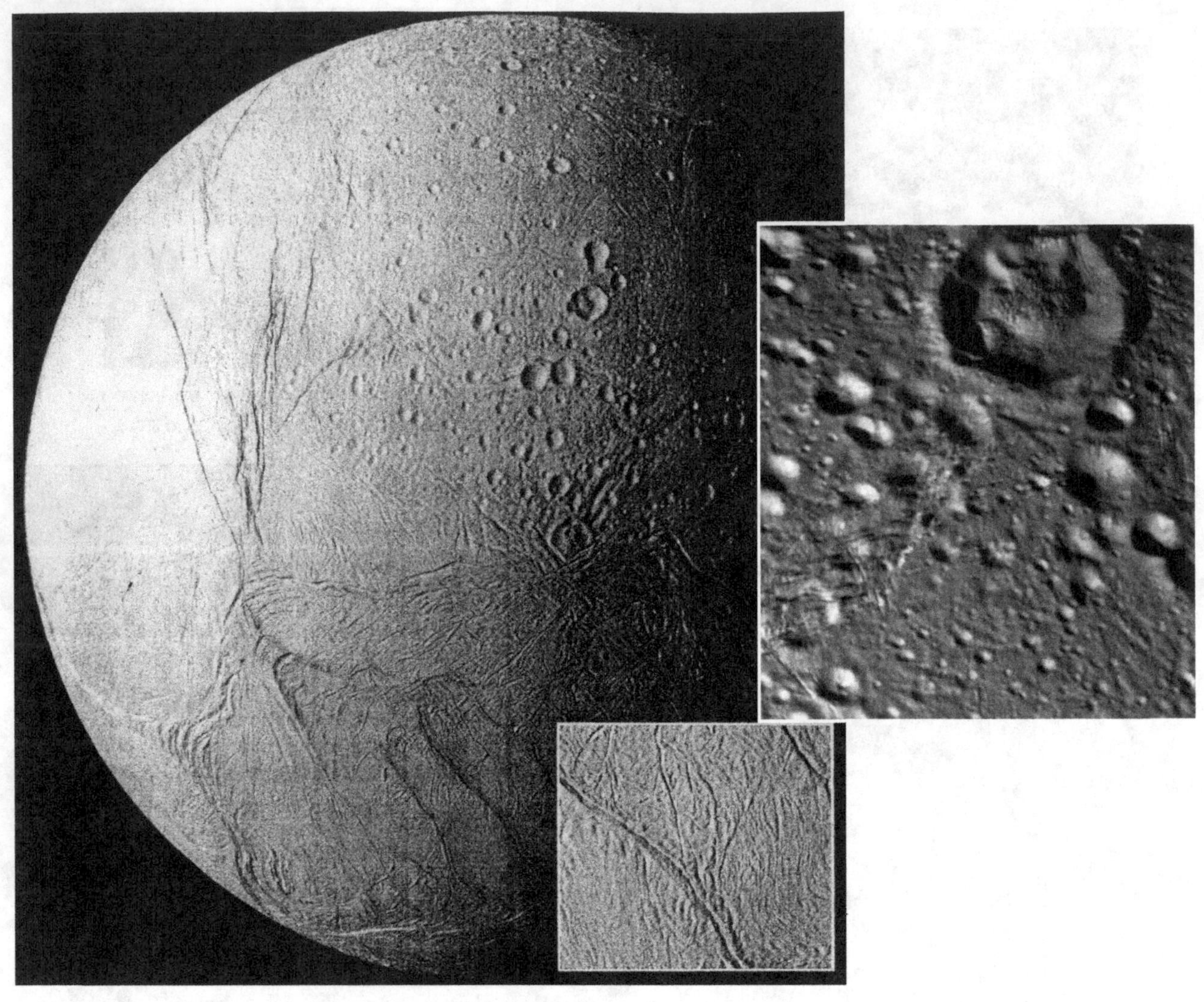

Cassini eyes youthful-looking Saturnian moon

On July 14, 2005, the Cassini spacecraft recorded images of Saturn's tiny moon Enceladus, whose diameter barely exceeds 300 miles. Images taken at close range show a large area of faults, folds and ridges devoid of craters, and a smaller area in which numerous craters do still exist. This pattern suggests that old pockmarks have been erased by recent geological activity in later times, perhaps only tens of millions of years ago. In the Big Bang (BB) concept, it is still a mystery why this tiny moon exhibits such diverse topographic features. But the mysterious aspect quickly vanishes in the perspective of the LB/FLINE model of universal origins. Craters made via ejecta processes, including out-gassing during earlier stages of the moon's evolution, once did exist over the entire surface before most were erased later by the geological activity revealed by the smoother terrain of faults, folds and ridges, all in full accord with the principles of Internal Nucleosynthesis (IN, from energy to matter) that drives Evolution (E) in all internally active bodies throughout the Universe. Cassini's images of Enceladus offer powerful support to the IN-E principles of the LB/FLINE model of universal origins. IN and E are inseparable; one cannot exist without the other.

Throughout the Universe, such smaller masses are simply ejecta from larger energy masses. Jupiter's small satellite, Io, is another, but much larger, example of a most recently ejected mass (from Jupiter's energy core) that's still very active, slowly evolving through the fourth of its five stages as the youngest big moon circling the giant planet that gave it birth. Cassini's findings add excitingly powerful evidence in support of the LB/FLINE model of universal origins and evolution (1973-1996).

To: Glenn Strait, Science Editor. **May 6, 1999**

The article, *Plate Tectonics....on Mars* (*Science News, The World & I*, Washington, DC, May 1, 1999, p 284), discussed in our most recent correspondence, arrived yesterday in my issue of *Science News* with more details of the discoveries on Mars. As usual, I am somewhat amazed, but not surprised, at how accurately they match my descriptions of planetary evolution detailed in the FLINE model. The authors, as usual, sound as if my book was used as a guide to interpret their findings, except that they stop just short of the Internal Nucleosynthesis-Evolution (IN-E) connection by merely describing it as the heat within. But their findings bring them ever closer to the FLINE paradigm.

To quote Maria Zuber of MIT: "The thermal state of the interior of the planet is something that you really need to understand to get to the early climate history of Mars. It's a very critical piece of the [FLINE] puzzle." Sufficient corroborating evidence for projecting backwards through the first four stages of planetary evolution is available for clearly explaining how the surface features of Mars formed. Inevitably, the answers, as with all planets, will lead directly to the IN-E connection in which E is not possible without IN (1975-1996).

But how long before this understanding becomes acceptable and useful knowledge in the scientific community remains questionable. Mars is prelude to Earth's distant future—an example of what our planet will look like in a billion years, more or less. And Earth is a guide to what Mars was like in the distant past. Each and every planet evolves through the five common stages of planetary evolution in full accord with size. Each eventually passes through the pattern of irregular patches of weak magnetic activity when it enters the fifth and final stage of evolution as the energy core becomes depleted.

Jack Connerney of NASA's Goddard Space Flight Center in Greenbelt, MD correctly states that the most likely explanation for the magnetic features is that "they are relics of a global magnetic field active on Mars long ago." After all, electromagnetism is inherent in the rotating nuclear-energy masses in which matter is created. Mario Acuna, also of Goddard, closes in on the FLINE model when he states, "The planet started off very hot and cooled very fast (relatively speaking), which you would expect for a small object." To quote from the article, "Scientists believe that the Martian field arose the same way that Earth's still-active field did, through the action of a dynamo generated by the rise and fall of molten material deep within a rotating planet."

According to the FLINE model, the two fields did arise the same way, but via the nuclear-energy core within each rotating planet, rather than via the prescribed "dynamo action". Connerney comes even closer to the FLINE model when he speculates that the fading of the global magnetic field and plate tectonics are intimately linked. This connection is explained clearly in my writings of the past 24 years, especially in my latest book, *The Spacing of Planets: The Solution to a 400-Year Mystery* (1996). As the planet enters the fifth and final stage (inactive) of evolution, the energy core becomes depleted, and all actions of electromagnetism and plate tectonics will cease.

The FLINE model offers assurance to Zuber *et al.*, that ancient Mars did indeed undergo a period of plate tectonic activity. And each of its many facets interlock precisely into the completed jigsaw picture-puzzle that provides a clearer understanding of the origins of solar systems and the orbital spacing and evolution of planets, thereby explaining the scientific reasons behind this former activity of Mars.

To: Dr. Neil E. Munch, Montgomery Village, MD 20886-3132. **May 14, 2003**

In follow-up to my letter of December 13 to Francisco, and as promised in our telecom of 5-13-03, I am listing 11 subjects for possible agreement by NPA members during the evening meeting on June 13. I had not realized the need to be more specific (detailed). I regret having to miss the meeting; perhaps some volunteer could pose some or all of them to the group.

Points to agree on:

1. The statements by Sagan and Hawking concerning, respectively, the myths of science and the conjectural nature of cosmology's prevailing hypotheses.
2. The structure of the Universe appears the same in all directions, perhaps indicative of its spherical structure; i.e., expansion in a vacuum is equal in all directions.
3. The distance of visible objects in our Universe is limited by the speed of light in a vacuum; i.e., astronomers might never be able to see what is happening beyond the point at which such distant light has not yet arrived at their telescopes.
4. Because of #3 above, mankind may never know for certain the size of our Universe, or whether it had a beginning or will have an ending.
5. The farther outward astronomers look, the faster the Universe is accelerating: an inherent, predicted and explainable phenomenon of the Little Bangs (LB)/FLINE model of origins and evolution—and just the opposite of the Big Bang prediction of a decelerating Universe.
6. There are only two fundamental substances in the Universe: energy and matter.
7. Energy and matter are interchangeable: $E=mc^2$.
8. The ongoing, never-ending reactions of the Universal Law of Creation of Matter (ULCM): from energy to atoms to molecules, usually in successive stages of gaseous, liquid and solid.
9. The ongoing evolution of planetary and other universal spheres is not possible without a nuclear-energy core to drive it forward.
10. The ultra-powerful explosions of Comet SL9 (1994) on Jupiter confirm the true nature of fiery comets: nuclear energy (not "snowballs"), accurately predicted and explained solely by the LB/FLINE.
11. The Five Fundamental Principles of Origins and Evolution (E): The FLINE model.
    a. The heliocentric Solar System by Copernicus.
    b. The First Three Laws of Planetary Motion by Kepler (FL).
    c. Gravity by Newton.
    d. Internal Nucleosynthesis (IN) via $E=mc^2$ by Einstein.
        (Note: In 1644, Descartes described Earth's interior as "Sun-like".)
    e. The Fourth Law of Planetary Motion (FL) explaining the mathematical spacing of planets: the final piece of the puzzle that melds together the First Four Principles, bringing the Copernican Revolution (a definitive understanding of the origin and evolution of our Solar System) full circle.

Thanks for all your help—past, present and future.

To: Dr. Cynthia K. Whitney, Editor, Galilean Electrodynamics. **July 28, 2003**

Re: 1. Bringing the Copernican Revolution Full Circle via the New Fourth Law of Planetary Motion. (Six pages of text, plus two pages of illustrations: the Seven Diagrams of the Origin of the SS)
2. Principles of Planetary Origins and Evolution. (14 pages of text, plus one page with five illustrations)

Here are the two manuscripts written with the continuity suggested in your recent letter. I'm sorry it took longer than anticipated. Much time was spent in refining the labeling of the seven Diagrams and five illustrations, plus making the explanations simpler to understand, and explaining why the Fourth Law is so important: it is truly the key to a definitive understanding of the dynamic origin and evolution of our SS and its history of changes. Along with Kepler's Three Laws, it details how the planets attained their orbital spacing (the solution that first eluded Kepler in 1595), eccentricities, ecliptic inclinations, etc., etc., etc. The LB/FLINE model is the highway to definitive solutions to all other planetary anomalies, including origins and, subsequently—along with Einstein's $E=mc^2$—planetary evolution. After 30 years of researching and developing this factual model from the ground up, along with keeping close tabs on Big Bang (BB) conjectures, I'm absolutely convinced—along with a number of other scientists—of its destiny to displace the BB early in this century. Its definitive evidence is overwhelmingly persuasive. It is the Golden Key that unlocks the floodgates to definitive solutions to planetary origins and evolution—past, present and future—and, ultimately, by putting science on the right track, the Key to all universal enigmas.

Because every facet of it interlocks precisely with all other facets, the new model is truly exciting and far more fruitful than the BB, which I now consider the greatest obstacle to genuine progress in the planetary and astronomical sciences. I feel sure you are familiar with Thomas Kuhn's book, *The Structure of Scientific Revolutions*, which accurately predicts the types of resistance to change experienced, and being experienced, unfortunately, by this revolutionary model. Any additional help you can render in bringing the Copernican Revolution—actually initiated by the work and beliefs of the Pythagoreans around 500 B.C.—full circle will be deeply appreciated by many scientists interested in accurate interpretations of their findings, which inevitably must be based on these principles of planetary origins and evolution. Should I submit the manuscript to any other editor(s) any time soon, and if so, to whom? The Royal Astronomical Society?

To: Eominic Feltham, Publishing Director, New Scientist, London. Sep. 20, 2005
Re: (1) Nearly a Planet (New Scientist, 10 Sep. 2005, p 19), and
(2) Tails of the Unexpected (New Scientist, 10 Sep. 2005, p 32)

During the past three decades, many scientists have openly recognized that mysteries of planets and planetary systems viewed via the Big Bang (BB) hypothesis are destined to remain "conjectural" (Hawking terminology, 1999), with no possibility of valid scientific solutions to problems such as those encountered in the subject articles. In reality, the BB gives no choice to scientists other than forcing them into awkward fairy-tale reasoning about their brilliant discoveries.

However, definitive solutions now are being found via the new well substantiated alternative to the Big Bang hypothesis: the LB/FLINE model outlined in the enclosed article embracing 12 giant steps for mankind (some little known) during the past 2500 years. Initially, around 500 B.C. the Pythagoreans gave us the first crucial tool for learning Nature's deepest secrets of its beauty, power and precision.

Since Copernicus in 1543 pointed science in the right direction, much brilliant work by a number of scientists has brought the Copernican Revolution nearly full circle. These findings obviate the need for conjecture, and open the floodgates to comprehending definitive origins and evolution of planetary systems, solar systems, galaxies, and ultimately, the functioning of all systems of our Universe in which everything is interwoven with everything else. Rather than conjectural science, substantiated facts should prove far more exciting to your readers, and much more effective in advancing the cause of science. They also prove Einstein right when he stated (paraphrasing): "The right answers will be simple and beautiful."

In reality, a plethora of substantiated evidence now strongly supports E. Chladni's work in 1794 that argues for the cosmic origin of fireballs and the generic connection between fireballs, meteorites, and meteors. When viewed today in the proper perspective, there's really nothing confusing about comets—or about Ceres. I urge you to give this revolutionary literature your most serious consideration for publication. If you would like more details, please let me know, and I will send pertinent papers and books written about these well researched subjects during the past 32 years, but yet remain ahead of their inevitable time.

To: Letters Editor, Discover, New York, N.Y. 10011. **May 22, 2006**

Jupiter Rules (*Discover*, June 2006) by Bob Berman raises several issues that warrant responses. The moot question of whether our Solar System has eight or ten planets is clarified by Kepler's First Three Laws of Planetary Motion, and the new Fourth and Fifth Laws dealing with the geometric spacing of the ten planets and Asteroids, along with the debris beyond. (Scarborough, 1995) (Proceedings, NPA International Conference, April 2006).

The same Laws reveal why Jupiter is the largest planet, along with substantiated solutions to more than three dozen SS anomalies. (*Journal of New Energy*, Fall 2003: *New Principles of Origins and Evolution*, Scarborough, 2002). Jupiter's "stripes of darker clouds called belts and lighter clouds known as zones," each running parallel to the equator, are readily explained via Jupiter's second (gaseous) stage of the five-stage planetary evolution common to all planets; e.g., Earth is in its fourth (rocky) stage of evolution.

Evidence clearly shows that the rate of planetary evolution is a function of planetary mass and time. Jupiter's stripes are separated via the same principles that separate planetary rings. And I agree with Del Genio: "Jupiter's belts crackle with lightning, and only water can make thunderstorms."

The Moon landing in 1969, the Voyager 1 probe in 1979, and the several probes to Jupiter and other planets since, have furnished sufficient data that, when correctly interpreted, give scientists the capability of fully understanding our Solar System's past, present and future. But, as often illustrated during the past few decades, this will not be possible until concepts other than the Big Bang illusion are granted the consideration they warrant.

Dear Kay,

There are actually two references relevant to your father's work: (1) Robert J. Heaston, "The Characterization of Gravitational Collapse as a Mass-Energy Phase Change", Proceedings of the Natural Philosophy Alliance, ISSN 1555-4775, Vol. 1, No 1 pp 33-42 (Spring 2004). Presented at Joint AAAS-SWARM-NPA Conference, University of Denver, Denver, CO (7-10 April 2004); and (2) Robert J. Heaston and Peter Marquardt, "The Constant Gravitation Potential of Light—Part 1: Theory". Presented at Joint AAAS-SWARM Conference, University of Tula, Tulsa, OK (3-7 April 2006).

The relevance of my work to your father's work is this. As I understand it, his LB/FLINE model includes "Little Bangs" and "Internal Nucleosynthesis", which both involve mass-transition to energy in both mechanisms. Normally this transition is accomplished by fission or fusion. A rather detailed analysis of the Einstein field equations of general relativity indicates that there may be a third means of transition from mass to energy. This third process is based upon a phase change from mass to energy from gravitational collapse. I do not know what mechanism your father uses, but my mass-energy phase change from gravitational collapse forces could possibly be a mechanism in your father's model. I am analyzing the extensive literature that reports on progress made from 1907 to 1916 when Einstein was deriving his field equations to see if he may have overlooked some significant options.

Good luck on the publication. Sorry for the delayed response. Your message got lost in 110 meaningless ones.

Regards,
Bob Heaston

Powerful Confirmation of Ejecta Craters and Nuclear Cores, Discover, Feb, 2008

The moon's ejecta craters and pushed-up mountains are typical results of the tremendous internal pressures created by nucleosynthesis within the cores of planets and moons during their second, third, and fourth stages of creative evolution into the fifth (inactive, dead) stage.

All spheres of the Universe (e.g., stars) are powered by identical cores of the nuclear energy supplied by black holes (actually spheres) created continuously as the space-time energy that powers the Universe, perhaps forever.

CHAPTER V

A BIG BANG OR LITTLE BANGS?

"When we try to pick out anything by itself, we find it is hitched to everything else in the universe."
John Muir

The Three Key Observations

Three observations provide the fundamental basis for the standard cosmology featuring the Big Bang (BB) theory:

1. The observed expansion of the Universe (usually interpreted in the framework of relativity as an expansion of the metric of space).
2. The 2.726 K cosmic background radiation (CBR), interpreted as a remnant of the BB.
3. The apparently successful explanation of the relative abundance of the light elements.

In reality, the same observations serve equally well as the fundamental basis for a different concept: the Little Bangs (LB) theory that intermeshes precisely with the geometric origin of the SS and the evolution of planets via nucleosynthesis. Pieced together during the past 31 years (1973-2004), this revolutionary perspective surprisingly offers answers to the many questions that pose challenges to the BB theory. The two concepts need to be examined closely with open minds.

Questions about an Expanding Universe (1992)

Hetherington's *Encyclopedia of Cosmology* summarizes the status of the BB theory, "Problems have been numerous and not all of them are solved to the satisfaction of all cosmologists. In particular, the large-scale homogeneities observed in the 1980s seem to indicate a structured Universe, which may contradict one of the foundations of the BB cosmology, the uniformity postulate (or cosmological principle). This and other problems have recently caused some cosmologists to declare the BB theory in a state of crisis."

"However, since no plausible alternative exists, the almost universal belief in the BB model has not been seriously shattered." The reason for the fallacy of the last statement is best summarized in the article, *Why Only One Big Bang?* (*Scientific American,* Feb 1992) by Geoffrey Burbidge, a professor of physics at the University of California, San Diego. To quote, "Those of us who have been around long enough know that peer review and the refereeing of papers have become a form of censorship. It is extraordinarily difficult to get financial support or viewing time on a telescope unless one writes a proposal that follows the party line. A few years back, Halton C. Arp was denied telescope time at Mount Wilson and Palomar observatories because his observing program had found, and continued to find, evidence contrary to standard cosmology. Unorthodox papers often are denied publication for years, or are blocked by referees [Amen]. The same attitude applies to academic positions. I would wager that no young researcher would be willing to jeopardize his or her scientific career by writing an essay such as this."

"This situation is particularly worrisome because there are good reasons to think the Big Bang (BB) model is seriously flawed," Burbidge continued. "One sign that something is amiss is the time-scale problem. The most favored version of the Big Bang model yields a universe that is between 7 and 13 billion years old. The large range of possible ages derives from uncertainty regarding the rate at which the universe is expanding, a value known as the Hubble constant."

"Comparison between observation and calculations of stellar evolution implies that the oldest known stars are 13 to 15 billion years old, with an uncertainty of plus or minus 20 percent. The estimated age of the elements in the solar system based on measurements of heavy radioactive elements is about 15 billion years, again including some uncertainty. If one accepts a high value of the Hubble constant, and hence a low-age universe, the simplest Big Bang model clearly fails, because the universe cannot be younger than the objects it contains. If one chooses a low value for the Hubble constant, it is touch and go," Burbidge concluded.

Recently, the Hubble Space Telescope (HST) provided new data used in revising the Hubble constant to 81-85 km/sec/Mpc. This figure proved disconcerting to advocates of the BB, because any result over 50 km/sec/Mpc reveals a serious flaw in the standard model. Further, there is an inescapable problem of creation in both the BB and the steady state cosmologies with no scientific solution. How did the singularity or primeval atom originate? Could all matter of the Universe really have been packaged in such a small unit and at trillions of degrees temperature? If an explosive expansion were nearly instantaneous, why is the Universe now expanding at merely the speed of light? Why wasn't gravity an impending factor during the time of instantaneous expansion? With the Universe expanding at or near the speed of light (as extrapolation of the factors of the Hubble constant indicates), why are extrapolations backward to the very beginning done on this basis instead of the basis of instantaneous expansion (an important basis of the BB theory)? Hasn't the Universe always expanded at the speed of light?

Quoting Burbidge again, "Rather than considering alternatives to the Big Bang, cosmologists contort themselves and propose that the rate of expansion is just small enough to accommodate the oldest well-documented stellar ages. Alternatively, they vary the Big Bang model by invoking an arbitrary parameter called the cosmological constant. In this version of the story, the initial Big Bang was followed by a waiting period and then a further expansion." In the LB alternative, no cosmological constant is needed to explain the accelerating rate of expansion of the Universe.

The Cosmic Microwave Background: The 2.7 K Radiation

The second fundamental basis of the BB theory deals with the 2.7 K radiation identified as a relic of the big event. While this interpretation makes a good story, it does raise serious questions that leave the door open to other possibilities. But first, a little background would be helpful. In 1965, two young scientists, Amo Penzias and Robert Wilson, decided to use sensitive microwave antenna in radio astronomy. Much to their chagrin, they discovered an irremovable background noise in the antenna. Frustrated, they sought help at Princeton University, where they were informed that the persistent "noise" was probably the most important radio signal ever to be received from outer space. This 2.7 K radiation, identified as the afterglow of the Big Bang, suffuses the sky in all directions at microwave frequencies. The cosmic microwave background (CMB) of 2.7 K radiation was interpreted as providing direct evidence of how radiation was distributed throughout the Universe when it was less than one million years old.

In 1989, the first Cosmic Background Explorer (COBE) brought back recordings of uniformity not varying more than one part in 100,000 in all directions. Advocates of the BB theory immediately interpreted these results, along with the thermal nature of the cosmic microwave background as evidence of its primordial origin. In the same 1992 article, Burbidge wrote, "The pervasive cosmic microwave background was predicted by the Big Bang theory, and is still considered to be one of its strongest pieces of supportive evidence. Measurements now, however, show that the background radiation is extremely smoothly distributed. Maps of galaxies, on the other hand, show structure on all scales."

"According to the standard version of the Big Bang theory, matter and radiation were strongly coupled together in the early universe, and only later did the two go their separate ways. If this were so, the cosmic microwave background would show some imprint from the lumpy matter distribution that led to the formation of galaxies. In actuality, however, the cosmic microwave background appears smooth to at least one part in 100,000, so close to the level at which the Big Bang must be abandoned or significantly modified."

However, not to worry: The COBE Satellite had been sent aloft again, and recorded another cosmic microwave background of 2.7 K radiation. Just two months after the Burbidge publication, the interpretation of this COBE probe became public. Incredulously, the presence of miniscule ripples—small temperature fluctuations of slightly more than one part in one million—was announced, and advocates of the BB breathed a huge sigh of relief. The tiny ripples were interpreted as fluctuations in the density of matter and energy in a very early phase of cosmic history—the clumping of matter into large structures such as galaxies.

In the same article, Burbidge had written, "Within the framework of the hot Big Bang, there is no satisfactory theory of how galaxies and larger structures formed. Galaxies cannot form by gravitational collapse in an expanding universe unless one assumes, without explanation, that large density fluctuations were present in the early universe. Under the influence of particle physicists, cosmologists are now proposing that these fluctuations occurred at an early stage of the Big Bang, or else, were caused by exotic entities such as cosmic strings. None of these ideas can be directly tested."

"The inflationary model, a pet idea of the past decade, holds that a period of extremely rapid expansion in the early universe accounts for both the smoothness of the cosmic microwave background and for the amount of matter present in the universe. But again, inflation is a non-testable addition to the lore of the Big Bang." Further, if "a period of extremely rapid expansion in the early universe" did happen, wouldn't this disrupt calculations of extrapolations backward to the time of the BB? To quote Burbidge again, "This form of inflation is arbitrary, and our successors will wonder when it goes out of favor, as the history of science suggests it will, why it was so popular?" Whether or not the new finding of miniscule fluctuations in the background radiation eliminated the necessity of the inflationary model is not clear in many minds. But when we take a closer look at the CMB radiation, the question becomes moot.

A Closer Look at the CM Background Radiation

While the cosmic microwave background is identified as a relic of the big event, the true source of the 2.7 K radiation remains debatable. The major problem with this pervasive radiation resides in its interpretation. In reality, it is blackbody radiation, and such radiation is characteristic of all Solar System objects. (Ref: *Encyclopedia of Physics*). Consider the radiation in the space enclosed within a hollow steel ball, which is maintained at a constant white heat. The radiation within such a sphere is called "blackbody" radiation because the radiation coming from each unit area of the walls of the enclosure is the same as that which comes from unit area of a perfect "blackbody" maintained at the same temperature of the walls and not by the substance of which the walls are composed. In such an enclosure, the intensity of the radiation falling on a unit area of the walls equals the radiation coming from the unit area. In principle, the same thing applies to a planetary sphere in which the white heat equates to its nuclear-energy core and the mantle/crust equates to the walls. The radiated energy originates from the internal energy associated with atomic and molecular motion and the accompanying accelerations of electrical charges with the object (sphere). Scientists realized that the results of the discovery of Penzias and Wilson were limited to only a few wavelengths clustered at one end of the Planck curve. Other explanations of the background radiation, such as a combination of radio sources, could explain those data points. It was not until the mid 1970s that enough measurements at different frequencies had been made to prove that the background radiation actually follows Planck's law.

Radiation is both absorbed and emitted by the walls of the enclosure, and by any gas, which might be in the enclosure. Planck suggested the idea that when radiation is emitted or absorbed, it is emitted or absorbed in multiples of a definite amount, identified as a quantum. The energy of a quantum of radiation of frequency v is proportional to v and so is equal to hv, where h is a proportionality constant known as Planck's constant. The full curve of Planck's law is the spectrum of the continuous radiation given out by a hot "blackbody". The plotted curve shows the distribution function with respect to wavelength and/or frequency. It applies to any planetary mass containing a hot nuclear core.

No problem would exist here if scientists could accept the fact that the spheres of the SS are evolving from nuclear masses, and the CBR is, in reality, the radiation emanating from these internal nuclear reactions. Simply put, the relationship among the three characteristics of these internal nuclear reactions is best shown by the full curve of Planck's law: the spectrum of the continuous radiation given out by these hot black bodies. Proponents of the BB theory claim evidence that the CBR originated at extremely high redshift. The evidence comes from the production of light elements. Knowing the present CBR temperature, and <u>assuming</u> entropy was nearly conserved back to high redshifts, they trace the thermal history of the Universe back in time to <u>temperatures high enough to have driven thermonuclear reactions</u>. Computations of the nuclear reactions, under <u>three other (conditional) assumptions</u>, predict that most matter comes out of the "hot BB" as hydrogen, with about 20% by mass, helium, and significant amounts of deuterium[3], helium[4] and lithium[7]. The computation, when using a mean baryon number and consistent with what is observed (supposedly) to be present in and around galaxies, yields computed abundances concordant with what is seen in old stars. Utilizing the same information except for the assumptions necessary in the BB theory, similar reasoning in the perspective of the Little Bangs (LB)/FLINE concept allows one to reach another confirmatory conclusion.

Rather than looking to the BB for answers to the origin of light elements, the computation of the nuclear reactions can be applied most assuredly to the nucleosynthesis common to all planets and stars. No complicated assumptions are necessary. As shown in previous chapters, the production of all elements, therein, will be in the same proportions predicted in the calculations used in the BB perspective. Planck's law applies to all thermonuclear reactions, no matter in which theory it is utilized.

The above information ranks high among the many crucial clues that reveal the true nature of planetary origins. While negating any claim that the CBR exclusively supports the BB theory, the facts stand firmly as powerful confirmatory evidence of the "blackbody" nature of the spheres of our SS—<u>the true source of the cosmic background radiation</u>. Consequently, any COBE search within the realm of our SS cannot escape the effects of the blackbody radiation emanating from objects comprising the system. Isn't it reasonable, then, to conclude that no matter where Penzias and Wilson pointed their microwave horn antenna, they inevitably would record the same noisy emissions in every direction?

The 2.7 K Background Radiation: A Question of Interpretation

The LB/FLINE model offers a definitive solution to the mystery of the 2.7 K background radiation. Quoting the *Encyclopedia of Physics* (1991): "Every object in the universe continuously emits and absorbs energy in the form of electromagnetic radiation ... [which] originates from the internal energy associated with atomic and molecular motion and the accompanying accelerations of electrical charges with the object." The LB model attributes this internal action to the Internal Nucleosynthesis within closed systems (e.g., planets) as perfect emitters and absorbers that radiate into a vacuum: i.e., 2.7 K blackbody radiation.

Planck's distribution applies to all thermonuclear reactions, no matter in which theory it is utilized. The above information ranks high among the many crucial clues that reveal the true nature of planetary origins. While negating any claim that the CBR exclusively supports the BB theory, the facts stand firmly as powerful confirmatory evidence of the blackbody nature of the spheres of our SS and all active spheres of the Universe—the true source of the cosmic background radiation. (4)

Consequently, any COBE search within the realm of our SS cannot escape the effects of the blackbody radiation emanating from objects comprising the system. Isn't it reasonable, then, to conclude that no matter where Penzias and Wilson pointed their microwave horn antenna, they inevitably would record the same noisy emissions in every direction? (4) (5)

Even a probe escaping the effects of our SS would never be able to escape the "noise" of other suns (stars) undergoing transformation in the normal course of evolution. Starting with our own planet, it can be heard as the "noise" of nucleosynthesis—the hum of creation—that does indeed pervade throughout the Universe. As long as there is an active, expanding Universe, the hum will not cease. Certainly, the eternal Hum of Creation is a suitable name for this 2.7 K blackbody radiation. Rather than a remnant of the BB, it is the humming of ongoing creation by which our planets evolve through the five stages of evolution and our Sun slowly sacrifices its mass in making the system possible. However, keep in mind that this persistent hum of the spheres bears little relevancy to Kepler's long-sought, elusive music of the spheres. That tune dwells within the beautiful Phi geometry of the SS.

Misinterpretation of the significance of the 2.7 K radiation is a prime example of many discoveries that have been force-fitted into the BB theory in efforts to advance the cause of a concept already fatally flawed. An excerpt from a letter published in *Science News* (27 Jul 1991) reads: "When the Big Bang proponents make assertions such as 'a whole bunch of observations that hang together', they overlook [misinterpret contradictory] observational facts that have been piling up for 25 [now 35] years, and that now have become overwhelming." (5)

These facts expose the sophistry of the BB/Ac myths, and establish the CMB 2.7 K radiation as a product of ongoing nucleosynthesis throughout the Universe.

Responses: These topics raise questions about the validity of assumptions made on the shoulder of previous assumptions. Each warrants an examination in the perspective of the LB/FLINE model:

1. In the arena of Internal Nucleosynthesis (IN) in the evolving spheres of the Universe, scientists do not yet understand the parameters governing the type and quantities of any element produced in a specific time-frame during the cycles of any sphere's evolution; e.g., the extreme internal conditions responsible for the helium-to-hydrogen ratios in our sun, stars, planets, etc. Or why our small central star produces a number of additional light elements, but none heavier than iron. Larger stars can produce a number of the heavier elements, while Earth can produce the full spectrum of 92 elements, including the heaviest one: uranium. This tells us that Earth's heavily encapsulated energy core builds up far more severe conditions of extreme temperatures and pressures than our Sun's interior (which is open to the severe cold of the

vacuum of space). Since hydrogen and helium are the easiest elements to produce in spheres, how can one ever expect to find an object whose helium abundance is far below the amount predicted from the Big Bang—23%—or from the LB/FLINE model?

2. Since all actively evolving spheres are "blackbody" or thermal form, the cosmic background radiation (CBR) could not have turned out to have a spectrum that differed from the "blackbody" form. In view of the scattered positions of spheres, the radiation temperature could not have been perfectly smooth; its unevenness precisely meets the expectations of the LB/FLINE model. The CBR is the 2.7 K "blackbody" radiation emanating from active sources: galaxies, stars, planets, young moons, etc.
3. If a neutrino is found to be the smallest unit of energy, its mass should be the equivalent of one divided by c^2 (the speed of light squared). If so, they are indeed far too small to be important contributors to dark matter. Perhaps, they are the most basic building block of all energy forms?
4. The low deuterium abundance is consistent with the LB/FLINE model's expectations, and is irrelevant to a Big Bang nucleosynthesis that never happened. Nucleosynthesis is ongoing, eternal—not a one-time Big Bang occurrence. It is the driving force that's both inseparable from, and crucial to, the evolution of active spheres and life itself.

Although these might-have-been discoveries that would have invalidated the Big Bang have not been made, they are of a speculative nature based on doubtful assumptions, and offer no positive evidence in support of the Big Bang. But many other discoveries do support definitive concepts that challenge its scientific validity while pointing clearly to a visible alternative. For example, recent discoveries of some 50+ extra-solar systems present enigmas to the Big Bang/Accretion hypotheses, but present no problem to the LB/FLINE model of dynamic origins and evolution that both predicted and explains their mysteries.

When viewed from the LB/FLINE perspective, the four absent observations do not appear to justify pursuit of the Big Bang hypothesis at the expense of ignoring the newer concept that provides a well-substantiated foundation on which to accurately interpret all relevant scientific discoveries. One example: The accelerating expansion of the Universe is the exact reverse of the slowing-down of the expansion as predicted by the Big Bang; however, it is an inherent principle of the LB/FLINE model, which predicted and explains this phenomenon without the need of a cosmological constant.

Should we believe the Big Bang/Accretion scenario or the LB/FLINE alternative? Is the polka-dot beginning of the Universe as misleading now as Ptolemy's antiquated Earth-centered universe was for 1400 years? Will future generations question how the Big Bang hypothesis (first recorded by Edgar A. Poe in 1848) could have survived so long in the face of so much powerful evidence against it, and no genuinely substantiated evidence to support it? Will they look to the Copernican-Galilean experience and ponder why history was permitted to repeat itself? Time, as always, will render the best judgment.

The Relative Abundance of Elements via the Big Bang (BB)

In the 1990s, scientists discovered that galaxies recede from one another in which the more distant ones recede at ever greater rates. The discovery of this ongoing expansion was interpreted to imply that the cosmos had once been concentrated as a very small unit that exploded into our universal system. To make the system work, scientists had to start with a tiny unit of mass at a temperature measured in trillions of trillions degrees. The tiny unit exploded instantaneously, spreading in all directions as a hot cloud of energy. Nuclear physics provided the tools for modeling the synthesis of the elements from fundamental particles within this hot cauldron.

The initial quantitative theory was developed by George Gamow, a Russian-born physicist, who had built a reputation by explaining radioactive decay. In the early 1930s, scientists realized that most stars consist of hydrogen and helium. Since the hydrogen nucleus contains only one proton, it seemed reasonable to assume that it was the first element to form, or precipitate, from the hot expansion as it began cooling down.

Helium, consisting of a nucleus containing four protons and four neutrons, was next in the line of elemental formation created by the fusion of hydrogen. However, the fusion of protons requires powerful forces to overcome the immense repulsion between them.

Many scientists believed that the tremendous heat and pressure required in the processes of fusion could be furnished only by a primordial event or the interior of a star. The prevailing theory of the nuclear physics of stars has been established by Hans Bethe in his explanation of how the Sun shines: nuclear fusion in stellar interiors converts mass into energy (and vice versa). This led to Hoyle's solution in which carbon is produced from three helium nuclei under the severe conditions in star cores.

By 1957, a scheme explaining how stars might have synthesized most of the elements from hydrogen and helium had been worked out by Fowler, Hoyle, Margaret Burbidge and Geoffrey Burbidge. A.G.W. Cameron in Canada did the work independently. But the cosmic abundance of helium remained a mystery, which was [allegedly] solved by Gamow. He suggested that the elements might have formed in an extremely hot, dense gas of neutrons, even before the stars came into existence. Some of the neutrons decayed into protons and electrons—the building blocks of hydrogen. According to this theory, larger nuclei formed in the primeval inferno when smaller ones, beginning with hydrogen, grew through the successive capture of neutrons. This process continued until the supply of neutrons was exhausted, the temperature dropped and the particles dispersed.

Hoyle has the distinction of calling this the Big Bang theory, a concept of creation of the elements that prevails today. Initially, the BB theory failed to explain the formation of the elements beyond helium, which has a mass number of four. Because there are no stable isotopes having mass numbers of five and eight, elements heavier than helium cannot be made by adding neutrons to them one at a time. Only invoking the stellar nucleosynthesis of Hoyle, associates solved this problem.

In a strange twist of fate, the BB theory and the works of Hoyle *et al.* have proven to be important steps in confirming the evolutionary process in which elements are also made in planets via nucleosynthesis. As evidence mounts against the BB theory, it grows more overwhelmingly in favor of the Little Bangs theory of the late 1970s. While it is true that elements are made via nucleosynthesis in stars, our Sun and in supernovae explosions, scientists have somehow managed to overlook an obvious source of creation of elements: nucleosynthesis in the SS spheres and other planets, moons, etc.

Here the BB can take much of the blame, which, in turn, can be passed back to the Kant-Laplace era in which the hypothesis of the formation of our Solar System from a gaseous dust-cloud originated. Scientists conveniently found a source of this material in the BB theory. This, in turn, led to erroneous conclusions that all universal spheres formed via condensation processes. Evidence shows that nothing could be further from the truth.

Quoting Burbidge again, "Big Bang cosmology is probably as widely believed as has been any theory of the universe in the history of Western civilization. It rests, however, on many untested, and in some cases, non-testable, assumptions. Indeed, Big Bang cosmology has become a bandwagon of thought that reflects faith as much as objective truth." Burbidge closes with this mutual thought: "Why, then, has the Big Bang become so deeply entrenched in modern thought? Everything evolves as a function of time except for the laws of physics. Hence, there are two immutabilities: the act of creation and the laws of physics, which spring forth fully-fashioned from that act. The Big Bang ultimately reflects some cosmologists' search for creation and for a beginning. That search properly lies in the realm of metaphysics, not science."

The Fiery Little Bangs Theory: A Plausible Alternative

In 1993, the author found a copy of *Einstein's Universe* (1979) in which Nigel Calder wrote an opinion on the BB theory: "It may be that no one has yet thought of a better and truer scheme for the Universe." At that time, Calder was among the vast majority who were unaware that a more logical scheme had begun to take shape in the author's mind in 1973 and was being stubbornly pursued, persistently developed, and to a very limited degree, recorded in the scientific literature. The Little Bangs (LB) theory of 1979 became a vital link in the FLINE concept that finally came to full fruition in 1995 when the Seven Diagrams of the SS evolved into the Fourth Law of Planetary Motion as the final piece of the giant puzzle of our origins.

Even at this writing, new discoveries continue to add more and more exciting evidence, each bit driving another nail into the coffin of the BB theory while adding substantial support to the LB concept that ties everything together. Calder's book seemed to vindicate my findings of the previous 20 years, while giving incentive to push onward. The initial clue to the LB was rooted in Hubble's discovery that galaxies are moving rapidly away from us and from each other. The BB interpretation that an explosive force imparted such vast momentum to the matter now comprising the distant galaxies did not make sense, especially since they were, and still are, moving aggressively, after billions of years, against strong forces of a centralized gravity.

Further, astronomers now know that galaxies at great distances are significantly younger than nearby galaxies. And on balance, present evidence favors never-ending expansion (not possible in the BB version) and the dire necessity of a continuity in nature. The youthful ages of distant galaxies furnish that continuity. In my mind, the utilization of powerful forces at the spherical perimeter of the expanding Universe was essential in keeping the system functioning in a perpetual motion manner. Black holes and quasars could be the sources of the powerful energy so essential to the system.

The concept fell together in 1979. As the white energy of light reaches the spherical Universe at its perimeter, it is slowed somewhat, and collapses into the form of tiny black holes, which continue onward at nearly the speed of light while continuing to gather energy, also, from the darkness of the space it enters. A black hole eventually becomes a quasar, shining brilliantly because of the subsequent infalling light crashing violently and noisily onto its perimeter. It seemed reasonable to believe that the speed of the infalling light increases dramatically as it is pulled into the new energy mass—perhaps attaining a velocity as high as the c^2 speed figure in Einstein's famous formula, $E=mc^2$. Perhaps this formula is telling us that energy is created by the infalling mass traveling at the speed of light squared.

In the same book, Calder wrote: "Physicists working with atom particles found that light of sufficient energy—very energetic 'gamma-rays' to be precise—could make fresh atomic particles. The energy of light was transformed into matter." At the time, I was unaware of both Calder's book and the matter-antimatter reaction. By sheer coincidence, we had reached essentially the same conclusion through similar processes of thought during the same year.

The LBT formulated at that time predicted that light-energy existing at the perimeter of the spherical Universe coalesces into "black hole" energy that later transforms into bright, noisy quasars that eventually erupt into a series of galaxies, thereby explaining the speed-of-light expansion of the Universe (galaxies and quasars moving at that speed at the perimeter). Calder continued the coincidence: "When a black hole is continuously fed with matter, its surroundings can glow extremely brightly. The falling matter exudes the energy, as if in a dying shriek, before it disappears forever. Quasars are the small cores of violently exploding galaxies, far outshining the ordinary stars."

This description was very similar to the LBT description of a quasar in describing its formation from a black hole, and later exploding into a series of galaxies—written the same year, but only discovered 14 years later. It served as a vital part of the LB theory. The manuscript was copyrighted, but unfortunately no editor accepted it for publication. It remains a good dust collector.

Quoting Calder again "... when a particle meets its anti-particle, they annihilate one another and disappear from the Universe. All that remains is a 'puff' of gamma rays--a reversal of the process of creation." Further, "All that is required to create a black hole is that light should feel the effects of gravity... Space cannot exist separately from 'what fills space', and the geometry of space is determined by the matter it contains."

At this time, it would be worthwhile to consider another, but similar opinion. Edward Tyron, City University of New York, has boldly stated that the Universe could have been created out of absolutely nothing without violating any of our physical laws, if it has certain properties. He wrote, "Every phenomenon that could happen in principle actually does happen occasionally in practice, on a statistically random basis. For example, quantum electrodynamics reveals that an electron, positron and photons occasionally emerge spontaneously from a perfect vacuum. When this happens, the three particles exist for a brief time, and then annihilate each other, leaving no trace behind. (This) is called a vacuum fluctuation, and is utterly commonplace in quantum field theory." Continuing with the LBT, after ages of time the brilliant quasar explodes, sending brilliant fireballs of energy embedded in clouds of energy-matter that spread in whirling finger-like formations. Out of this fiery soup, a group of galaxies is created in spherical bubble-shaped formation around the central point of the explosion.

Such bubble-shaped formations are observable in the distant sky. Closer examination shows a clear hierarchy of galaxies grouped into clusters and clusters into superclusters. Such formations do not seem possible in the time frame and perspective of the BB theory—in spite of the tiny fluctuations in the COBE recording. Within each whirling galaxy, the fireballs react with cooler enveloping clouds. Fiery masses near enough to each other interact to form binary-star systems, a common sight to astronomers. When some of these systems meet the right conditions of relative sizes, velocities and distances apart, the result can be a solar system like the three recently discovered: 51 Pegasi, 70 Virginis and 47 Ursae Majoris. In very rare cases, they go a step further and meet the special prerequisites that cause the smaller, faster mass to break up at the points dictated by Nature's Phi geometry (Chapter III). The result is a SS like ours.

In his excellent article, *Monsters at the Heart of Galaxy Formation* (*Science*, 1 September 2000), John Komendy expounds on the history and most recent findings concerning the relationship between black holes and galaxies as viewed from the current perspective of the Big Bang (BB) concept of origins. He states that massive black holes "were first invoked in the 1960s to explain the enormous energy output of active galactic nuclei (AGNs) such as quasars. These super massive black holes (BHs) stand in sharp contrast to ordinary BHs, which have masses of only a few solar masses and which are well known to form when massive stars die. The origin of super massive BHs is unknown, and their existence long remained a hypothesis. By the mid-1980s, BH 'engines' had become part of the theoretical framework for understanding AGN activity, but evidence for their existence was still lacking." "Surveys with the Hubble Space Telescope (HST) are finding BHs in every galaxy that has an elliptical-galaxy-like 'budge' component. These observations.... indicate that BH growth and galaxy formation are closely linked. These results have profoundly changed astronomers' views of BHs ... BHs are becoming an integral part of our understanding of galaxy formation." The recognition of this BHs/galaxies connection is a giant step in the direction of the Little Bangs (LB) concept (1979-1980) of the origin and ongoing creation of the Universe. In this revolutionary concept, BHs are created as the densest form of energy at the spherical perimeter of a Universe expanding at the speed of light. Eventually, the massive BH is triggered explosively into a small series of fragments in a bubble-shaped formation around the center of the system. Each fragment of the shattered BH spews its energy/matter from its two opposite poles, thus giving the pinwheel effect commonly seen in galaxies.

In this LB perspective, BHs grow first, then later spawn the galaxies consisting of gases, dust, and the fiery masses of energy we call stars, all moving at great speeds. This scenario provides the ideal situation for dynamic interactions of these energy masses which can, and do usually, form into various combinations of solar systems, both binary and multi-planetary. The smaller masses eventually evolve into the planets and exo-planets now being discovered at an ever faster pace. Since 1980, in this LB scenario, spheres always begin as fiery masses of energy and slowly evolve via nucleosynthesis into the spheres we see today. Thus, the basic principle of the galaxies/BHs relationship described in the LB concept of 1980 is being confirmed by these recent discoveries.

But rather than viewing their findings in the perspective of the LB, astronomers still attempt to interpret them in accord with the BB. This raises questions concerning the nucleosynthesis/evolution relationship. In Nature, everything evolves. Since the Evolution of any sphere is not possible without the Internal Nucleosynthesis that drives it forward, how is it possible that nucleosynthesis of all the elements occurred during the first moments of the BB rather than in the individual spheres now undergoing evolution?

The answers to this question remove any doubt about the fallacies of the BB/Accretion concept, while offering powerful substantive evidence for the definitive LB/FLINE concept. Conclusive evidence will stem from future discoveries about the true nature and sources of comets and asteroids during the first decade of the 21st century.

More on Galaxies

Among the recent findings of the Space Telescope was a distant elliptical galaxy billions of light years away that looked too modern at a time when the Universe was supposedly a tenth of its present age. Hubble images show that the spiral shape of galaxies is far more common among these youthful galaxies than in ones closer to us in space and time. Many astronomers had postulated that elliptical galaxies would form more slowly than spiral ones. And all of them had formed so much earlier than astronomers expected galaxies of any kind to form. Yet, going back 12 billion years ago, nine-tenths of the way back to the BB, Hubble had photographed a fully formed elliptical galaxy. When Duccio Macchetto of the Space Telescope Science Institute in Chile aimed the Space Telescope at the mass in the sky, he saw the unmistakable image of the elliptical object.

Proponents of the BB do not believe that the presence of the full-fledged elliptical so soon after the Universe is supposed to have formed casts doubt on the reality of the BB, but it does raise questions about the true nature of the beginning. One favored variant of the BB assumes the spawning of a Universe containing a density of matter high enough to eventually halt its expansion. But that model also predicts that primordial matter was distributed so evenly that galaxy formation was delayed. Consequently, Macchetto's discovery may force cosmologists into a version of a less dense, ever-expanding, open Universe, in which galaxies could take shape earlier. If so, it will be another giant step in the direction of the LB concept.

In *Asimov on Astronomy* (1974), Asimov made the following prediction: "In fact, even in a finite Universe, with a radius of 12 billion light years, there might still be an infinite number of galaxies, almost all of them (paper-thin) existing in the outermost few miles of the Universe-sphere." This prediction might not be too far off. The photograph made by the HST in December, 1995, is the deepest archaeological dig in the history of galaxies.

The great abundance of faint galaxies some five-sixths of the way to the alleged edge of the visible Universe presents a problem. Much interest will focus on the 1500 to 2000 bluish dots, which many astronomers believe are young galaxies in the distant Universe. The implications extend well beyond one tiny patch of sky. Based on it, a survey of the entire sky to the same depth would reveal a total of 50 billion faint objects. To make such a map would require a million years. At these great distances, most astronomers expected galaxy numbers to decrease. The fact that they don't decrease expands the population of known galaxies and leaves an uncomfortably short time for them to form after the BB.

Further, radio galaxies and quasars are more densely packed at distant phases of the Universe than near at hand. Here again, the evidence mounts against the BB theory, while adding powerful support to the LB theory of expansion and growth of the Universe at its spherical perimeter. Not surprisingly, all things in Nature grow in similar manner; e.g., a tree grows upward and outward by adding new matter to the ends of all the branches and trunk.

The immensity of the known Universe continues to grow by leaps and bounds. These vast quantities begin to cast more doubt that they all could have been stored inside a pinhead. Simultaneously, they add significant credibility to the LB theory in which they are created continuously. An interesting comment on the BB theory was made recently in the weekly *Ask Marilyn* column. Responding to an inquiry, she wrote: "I think that if it had been a religion that first maintained the notion that all matter in the entire universe had once been contained in an area smaller than the point of a pen, scientists probably would have laughed at the idea."

In expressing his cynical view of science, Planck also gave hope for the future. Writing in his *Scientific Autobiography and Other Papers* (1949), the great physicist argued, "A new scientific truth does not triumph by convincing its opponents and making them see the light, but rather because its opponents eventually die, and a new generation grows up that is familiar with it."

In the article, *Is There a Super Way to Make Black Holes?* (*Science News*, Sep. 11, 1999), scientists discuss theories and new findings on the origins of black holes. In one theory, massive stars succumb to gravity after burning up their nuclear fuel, and collapse under their own weight to become black holes. In the second model, the star explodes as a supernova, hurling its outer layers into space and leaving behind a dense, burned-out remnant called a neutron star. Debris from the explosion then falls back onto the hole, and is offered as support of the supernova model of black hole formation (Sept. 9, 1999, *Nature*). The research team found that the star's outer layers contain oxygen, magnesium, silicon, and sulfur in abundances six to ten times those found in the Sun. They were puzzled because of the belief that the lightweight star, only twice the mass of our Sun, would never have reached an internal temperature high enough (believed to be greater than three billion K) to forge high concentrations of these elements. They believe that the star's massive companion could have generated them by exploding as a super-nova, ejecting the elements into space before collapsing into a black hole. Scientists were in general agreement that this is the only way that could have enhanced the four elements. These findings, assumptions and conclusions raise issues and questions.

Ranging between atomic numbers, 8 to 16, on a scale of 92 for uranium, these four elements are lightweight and relatively easy to forge in the Internal Nucleosynthesis of all active spheres of the Universe under specific combinations of high temperature and pressure that determine the types and quantities of each element produced. If, according to prevailing beliefs, these lightweight elements were formed solely in the Big Bang (BB), could they have been forged in the star's massive companion before the supernova explosion? Isn't this assumption actually confirming the Internal Nucleosynthesis/Evolution phases of the FLINE model of planetary origins while presenting a good argument against the BB? Why would such explosions not create significant quantities of the heavier elements credited as being created in super explosions? For example, elements 99 and 100 are produced in explosions of hydrogen bombs. At best, these findings do add strong evidence to the FLINE model of creation of atoms via Internal Nucleosynthesis under extreme conditions of temperature and pressure, while raising genuine doubts about the creation of all lightweight atoms (and as originally believed, all heavier elements) in the BB.

At this time, we cannot reasonably conclude that black holes can be made only in the manners described in the article. Open minds can lead to other possibilities. We must question what happens at the ever-expanding perimeter of the Universe when its exiting energies collide with the absolute-zero temperature where all atomic motion ceases. Could these energies collapse into the densest form of energy, black holes that grow rapidly into powerful gravity masses? Would such a mass later burst into spinning twin fountains spewing masses of fireballs in a spiral configuration we call a galaxy? This likely scenario was aptly named the Little Bangs Theory (LBT) in 1979, the year of its conception. Unlike the BB, the LBT answers more questions than it raises. It blends precisely with the FLINE model; both concepts need to be explored in depth by the scientific community. Scientists need to rethink prevailing beliefs about the BB, solar system origins and planetary evolution. Determining the precise combinations of temperature and pressure necessary to create each of the atomic elements via Internal Nucleosythesis within each active sphere during each stage of its evolution is a good starting point.

How Stars Form (1995)

On April 1, 1995 the Hubble Space Telescope snapped a photograph showing about 50 stars inside monstrous columns of dust and dense, molecular hydrogen gas. The gaseous towers, six trillion miles long, resemble stalagmites rising from a cavern floor. At their edges can be seen finger-like protrusions, each with tips larger than our SS, in which the stars are embedded. The information released to the public six months later interpreted the photograph in the perspective of the BB theory; the stars supposedly are created when "the gas collapses under its own gravity". As they grow, the massive stars produce huge amounts of intense ultraviolet radiation that hollows out a cavity around them by heating their surroundings, thereby making them visible. As the cloud gets boiled away, it uncovers stars buried there.

Interpreted in the LB perspective, the monstrous columns of gas and the fiery stars are products of a tremendous explosion (or a series of explosions) that imparts the momentum essential for interactions that can result in active binary systems while scattering them eventually over the six-trillion-mile distance. The stars were created as fireballs in the explosion(s), rather than from the explosion's clouds in which many remain hidden from view. However, the two viewpoints do agree on the cavity formation aspect of the theory.

Cosmic Misfits Elude Star-Formation Theories (Science, 2 March 2001)

To quote from the subject article, "Astronomers have become increasingly perplexed over the last few years by a strange new class of celestial body. Too small to fit conventional definitions of brown dwarfs, they, nonetheless, move through star-forming regions in a manner that separates them from planets orbiting a star. Once seen as anomalies, their growing numbers are forcing astronomers to sit up and take notice. On 14 February, a Japanese team raised the stakes by reporting its discovery of more than 100 of these objects in a star-forming region known as S106." The objects do not neatly fit any conventional definitions. "This poses a big challenge for the standard picture of star formation," says Shuichiro Inutuska, a theorist at Kyoto University.

In addition to hundreds of brown dwarfs, the team spotted more than 100 fainter free-floating objects with estimated masses five-to-ten times that of Planet Jupiter. Their infrared emissions placed the objects within the region. The discovery sheds new light on the ubiquity of isolated planetary-mass objects. One astronomer, Joan Najita, stated, "I think these kinds of results show that the process that makes stars can also make things that are sub-stellar."

Quoting again, "Most astrophysicists believe brown dwarfs and stars condense directly out of vast molecular clouds, whereas planets form in disks of matter swirling around nascent stars. Small lone bodies, however, don't mesh well with either scenario." "Two theories about the origins of planetary objects shed light on the elusive creations, but fall short of supplying a complete answer. One proposes that they are ejected from young stellar systems, the other that they form from molecular cloud cores with masses too low to give birth to stars. But Inutuska says neither idea can account for the large numbers of smaller objects spotted in S106."

The first proposal is very close to the teachings of the LB/FLINE model which predicted and explains the inevitable broad range of sizes of stellar bodies as ejecta initially from much larger masses of energy, perhaps quasars, black holes or large stars during the explosive birth of galaxies. Many of the ejected masses end up as free-floating singles, many as binary systems, and occasionally, but rarely, as a multiple-planet system like ours. In the chaos of such powerful energy systems, there seems to be no rules other than the Four Laws of Planetary Motion to govern sizes or combinations of interacting spheres. Within the realm of the universal ongoing LB/FLINE model in which all things appear to intermesh precisely, there seems to be no room or need to believe in a Big Bang, the primary cause of the perplexity of the astronomers.

The Irony of Iron in Stars and Planets

The news article, *Study reinforces Earth-not-alone theory* (*The Atlanta Constitution*, Feb. 21, 2001) by Alexandra Witze, carried the subtitle, *Amount of iron in stars suggests planets like ours exist*. To quote, "Rocky planets like Earth may orbit most of the 100 billion stars in the Milky Way, a new study suggests. Many astronomers have suspected planets may be common in the galaxy and that extraterrestrial life may have gotten started on other planets like Earth. But the new research increases the likelihood that Earth-like planets really do exist."

The findings were announced by astronomer, Norman Murray, at the annual meeting of the American Association for the Advancement of Science (AAAS). At this writing, astronomers have discovered about 55 planets around stars other than our Sun; however, rather than rocky like Earth, they are huge and gaseous like Jupiter. Continuing the quote, "The new study's findings rely on a statistical analysis of how much iron stars contain, which [allegedly] is a potential measure of how much rocky debris has been incinerated in the stars' upper atmospheres. A lot of such debris could mean enough of it also existed to clump into planets."

"Murray's team scoured scientific studies for data on the chemical makeup of 466 stars. The analyses showed most of the stars contain an amount of iron equivalent to half Earth's mass. The iron probably got there as meteorites burned up in the star," Murray said. "And where there are meteorites, there are probably rocky planets. I can't point at a single star and say that one's got it (a planet). But on average, the stars contain enough iron that they are probably accompanied by Earth-like planets."

It is unfortunate that researchers currently have no choice but to interpret their findings solely in the perspective of the erroneous Big Bang/Accretion concept. If viewed in accord with the teachings of the LB/FLINE paradigm, each and every nuclear energy mass continuously undergoes nucleosynthesis that produces its own atomic elements, especially the lighter third of our universal elements ranging from hydrogen and helium up through iron (and beyond). The 466 stars are no exception to this ubiquitous rule. Their persistent self-sustaining production of elements negates the need to hypothesize erroneously that iron came from meteorites that burned up in the stars—speculation that would be hilarious if not for its tragic consequences. The giant gaseous planets simply are relatively small masses of nuclear energy self-encapsulated with huge volumes of gaseous matter of their own making: the second stage of all planetary evolution.

The most basic principle of the FLINE model is that Internal Nucleosynthesis and Evolution are inseparable: one cannot exist without the other. Nucleosynthesis is the internal nuclear engine that drives the evolution of all active spheres universal. Planets do not depend on stars for their sources of iron; each of the two types of spheres creates its own iron. Thus, there is no factual basis for the conclusion that the amount of iron in stars suggests planets like ours exist. The FLINE model both predicted and explains why such planets do exist, and makes it easy to understand definitively the origins and evolution of the countless billions of planets. The new Fourth Law of Planetary Motion assures us that Earth-like planets like ours might exist, but in an extremely low ratio to the total number of planets in galaxies throughout the Universe.

Redshifts: A Shaky Measuring Rod for Astronomical Distances?

As reported in the article, *Radical Theory Takes a Test* (*Science*, 26 Jan. 2001, p. 579), nearly 3000 scientists and educators gathered in San Diego for the 197th biannual meeting of the American Astronomical Society, held in conjunction with the American Association of Physics Teachers during the second week in January, 2001. A radical theory presented by Margaret Burbidge, a noted astronomer, described how she and two collaborators—Halton Arp and Yaoquan Chu—had found a pair of quasars flanking a galaxy known as Arp 220. Redshift measurements reveal the galaxy to be only 250 million light-years away, while the redshifts of the quasars indicate that they are about six billion light-years away. This brought up questions of whether the finding is a chance alignment of two distant quasars flanking the closer galaxy, and whether redshifts are a valid and dependable method of measuring astronomical distances.

Most astronomers believe that redshifts are a result of the rapid outward motion of the quasars as the Universe expands: the faster the outward motion, the greater the distance away, and thus, the greater the red-shift. "So far," Burbidge stated, "11 nearby active galaxies with high-red-shift 'paired quasars' have turned up since the discovery of the first one four years ago." She believes that in each case, the paired quasars and the associated galaxy might be equally close to observers on Earth.

This naturally brings up profound questions of the odds and significance of finding an X number of such paired systems among the billions of galaxies and quasars. Quoting the article, "Arp goes further. He thinks the quasars originated inside the galaxies, as clumps of new matter created billions of years after the Big Bang. Arp and a handful of other cosmological dissidents believe that matter is still coming into being in some [most] parts of the universe, including the cores of active galaxies [and stars, planets and moons]. The newly created matter [usually] is flung out in two opposite directions, just like the radio-emitting jets of high-energy particles that stream out of many active galaxies [and spheres]. The high redshift of the ejected matter, they say, may be due to its youth (an idea developed by the Indian astrophysicist, Jayant Narliker) or to relativistic effects." The answers to the posed questions remain indeterminate at this time; however, we can look to the Little Bangs concept for some plausible explanations. As explained earlier, galaxies are created when fiery spheres (stars) of all sizes, along with voluminous gaseous dust-clouds, are forcibly ejected from the two poles of a spinning mass of energy, probably a quasar or a black hole, thereby creating the common pinwheel-galaxy effect revolving around the remaining central energy core. These huge, newly-created energy masses, continuously forming while moving outward at the speed of light in the vacuum at the ultra-cold perimeter of the Universe before and during birthing of the galaxies, explain why astronomers observe these massive galaxies moving at an ever greater speed as they look ever farther into space. Naturally, the farther out they look, the younger the observable galaxies. This concept eliminates the need for a cosmological constant. Astronomers will never be able to observe the limits of the Universe, simply because the light from these new systems cannot reach Earth before a vast number of ever newer and unobservable systems already have been created even farther out.

In the paradox of redshifts, one must consider the beautiful correlation between the distance and age of stars—in accord with the Little Bangs theory. Since greater distances equate to younger systems, perhaps redshifts can be attributed to both distance and age. Quasars form near the perimeter of the expanding Universe a relatively short time before ejecting their galactic systems, so it does not seem logical that the findings of Burbidge, *et al.*, present a sound argument against the validity of the red-shift method for measuring astronomical distances. Their findings, at this time, do appear to be chance alignments of more distant quasars and closer-in galaxies. Or perhaps, as a good probability, the 12 closer galaxies in the paired systems were ejected into existence in the opposite direction of expansion and, thereby, slowed sufficiently to account for their apparently slowed rate of outward movement—as indicated by their redshifts—away from observers on Earth. Certainly, more evidence is needed.

Should We Believe the Big Bang Scenario? (Science, 8 December 2000)

This short essay reads as follows: "The extrapolation by astrophysicists and cosmologists back to a stage when the Universe had been expanding for a few seconds deserves to be taken as seriously as, for instance, what geologists or paleontologists tell us about the early history of our Earth: Their inferences are just as indirect and generally less quantitative. Moreover, there are several discoveries that might have been made over the last 30 years which would have invalidated the Big Bang hypothesis, and which have not been made—the Big Bang theory has lived dangerously for decades and survived. Here are some of those absent observations:

1. Astronomers might have found an object whose helium abundance was far below the amount predicted from the Big Bang—23%. This would have been fatal because extra helium made in stars can readily boost helium above its pre-galactic abundance, but there seems no way of converting all the helium back into hydrogen.
2. The background radiation measured so accurately by the Cosmic Background Explorer satellite might have turned out to have a spectrum that differed from the expected 'blackbody' or thermal form. What's more, the radiation temperature could have been so smooth over the whole sky that it was incompatible with the fluctuations needed to give rise to present-day structures like clusters of galaxies.
3. A stable neutrino might have been discovered to have a mass of 100 to 10 electron volts. This would have been fatal because the hot early Universe would have contained almost as many neutrinos as photons. If each neutrino weighed even a millionth as much as an atom, they would, in total, contribute too much mass to the present Universe—even more than could be hidden in dark matter. Experimental physicists have been trying hard to measure neutrino masses, but they are seemingly too small to be important contributors to dark matter.
4. The deuterium abundance could have been so high that it was inconsistent with Big Bang nucleosynthesis (or implied an unacceptably low baryon density). The Big Bang theory's survival gives us confidence in extrapolating right back to the first few seconds of cosmic history and assuming that the laws of microphysics were the same then as now."

More Validation of the Little Bangs (LB) Concept (1973-2002)

In reality, the alleged evidence gained from recent findings argues against the Big Bang (BB) hypothesis while simultaneously adding strong support to the LB concept. "...filaments of these gases, whose temperatures are so hot that they are invisible to optical, infrared or radio telescopes" raises questions about the scientific validity of the interpretations: how can such temperatures exist except via creation of the "dark matter" in more recent times via the LB concept? What is the relationship between such high temperatures and the 2.7 K background radiation allegedly left over from the BB? Why has "dark matter" previously been known only as "cold dark matter" in the BB hypothesis? Do such interpretations based on conjectural myths only add to the mythical conjecture of the BB?

At this time, there is no way for scientists to determine whether the Universe, beyond which mankind will never be able to see, even with the most powerful telescopes, gives assurance of this fact. Until their light rays finally reach us, newer galaxies beyond the newly visible ones will remain invisible, and even newer ones will have been created far beyond the newest invisible ones.

Evidence against the BB/Accretion hypotheses mounts with every relevant discovery, adding strength to the plethora of evidence clearly exposing these "myths" as the grandest scientific hoax ever perpetrated on a naive public. Simultaneously, it strongly supports the LB/FLINE concept as the only alternative underpinned with sufficient substantiated evidence to qualify as a valid scientific concept. According to a basic principle of the Little Bangs (LB) concept (1979-1980), black holes, the densest form of energy, form continuously at the spherical perimeter of the speed-of-light expanding Universe. In time, each is triggered into continuous exhalation (ejection) of dust and brilliant stars from its two poles as it spins to form pinwheel galaxies. During their two stages of noisily scooping "energy/material" into their gravitational gullet, and eventually exhaling stars from their poles, their brilliance (as quasars) is indescribable. A remnant of the birthing black hole remains at the center of each spinning galaxy, holding it together via its strong gravity.

Turkevich's finding that the Moon's surface is volcanic-rock basalt with a high degree of titanium is powerful confirmation of the five-stage evolution of planetary spheres, all in full accord with the LB/FLINE concept. This volcanic method of forming spherical crust is common to all planetary/moon spheres, although the composition of each one's crust and atmosphere will vary in accordance with the variable conditions within each sphere. Such conditions include the nuclear energy core's variables of temperature and pressure that dictate the types and quantities of each element comprising the composition of the surface of each sphere. In turn, these variable conditions are functions of the size (mass) of each sphere, its distance from a central sun, and time. Each sphere is a self-sustaining entity that creates its own compositional matter as it evolves through the five common stages of planetary/moon evolution. Any conclusion predicated on the mythical BB/Accretion hypotheses will itself be mythical and inaccurate. Universal elements are products of ongoing nucleosynthesis within all active spheres of the Universe.

Binary-star systems in which the two stars interact continuously are known to exist in large numbers, and are best explained via the LB/FLINE concept, rather than by the BB/Accretion hypotheses. Galactic stars gain their velocities via the explosive exhalations from the two poles of spinning black holes during the formation of galaxies. As smaller stars pass in the vicinity of larger stars, they lock into orbit around each other.

Imagine a potential binary-star system in which the smaller, faster mass zips past the larger mass at a specific distance apart. If the two masses are of proper relative sizes and have proper relative velocities, the result will be a breakup of the smaller mass into fragments that split off at points determined by Nature's ubiquitous GR (Phi) geometry. With this special scenario in mind, one begins to realize why and how the planets of our SS were placed precisely in geometrically spaced orbits. By combining this realization with the mysterious Bode's Law (BL) for the current spacing of planets, one can derive a plausible Fourth Law of Planetary Motion in which significant changes in the orbits and velocities of planets are revealed and future changes accurately predicted.

In a related manner, the slight variations in the 2.7 K radiation charted by the COBE satellite are best explained via the LB/FLINE concept. Ref: *The Spacing of Planets: The Solution to a 400-Year Mystery* (1996), A.A. Scarborough; *New Principles of Origin and Evolution: Revolutionary Paradigms of Beauty, Power and Precision* (2002), Scarborough.

Abstract for NPA Meeting (9-13 June 2003)

Sagan, Hawking, Burbidge, vos Savant, Mitchell (NPA) and many others recognize the absence of substantiated evidence in the mythical structure of the Big Bang/Accretion hypotheses. Such insight justifies an imminent revolution in scientific beliefs. "If anything is clear in Thomas Kuhn's familiar book, *The Structure of Scientific Revolutions*, it is that no revolution has occurred, nor will occur in the future unless there is a proposed alternative theory to replace the older one." (*NPA Newsletter*, Dec. 2002). *Origins and Evolution* presents the prerequisite alternative that displaces older beliefs, and thereby launches an imminent scientific revolution.

The revolutionary Little Bangs (LB)/FLINE model, underpinned with a plethora of established facts, offers valid scientific alternatives to the Big Bang/Accretion hypotheses. This revolutionary model, featuring the new Fourth Law of Planetary Motion, melds together relevant works of Copernicus, Kepler, Newton, Descartes, Einstein and other philosophers, geologists and astronomers in a continuity of voluminous substantiated evidence. Consequently, the Copernican Revolution (a definitive understanding of origins and subsequent evolution of planetary systems) is brought full circle. The Fourth Law details how energy masses attained their orbital spacing around our Sun before evolving into planets.

Precious Sun Dust

Quoting from the article, *Spacecraft lands with a $260 million thud* (*AJC*, Sep. 9, 2004), "A parachute failed to open as NASA's Genesis space capsule plunged back to Earth on Wednesday, causing it to take a tragic tumble from the heavens that buried it in the desert sands of western Utah, perhaps seriously damaging precious cargo revealing the origin of the solar system... Scientists said the crash that abruptly ended the six-year, $260 million mission breached a canister containing more than 200 ceramic tiles within the 450-pound capsule, exposing the payload to atmospheric contamination and likely reducing the tiles to a mishmash of shattered glass." Further, "There is still hope for a science result from this mission."

"Genesis, launched three years ago, spent 850 days with the tiles exposed to the solar wind to collect atoms representing all the elements and isotopes of the periodic table. The particles—a few micrograms of the Sun's primordial stuff—were expected to tell scientists the composition of the Solar System when it formed 4.5 billion years ago." The Genesis project is a prime example of how belief in the Big Bang myth can misguide scientists into embarrassing and costly situations. Regardless of the final analysis of the solar-wind sample, it cannot tell scientists the composition of the mythical gaseous dust-cloud from which our Solar System allegedly formed 4.5 billion years ago; myth begets myth.

Transformations of energy-to-matter are common to all suns, stars, and all still-active planets, moons, comets, etc.—a fundamental principle of the revolutionary LB/FLINE model of origins and evolution (E) of all universal systems. IN and E are inseparable; one cannot exist without the other.

Big Bang or Little Bangs?: Crucial Questions, Simple Answers

Oct. 5, 2006

The 2006 Nobel Prize in physics was awarded to John C. Mather and George F. Smoot for work that allegedly "helped cement the BB theory, and deepen understanding of the origin of galaxies and stars." While their work has not proven the BB theory, it has given it very strong support, according to the Nobel Committee making the award.

Their discovery: the nature of blackbody radiation (CBR) believed to stem from the BB at the birth of the Universe. Readers may recall the discovery of the CBR by Penzias and Wilson in 1965, for which they, too, were awarded the Nobel Prize. But, is the CBR a product of the BB, or can it be more logically explained by another scenario? Perhaps this can best be answered by asking questions of the two competitive theories:

Question	BB	LB/FLINE
1. Does theory have any substantiated support?	No	Yes
2. How much substantiated support (on 0-10 basis)?	0	8
3. Could microwaves have existed for 14 billion years after BB?	Yes	No
4. Does theory explain ongoing creation of short-lived microwaves?	No	Yes
5. Does radiated energy originate from internal energy associated with atomic and molecular motion, and acceleration of electrical charges within active universal spheres?	Yes	Yes
6. Does every object in the Universe continuously emit and absorb energy in the form of electromagnetic radiation?	Yes	Yes
7. Is the electromagnetic energy emitted by a "blackbody" called blackbody radiation?	Yes	Yes
8. Can theory explain the ongoing source of blackbody radiation (the 2.7K CBR)?	No	Yes
9. Does theory explain self-accelerating universal expansion?	No	Yes
10. Assuming the BB happened, should universal expansion be slowing down?	Yes	No
11. Can the five laws of planetary motion be explained definitively?	No	Yes
12. Can the mathematical spacing of planets be explained definitively?	No	Yes
13. Can Pluto be definitively shown as the original 10th, and now the 9th, planet in SS?	No	Yes
14. Is the ongoing five-stage evolution of planetary spheres well-substantiated?	No	Yes
15. Does theory definitively explain our SS's origin, evolution and orbital changes?	No	Yes
16. Does theory definitively explain why the orbits of Uranus and Neptune are the only two drifting slowly away from the Sun, while all others are drifting closer?	No	Yes
17. Does definitive evidence reveal asteroids as remnants of a disintegrated planet (Olbers,1802)?	No	Yes
18. Does theory enable scientists to decipher the history of our SS and predict changes?	No	Yes
19. Does theory definitively explain the basic cause of volcanism, earthquakes, etc?	No	Yes
20. Does theory definitively explain ongoing origins and evolution of all SSs and planets of all sizes, including giant gaseous planets; e.g., HAT-P-Ib and HD 209-458b?	No	Yes
21. Does theory definitively explain existence of gaseous giant planets too close to a central star to have formed there via accretion?	No	Yes
22. Does theory definitively explain ongoing origins and evolution of galaxies?	No	Yes
23. Does theory explain the ongoing source of superstrings, the fundamental building blocks, at the interface of the Universe's spherical perimeter and Infinite Space beyond?	No	Yes
24. Is $E=mc^2$ a huge, ongoing, crucial factor in the theory?	No	Yes
25. Does the theory violate basic laws; e.g., thermodynamics?	Yes	No
26. Is the theory in sync with Einstein's relativity theories?	No	Yes
27. Has the theory advanced beyond the purely speculative stage?	No	Yes

The search for Truth is life's noblest cause; its discovery, the greatest joy. Author

Facts Inherent in the LB/FLINE Model: Challenges to the Big Bang

September 20, 2006

1. The self-accelerating, speed-of-light expansion of the spherical Universe.
2. The ongoing origin and evolution of the Universe at its spherical perimeter via the fiery birth of majestic pinwheel galaxies.
3. We live in a "subdivision" of an expanding Universe in which its beginning, size, and ending may be indeterminate.
4. The Universe gives the illusion of being static, no matter the galaxy from which it is viewed.
5. Our Milky Way Galaxy appears to be at the center of the Universe—an illusion that holds true for all other visible galaxies when viewed from any one of them.
6. Electromagnetism and gravity are inherent in superstrings, Nature's fundamental building blocks comprising all universal energy and matter. They are what hold everything together at all stages of universal origins and evolution.

The Five Laws of Planetary Motion (FL):

7. The Five Laws of Planetary Motion govern the origin, movements, and mathematical spacing of the nine planets in our SS. Three sets of SS diagrams plus SS-8 corroborate each other.
8. Heinrich Olbers: realization of the disintegration of Planet Asteroid (originally the fifth planet in our SS) into countless thousands of asteroids, recorded in 1802.
9. While confirming Pluto as currently the ninth planet, the Five Laws offer the capability of deciphering the history of our SS, and predicting future changes in orbital positions vs. time.

Internal Nucleosynthesis/Evolution (IN/E):

10. The ongoing, observable five-stage evolution of planets and spherical moons via Internal Nucleosynthesis enables determination of rates of evolution through all five stages.
11. The 2.7 K MBR (universal "noise") emanating from all active spheres of the Universe is short-lived microwave background radiation produced continually via IN/E.
12. Every active sphere has a core of nuclear energy that drives its evolution, and is the basic cause of planetary catastrophies (volcanism, earthquakes and extinctions).
13. Every universal sphere is a self-sustaining entity that creates its own matter (planetary atmospheres, land, water, minerals, iridium) via IN/E.

The LB/FLINE Model of Universal Origins:

14. We live in an ongoing clockwork Universe governed by inviolate laws of physics and chemistry that connect everything to everything else. (John Muir was right.)
15. The LB/FLINE model appears to be a testable theory at every stage of universal progression described above.

Conclusion

The new model is in sync with Einstein's relativity theories, and fulfills his belief that the right answers will be beautiful and simple. These answers provide a serious challenge to the BB hypothesis that remains untested at every stage of its progression.

A Change in Direction of Scientific Thought

Eight Major Dilemmas in the Big Bang Hypothesis

1. The Horizon Problem
2. The Smoothness Problem
3. The Flatness Problem
4. The 2.7 K MBR Problem of short-lived microwave radiation.
5. The accelerating rate of expansion of the Universe.
6. Limitations imposed on evolution of planetary atomic/molecular matter.
7. The spacing of planets in all solar systems.
8. Singularity Problem

These dilemmas (8) pose no major problems in the LB/FLINE model of universal origins and evolution. Additionally, the new model is in perfect tune with the principles of relativity and natural laws. Although lacking any substantiated evidence, the Big Bang has thrived in the absence of a valid alternative—a situation that gives scientists no choice other than working within the limitations of the BB hypothesis.

Today, the opportunities to work beyond these limitations are greatly expanded by the revolutionary LB/FLINE model of universal origins (1973-1996). To paraphrase Thomas Kuhn in his *Structure of Scientific Revolutions:* Before there can be any scientific revolution, there must be a reasonable concept to replace prevailing beliefs. The Little Bangs (LB)/FLINE model of universal origins meets Kuhn's requirement for a valid scientific revolution, and offers a definitive change in direction of scientific thought. A recent journal-published paper presents a mathematical solution to the enigmatic Fourth Law of Planetary Motion (3). Together with Kepler's First Three Laws, the Four Laws (FL) present a definitive explanation of the geometric spacing of the planets during the dynamic origin of our Solar System (SS) some five billion years ago (4) (5) (6)—an impossibility via the BB model, as proven via computers by George Wetherill (*Science*, Aug. 2, 1991).

A follow-up paper in the same *Journal of New Energy* presents a plethora of substantiated evidence detailing the five-stage evolution (E) of planets and moons via Internal Nucleosynthesis and subsequent polymerization. The same principles in the FLINE model of origin and evolution apply to all matter comprising planetary systems (6) (7). Working backwards, the evidence contained in the two papers (and books) provides strong clues to the origin of our Milky Way galaxy, and, subsequently, to the billions of galaxies in our Universe. The result: the 1980 Little Bangs (LB) model for the origin and ongoing functioning of our Universe.

Combined, the two revolutionary models blend into one: the LB/FLINE model that offers valid scientific solutions to many mysteries that remain unsolvable in the Big Bang model of universal origins (3) (4) (5) (6). The new model needs neither dark energy nor dark matter to explain these mysteries. Nor does it need the BB's cosmological constant to explain the rapid expansion of our Universe, nor gravitons to explain gravity. Along with these explanations, the LB/FLINE model explains the 2.7 K radiation throughout the Universe as a product of the ongoing nucleosynthesis in stars and active planets in all galaxies; such radiation is relatively short-lived, and must have a constant source that produces it continuously (6).

The Three Fundamental Infinities

To understand the LB/FLINE model of universal origins, we begin with the three plausible infinities: Space, Energy and Time (SET). Space is saturated with frozen, inactive energy particles, and remains totally dark at absolute-zero temperature until exposed to energies, usually in the form of light, an event that is happening continually at the spherical perimeter of our Universe. The exiting energies, moving at the speed of light, crash into the absolute-zero coldness of space where all atomic motion ceases. The result is the collapsing of energies into the densest form of energy, tiny black holes that grow ever larger by accretion of more and more energies as they move outward at the speed of light on the perimeter of our spherical Universe (6). Meanwhile, Time stands forever still, while everything in Energy-packed Space moves forever onward.

Eventually, due to the pull of universal gravity, each black hole (BH) is slowed somewhat, warmed slightly and triggered into a continuous ejection of dust and brilliant stars from the two poles of the spinning black mass, creating a beautiful pinwheel galaxy. This process continues until it reaches a balance between each black hole and its surrounding masses of billions of stars buried in gaseous dust-clouds (6). As determined recently by Carl Gebbardt *et al.*, this balance occurs when each black hole is reduced to a mass equal to 0.5% of the (ejected) mass. The remnant of the original BH remains at the center of each galaxy to hold the system together; no other forces such as dark energy or dark matter are needed to hold the masses in orbit around the BH (6). This ongoing dynamic formation of speeding galaxies explains Hubble's discovery that the farther out astronomers look in any direction, the faster the galaxies are moving away. The LB/FLINE model, when visualized in the manner explained above, needs no cosmological constant or other forces to explain this phenomenon (6).

According to the new model, nuclear-energy cores of all active spheres of the Universe consist of quark-gluon plasma (qgp); it is the energy source of their Internal Nucleosynthesis, the ongoing transformation of nuclear energy into all forms of radiation and atomic/molecular matter found in varying quantities in all stars and all active planets, moons, comets, etc. It is what makes the stars shine, and spheres evolve. It is the key to understanding the interior of atomic nuclei and gravity; in all forms of atomic energy and matter, gravity is an inherent property that holds everything in orderly fashion (6).

But rather than this continuous process in the ongoing, expansive universal creation in which all data, both old and new, fit precisely into the new model, Big Bang advocates interpret qgp as a one-time primordial product produced at the birth of the Universe some 13 billion years ago—a belief that puts undue limits on our expanding, evolutionary Universe—and forces scientists into speculation and misinterpretations of their excellent discoveries and subsequent data (1) (2). All universal systems continually evolve from energy into matter; nothing is as it once was, and nothing is where it was before—an impossibility within the rigid confines of the BB concept that places undue limits on the ongoing creation of atomic elements comprising universal systems.

Evidence clearly shows that in the absence of Internal Nucleosynthesis (IN) via qgp in planets, Evolution (E) of planetary systems (comprised of virgin atomic matter) via IN would not be possible: qgp, IN and E are inseparable; one cannot exist without the other (3) (4) (5) (6). The ongoing, never-ending source of the qgp via an ever-expanding Universe is the key to understanding universal origins and the integrated functions of universal systems (6).

Perhaps recognition of the LB/FLINE model will come when scientists learn the true nature and sources of asteroids, comets and the cosmic microwave background (the 2.7 K radiation) early in the 21st century (4) (6) (8) (9)—findings that will definitively expose the sophistry of the BB/Ac myth.

References:

1) Thomas Kuhn, *Structure of Scientific Revolution,* (1956).
2) William Mitchell, *The Cult of the Big Bang. Was There a Bang?* (1995). (Cosmic Sense Books, Carson City, Nevada 89702).
3) Alexander Scarborough, *Bringing the Copernican Revolution Full Circle via a New Fourth Law of Planetary*

Motion, Journal of New Energy, Fall 2003. Proceedings, Natural Philosophy Alliance, International Conference, June 9-13, 2003.

4) Alexander Scarborough, *The Spacing of Planets: The Solution to a 400-Year Mystery,* (1995), 8th Edition, Energy Series. (Ander Publications, LaGrange, Georgia 30241).
5) Alexander Scarborough, *Origin and Evolution of Planetary Systems, Journal of New Energy,* Fall 2003. Proceedings, NPA, International Conference, June 9-13, 2003.
6) Alexander Scarborough, *New Principles of Origins and Evolution: Revolutionary Paradigms of Beauty, Power and Precision,* (2001), 9th Edition, Energy Series. (Ander Publications, LaGrange, Georgia 30241).
7) R.A. Karam, J.M. Wampler, Neely Nuclear Research Center, Georgia Institute of Technology, Atlanta, Georgia. *Characterization of Natural Gas, Oil and Coal Deposits With regard to Distribution Depth and Trace Elements Content,* (1986).
8) R.G. Lerner, G.L. Twigg, *Encyclopedia of Physics,* Second Edition. (VCH Publishers, Inc. NY).
9) N.S. Hetherington, *Encyclopedia of Cosmology,* (1993). (Garland Publishing, Inc., NY).

Solutions to Anomalies: A Summary List

The ramifications of the geometric solution to the origin of the SS are immense. As the final phase of the FLINE model, it completes the continuity of evidence that weaves together all the anomalies of our SS. Most of the anomalies listed below have been discussed in this book or in previous writings. All are either solved or solvable in this new perspective.

1. Why planets obey Kepler's First Three Laws of Planetary Motion.
2. The complete geometry of the SS's origin.
3. The original and current spacing of planets.
4. The valid explanation of the enigmatic Bode's Law of the spacing of planets.
5. Why Uranus, Neptune and Pluto do not obey Bode's Law.
6. Why planetary orbits are inclined to the ecliptic.
7. Why the displacement (AU) of each planet from its original orbit.
8. The highly elliptical and large inclination of the orbit of Pluto.
9. The speeds and directions of rotation and evolution of planets.
10. The slow spin of Venus.
11. The huge size of Jupiter (Geometric Mean of the SS).
12. The tapered sizes of the planets.
13. How large moons formed and evolved separately or contemporaneously with planets.
14. Why the Sun's equatorial mass has a 7-degree inclination to the ecliptic.
15. Why the Sun's equatorial mass rotates faster than the remaining mass.
16. The huge discrepancy in angular momentum of the Sun and the planets.
17. The nature of Earth's core: nuclear energy (ten clues).
18. The abiogenic origin and evolution of hydrocarbon fuels (1973-1980).
19. The origin of surface features: water, land, salt mines, etc.
20. The expanding Earth: increases in sea level and land.
21. How the Asteroids came into being: Olbers was right.
22. The craters on moons, asteroids, comets and planets: Mills was right.
23. How planets evolve through five common stages in accordance with size.

24. The physical and chemical differences among planets and moons.
25. Why species come and go: the extinction of the dinosaurs.
26. Why Earth and other planets are self-sustaining entities.
27. Plate tectonics: from highly mobile to barely movable.
28. Why the electromagnetic strength of each planet is different.
29. The unexpected powerful explosions of Comet SL9 on Jupiter.
30. The origin and evolution of comets.
31. The origin and fate of planetary rings.
32. The cometary moons of Mars: Phobos and Deimos.
33. How the discoveries of space probes fit into the FLINE concept.
34. The 2.7 K radiation throughout the Universe: not from the Big Bang.
35. The erroneous notions of the Big Bang.
36. The too-high Hubble constant revealed by the Hubble Space Telescope.
38. How and why the Universe expands at its spherical perimeter—forever.
39. The 51 Pegasi system discovered by the HST.
40. The large extra-solar planets: 70 Virginis and 47 Ursae Majoris.
41. The origins of known extra-solar systems (50+) and binary systems.

New Principles of Origins and Evolution (2001)

The full significance of the Fourth Law of Planetary Motion yet lies camouflaged within the mysteries of the SS and among the enigmas of the Universe. Einstein was dogmatic about continuity in Nature, and rightly so. Nigel Calder wrote, "The uncomprehending antagonism evoked by Einstein's idea is a sign that the old conflict between scientific inquiry and dogma is far from dead." Perhaps it always will be so. Quoting Calder again, "... great syntheses are made in individual minds." (Not by committees, as has been shown many times in the past.) "...all that is lacking is a new comprehensive insight... That must come soon."

To the credit of science, the majority of scientists have significant doubt about the Big Bang theory. However, as long as the concept prevails in the news media, any new insight will have a tough uphill battle just to establish a foothold with that majority. At this writing, the revolutionary LB/FLINE concept has not been permitted to establish that foothold, in spite of the many substantiated facts comprising its sturdy structure. One day in September of 1992, in frustration of the situation, I wrote: "Even as the 21st century approaches, the science of planetary origins and evolution remains bogged in a quagmire of egoistic bias, snugly entrenched in an ivory tower built of cards on a foundation of sand, with shades drawn against the light of new ideas and factual knowledge, reveling in the fantasies of hypotheses that would burst like a bubble with the singular probing of the finger of question."

While it did relieve the frustration somewhat, the situation has shown no perceptible sign of change. Perhaps some day the system will improve, but right now, there is no light at the end of the tunnel. Perhaps this ninth edition of forced self-publications will add enough substance to the already overwhelming evidence against current dogma to force a change in the direction of scientific thought on the origins of our SS and evolution of its planetary systems. In view of the continuity of evidence herein, certainly the LB theory is a plausible alternative to the BB theory. The great beauty of it is that no changes in established evidence should be necessary; however, it is essential, even crucial, to the advancement of the planetary sciences and astronomy that the perspective in which this evidence is interpreted be examined closely. In a battle for truth, one cannot afford to tread too lightly—or too heavily.

More Signs Black Hole Exists in Milky Way Center

September 7, 2001

An interesting article on the black hole at the center of our Milky Way appeared in the journal, *Nature*, on September 6, 2001. Scientists using the powerful new Chandra X-ray telescope garnered evidence that apparently clinches the case for the existence of a super massive black hole at the center of our galaxy. Through NASA's $1.5 billion telescope, astronomers observed a flare of X-ray energy produced where the lip of the invisible black hole should be. The clear-cut image of the flare was the first of its kind; it dimmed and brightened over ten minutes, the time it would take for light to travel about 93 million miles around the lip of a black hole.

Scientists calculated the mass of the black hole to be 2.6 million times that of our Sun, all packed in a small mass. "We are now able to say that indeed all of the mass, by implication, is within that small region, and there is nothing we know that can be that dense and not be a black hole," Frederick Baganoff said. The black hole is 24,000 light-years from Earth. Scientists now believe there are billions of black holes throughout the Universe, varying greatly in mass and luminosity.

More good news came that same evening from TV's Stephen Hawking's *Universe*. One astronomer stated that most scientists now agree that the Universe will expand forever. Thus on the same day, two stunning releases provided powerful support of the Little Bangs Theory in which black holes endlessly form at the perimeter of the Universe, moving initially at the speed of light, gradually slowing while spewing out galaxies of fiery masses and hot gaseous dust, thereby accounting for the observed accelerating expansion of the Universe in every direction, as well as for other anomalies of astronomy. The LBT, thus, quietly continues to provide answers to critical questions raised and unanswerable by the Big Bang hypothesis.

Einstein's general theory of relativity expands the time and space proposals of the special theory of relativity from the areas of electric and magnetic phenomena to all physical phenomena, with emphasis on gravity. Black holes, the densest form of electric, magnetic and gravitational phenomena, are the medium through which energy is created at the perimeter of the Universe and distributed forcefully into galactic systems, each consisting of billions of brilliant stars enshrouded in gaseous dust-clouds. Our Milky Way is but one among billions of galaxies transformed in like manner, each from a black hole, a remnant of which normally remains at the galactic center.

The new X-ray discovery is another confirmation of the LB concept in which our Universe is continually transformed into black energy destined to become brilliant white galaxies in which solar systems form primarily as binary systems and, very rarely, as multiple-planetary systems similar to ours. In this manner does the LB/FLINE paradigm blend with the theories of relativity that expand the time and space proposals from the areas of electric and magnetic phenomena to all physical phenomena that eventually evolve into spheres we call planets, moons, comets, and gaseous dust-clouds.

Letters and Memos

Memo: 01-19-98

As scientists gain opportunities to closely examine the new LB/FLINE model of universal origins, they will discover a valid scientific concept fully capable of providing the substantiated answers not possible in the BB/Ac hypothesis. Although lacking any substantiated evidence, the BB has thrived in the absence of a valid alternative—a situation that gives scientists no choice other than working within the limitations of the BB hypothesis. Today, the opportunities to work beyond these limitations are greatly expanded by the revolutionary LB/FLINE model of universal origins (1973-1996).

All universal systems continually evolve from energy into matter; nothing is as it once was, and nothing is where it was before—an impossibility within the rigid confines of the BB concept that places undue limits on the ongoing creation of atomic elements comprising universal systems. Evidence clearly shows that in the absence of Internal Nucleosynthesis (IN) via qgp in planets, Evolution (E) of planetary systems (comprised of virgin atomic matter) via IN would not be possible: qgp, IN and E are inseparable; one cannot exist without the other (3) (4) (5) (6). We live in a self-perpetrating universal ocean of electromagnetism, a most critical fundamental of everything—both inanimate and animate—in our Universe. Electromagnetism is inherent in elementary particles of energy, whether in the nuclear stage or comprising the atoms of all matter. One might question whether electromagnetism and gravity are one and the same; logic and evidence indicate they are indeed the same. If so, the EM-G entity could well be the attractive force that brings and holds together all the components of universal systems throughout their complete cycles. In the EFT, the Universe operates entirely through electromagnetic force. Several cited factors explain the wide variations in electromagnetic field strengths of planetary spheres in accord with the LB/FLINE model of universal origins. This concept permits definitive solutions to universal anomalies; e.g., origin of solar systems, the Fourth Law of Planetary Motion, the spacing of all planets, the 2.7 K radiation, lightning, etc.

To: Glenn Strait, Science Editor, The World & I Magazine, Washington, D.C. 20002. Jan. 16, 1998

Thanks for the fax of January 14 concerning two press releases about uranium-prospecting with neutrinos and rapid shifting in tectonic masses. You do know how to ask some tough questions in the realm of physics. I'm not a physicist, so I will respond only from the perspective of the FL/IN concept. Whether neutrinos emanate from encapsulated nuclear energy cores at the highly-elevated conditions of temperature and pressure, therein, is beyond my knowledge. My guess is that they do, but in quantities that vary considerably from those calculated (or recorded) from the Sun (relatively speaking).

I do not understand how one can determine that 40% of the terra watts of energy leaving Earth's surface comes from radioactive decay of uranium - 238 and thorium - 232 (products from the core), rather than directly from the nuclear core. In my mind, it seems difficult, if not impossible, to differentiate the sources with any degree of accuracy, and neutrinos do scatter in all directions. So it seems unlikely that geophysicists will "map the deposits more accurately" by the neutrinos method—but perhaps I do not understand the situation. Further, the belief that "radiogenic heat has had a large role in determining the dynamics of our planet's interior" is a misleading assumption: the taproot reason for the failure of scientists to recognize the true source of Earth's heat. Radiogenic heat is a relatively cool by-product of the core's nuclear reactions; it plays little, if any, part in heating Earth's interior. Neither presentation described the original source of the elements providing the radiogenic heat.

To the second question: If, "from the perspective of (my) ideas, would (I) interpret the evidence in the same way as is done here?" I'm excited to report that these findings and their interpretations fall right in line as additional supportive evidence of the FL/IN paradigm—just as the findings on Mars and of other space probes do—and the basic reasons for these dramatic changes in Earth's evolution can be traced to the nuclear-energy core. Without this energy source to drive evolution, such changes would not be possible. All new findings continually pull scientists ever closer to the new paradigm. The next giant leap in progress will come when the truths of the FL/IN concept are recognized, accepted and utilized.

David Evans is correct in stating, "We might expect the Cambrian example to have been caused by some other tectonic process not involving such rapid sliding of tectonic plates over the upper part of Earth's mantle...a process known as true polar wander, caused by an imbalance in the mass distribution of the planet itself, which is forced by the laws of physics to equalize itself in comparatively rapid time scales."

The next step is to find the basic reasons such imbalances occurred, and are still occurring. This is where the evolution of planets through five stages of evolution via Internal Nucleosynthesis fits into the big picture. To conserve writing time, I've enclosed copies of pages 32-33 from my book, *The I-T-E-M Connection: How Planet Earth and Its Systems Were Made by Means of Natural Laws* (1991) that explain the processes involved. Briefly put, separate islands of earthly mass did form and grow gradually, and drifted dramatically for eons before the entire encapsulation of the core mass/mantle was completed. Land drifting did continue at an ever diminishing pace until the entire crust solidified and stabilized, locking the masses in place. Meanwhile, mountains rose, while waters flowed to the lower regions to form oceans and lakes. Thus, plate tectonics changed from the loose, drifting stage to the current "stabilized", or locked-in, stage. The movements scientists detect today are due primarily to the very gradual expansion of Planet Earth as the core's nuclear energy continually transforms into matter—a process that results in tremendous internal pressures (and temperatures) caused by the powerful expansions of the new matter from energy.

Consequently, Earth's circumference gradually expands, cracking the crust at the weak points (e.g., the rim of fire). The expanding virgin lava is forced up the paths of least resistance, continuously from undersea seams and intermittently from some 1600 volcanoes around the globe, always building more land and larger, deeper oceans. Like spots on an inflating balloon, the continents continue to drift apart. All events can be understood better in this perspective. For example, when huge volumes of matter accumulate at a given location (e.g., Deccan Traps) while other volcanic sites remain dormant, an imbalance on Earth's surface can occur—an imbalance that could result in "true polar wander".

I strongly suspect that Uranus, Pluto, and perhaps other planets, to a lesser degree, are other examples of this occurrence. Of course, as Earth's crust thickens and stabilizes more and more over time, scientists should observe an ever-diminishing "polar wander"—perhaps until close to the time of exhaustion of the core's nuclear energy some three-to-four billion years from now (theoretically speaking).

You did not ask, but I would like to comment on the "evolution of life forms" before closing. As Earth evolved (and still evolves) through its five stages of evolution, species come and go in compliance with the demands of the ever-changing environment. The dinosaurs were no exceptions; they died out over a period of 500,000 years, due to the persistent environmental changes wrought by the core's nuclear energy. The pattern of the layers of iridium is a clue to their extinction; however, the patterns clearly indicate Earth's nuclear core as the source of all iridium layers, each layer occurring in different eras of time via volcanism whenever internal conditions dictated the production and ejection of iridium.

Thanks again for your interest and considerations.

P.S. Several articles in the current issue of *Astronomy* magazine (Feb. 1998, pp 37-56) entitled, *On the Origin of Everything,* are interesting. They present clearly the current views of the origins of our Solar System. By coincidence, I watched *A Science Odyssey* on PBS TV last night on the origin of Earth—the same version described in the library magazine earlier in the day. In neither instance was I able to detect even one statement of valid scientific evidence supporting these views.

To: Glenn Strait, Science Editor, The World & I, Washington, D.C. 20002. 02-09-98

I feel compelled to comment on four new writings in the scientific literature received this week. James Glanz's article, *Exploding Stars Point to a Universal Repulsive Force* (*Science*, 30 Jan 1998, pp 651-652), states, "Not only did the results support the earlier evidence that the (cosmic) expansion rate has slowed too little for gravity ever to bring it to a stop; they also hinted that something is nudging the expansion along." Perlmutter concluded: "That would introduce important evidence that there is a cosmological constant." Michael Turner added, "What it means is that there is some form of energy we don't understand." The article continued, "Other observers had already found signs that the Universe contains far less mass than the mainstream theory of the Big Bang (BBT) predicts, which left open the possibility that some form of energy in empty space could be making up the deficit ... But the mass just doesn't seem to be there... Both teams concluded that the expansion had slowed so little that it will probably go on forever." (*Science,* 31 October 1997, p 799).

While these findings present even more problems for the BBT, they fit precisely into the Little Bangs Theory (LBT) of 1979-1980—described in my earlier writings—in which a cosmological constant is not needed as a repulsive force to complete the big picture. Einstein was the first to propose a universal repulsive force, which he later—and according to the LBT paradigm—correctly abandoned. The basic principles of the LBT in which the Universe expands at its spherical perimeter at the speed of (exiting) light (via energy/matter interactions) are further enhanced by another article: *Gamma Rays Create Matter Just by Plowing into Laser Light* (*Physics Today*, Feb 1998, pp 17-18). To quote: "But the SLAC experiment was the first direct observation of material particles produced by nothing but protons," which "can also be thought of as 'the sparking of the vacuum', an exotic prediction of quantum electrodynamics at extremely high electric field intensities."

I believe Dirac initiated this sparking concept in the early 1920s, and other name-scientists have added to the concept. In *Comet Shower Hit, But Life Didn't Blink* (*Science*, 30 Jan 1998), "researchers tapping a new sort of geologic record find solid evidence that at least one comet shower did pelt Earth—but without apparent effects on life ... The Massignano sediments yielded a two-million-year-long surge in helium-3. It peaked about 35.5 million years ago, right at the time of the two major impacts that allegedly caused the extinction of the dinosaurs." The evidence, per se, is in perfect harmony with the Energy Fuels Theory (EFT) of 1973 in which significant quantities of helium-3 show up with gas from deep-Earth drillings. The new EFT reveals that all elements comprising these deposits were (and still are) created via Internal Nucleosynthesis.

Thomas Gold presented an excellent paper at the AAAS Meeting in 1985 on the gas/helium-3 association in deep-Earth drillings—which added strong evidence to the EFT. Unfortunately, an incorrect version of the 1973 EFT became known as Gold's theory in the 1980s. (See enclosed copy.) Next, *The New Gamma-Ray Astronomy* (*Physics Today*, Feb 1998, pp 26-28) presents a good discussion on nucleosynthesis sites. "The modern era of theoretical nucleosynthesis sprang from a classic 1957 paper by Burbidge *et al.",* in which heavier elements are created in stars, supernovae and novae rather than being components of primordial gas from the Big Bang. The next logical step came from my realization in 1973 that planets evolve(d) from smaller masses of star-like energy—each one creating all elements comprising its atmosphere and/or crust via Internal Nucleosynthesis (IN)—until the nuclear core is depleted. Ironically, this concept inevitably led to the solution to the proposed Fourth Law of Planetary Motion (1980-1995) (now being referred), the last phase that finally brought the longtime evolving FL/IN paradigm full circle.

Now, as with the four papers discussed herein, all relevant articles and discoveries continually add more supportive evidence to this new perspective. The fourth article, *The Origin of Chondrules at Jovian Resonances* (*Science*, 30 Jan 1998, p 681) states: "Isotopic dating indicates that chondrules were produced a few million years after the solar nebular timescales for the formation of planetesimals." The authors present a tentative explanation of the puzzling discrepancies by attributing them to "Jovian resonances that cause collisions and melting of dust by bow shocks in the nebular gas"—an explanation based more on erroneous speculation than fact. The basic problem here is the outmoded hypothesis that planetesimals accrete, or accreted, from a nebula of dust and gas.

The findings can be interpreted more accurately in the perspective of the FL/IN paradigm in which planetesimals are remnants of the (three) Asteroids planets that exploded—just as Olbers concluded in 1802. The proposed Fourth Law of Planetary Motion and Kepler's First Three Laws clearly reveal how the original Asteroids' masses of energy attained their orbital positions around the Sun before beginning the five-stage evolution common to all planets. The ages of chondrules should indicate the time(s) the disruptions (perhaps explosions) occurred. Chemical analyses of their compositions should reveal much about both the internal and explosive nucleosynthesis aspects of these evolutionary events. For example, there are direct connections between the nickel-iron contents in some asteroids and meteorites and the results discussed in *The New Gamma-Ray Astronomy* article above.

Finally, the findings described in the four articles can be tied neatly together via the FL/IN paradigm. These four examples illustrate a crucial point: it is highly probable that interpretations in the perspective of the FLINE paradigm will provide correct answers to the many problems now existing throughout the prevailing beliefs about the origins and evolution of planetary systems.

The same thing can be said for the LBT vs. the BBT. At the frontier of research, these new ideas seem destined to stimulate new directions for research; for example, in the basics of Internal Nucleosynthesis and its connections with El Nino and all other phenomena (on Earth, other planets and moons). Unless someone is able to find a flaw in the LBT/FL/IN paradigm, my faith in its scientific validity remains steadfast. Thanks for taking the time to read this long letter, and for your other considerations.

P.S. - Most of the evidence in C. W. Hunt's *Anhydride Theory: A Theory of Petroleum and Coal Generation* (1997), herewith, is in agreement with my Energy Fuels Theory (EFT) (1973). However, as with Thomas Gold (1980), Hunt misses the crucial starting point: methane (C & H molecules) created via Internal Nucleosynthesis, followed by polymerization and cross-linking into petroleum and coal.

P.P.S. - Publication of the EFT and the LBT/FL/IN paradigm will lead to a provable understanding of the origins of solar systems and the evolution of all planetary systems.

To: Glenn Strait, Science Editor, The World & I, Washington, D.C. 20002. 04-15-98

Thanks for your interesting Internet article of 98-04-10, concerning the vast amounts of water found in Orion. You asked how the explanation, therein, for the origin of water compares with my explanation. In the FL/IN/E paradigm, water and all other planetary matter are end-products of Internal Nucleosynthesis (IN) that drives Evolution (E) in all active spheres of the Universe. IN and E are inseparable; one cannot exist without the other (a fact that deftly eludes advocates of the Big Bang) (BB). The Four Laws of Planetary Motion (FL) irrefutably support the fiery origin of solar systems and the evolution of the orbiting energy masses that gradually evolve into planetary matter through five common stages of evolution in accord with size and in full compliance with all natural laws.

This also explains the anomalies of exo-planets (e.g., most are too close to the parent star to have condensed from a gaseous dust-cloud and/or small planetesimals). The Universal Law of Creation of Matter via IN applies equally to the Orion water—an end-product of IN—expelled into space in a manner yet to be detailed. With the headline that a "Factory found in Orion would fill Earth's oceans 60 times per day", BB advocates appear to be stepping gingerly into the FLINE paradigm. Are they saying that water is still being created in cold space, and with or without Internal Nucleosynthesis? What is the source of the water's oxygen? I am under the impression that all water (and all other matter) had been created during cool-down from the BB, and everything coalesces via gravity.

In any case, the interpretations expressed in the article appear to be far-fetched and force-fitted to the occasion, while the aforementioned fact that IN and E are inseparable, and one cannot exist without the other, is ignored. Even though the BB admittedly is highly speculative and contains a number of fatal flaws, many advocates remain adamant in their refusal to consider alternatives—as was illustrated in the Internet article. The problem with the interpretations stems from this situation.

To: Dara Horn, Harvard University, Cambridge, MA 02138. June 4, 1998

Congratulations on your excellent and perceptive article, *The Shoulders of Giants* (*Science*, 29 May, 1998). During 63 years of study and research in various fields of science, I have never read a more tactfully perceptive description of the way things are in real life. However, I must admit my genuine surprise that such an exposé could be written by anyone so young, and that it could get past peer reviewers for publication in this magazine. Cecilia Payne-Gaposchkin's fate is a sad commentary on science. Some years ago, I read of a similar situation in *The Cry and the Covenant* (author, Thompson) in which Dr. Samuel Semmelweiss met an even sadder fate for his perceptive contribution of cleanliness during childbirth in the mid-to-late 19th century. If you have not read the book, I strongly recommend it to you for further understanding of real life in the sciences. Many scientists have met, and still do meet, similar undeserving fates.

My most familiar example is my present situation. For the past 25 years, my research in the field of planetary origins and evolution inevitably led to the proposed Fourth Law of Planetary Motion (1995) that first eluded Kepler (the discoverer of the First Three Laws) in 1595. The geometric solutions to the new Fourth Law (FL), together with Kepler's First Three Laws, detail how the planets attained their current spacing around the Sun—eventually to evolve via Internal Nucleosynthesis (IN) (from energy to atoms) into their present stages of Evolution (E)—all in full accord with size.

In our Solar System, the three facets of this FL/IN/E paradigm are inseparable (until the energy-source core is depleted); one is not possible without the other two. Powerful supportive evidence continuously mounts; every relevant discovery serves as additional corroborative proof requiring no speculation, while leaving little, if any, room for doubt. Currently the major problem is the one you described so well concerning Cecilia's situation. As with Copernicus, this revolutionary concept upsets prevailing speculative beliefs about solar systems and the evolution of planets. As such, it cannot get past peer reviewers schooled in current dogma.

To preserve it for posterity, I was forced to *self-publish at great sacrifice to my budget. Grants, obviously, were unattainable. For the past three years, most of my abstracts and papers have been rejected without cause, reason or explanation by members of my science organizations—censorship at its finest. The irony is that the total package of the FLINE paradigm explains why the stars are "amazingly uniform" in their composition and why "hydrogen is millions of times more abundant than any other element in the universe". Also, why and how light and heavy elements are created within Planet Earth.

Unfortunately, we both have been disillusioned: science is not the pure and objective pursuit of knowledge; rather, it has evolved into a political big-money game rift with the emotions best described in your perceptive article. The question now is what can be done by younger people like you to bring science back to where it should be. For the sake of science and posterity, I hope you will continue to use your tremendous talent as a writer always to strive towards this goal.

*Ref: *The Spacing of Planets: The Solution to a 400-Year Mystery*, (1996).

To: K.C. Cole, Journalist, Los Angeles Times, Los Angeles, CA 90053. 1998

Your excellent article headlined *33-year-old changing way astronomers see universe* (*Atlanta Journal-*

Constitution, Sept 13, 1998) featured Andrea Ghez's presentation in August at Rutgers University in New Jersey. Her paper adds corroborative evidence to a revolutionary concept that has evolved during the past 25 years: the little-known FLINE paradigm of origins of the Universe and its solar systems, and the orbital spacing and subsequent evolution of planets. The lack of familiarity with this valid scientific concept, perhaps, stems from its serious threat to current beliefs about such origins.

The FLINE paradigm explains why "a massive black hole sits at the center of the Milky Way" and at the center of all galaxies—and why a black hole is essential to the formation of each and every galaxy. In the perspective of the Little Bangs Theory (LBT) of 1980—a vital part of the FLINE paradigm—the explosive black hole spews out, in pinwheel fashion, the dust, gases and fiery masses of energy we call stars. Obviously, such close initial proximity accounts for Ghez's "discovery that most new-born stars appear to be twins". Seldom can they remain single stars. Clearly, they usually were born double; however, some can drift apart later as single stars. In similar manner since 1979, the evolving FLINE paradigm has contradicted the prevalent theories of how stars form from dust and gases—which, in reality, are by-products created simultaneously with the fiery stars comprising the galaxy ever since their explosive ejections from the twin outlets of a central black hole.

Obviously, this concept adds dramatically to the large amount of evidence already contradicting the Big Bang hypothesis of the origin of our Universe. The new paradigm provides the perfect (and perhaps the only logical) mechanism to explain the formation of double stars, including the recently discovered exo-planets. Ghez's belief that the current theory "very nicely produces our sun and planets" continues to ignore the Four Laws of Planetary Motion (FL), detailing how the planets attained their orbital spacing around the Sun. The Four Laws clearly reveal how our dynamic Solar System formed as a special and rare multi-star system via the same explosive forces that formed all binary-star systems.

Planetary Evolution (E) is a fact of Nature, but such ongoing changes are not possible without a source of energy to drive them forward through the five stages of evolution common to all planets: from energy to gaseous to transitional to rocky to inactive (dead) spheres. In scientific terms, these processes are known as Internal Nucleosynthesis (IN) (the creation of atoms from energy, per $E=mc^2$) and polymerization (the combining of atoms to form all planetary matter). The complete details are found in my writings of the past 25 years. In further confirming the FLINE paradigm, Ghez clearly reinforces this revolutionary concept to the extent that it now seems destined to displace prevailing beliefs about planetary and universal origins early in the 21st century. Perhaps her paper and many corroborative findings of space probes will encourage scientists, finally, to give more serious consideration to the FLINE paradigm of 1973-1998. If you care to delve deeper into this new concept for more stories, I would be happy to work with you.

To: Glenn Strait, Science Editor, The World & I, Washington, D.C. 20002. March 9, 1998

Your e-mail on *Electric Space* does stimulate thinking. I'm convinced that there is a lot of truth in the plasma model of the Universe, and that it does offer significant challenges to the Big Bang. However, it does fit well in the Little Bangs Theory (LBT).

As far back as science is able to comprehend, everything in the Universe begins with "electric space". We may never know how this basic form of energy came into being; however, the LBT describes its continual creation out of empty space at the rapidly expanding spherical perimeter of our Universe.

Nature's next step is the transformation of "99.999 percent" of the basic electric energy (BEE) (which some scientists call ether) into other forms of energy (i.e., black holes and quasars) that form at the perimeter of the ever-expanding-at-the-speed-of-light Universe. From these sources springs the explosiveness that creates the galaxies of plasma comprising their stars, dust, gases, solar systems, etc. In this explosive mix, the observable plasma energy forms are usually misinterpreted in the perspective of the BB as stars created via condensation of the dust and gases.

In reality, the starry fireballs of plasma energy are by-products of the powerful explosions, or outpourings, of these forms of energy. Under the proper conditions, smaller masses become locked in orbit around larger masses in the initial phase of binary and—in rarer cases—multi-bodied "solar systems". This revolutionary concept is corroborated strongly by the Four Laws of Planetary Motion—which are vital factors in explaining the anomalies of our Solar System and exo-planets. Once placed in orbits around a Sun, each fireball of plasma begins it transformations through the five stages of evolution common to all planets and moons—until the energy source is depleted (totally transformed into planetary matter). Evolution ceases, and everything eventually grinds to a halt at the end of the long journey of transformations from the original plasma energy into its ultimate destination: atomic matter. The fact that "plasmas generate electromagnetic fields" clearly confirms the true nature of planetary cores, thereby providing convincing evidence that no other source (i.e., the geodynamo) is necessary to create planetary electromagnetism. The fact that "plasmas are also prodigious producers of electromagnetic radiation, from very low frequencies, to microwaves, to very high frequencies such as those associated with cosmic rays" is interpreted in the FLINE paradigm as an explanation of the 2.7 K radiation and the recently discovered infrared glow throughout the Universe, both of which have been identified erroneously as leftovers from the BB.

The FLINE interpretation renders a better understanding of the significance of Planck's curve of "blackbody" spectrum that applies to all thermonuclear reactions—which further explains why the cosmic background radiation can be recorded so smoothly in every direction (as detailed in my latest book, *The Spacing of Planets*). In confirming my writings on the subject, this press release brings researchers another giant step closer to the FLINE paradigm and to "a revision of our understanding physical processes in space as far-ranging as the formation of planets to the sources of high-energy particles and radiation".

To: Glenn Strait, Science Editor, The World & I, Washington, D.C. 20002. **09-25-98**

Thanks for the information and suggestions in your e-mail of 98-09-15 under the subject of *Common Sense Science*. The address you requested has been mailed in a previous response to another subject, but I would like to make some comments. I feel at fault in not being able to convey the connection between the Little Bangs Theory and the FLINE paradigm for the origin of the Solar System and the evolution of planets. The link resides in the necessity of continual creation of energy and its subsequent explosive manner of conversion into lesser forms of energy systems such as galaxies of binary and multiple-star systems of fiery masses that subsequently are transformed into the atomic form of matter. The LBT is to the FLINE paradigm as the Big Bang is to the Accretion concept of planetary origins. The basic principle is the continual transformation of universal energy from one form to another in full accord with the prevailing conditions imposed upon it. But I will leave the cosmology details to others, as you suggested.

The FLINE paradigm has evolved slowly during the past 25 years of carefully piecing together as many relevant and factual bits of information as feasible from the scientific literature. The primary objective was to separate facts from myths. I feel that this guiding principle has forced me into meeting that goal. The really surprising thing is the huge number of facts that have accumulated during those years of research and how well they interlock throughout without the need for assumptions and speculations.

The best analogy for me: the finished jigsaw puzzles of many years ago in which it was essential to interlock each piece into its correct position. If a piece did not fit into a position, it could never be force-fitted. When finished, all the pieces fitted precisely in place to reveal the big picture. My weaker point does reside in the field of sub-atomic physics, even though I have gained some knowledge of it through the years. So I must leave the details of the Internal Nucleosynthesis of stars and planets to the better-trained physicists. However, the finer details really are not crucial to the structure of any valid concept as long as there is sufficient evidence, otherwise, to corroborate its every facet.

The vast quantity of such evidence corroborating the FLINE paradigm is truly surprising. Its IN/E facets forced me into finding the final piece of the puzzle that ties everything together: the solution to the Fourth Law of Planetary Motion (1980-1995) that first eluded Kepler in 1595. As to the neutrino puzzle, my best guess, at this time, is that the answer might reside in the differences between the parameters of the two sources you mentioned: our Sun and planets. One is a huge, open system; the other is a very small closed system. My chemical engineering professors at Georgia Tech stressed that the differences in two similar systems will account for differences in the end results. This lesson proved to be crucial to understanding the differences in elemental compositions of planets, caused by the progressively changing conditions (e.g., encapsulation, pressure, temperature, concentration, etc.) within nuclear-energy cores as functions of time and planetary size—which explains why Earth can produce elements up through uranium, while the Sun can produce elements only through iron. Allow me to pose two questions: How does one distinguish between a neutrino from the Sun and one from our planet? Can neutrinos be detected and identified as emanating from Jupiter (in only the second stage of evolution)? Since I don't have the answers, any other suggestion of mine would be in the realm of pure speculation.

To: Christine Gilbert, Letters Editor, Science, Washington, D.C. 12-29-98

In his excellent article, *Cosmic Motion Revealed* (*Science*, 18 Dec. 1998), James Glanz discusses how "astronomers peered deep into the universe and found that it is flying apart ever faster, suggesting that Einstein was right when he posited a mysterious energy (lambda) that fills 'empty' space". Without lambda, this discovery conflicts with the Big Bang Theory (BBT) in which just the opposite (a slowing down at these great distances) is more likely. Quoting Glanz, "Back in 1917 when Einstein proposed the (lambda) constant, he thought the universe was static, neither expanding nor collapsing. He put the cosmic repulsion (lambda) into his equations to prevent the universe from collapsing on itself from the gravitational pull of the matter inside it."

Years later, Einstein reasoned that if the expansion was a relic of a primeval explosion, the cosmological constant—which he felt made the equations unaesthetic—wasn't needed. He withdrew the idea and called it (correctly) his "biggest blunder". In the BBT, the observation that distant matter is flying apart ever faster must, of necessity, utilize a lambda factor. However, there are caveats.

So far, "Calculations suggest that such a lambda should be many orders of magnitude larger than the supernova groups have seen. That puzzle has launched a search for new physics principles ... At this point the cosmological constant remains in the realm of theory; no one yet knows the precise nature of the energy causing the universe to fly apart ever faster." To exacerbate this odd situation, volumes of relevant substantiated evidence against the BBT continue to be ignored.

Back in 1980, a revolutionary concept of universal origins accurately predicted the findings of the supernova groups, and clearly explained the reasons the Universe should be flying apart ever faster as astronomers peer deeper and deeper into space. The definitive concept was named the Little Bangs Theory (LBT). In this version, energy is created at the perimeter of the Universe expanding outwards at the speed of light. From these energy masses (black holes and quasars) come the ever-expanding numbers of limitless galaxies. In the perspective that everything is connected to everything else in the Universe, the LBT led to a FLINE paradigm of planetary origins structured with the Four Laws of Planetary Motion (FL) and Internal Nucleosynthesis (IN) processes without which Evolution (E) is impossible.

But the LBT concept can hold true only if the Universe does not fly apart faster than the speed of light; if that speed limit is found to be exceeded, then the necessity of a lambda constant might be likely. But as of now, Einstein appears to be correct in his "biggest blunder" assessment; a lambda constant is not needed here. Although only a small facet of the concept, this exciting "breakthrough of the year" does corroborate the predictions and explanations of the LBT. Rather than searching for lambda or for new physics principles, perhaps a close examination of the 1980 concept would be more productive.

To: William C. Mitchell, Institute for Advanced Cosmological Studies. 3-5-99

Thanks for the copy of your interesting paper, *The Recycling Universe*. I enjoyed reading it, especially the quotes of some great minds of past and recent times (Ref: Jeans, Harwit, Rees, Silk, Davies, etc.). All of the statements fit precisely into the Little Bangs Theory (LBT) of 1980. I could imagine that each of them was speaking of a specific facet of the LBT. And your comments on galaxies fit equally well. In my mind, galaxies consisting of fiery stars, dust-clouds and gases are products of explosions of much larger masses that often place them in bubble-shape formation like points on a spherical perimeter.

To: Physics Today, American Center for Physics, College Park, MD. 05-24-99

In publishing the two opposing views, *A Different Approach to Cosmology* (*Physics Today*, April 1999) by Burbidge *et al.*, and the reply by A. Albrecht, you have done a great service by encouraging open debate in the sciences pertaining to origins and evolution of our Universe. While both articles present interesting viewpoints, the different approach offers the more convincing arguments. The most surprising and pleasant aspect, however, is that these frank views were brought clearly into open debate in the same magazine issue—a good idea concerning universal origins that too often seem biased in favor of the BB model.

While the quasi-steady-state universe (QSS) supplies the different and more logical approach, both articles raise questions that appear unanswerable by either of the two models. These questions warrant consideration of a compromising third model capable of supplying these answers. The Little Bangs Theory (LBT), initiated in 1979, offers valid explanations of the controversial issues of a cosmological constant (not needed; Einstein was right), the microwave background, its fluctuations and its connection to ongoing nucleosynthesis throughout the Universe (not really possible at any time in the BB concept), the creation of energy and its transformation via nucleosynthesis into the matter comprising all galaxies (the dynamic ejection method) along with many other relevant anomalies.

The principle points of the QSS and the LBT are in general agreement except for the status of the Universe. The LBT teaches about an ever-expanding Universe, growing at the speed of light at its perimeter where energy is created, rather than being created within a steady-state universe. These forms of energy eventually are transformed into the matter comprising all galaxies, which are created, and grow, via the ejection manner described in the QSS version. The discovery of the proposed Fourth Law of Planetary Motion, explaining the spacing of planets in solar systems, was accomplished in the years of 1980-1995. Together with Kepler's First Three Laws, the Four Laws (FL) and the Internal Nucleosynthesis (IN) that is responsible for all planetary Evolution (E) offer valid evidence of how our Solar System came into its current stage of existence.

All findings on Mars (and other planets) corroborate this FLINE concept of planetary origins and evolution in which the FL, IN and E are chronological, inseparable and ongoing realities of dynamic solar systems. They clearly reveal the necessity of ejection processes in the formation of binary and multiple-planet systems while arguing strongly against the condensation concept. Much corroborative evidence in the QSS and the LBT bear powerful witness to this conclusion.

All relevant discoveries of the space probes and of Earth continue to interlock precisely into this revolutionary LBT/FLINE concept (initiated in 1973). The primary hope of this letter is to keep the debate on origins out in the open until we reach the point of absolute proof of how everything came into being. At this writing, the evidence continues to mount against the BB, and in favor of the LBT and much of the powerful evidence presented in the article by Burbidge *et al.* Mainstream cosmologists cannot, and must not, yet, conclude that we understand the origins and evolution of galaxies, solar systems, planets, etc., via the BB and condensation concepts.

To: Glenn Strait, Gordon Rehberg, William Mitchell, File. 07-29-99

The Heart of the Matter (*Science*, 2 July 1999, pp 55-56), a book review by Roger Blandford, contains information relative to the basic principles of the Little Bangs Theory (LBT) (1973-1980) which state that every galaxy is spawned from the two poles of a central black hole of energy previously created at the spherical perimeter of the Universe during its speed-of-light expansion. The ejecta include all the forms of energy and matter comprising each galaxy. From such explosive actions come the huge masses of white energy we call stars. Subsequent interactions of these speeding masses result in many formations of solar systems, usually of a binary nature. In each case, the system must comply with Kepler's Three Laws of Planetary Motion.

In the much rarer case of a multiple-planet system like ours, two energy masses meet the precise prerequisites of relative masses, relative speeds and precise distance apart to effect a breakup of the smaller, speedier mass into smaller masses as they space themselves in full accord with the Golden Ratio in orbits around the larger mass. Such systems must comply with all Four Laws of Planetary Motion. Whether all systems must comply in some way with all Four Laws remains to be determined. With this background, we can understand why Blandford's review of *Active Galactic Nuclei: From the Central Black Hole to the Galactic Environment* (Princeton University Press, 1999), authored by Julian H. Krolik, is an exciting confirmation of the LBT.

Krolik's two confirmations add powerful support to the LBT, yet he appears to cling to the ludicrous Big Bang hypothesis. Herewith is a copy of the review for your perusal and file. (Note: The second page contained only one-half paragraph which is not essential.) Any feedback in the form of questions and comments would be appreciated.

To: Glenn Strait, Editor, The World & I, Washington, D.C. 20002. 08-31-99

Some time ago, you inquired about my terminology of the composition of the energy cores in universal spheres, specifically of Earth's core. As stated then, whether we can say that such cores consist of energy particles or plasma or a mixture thereof still remains debatable, and scientists have only to look to these cores to understand the mysteries of how atoms are created within these spheres under severe conditions of high temperatures and pressures.

By reversing Nature's processes, scientists are now very close to finding the answers to creation as defined in the LBT/FLINE model. An exciting article, *Making the Stuff of the Big Bang* (*Science*, 20 August 1999, p 1194), by David Voss, unwittingly sheds some light on active core compositions. Voss writes, "If all goes well later this year, physicists (at the Brookhaven National Laboratory) ... will create miniature copies of the Big Bang by smashing together the bare nuclei of atoms traveling at nearly the speed of light. Reaching temperatures a billion times hotter than the surface of the Sun, the protons and the neutrons will melt into their bizarre building blocks: quarks and gluons that hold them together. Out of the inferno will come an exotic form of matter called quark-gluon plasma, primordial stuff that may have been the genesis of all the normal matter we see around us. And most profoundly, the very vacuum, what we think of as empty space, will be ripped apart, revealing its underlying fabric."

The quark-gluon plasma (QGP) concept (with possibly even smaller particles) is what I have always visualized—but until now did not know how to express clearly—as the true composition of the energy cores of all active, self-sustaining spheres of the Universe. The researchers are attempting, albeit, unwittingly, to duplicate this core composition. By reversing Nature's processes for creating atomic matter, they will be re-creating the QGP from which the atoms were, and are, made within all active spheres, thereby striking at the very heart of ongoing creation. But to move on to the next step in understanding planetary anomalies, scientists must recognize that the QGP processes will go on forever; they did not happen, and could not have happened, in the alleged Big Bang. They are indeed the "primordial stuff that may have been (is) the genesis of all the matter we see around us".

But rather than from a BB, the origin of the QGP can be best explained via the Little Bangs Theory (LBT) in which the "primordial stuff" will be forever created from "the very vacuum, what we think of as empty space", at the spherical perimeter of the Universe that is forever expanding at the speed of light. When the QGP-to-atoms concept of creation of atomic matter—the inseparable Nucleosynthesis/Evolution relationship of the FLINE model—inevitably is recognized and accepted as Nature's basic principle of ongoing creation, the solutions to all planetary anomalies are sure to follow in rapid sequence. The work described by Voss brings scientists to the very forefront of the LBT/FLINE paradigm. After more than two decades of frustration, the exciting breakthrough cannot be too far away!

To: Marilyn vos Savant, c/o Parade, New York, N.Y. 01-11-00

Your response to Mike Berman (*Parade*, Jan. 9, 2000) concerning current scientific theories was a pleasure to read. As a scientist (retired), I agree wholeheartedly with the questions raised and the comments made about scientific theories; thanks for sharing these insights. You "received mail from furious scientists who proceeded to cite every known argument in favor of the Big Bang (BB) theory except that one to which (you) specifically referred—the argument in favor of the entire cosmos once being smaller than a polka dot."

That is why I chose it, and its absence from those letters supports my point. I also think we must be careful not to teach theories as fact. It slows scientific progress immeasurably." Comment: Furious responses often stem from uncertain consciences. Not all scientists believe in the BB; many of us are fully aware of its highly speculative structure that stems from evidence based on erroneous assumptions; e.g., a small polka-dot beginning.

Even one false assumption can lead to many false conclusions. And that is precisely the awkward position in which advocates of the BB have placed science. If the data do not fit an anticipated conclusion, advocates often force-fit it into the concept, softened by "might", "maybe" or "could-be" speculations.

One example: In recent times, the acceleration of the Universe has been attributed to a repelling force, lambda, a fudge factor to support the erroneous current belief; this, after Einstein called his expansion force the "biggest blunder" of his life. And the idea that all elements were created in the BB is pure speculation that has no basis in fact; a more logical and provable creation of elements is provided later in this letter.

The BB prevails in spite of its readily refutable evidence and the powerful substantiated evidence against it (Ref: *The Cult of the Big Bang*, by William C. Mitchell). It does so in the face of a more logical concept that's supported by powerful substantiated evidence, requiring neither a lambda factor to explain the acceleration of the Universe nor a polka-dot beginning. These two fallacies pose no problem in the Little Bangs Theory and the subsequent ongoing processes of nucleosynthesis within every active sphere of the Universe.

The Internal Nucleosynthesis (an idea first hinted at by Descartes in 1644) is the driving force behind the evolution of all universal spheres: the two are inseparable—one cannot exist without the other. Many substantiated facts firmly establish these ongoing relationships. Kepler's Three Laws of Planetary Motion and the new Fourth Law, explaining the geometric spacing of planets, combine their powerful weight against the Planetary Accretion concept, which, in turn, undermines the BB. Together, the Four Laws (FL) and the Internal Nucleosynthesis (IN) that drives all Evolution (E) form the irrefutable FLINE model of the origins of solar systems and the evolution of planets.

Advocates of the BB should heed your warning that "we must be careful not to teach theories as fact. It slows scientific progress immeasurably" (and tragically). As long as the BB and the Accretion hypotheses prevail, a true understanding of universal origins and evolution remains highly improbable.

To: Program Chair, 2001 Meeting, AAAS, Washington, D.C. 20005. 06-29-00
Re: Origins of Solar Systems and the Evolution of Planets

Science, today, finds itself in a situation that should be, and must be, resolved before any truly significant progress can be realized in understanding planetary origins and evolution. The tendency toward teaching theories as fact must be carefully avoided while simultaneously keeping open minds to new ideas. As history teaches, when facts displace speculation, new ideas inevitably displace old ones. Your help in bringing these new ideas before the AAAS membership is essential.

I respectfully request that you seriously consider permitting the introduction and open discussion of my work on either one or both new concepts into the main program at the next AAAS Annual Meeting in San Francisco. One pertains to the origin of the Universe; the other to the origins of solar systems and the subsequent evolution of planets. The two concepts are intertwined; both remain within the realm of provable facts and sound logic without the need for speculation. Both, in conjunction with current beliefs, were painstakingly researched, developed and carefully evaluated during the past 27 years. The findings have been summarized in the two pages entitled: (1) *Origin of the Universe;* and (2) *Origins of Solar Systems and Evolution of Planets*. The listed claims of both the FLINE model and the Little Bangs model are backed by substantive evidence; both present strong arguments favoring them over current beliefs.

Additionally, my paper, *On the Spacing of Planets: The Proposed Fourth Law of Planetary Motion*, has been in peer review at the Royal Astronomical Society, London, for over two-and-one-half years, apparently without a flaw to justify its rejection. If accepted, of course, it would cause scientists to rethink current beliefs about the origins and evolution of all planets in all solar systems.

Since completion of the proposed Fourth Law (1980-1995) as the final link in the FLINE model, I have been able to reach a number of people who have taken the time and interest to learn and accept its factual message, and consequently, have become enthusiastic advocates. But the greater scientific community, who stand to benefit most, remains unaware of its great potential simply because its new ideas have not yet been permitted into the system other than via ineffective poster sessions. These new ideas offer valid and definitive scientific alternatives to the Big Bang and Accretion concepts.

Please examine these findings carefully for any fatal flaw or for any question about the scientific validity of these concepts. If you feel they warrant open discussion and would make an interesting topic at the meeting, please put them on the main program as an invited paper in the planetary sciences. If doubts or questions arise, the feedback from you would be deeply appreciated and helpful.

To: Michael S. Strauss, Program Director, 2001 AAAS Meeting, Washington, D.C. **9-4-00**

For the past quarter-century, I have carefully researched the origins of solar systems and the evolution of planets from the perspective of prevailing beliefs versus newer concepts. The result is a non-speculative, factual theory. As each facet of the evolving concept became incontrovertible, it was presented (usually in poster sessions) at various annual meetings of the AAAS, AGU, GSA, ACS, GAS and other organizations.

The final link that ties everything together and brings the definitive FLINE concept full circle is the solution (1980-1995) to the enigmatic Fourth Law of Planetary Motion, clearly revealing how the planets attained their orbital spacing around the Sun. This solution first eluded Kepler in 1595, and until 1995, it had remained the unknown key to understanding the dynamic origins of solar systems. In the sense that it is destined to change scientific beliefs about the way solar systems came into existence, this revolutionary discovery is equivalent to the Copernican idea of a Sun-centered Solar System. My efforts to bring this important discovery to the attention of the greater scientific community via mainstream scientific programs, other than ineffective poster sessions, have been thwarted for the past five years by letters that failed to cite a valid reason for the rejections.

To circumvent the anticipated rejections, I published my findings in book form to preserve them for posterity. Herewith are two pieces of literature that help convey the significance of these quarter-century findings. In allowing more logical and accurate interpretations of discoveries in the planetary sciences, the FLINE and LB models have the potential of tremendous benefits to scientists. Poster sessions have proven ineffective, so I see no reason to put these models through another one. For these reasons I responded promptly to your published call for suggestions for the 2001 AAAS Meeting.

With all due respect, I urge you and the program committee members to become familiar with the three fundamental and incontrovertible principles of the FLINE model, so that when its time inevitably arrives, you will permit a symposia presentation to the scientific community by a number of scientists who already understand and accept this revolutionary concept. Now at age 77 and with cancer, I would like to see this come to pass in the near future.

From: Brig. Gen. Gordon C. Carson III (Ret.), Livingston, TX. **12-26-00**

I am the son of Gordon and Eleanor Carson whom you met through Opal Falligant. When I was home for Christmas, my parents showed me a copy of your book, *The Spacing of Planets*. During my stay in Savannah, I read it several times and find it to be the most brilliant and fascinating work I have seen in many years. The Big Bang Theory has never been acceptable to me. In my opinion, the Universe has always been here and will always be here as a perpetual motion-machine. The age of the oldest galaxies may simply represent the normal life span of such bodies.

And as you point out, millions of new galaxies are continuously being created at the perimeter of our expanding Universe. Another point in which I agree is the solar origin of the planets and moons. The iron-core theory has not been convincing, and your belief of the internal nuclear furnace makes perfect sense, owing to the composition of the Sun. There are many other points which I find intriguing and would enjoy discussing with you, but the reason for my letter is to ask how I might buy four copies of your publication ... one for myself, and one each for three friends. Please send purchasing information to my Georgia address below.

To: Dr. William C. Mitchell, Chairman, Institute for Advanced Cosmological Studies, Carson City, NV. 06-25-00

Just in case you missed the recent article, *The Universe in a Sphere*, in *Parade Magazine*, I'm sending a copy. After you have read it, I'm sure you'll understand why I felt compelled to comment on the brand new Hayden Planetarium that is its centerpiece. Exhibits depict the development of the Universe from the "Big Bang" (BB) on. Its Center is planning extensive nationwide educational programs via various news media, thereby making scientific findings, data, interpretations, etc., available to classrooms and communities across the land.

It is unsettling to learn that so much effort and expense is being put forth to advance the BB as a valid scientific concept, especially when there is far more substantiated evidence against the BB, and far too much speculative interpretations used in support of it. What a terrible waste! How embarrassing it will be to science in the not-too-distant future when astronomers learn the true nature and origins of comets and asteroids.

Revved-Up Universe: Astronomers check out an expansive finding (*Science News*, Feb. 12, 2000) states, "Just two short years ago, two teams of astronomers presented the first evidence that we live in a runaway universe, driven to expand at a faster and faster rate." You may recall my letter of February 20, 1998, in which I explained that this accelerating expansion was first predicted and explained in my 1980 version of the Little Bangs Theory (LBT) in which the Universe is expanding at its perimeter at the speed of light. This naturally explains why the Universe appears to astronomers to expand at a faster and faster rate as they look farther and farther out.

"That finding is in direct conflict with the simplest version of the BB," the article continues. "According to that theory, the Universe has expanded ever since its explosive birth, but gravity has gradually slowed the expansion. Even if the universe grows forever [a principle of the LBT], the theory predicts that it should do so at a steadily decreasing rate."

Recent observations of exploded stars, however, suggest that the Universe's rate of expansion is, in fact, increasing (another principle of the LBT). Over the past year (and as predicted by the LBT), new data appear to corroborate those findings. However, the reason given by astronomers for the apparent increase in the rate of expansion requires a "fudge" factor, the cosmological constant that Einstein discarded in later years by calling it his "greatest blunder". There is no need for the cosmological constant in the LBT; the apparent increasing rate of expansion is a natural consequence, a basic principle, of the LBT.

The reasons I'm writing this, other than to be sure you are aware of the new Hayden Planetarium and its educational programs, is to get this load off my shoulders and into the written record for future generations; also, to record that not all scientists believe in the unsubstantiated BB. To me, it does seem wrong to push the BB into people's minds without more definitive evidence to support it. But, of course, such evidence will always be elusive.

P.S. - Astronomers appear to be very familiar with the LBT; they seem to be headed in that direction.

To: Brig. Gen. Gordon C. Carson III, Conyers, GA 30013. 12-29-00

Thank you for the tremendous compliments expressed about my latest book, *The Spacing of Planets*, and for the four-book order in your letter of December 26th. By the time you receive this response, you should have received the books mailed to your Georgia mailing address, given by Eleanor by phone the evening prior to receiving your letter. You may already know that the inspiration for the research efforts that ultimately led to a series of books on origins and evolution stemmed from a conversation with your father in the early 1950s, speculating about the origin of our Moon. This occurred at the Falligants' (Claude & Opal) home on Talahi Island between card games.

After almost 20 years of cogitating and rearing a family had passed, the pieces began interlocking precisely into place. In-depth research soon proved, beyond doubt, the abiogenic origins and intimate relationship of the hydrocarbon fuels, first self-published in 1973 as *Fuels: A New Theory*. This revolutionary concept proved to be the crucial key to unlocking the mysteries of how all planetary systems came into being, and to the factual understanding of planetary anomalies. In turn, this led to the enigmatic solution to the Fourth Law of Planetary Motion that first eluded Kepler in 1595.

The new Fourth Law (1980-1995) was the final link that brought the FLINE model full circle. As it turns out, my real work had only begun. During the past two decades of the evolving model, each new facet was presented in various science meetings to national and international groups of scientists. Since its completion in 1995, it still accumulates rejections by peer reviewers schooled in current beliefs based on the erroneous Big Bang (BB) hypothesis.

To put the BB situation in perspective, I'm taking the liberty of sending you a copy of *The Cult of the Big Bang: Was There a Bang?*, by William C. Mitchell. In my opinion, this is the most authoritative and factual book ever written on the subject. Its contents mesh perfectly with my Little Bangs/FLINE model. However, since revolutionary ideas are upsetting to BB believers who exercise control over scientific content (press releases, etc.), we both experience the same rejection problems. So, we join with many other scientists in attending the Natural Philosophy Alliance meetings, an international group who meets one or two times a year to discuss, with open minds, all new ideas that challenge prevailing beliefs.

A copy of page 10 of the AAAS meeting, February 15-20, 2001, is enclosed. The four subjects check-marked will be based on the BB; other concepts are not permitted on their main program—only in ineffective poster sessions. Thus, progress in the planetary sciences will remain stymied. *The Spacing of Planets* now can be found on the internet at www.iuniverse.com, by typing an author's name. The updated version that embraces relevant findings of the past five years is due on-line in late January or February. You will find three of its pages enclosed. With your permission, I would like to use, as a promotional blurb, the statement from your letter: "I read the book several times and find it to be the most brilliant and fascinating work I have seen in many years."

To: Glenn Strait, Science Editor, The World & I, Washington, D.C. 20002. 03-07-01
Re: On the Spacing of Planets: The New Fourth Law of Planetary Motion

Because exciting breakthroughs within the realm of natural laws, along with the rapid accumulation of corroborative evidence, assure the scientific validity of this manuscript, I'm taking the liberty of sending it for your consideration for publication. If too long as is, its separate, but interlocked topics, with editing, can be conveniently published in any combination, thereof, to meet any specification of length of an interesting and historical article.

For example, the Abstract, Introduction and The Geometric Solution to the Fourth Law of Planetary Motion could be used as one article. The addition of *Extra-Solar Systems: How and Why They Differ from Our Solar System* would add more interest and only two additional paragraphs. The section headed *The Five Stages of Planetary Evolution* appears capable of standing alone, but could be enhanced by a connection with the manner in which the planetary nebulae attained their orbital spacing around our Sun before beginning their evolution through the five stages. The same reasoning could be applied to *Corroborative Evidence for Nucleosynthesis within Planets and Moons* and to *The FLINE Paradigm of Planetary Origins and Evolution*. These are only suggestions—probably superfluous—which you, as editor, can accomplish better than I could. At any rate, I hope you will give the contents the full considerations they warrant, and will feel free to edit as you wish.

As more people read about these breakthroughs, the list of excited advocates grows (see blurbs) along with my confidence that the concept contains no fundamental flaw—especially since no one has managed a valid rebuttal during the past 28 years of its evolution. But there is one caveat: my paper on *The Spacing of Planets* has been in peer review for slightly more than three years at the Royal Astronomical Society, which gives me even more confidence in its scientific validity.

Isn't it reasonable to believe that a fundamental flaw, if any, would have been discovered and exposed during the many years of the FLINE model's development and discussions with knowledgeable scientists? Its potential benefits to science are immense. It is time to move on to the next paradigm shift as recognized by Thomas Kuhn—one that finally offers provable solutions to the anomalies of solar systems and planets. Imagine the tremendous controversy in which your magazine could reap the benefits. Self-explanatory letters concerning its status accompany the manuscript. You have the option for first rights to magazine publication; however, I will retain these and all other rights if, for any reason, you decide not to publish it soon.

To: Dr. William C. Mitchell, Institute for Advanced Cosmological Studies, Carson City, NV. 03-16-01
Re: Space observatory shows black holes once dominated. (Atlanta Journal, 03-14-01)

This article discusses the findings of the Chandra X-ray telescope. To quote: "Huge black holes once dominated the Universe—sucking in gas, dust and stars, erupting with surges of X-rays that have journeyed since for billions of years across the heavens. That's a picture of the early Universe captured by the orbiting Chandra X-ray telescope in a study that focused on a small section of the sky for days-long exposure to capture faint X-rays streaming from more than 12 billion light-years away."

"The Chandra data show us that giant black holes were much more active in the past than at present," Riccardo Giacconi, a Johns Hopkins University astronomer, said Tuesday at a news conference. Added Bruce Margon, a professor at the University of Washington, Seattle: "If you look at the sky with X-ray eyes, you see almost nothing but black holes."

Proponents of the Big Bang (BB) believe that the study confirms theories by showing that the early Universe teemed with active black holes, spewing X-rays across the heavens, and that Chandra is looking at the X-ray Universe as it existed up to 12 billion years ago. But they do not say exactly which theories are confirmed. However, when the evidence is examined more closely, it presents very powerful arguments favoring the Little Bangs (LB) over the BB theory. The evidence is precisely as predicted and explained in my book, *The Spacing of Planets: The Solution to a 400-Year Mystery* (pp 111-126).

The fact of ever greater numbers of black holes seen upon approaching the Universe's rapidly expanding perimeter is a basic principle of the LB. They are the ultimate source from which all galaxies are spawned. This fact explains both the ever greater outward velocity of galaxies as astronomers look ever farther into space and the close relationship of greater distances to younger galaxies. All other relevant anomalies become vulnerable to easy solutions via this revolutionary LB/FLINE model—a concept solidly underpinned with the Four Laws of Planetary Motion. Its supporting evidence is incontrovertible.

Thus, the conclusion the Chandra data show us that giant black holes were much more active in the past than at present is not justified; one can conclude only that their ever-increasing numbers verify the rapid growth of the Universe via the creation of the densest form of energy (black holes) at its ever-expanding perimeter. When viewed in this light, the data bring astronomers ever closer to the LB/FLINE model of ongoing origins and evolution while pushing them ever further from the BB.

Please feel free to use any of this information in your next book.

To: Sen. Newt Gingrich, 1301 K Street, NW, Washington, D.C. 20005. 05-11-01

As a concerned R&D scientist (retired), I would like to offer some belated comments on your fine article, *An Opportunities-Based Science Budget* (*Science*, 17 Nov 2000), and hope that perhaps they are not too late to do some good. I am in total agreement with the five major increases in federal funding for scientific research that you feel should be seriously considered in the science budget. Your suggestion number three (3) is of special interest: "Money needs to be available for highly innovative, 'out-of-the-box' science. Peer review is ultimately a culturally conservative and risk-avoidance model. Each institution's director should have a small amount of discretionary money, possibly 3% to 5% of their budget, to spend on outliers. The history of plate tectonics should remind all of us that accepted wisdom can be wrong."

Proper use of discretionary money could alleviate some very costly situations in science. Most scientists realize that throughout a history beginning with the Copernican idea of a heliocentric Solar System, the accepted wisdom of the time has almost always been proven wrong. And today's beliefs already have been shown to be no exception to this great truth. Any suggested change in the direction of scientific thought, as with Copernicus, still meets with persistent and unfair peer resistance, almost always through peer review by advocates of prevailing beliefs. No new idea that challenges current beliefs, regardless of its scientific validity, can get past advocates in control of peer review and grant money. My personal experience during the past quarter-century certifies this point. History clearly reveals that dramatic breakthroughs always occur in the mind of one person rather than via consensus of many minds. Science finds itself in the situation of being, once again, at the crossroads of decision. Revolutionary evidence of new paradigms as irrefutable as that of Copernicus cannot get past peer review.

Your wise suggestion of discretionary money to spend on outliers seems to be the only hope for revolutionary ideas that can point us in the right direction of scientific thought in the planetary sciences. The remaining choice is to continue to spend billions of dollars teaching and trying to prove current speculative and erroneous hypotheses. You have touched upon the most crucial issue now facing the planetary sciences.

Enclosed are some pages of self-explanatory literature that illustrate the issue at hand. Thank you for the timely article; I hope this information is not too late to be of significance to the cause.

P.S. NASA space science chief, Ed Weiler, is right to put the freeze on the Pluto mission (*Science,* 17 Nov 2000)—at best, an ego trip. Scientists can learn nothing of real significance that's not already known about this planet; the $1.4 billion price tag would be close to a total waste. If acquiring highly significant knowledge is the goal, the money would be spent far more profitably on two other missions: to Europa and to the next nearby comet. The latter mission will reveal the most shocking and revolutionary results; perhaps, this could be the reason such a mission has been delayed as long as possible.

To: Donald Kennedy, Editor-in-Chief, Science, AAAS, Washington, D.C. 20005. 05-28-01
Re: The Five Fundamental Principles of Origins and Evolution of Planetary Systems

The literature herewith includes a brief synopsis of the definitive FLINE model embracing the five fundamental principles of origins and evolution of planetary systems. The fifth and final principle—the enigmatic solution to the new Fourth Law of Planetary Motion that first eluded Kepler in 1595—inexorably links with the first four principles of the revolutionary FLINE paradigm to bring it full circle to an irrefutable conclusion. This geometric solution clearly reveals, beyond any reasonable doubt, how the planets attained their orbital spacing around our Sun. The new model provides a solid foundation for a full and complete understanding of all anomalies of solar systems.

Recent examples of the FLINE model's potential for elucidation of new data are found in the seven articles on *Comet C/1999 S4 (LINEAR)* (*Science*, 16 May 2001). The strange behavior of the disappearing comet leaves many unanswerable questions, solely because the interpretations of all data are based on the prevailing belief (posing as fact) in the speculative "snowball" nature of comets. Interpretations of such data in the perspective of the realistic "fireball" nature of comets never encounter the problem of unanswerable questions. The FLINE model leaves no room to doubt the true nature and sources of comets, or why they disappear, and why the total mass measured following the breakup of Comet C/1999 S4 (LINEAR) is about 100 times less than the estimated total mass prior to its breakup.

Any of the intermeshed facets of the FLINE model can stand alone on its own merits. Each of them tells a factual story of crucial significance to the scientific and world communities; e.g., The Energy Fuels (Non-Fossil) Theory, The New Fourth Law of Planetary Motion, The Fiery Nature and Sources of Comets and Asteroids, The Dynamic Origins of Solar Systems, The Five-Stage Evolution of Planets, and the subject paper: *The Five Fundamental Principles of Origins and Evolution of Planetary Systems*. In conjunction with its many interlocked facets, the fifth principle predictably—as with the Copernican idea—will force dramatic changes in beliefs about our planetary systems in which all five of these inseparable principles, each dependent on the other four, already are playing equally significant roles.

A page of old and new blurbs adds assurance of public interest in, as well as the scientific validity of, these new perspectives, all carefully researched during the past three decades to eliminate myths and speculation. A manuscript of specified length on any facet of the new model is available upon request.

Thank you for your time and consideration to publish this short version of *The Five Fundamental Principles of Origins and Evolution of Planetary Systems*. I fully comprehend the situation it entails, and urge you to pursue these crucial findings for the sake of pure science, the AAAS creed, and posterity.

The Genesis Mission to the Sun: Scheduled for July 30, 2001 - Sept., 2004

The news article, *Spacecraft to Catch a Piece of the Sun, Bring it Home,* by Marcia Dunn, AP (*Atlanta Journal-Constitution*, July 15, 2001), stated, "A robotic explorer named Genesis is about to embark on an unprecedented journey to gather and bring back bits of the Sun. NASA hopes the specks of solar wind—equivalent to perhaps 10 grains of salt, if added up—will help to explain what it was like in the beginning. The very beginning, when the planets in the solar system were forming." "... As the solar wind streams past the spacecraft, microscopic traces of chemical elements will become embedded in materials on collector panels. These castaway elements are the same material as the original solar nebula, the disk of gas and dust from which all the planets and all other Solar System objects formed 4 1/2 billion years ago. The outer layers of the Sun, continually streaming into space as solar wind, provide the most feasible way to access this fossil record." The spacecraft is on schedule for a July 30 launch, and should return to Earth in September, 2004 with the small samples.

But, there are caveats. Stephen Hawking, a prominent physicist, recently made a stunning admission: "Cosmology's hypotheses are conjectural when pressed to the limit." True; they are purely speculative. As a matter of fact, definitive evidence against these prevailing concepts has become overwhelmingly persuasive; e.g., the five principles of origins and evolution revealing that planets did not, and do not, accrete from a disk of gas and dust. As divulged by the five principles, they began as moderate-sized masses of nuclear energy that had broken from a separate, but smaller, faster-moving, mass that passed close to the larger Sun, and had been forced into geometrically-spaced orbits around it. Most are still undergoing the processes of evolution via the transformation of their internal energy into matter, all in full accord with Einstein's ubiquitous $E=mc^2$, and with Nature's natural laws. Some smaller ones have exhausted their energy to become inactive (dead) planets (e.g., Mercury, Mars, Pluto).

The Four Laws of Planetary Motion assure that, like the planets, our Sun began as a nuclear-energy fireball, larger than now, but much as we see it today—a huge fiery mass of ejecta from a much larger energy mass (black hole or quasar). They clearly confirm that the dynamic interactions between the two energy masses gave birth to our Solar System. Here, no conjectural cosmology is needed; the plethora of supportive evidence appears incontrovertible.

Should the Genesis mission's objectives be viewed in this new perspective, rather than in the current one, the futility of the mission becomes obvious. The captured pieces of the Sun will be virgin material created within its massive interior and ejected into space, as is true with all fiery stars. This material will be the elemental products of Internal Nucleosynthesis, which occurs in all active spheres of the Universe—and it will be basically the same (selectively qualitative, but not quantitatively equal) as all other ejecta material from other active spheres. Its significance will have no relevance to the alleged Big Bang origin of our Universe or of the Accretion of planets; both appear to be erroneous conjectures. But perhaps, the results will prove beneficial in advancing scientific knowledge about the Sun's Internal Nucleosynthesis, and thereby, increase our knowledge about the first stage of evolution of all planets: the energy stage.

To: Joel Martin, Staff Writer, LaGrange Daily News, LaGrange, GA 30241. December 23, 2003

Here are examples comparing the interpretations of recent discoveries that illustrate the illogic of the Big Bang versus the logic of the LB/FLINE model. These were overlooked during the interview. While we can hope that some books will be sold, my primary objective is to alert the public to the myths and facts of science: to establish definitive truth in the sciences of origin and evolution of universal systems.

Galileo spent the last ten years of life suffering the consequences of defending the 1543 idea of a heliocentric Solar System that initiated the Copernican Revolution (Ref: *The World Book Encyclopedia*). In the mid-19th century, Samuel Semmelweiss spent 20 years trying to save the lives of mothers and their newborn babies by convincing the medical establishment that cleanliness is the essential ingredient.

Because of ego, they would not listen, the deaths continued unabated, and Semmelweiss ended up in the insane asylum (Ref: *The Cry and the Covenant* by Thompson).

Obviously, egoism, bias and censorship should have no place in science. In bringing the Copernican Revolution full circle via the definitive Fourth Law of Planetary Motion, I find myself in the same predicament as Galileo and Semmelweiss. Why is it that people will readily believe hearsay, gossip and myths, but find it difficult to believe definitive truth underpinned with substantiated evidence accumulated over a lifetime? Separating the myths and facts of science in order to establish truth in science has been the primary goal in my life.

Fortunately, time is always on the side of truth. Although the Fourth Law and the First Three laws clearly reveal the dynamic GR origin of our SS, thus making them the two most significant discoveries for solving the mysteries of SS origins, I would refrain at this time from calling them the greatest in the history of science; perhaps, the most significant for SSs would be more accurate. I don't believe either one of us can realize the full significance of these findings at this time. Certainly, you are writing one of the biggest news stories of all time. Good Luck! You'll need courage and foresight to see it through, and this could take a lifetime for coverage of this subject of universal proportions that grows with every discovery, should anyone want the task.

Thanks for the time and courage to do the story. If you need any clarification or have any questions, please feel free to give me a call.

To: Jane Selverstone, Technical Program Chair, GSA, Denver 2004 Meeting. January 12, 2004

Quoting from *GSA Today*, "Pardee Keynote Symposia (PKS) are special events of broad interest to the geoscience community. Topics appropriate for these symposia are those that are on the leading edge in a scientific discipline ...; address broad, fundamental problems; are interdisciplinary or focus on global problems."

Origin and Evolution of Planetary Systems meets all criteria for a PKS. Its definitive fundamentals underpin every geological presentation—past, present and future—while offering opportunities for geologists and other scientists in relevant fields to definitively interpret their new data.

In *An Alternative Earth* by W. B. Hamilton (*GSA Today*, Nov 2003, pp 4-12), the author could have avoided much speculation, while deriving tremendous benefit from the fundamentals in *Origin and Evolution of Planetary Systems*. Like the heliocentric idea of Copernicus, the definitive LB/FLINE model that brings it full circle bodes a truly exciting future for science. "The goal of the Technical Program Chair and the JTPC representatives is to provide presenters the best possible opportunity for communicating new scientific information rather than to dictate what can or will be presented." It is possible to reach this lofty goal, but only if new, well-substantiated ideas are given a fair chance to be heard, vetted and validated by the scientific community.

I believe you will agree that censorship in any form has no place in true science, and that substantiated facts should not be relegated to the level of irrelevancy, else we run the risk of delving in pseudo-science. Inevitably, science must recognize the futility of the BB/Ac hypotheses, and become willing to examine another model. If so, the LB/FLINE model should be given a high priority; it is testable, and should pass every test with ease and confidence. This PKS Proposal presents a great opportunity for the GSA to seize now.

To: Mrs. Pam Smrekar, LaGrange, Georgia 30240-9569. April 27, 2004

Many thanks for the cheerful note and picture from the *LDN*; it came at a time when sorely needed. Since your fine efforts on my behalf many years ago, much progress has been made in understanding the origin and evolution of universal systems. Nine editions of my book have been completed, and the ninth edition is now on www.authorhouse.com. I keep some on hand for immediate sale (*New Principles of Origins and Evolution*). All editions of this Energy Series (1975-2002) are cumulative. I have run into progressively more resistance from advocates of the Big Bang/Accretion (BB/Ac) hypotheses that Sagan called "myths" and Hawking called "conjectural".

Unfortunately, in science, BB/Ac advocates control publications, programs, grants, etc., and diligently protect their beliefs from anything that might pose a threat to them. The situation is somewhat analogous to the era of Copernicus and Galileo.

However, I remain a member of an international group of some 200 scientists who recognize the fallacies of established beliefs, and are struggling to bring open-mindedness and truth back into science where it belongs. We meet once or twice each year.

My book sales are slow, but improving; they need a good PR person. I'm trying to establish local sales outlets in this and other areas, while continuing to lecture to groups who have an interest in science. I would like to work with you again in these efforts; if the feeling is mutual, we can meet at your convenience to discuss the possibilities. It was good hearing from you again.

To: Alexander A. Scarborough, LaGrange, Georgia 30241. May 13, 2004

Many thanks for your letter of 26 April. Don't worry; your efforts are not in vain! Cosmology is generally in a period of way too much speculation. We can only hope that gathering more information over time will someday cause ideas to coalesce more sensibly.

From: Dr. Cynthia Kolb Whitney, Editor, *Galilean Electrodynamics*, Tufts University

To: Editor, New Scientist, UK.
Subject: Editorial—Not the end of the world; Does it matter that some cosmologists are finding fault with the big bang? (New Scientist, 2 July 2005). 18 July 2005

The reason the "huge array of variables of the Big Bang (BB) can be changed pretty much at will" resides in the fact that, invariably, all myths are flexible and adjustable via additional myth—as best illustrated in the fairy tale, *Alice in Wonderland*—no matter its absurdity. The BB myth is promulgated, understandably, by advocates who remain unaware, often for personal reasons, of a viable alternative in which to interpret their excellent findings. But there is now a valid scientific concept "showing us where our thinking may be flawed and forcing us to ask new questions"—the revolutionary Little Bangs (LB)/FLINE model of universal origins. The new model (1973-2002) forces science into a position like the one initiated by Copernicus in 1543.

The model is structured on a solid foundation of much substantiated evidence encompassing the works of Pythagoras, Copernicus, Kepler, Newton, Descartes, Einstein, and numerous modern-day scientists. The latter works include the enigmatic solution to the new Fourth Law of Planetary Motion (1980-1996) that first eluded Kepler in 1595, and remained dormant for 400 years.

Understanding the new Fourth Law, along with Kepler's First Three Laws, is crucial to comprehending the origin and evolution of all solar systems, and subsequently, galaxies, and other universal systems. While this may be bad news for BB devotees, it is truly exciting news for the future of science. Certainly, it is not the end of the world, and most assuredly, it does matter a great deal that some cosmologists are finding fault with the BB. Much substantiated evidence dictates that the time has come for a change in direction of scientific thought. Many thanks for your insightful and timely editorial.

To: Editor, Journal of the Royal Astronomical Society, Burlington House, Piccadilly, London, UK. 07-04-06
Re: A Really Bright Future for Science!

My manuscript briefly presents the new Fourth and Fifth Laws of Planetary Motion; fuller details are found in the References. The two laws underpin the continuity of interrelationships comprising the revolutionary LB/FLINE model (1973-2006) of universal origins and evolution, and are the primary objectives of this paper submitted for your journal. The revised 1995 Fourth Law presents a definitive understanding of the dynamic origin of our Solar System via the geometric spacing of planets around our Sun, all in full accord with the Golden Ratio.

The new Fifth Law defines and explains the gradual changes in orbits and velocities of all planets since the dynamic origin of our SS some five billion years ago; e.g., why Neptune and Uranus are the only two planets slowly drifting away from the Sun, while all others are drifting slowly toward the Sun. Both laws are in sync with the First Three Laws discovered by Kepler early in the 17th century.

Surprisingly, the Five Laws (FL), in conjunction with $E=mc^2$, open the floodgates to definitive understanding of the intimate relationship among SSs, galaxies, black holes (spheres), electromagnetism, gravity and space-time, all in accord with the LB/FLINE model. The new model offers valid scientific solutions to a number of the greatest mysteries of the Universe—e.g., the mathematical spacing of planets in our SS, and the accelerating expansion of the Universe (inherent in the model without the need for a lambda-type cosmological constant or a dark substance)—validating its supportive evidence to the point of making it seem incontrovertible.

Presentations at national and international science meetings during the past decade have added much assurance of its scientific validity. Although it remains ahead of its time, the model has gained a firm beachhead via substantiated evidence. Fortunately, every relevant discovery further corroborates its scientific validity. Originally too broadly encompassing, the Fourth Law (1995)—with the help of Dr. Robert Heaston—has been revised into the Fourth and Fifth laws in compliance with Kepler's style and for improved clarity. (Detailed graphs are included.) This revolutionary concept bodes a truly exciting future for science!

The most significant developments in BB cosmology of the past 25 years (Discover, Sept 2005) vs. Viewpoints of the new LB/FLINE model of universal origins ***August 12, 2005***

1. BB: The discovery that the expansion of the Universe is accelerating, rather than decelerating as had been expected (1998). But it needs assistance from dark matter or dark energy (Einstein's cosmological constant? admittedly his "greatest blunder"); both remain speculative and undiscovered. LB/FLINE: The accelerating expansion of the Universe is an inherent, unassisted factor in the new model of universal origins that needs no dark matter or dark energy (1980-1996).

2. BB: Supernova observations and other data led to a clear accounting of the composition of the Universe: about 71% dark energy propelling the acceleration, about 25% dark matter not composed of normal matter such as protons and neutrons, and only 4% normal matter. LB/FLINE: Dark energy and dark matter remain elusive and doubtful, not really needed for any known purpose. Throughout the Universe, Energy Strings are the ubiquitous fundamental substance from which all matter of the Universe is derived via ongoing nucleosynthesis/ polymerization and other miscellaneous processes. The composition percentages appear unrealistic.

3. BB: The discovery of fluctuations in the cosmic microwave background radiation, believed to be leftovers from the BB event. LB/FLINE: Microwave radiation cannot survive for billions of years; in reality, it is short-lived, and requires constant sources of supply throughout the Universe. The ubiquitous sources are the nuclear energies, primarily in stars, planets, comets, and perhaps other smaller masses with similar energy cores that drive their evolution, eventually, into matter, all in accord with $E=mc^2$. Microwave background radiation is shown as one small area on Planck's curve, which itself is a full graph of all thermonuclear energies currently emanating from nuclear-energy sources throughout the Universe.

Conclusion

The three selections by BB advocates appear to miss the target as qualifying contenders for the most significant discoveries of the past 25 years. The BB belief that all atoms were created during the BB event puts extreme limitations on the ongoing nucleosynthesis processes that drive all universal origins of all forms of energy and matter within our ever-expanding spherical Universe.

The new model depends heavily on continuous universal growth at a speed-of-light rate at its spherical perimeter via natural energy/atomic matter relationship ($E=mc^2$ and string theory), followed by processes of nucleosynthesis/ polymerization in which all universal spheres and their functioning systems freely operate by means of natural laws. Everything evolves; nothing is as it once was.

This new model is solidly underpinned with the Laws of Planetary Motion that governed the dynamic birth of our Solar System, and in conjunction with $E=mc^2$ processes, still governs its ongoing functions. As the key to understanding the dynamic origin of our Solar System, especially the mathematical spacing of its planets, the new Fourth Law presents a powerful argument as the single most significant development in the field of astronomy during the past 25 years. Until science forsakes the BB myth and pursues a more promising direction in scientific thought, the prospects for comprehension of universal origins will remain bleak.

Addendum:

To reach a specific station, a train must start out on the right track. Equally essential for comprehending universal origins is the right direction of scientific thought: the two fundamental keys, $E=mc^2$ *and the Laws of Planetary Motion (FL). One key opens doors to origins and evolution of planetary systems, the other to origins of solar systems. Subsequently, both keys will be necessary for unlocking the mysteries of galaxies and universes. Once on the right track, the ultimate destination will be within reach.*

CHAPTER VI

INTRODUCTION TO THE INFINITE SOURCE OF SUPER-STRINGS

Jan. 25, 2007

The revolutionary model of universal origins teaches that we live in a dynamic Universe undergoing constant change, constant motion, constant evolution, and constant expansions. Billions of galaxies, each containing billions of fiery stars of all sizes, each rotating while embedded in an arm of dust-clouds, all revolving around a central mass (a black sphere of dense energy, a.k.a. black hole) in beautiful pinwheel fashion. From their interactions, solar systems and planets form. Billions of new galaxies continuously form at the perimeter of our spherical Universe expanding at the speed of light. We can only know the size of the Universe that is visible through powerful telescopes. But no matter their ultimate power, it will never be possible to view the perimeter of the Universe because it always will be beyond the point at which its light can reach Earth—even while traveling at the speed of light. When, at any point in time, such light does reach Earth, the perimeter will have moved ever farther out of sight.

Such unimaginable numbers of stars, planets, and galaxies require an unimaginable amount of energy from a dependable source in order to sustain the forever-ongoing system. What could possibly be that source of infinite energy that sustains the origins of our universal systems?

The most probable source is the infinite Space of total darkness and absolute-zero temperature, the conditions at which no motion, atomic or otherwise, is possible, until energy in the form of light hits it. This happens at the rapidly expanding perimeter of our Universe; the result is the activation of dormant space-energy in the form of black spheres of super-strings, which ultimately serve as the most fundamental building blocks of all universal systems. These cold black spheres are the deepest form of energy, and the eventual source of all white-hot galaxies, each consisting of billions of stars of all sizes embedded in hot, gaseous dust-clouds, as will be detailed later on.

Superstrings fulfill the micro-prerequisites of the new LB/FLINE model of universal origins that fulfill its macro-demands. Together, the two concepts provide sufficient substantiated evidence—all in perfect sync with Einstein's relativity theories—to fulfill the prerequisites and results outlined by Laplace.

"An intelligence that, at a given instant, could comprehend all the forces by which nature is animated and the respective situation of the beings that make it up, if moreover it were vast enough to submit these data to analysis, would encompass in the same formula the movements of the greatest bodies of the universe and those of the lightest atoms. For such an intelligence, nothing would be uncertain, and the future, like the past, would be open to its eyes."

This quote speaks loudly and clearly of the new LB/FLINE model of universal origins and the superstrings theory.

The fulfillment of this prophecy via the BB-myth perspective is an impossibility.

The Five Laws of Planetary Motion provide a solid foundation for the new model. They enable scientists to decipher definitely the history of our Solar System, and to predict future changes in orbital positions of its nine planets. They remove any doubt about Pluto's original status as the original tenth planet, and now the ninth planet, (since the disintegration of the Asteroids planet, as correctly identified by Heinrich Olbers in 1802).

Little Bangs: Explosive Galactic Births via Black Spheres

July 2006

In response to a recent request for defining LB, the LB/FLINE model of Universal origins offers evidence revealing the relationship of superstrings, black holes (spheres), and galaxies, all in sync with Einstein's relativity theories.

Black spheres are produced in countless numbers at the spherical perimeter of a Universe expanding at the speed of light. At the interface of the Universe and the absolute-zero temperature of the dark Space beyond, sustaining actions occur. Mass in the form of collapsed superstrings and exiting energies (e.g., light) form continually in the small, black spheres that grow ever larger as functions of time.

According to Robert Heaston, "A rather detailed analysis of the Einstein field equation of general relativity indicates that there may be a third means of transition from mass to energy ..., based upon a phase change from mass to energy from gravitational collapse ..., a mass-energy phase change from gravitational collapse forces could possibly be a mechanism in [the new] model."

When this mechanism reaches its trigger point, huge volumes of fiery stars (nuclear energy) embedded in hot dust-clouds are spewed from the two poles of the spinning black spheres, forming the curved arms of beautiful galaxies—and the illusion that the embedded stars coalesce from the hot, gaseous dust-clouds. Gamma rays may be one result of this noisy action. Interactions among the stars result in many binary systems and lesser numbers of trinary and multiple-solar systems of stars of various sizes. Our Solar System is an excellent example in which the smaller orbiting energy masses have evolved into planets currently in various stages of the five-stage evolutionary cycles, at rates in accord with size—all via Internal Nucleosynthesis (IN). Each planet is a self-sustaining entity that creates its own planetary matter via IN.

In this scenario, black spheres are all the dark energy (six-to-one ratio) needed in driving the accelerating expansion of the Universe, and are responsible for the illusions that Earth is the center of the Universe, and no matter the galaxy from which everything is seen these views will be the same.

In summary, a Little Bang occurs via the explosive nature of each birth of a galaxy spewing from two poles of a spinning black sphere that had formed at the expanding perimeter of the Universe, thus feeding the ongoing Universe the energy needed for its perpetuity.

Is Jiggling Vacuum the Origin of Mass?

Quoting from the article, *Is Jiggling Vacuum the Origin of Mass?* (*New Scientist*, 13 Aug 2005) by Mark Anderson, "Where mass comes from is one of the deepest mysteries of nature. Now a controversial theory suggests that mass comes from the interaction of matter with the quantum vacuum that pervades the universe." In the 1990s, Rueda and Haisch "suggested that a very different kind of field known as the quantum vacuum might be responsible for mass."

NASA

A nearby galaxy is revealed in one of the first images from the Spitzer Telescope released Thursday by NASA.

1. The spiral necklace of surrounding stars and the clouds of glowing carbon dust are ejecta spewed from the two poles of the central black hole: a perfect example of the LB/FLINE model, verified further by the following two statements.
2. The stars and dust are original ejecta; stars do not form from dust! Solar systems and binary-star systems result from the interactions of two or more stars during their pell-mell flight away from their powerful black-hole source.
3. The glowing core of a comet is identical to the core of a star: nuclear energy, as proven via the LB/FLINE model. Ejected from a larger energy mass, a comet's coma usually exceeds one million miles in diameter.

Omitting their explanation for the sake of brevity, the revolutionary LB/FLINE model agrees that the quantum vacuum is responsible for mass, but differs in the manner of its transformation, which occurs only at the spherical perimeter of the speed-of-light-expansion of our finite Universe. When exiting energies crash into the absolute-zero interface of the Cosmic Strings Space beyond, where all atomic motion ceases, they collapse into the densest form of energy that we call black holes (actually black spheres) that grow ever larger by accretion of collapsed energies, including the jiggling cosmic strings that permeate the infinite Space beyond our expanding universal boundaries.

In time, the spinning black spheres release countless masses of fiery stars buried in voluminous gaseous dust-clouds, forming beautiful pinwheel-shaped galaxies, filling our skies with billions of stars, and our hearts and minds with awesome wonder.

Classical Electrodynamics

Perhaps the most exciting concept in modern physics is the Electromagnetic Field Theory (EFT): the idea that the Universe operates entirely through electromagnetic forces; i.e., the fundamental interactions of forces in the Universe are all electromagnetic in nature. *In a New Foundation for Modern Physics*, Charles W. Lucas offers powerful arguments and irrefutable logic, in strong support of this concept, which is in full agreement with the revolutionary LB/FLINE model of universal origins. His paper was first presented at the Fifth International Conference: *Problems of Space, Time and Motion*, St. Petersburg, Russia, 23 June 1998, updated and first published in *Common Sense Science: Foundations of Science* (Nov. 2001). The abstract and selected excerpts follow.

Abstract. Principles of logic and criteria for acceptance of theories in science are presented. According to logic, whose purpose is to guide science toward truth, Maxwell's Equations, Einstein's Special and General Relativity Theories, Quantum Mechanics, the Bohr and Dirac Theories of the atom, the Quantum Electrodynamics Theory of elementary particles and Newton's Universal Law of Gravitation are not acceptable theories for science.

A new foundation for modern physics is presented that is based on the fundamental empirical laws of Classical Electrodynamics for finite-sized elastic elementary particles in the shape of a toroidal ring and composed of plasma filaments. Using combinatorial geometry, three-dimensional physical models of the atom and nucleus have been developed that describe the Periodic Table of Elements and nuclear shell structure better than quantum theories. Blackbody radiation, the photoelectric effect, and atomic emission spectra are explained in terms of toroidal-shaped electrons. New spectra lines for hydrogen lines for hydrogen in the extreme ultraviolet (not predicted by quantum theories) are predicted and found experimentally by the Berkeley Extreme Ultra Violet Physics Laboratory from rocket-based experiments in space.

A gravitational force law obtained from corrected Classical Electrodynamics is superior to that of Newton (even with Einstein's general relativistic corrections incorporated) as shown by analysis of Solar System data.

Text

To the rules of logic that have been continuously held for thousands of years, Lucas adds one more: "All scientific theories must acknowledge in a self-consistent way the mutual interaction and interconnectedness or unity of all parts of the Universe—Mach's Principle."

"In 1966, Winston H. Bostick proposed that the electron is a string-like submicroscopic force-free plasmoid constructed by the self-energy forces of electric E and magnetic H vectors. He found that a string of charge that makes up the electron naturally assumes the configuration of a helical spring that is connected end to end to form a deformable ring of torus."

"The cornerstone of this completely electrodynamic model of the Universe was electric charge in the form of an extremely slender, electrically charged, electromagnetic fiber that is in stable equilibrium by its own self-forces—and whose electromagnetic energy is a 2.5 X 10 GeV. Bostick proposed this physical structure as the origin of all superstring effects and gave arguments that:

*all mass, momentum, and energy are electromagnetic in character;
*the strong force is due to the electromagnetic forces between two finite-sized toroidal particles; and
*the transverse deformation waves on the filament are equivalent to the De Broglie waves of Quantum Mechanics."

"In 1996-1997, Charles and Joseph Lucas explained the fundamental phenomena that established Quantum Theory; i.e., blackbody radiation, the photoelectric effect, and the emission spectra of atoms, in terms of the toroidal Ring Model of the electron. The explanation of these phenomena turned out to be logically superior to that of Quantum Physics—with none of its problems in logic. Furthermore, the new physical model of the atom predicted emission spectra lines for hydrogen and other atoms in the extreme ultraviolet."

"In 2000, Lucas extended his previous work to obtain a more complete electromagnetic force that includes the force of gravitation with angular and velocity dependencies that go beyond Newton's Universal Law of Gravitation and Einstein's General Theory of Relativity. An analysis of Solar System data confirmed these results." (See FLINE.)

"Summary: This paper references key experiments, theoretical developments, and arguments from logic that are to form a new working foundation for modern physics which is based on Classical Electrodynamics. Some long-standing errors in classical physics have been fixed so that the corrected theory acknowledges particles to have finite sizes in the shape of toroidal rings formed by helical filaments of charge. The new and improved version of electrodynamics appears to satisfy all the rules of logic that undergird the Scientific Method. In its current rudimentary form, this approach is able to give the first holistic description of the atom—including its nuclear structure and spins, blackbody radiation, the photoelectric effect, and the emission spectra of atoms. This new approach in physics is logically superior to relativistic Quantum Electrodynamics Theory. In addition, it explains the electromagnetic origin of gravity and gives many correction terms to Newton's Universal Law of Gravity without invoking the General Theory of Relativity."

Quoting from *Testing String Theory* by M. Kaku (*Discover*, August 2005), "Nobody has yet detected a gravitational wave, but not for lack of trying. The new Laser Interferometer Gravitational Wave Observatory, housed in two sprawling facilities in Louisiana and Washington State, went on-line in 2002. Scientists are still calibrating the equipment and increasing its sensitivity; they are hopeful that, in the coming years, the observatory will detect gravitational waves for the first time."

While advocates believe vibrating strings underlie every particle and every force in the Universe, many scientists question whether anyone will be able to prove it. Direct proof may well be impossible, made so by their permanent invisibility. But indirect evidence already has revealed their existence as crucial; their gravitational and electromagnetic natures exist within all energy and all matter—the reason everything attracts everything else in the growing Universe.

The LB/FLINE model teaches that gravity and electromagnetism are inseparable from the beginning and throughout the history of our dynamic Universe; they are one and the same, the fundamental building blocks of the Universe.

Interconnections with the LB/FLINE Concept (2002)

While inherently opposing Edgar Allan Poe's Big Bang (BB) idea of 1848, the Classical Electrodynamics (CE) concept enhances, and melds precisely with, the Little Bangs (LB) concept of 1980. Both obey Mach's Principle; both acknowledge in a self-consistent way the mutual interaction and interconnectedness, or unity, of all parts of the Universe via the presence of electromagnetic forces throughout; all are electromagnetic in nature, no matter the mass, momentum, or energy form. This does, of course, make it easier to understand the gravitational forces among all particles and spheres of the Universe, no matter how far apart.

In both concepts, the blackbody radiation of all atoms definitively tells us that the 2.7 K background radiation throughout the Universe stems from the uncountable numbers of atoms that get in the path of any, and all, recorders sensitive to the radiation, no matter in which direction the instruments are pointed. In the BB version, this radiation is erroneously identified as a leftover from the initial BB explosion. In the LB/FLINE model, it is more logically explained as emanating from all energy masses throughout the Universe, primarily from the ubiquitous energy masses undergoing transformation into atomic matter. In the LB/FLINE model, the ubiquitous 2.7 K radiation is a constant that has never, and will never, change, a universal constant forever in compliance with Mach's Principle: "All scientific theories must acknowledge in a self-consistent way the mutual interaction and interconnectedness, or unity, of all parts of the Universe."

The toroidal ring structure simplifies our understanding of the fundamental phenomena of the atom, making it easier to understand the forever-ongoing processes of transformation (within all nuclear masses) of nuclear energy into atoms, and how these atoms function so well in all molecular combinations. Such knowledge is crucial to solving the anomalies of solar systems and broadly, the mysteries of the origins and evolution of all systems universal. If viewed in the perspectives of the Classical Electrodynamics concept and the LB/FLINE model, all anomalies pertaining to origins and evolution eventually will shed their veils of mystery.

The Four (now Five) Laws of Planetary Motion (FL), comprising one-third of the FLINE model, clearly reveal the dynamic origins of solar systems as stemming from ongoing processes of the LB, whose origins are attributed to the electromagnetic nature of the ever-expanding Universe. Yet, there remain the two greatest mysteries that may never be solved outside the realm of metaphysics: the origins of Space and the dormant electromagnetic energy that fills that space.

In *Fit to be Tied: Impatience with String theory boils over* (*Science News*, Oct. 21, 2006), Peter Weiss states, "Although the string universe includes the point-like, elementary particles of conventional physics, such as quarks and electrons, those are just vibrations of the more fundamental strings. In string theory, there are also many yet-to-be-discovered particles that are partners to those already known. The hypothesis of partner particles arises from a concept called supersymmetry, so researches also refer to string theory as superstring theory."

"By going out on a limb with extra dimensions and extra particles, string theorists get glimpses of possible answers to major questions that conventional theory has left unanswered. For example: Are the four basic forces of nature—the electromagnetic force, the weak force that controls nuclear decays, the strong force that holds atomic nuclei together, and gravity—variations of a single, more fundamental force? Many physicists suspect that they are, but today's prevailing theory of particle physics has revealed intimate connections only between the electromagnetic and weak forces."

"Yet in string theory, simple behaviors of strings—for instance, how they vibrate and whether they break or form loops—generate not only all four forces but also all the elementary particles."

What Holds the Universe Together?

Jan. 2006

Electromagnetism, Gravity, or Dark Matter?

According to the LB/FLINE model, from fundamental superstrings to universal systems, electromagnetism is the glue that holds everything together. Gravity is inherent in electromagnetism—from superstrings to all universal systems—else why would every universal body be attracted to all other bodies of every size at all stages of their evolutionary cycles? Electromagnetic superstrings in which gravity is an inherent characteristic are the glue that holds the Universe together, and negates the need for elusive dark matter to keep galaxies, solar systems, planets, and everything else from flying apart.

Dark Energy: Source and Function in the Expanding Universe

November 30, 2006

Quoting from *Dark Fingerprints* (*Science News*, Nov. 18, 2006), "The mysterious cosmic push that's tearing apart the universe began revving up about five billion years ago. But a new study reveals that several billion years earlier, the bizarre, elastic substance [dark energy] that fuels this push was lurking in the shadows and already beginning to fight gravity's tendency to pull things together."

The article further speculates that dark energy "might emanate from the cosmic vacuum and have a constant density ... resembling the cosmological constant" [that Einstein later repudiated as his greatest blunder]. These "findings" are confirmation of, and good attempts to explain, the 1998 discovery of the accelerating expansion of the Universe; all are based on studies of exploding stars called 1a supernova.

According to the newer LB/FLINE model of universal origins, the only dark energy needed to drive the accelerating expansion of the Universe originates as black-energy spheres at its spherical perimeter. Here, the exiting white energies moving at light-speed collide with dormant superstring energies residing in the absolute-zero coldness of the infinite Space enveloping our Universe.

As these energies coalesce, the black spheres grow ever larger and denser, while maintaining their original light speed. Their increasing gravity and rapid outward momentum are the fundamental reasons behind the accelerating rate of expansion discovered in 1998.

From the two poles of each spinning black sphere, a beautiful pinwheel galaxy is spawned: stars of all sizes buried in hot gaseous dust-clouds—a fact readily confirmed via close observation that's underpinned with substantiated evidence. The smaller star masses eventually evolve into planets undergoing five stages of evolution at rates in accord with size.

Rather than tearing apart the Universe, the dark energy spheres are the crucial source of energy needed to continue building an expanding and precisely functioning Universe.

Dark Fingerprints is another example of how the BB hypothesis forces scientists into speculations and misinterpretations of their spectacular discoveries.

Dark Matter or Dark Energy/Matter?

Oct. 20, 2006

According to Robert Caldwell at Dartmouth College, "Dark matter is a proposed solution to an as yet unsolved phenomenon—the mismatch between measurements of the gravitational mass and the luminous mass (the mass contributed by light-emitting matter) of galaxies and clusters, gravitationally bound groups of galaxies. This disparity suggests the presence of matter in the universe that does not efficiently produce light—hence it is invisible, or dark."

"This mass-to-light comparison indicates that the gravitational mass of galaxies and clusters far exceeds the luminous mass ... dark matter outweighs normal matter by a factor of 6 to 1." (*Scientific American*, Nov 2006). The prevailing BB model predicts our Universe dwells in a sea of dark matter with a density as high as roughly 105 particles per cubic meter. What this accomplishes can only be a matter of speculation.

In the LB/FLINE model, dark energy/matter in the form of black spheres is found at, or near, the perimeter of the speed-of-light expanding Universe where these invisible masses form via accretion of superstrings energy at its interface with the absolute-zero coldness of the infinite Space beyond. Here, the huge, dense spheres continue growing via accretion, remaining invisible until triggered into spewing from the two poles of each spinning black sphere brilliant galaxies in pinwheel fashion: fiery masses (stars) embedded in hot gaseous dust-clouds. A small remnant (0.5% of the ejected mass) remains at each galactic center to hold the system together.

According to the new model, the dense, uncountable black spheres may be the sought-after dark matter of the Universe, apparently in the proper 6-to-1 ratio, and fulfilling all crucial functions of the Universe; e.g., its ongoing accelerating expansion. Certainly, this ongoing process is the most logical source of the new energy needed to maintain our very active universal systems in the manner in which they are accustomed.

Other than infinite energy, this process gives assurance that superstrings are indeed Nature's most fundamental building blocks—an inherent factor in the LB/FLINE model of universal origins (1980-2004).

Addendum: *The Hawking Paradox* (Oct. 20, 2006, *AJC*): "In 2004, Stephen Hawking admitted that he'd made a mistake. The discoverer of black holes was now claiming that his entire theory about them was wrong. *What Hawking said, and how the scientific world reacted*. 9:00 tonight, *Science Channel*." [Oct 22, 2006]. This author was unable to get *Science Channel* 202; however it went, it should be good news for the LB/FLINE model of universal origins.

A Revolution in Scientific Thought

Oct. 2006

In 1848, Edgar A. Poe published the first version of the Big Bang hypothesis, which immediately became subject of laughter. But the hilarity slowed dramatically in 1917 when scientists first accepted it as the only available concept that offered a foundation, however shaky, on which to establish a scientific theory of universal origins that would include Laplace's Accretion hypothesis of 1796. Over the ensuing years, the BB/Accretion hypotheses became more firmly entrenched in the scientific community.

Scientists, then, were forced into attempts to interpret their discoveries within the context of the BB/Ac hypotheses, which usually proved difficult, if not impossible, to do without the use of pure speculation. Before his untimely death from cancer, Carl Sagan, the brilliant spokesman for science and the famous narrator of *Cosmos*, the TV serial on universal origins, called the BB a myth. In the summer of 1999, Stephen Hawking, the eminent physicist, stated, "Cosmology's hypotheses are conjectural when pressed to the limit."

Since myth begets myth, the validity of the BB now is brought sharply into question. After 30 years of researching universal origins, and unable to find one iota of substantiated evidence for a BB, the author agrees with the conclusions of Sagan, Hawking, and a large number of scientists who feel that the BB myth is science's equivalent of *Alice in Wonderland*.

In his 1956 book, *The Structure of Scientific Revolutions*, Thomas Kuhn makes it crystal clear that no scientific revolution can occur until there is a valid concept to replace it. *Origins of Universal Systems* meets Kuhn's prerequisite. Its new LB/FLINE model of universal origins is based on the solid foundation of the Five Laws of Planetary Motion that explain precisely how our Solar System came into being, why its planetary orbits are geometrically spaced and at various angles of inclination to the ecliptic, and how the planets are evolving through five stages of evolution at rates in accord with mass (size), plus many other anomalies that remain impossible to explain via the BB/Ac hypotheses.

An understanding of the Five Laws and our Solar System (and all SSs) opens pathways to comprehension of their galactic origins via black holes (actually spheres). The spewing of a galaxy from a spinning black sphere—a remnant of which remains at the center to hold it together—explains the fiery formations of many types of solar systems and single stars embedded in dust-clouds, all components of pinwheel galaxies.

Perhaps the most perplexing question in BB physics is the accelerating expansion rate of the Universe. It presents a major problem for the BB concept that predicts exactly the opposite: a gradually slowing rate of expansion. But the answer to this question is inherent in the LB/FLINE model in which the black spheres are seeded and, via accretion, grow large and extremely dense at the spherical perimeter of the speed-of-light expansion of the Universe into infinite Space.

The powerful gravity and outward light-speed momentum of the billions of black energy spheres are the fundamental reasons for the accelerating expansion of our Universe: a continual revitalization via this inexhaustible source of universal energy. In this process, new subdivisions of galaxies come into existence as fast as older ones evolve into inactive spheres, later consumed by larger masses. Thus, the Universe evolves, forever replicating itself, onward and outward.

While our spherical Universe occupies an extremely big space that grows ever bigger at a speed-of-light rate, it remains but a speck among countless other replications in the dark absolute-zero coldness of infinite Space, the inexhaustible source of all energy. This empty Space energy that drives the accelerating expansion of universes is absurdly improbable via BB principles, but remains absurdly simple via the LB/FLINE model.

The primary inference here is that all universes abide by identical laws of physics and chemistry, and by the homogeneity and symmetry governing the beauty, power and precision of Nature, all in accord with the LB/FLINE model of universal origins and evolution.

Note: Historical findings and relevant correspondence letters of interest are dated for the sake of more accuracy in the records.

The Shape of Things to Come

The Shape of Things to Come (*New Scientist*, 30 July 2005) by Davide Castelvecchi discusses insurmountable problems when viewed in the Big Bang (BB) perspective, and poses the question: Must we abandon the aims of physics? Quoting Raphael Bousso's conclusion, "Physics as we know it is impossible in the universe we live in."

"The mysterious dark energy that seems to be blowing the universe apart will always prevent us from discovering the ultimate physical truths." Fortunately, this is not exactly true! As many scientists already recognize, many such problems simply vanish in light of physics viewed in another perspective. The off-track culprit in this case is the mythical BB, not physics.

According to the new LB/FLINE model of universal origins, we live in a spherical, open universe that is expanding and growing forever via births of black holes—the densest form of energy and the eventual source of galaxies—moving at the speed of light at the interface of its spherical perimeter and the absolute zero of infinite Space where all atomic motion ceases (and where the string theory and its inherent gravity come into play).

This accounts for the accelerating expansion of the Universe as viewed from Earth's position—or any position—in the cosmos in which the farther out astronomers look in any direction, they see galaxies moving apart, apparently at accelerating speeds that can never exceed the original speed of light of their central black holes. In reality, the galaxies are slowing down very gradually and in sync, while maintaining their moving-apart paces and relative positions, perhaps forever.

There is no need for dark energy, or lambda, or dark matter to explain this self-perpetrating phenomenon; it is an inherent factor predicted and explained via the LB/FLINE model.

Wrenching Findings

Quoting from the interesting article, *Wrenching Findings: Homing in on Dark Energy* (*Science News*, Feb. 28, 2004), "Something is pulling the universe apart, causing galaxies to flee from each other at an ever-faster rate. ...researchers have taken one of the first steps toward identifying this bizarre influence, often known as dark energy....dark energy might be an intrinsic property of space itself."

Prevailing belief in the Big Bang (BB) myth forces scientists into the need for a force to explain how the Universe is being pulled apart. The idea of dark energy as an intrinsic property of space itself is only the newest one that furthers the conjectural nature of the mythical BB; myths breed myths.

In the LB/FLINE model of origin and evolution that definitively explains why galaxies flee from each other at an ever-faster rate, there is no need for the elusive dark energy. When astronomers first discovered the ever-faster rate of universal expansion, it was immediate confirmation of the new model's explanation—an inherent and unavoidable part of the new model—all without the need for dark energy or any other factor.

BB science is at the crossroads. A plethora of substantiated evidence warrants a change. Why not now?

Dark energy, matter mystery baffles experts

By Robert S.Boyd, June 20, 2003

WASHINGTON—"According to a batch of reports to be published today in a special *Welcome to the Dark Side* issue of the journal, *Science*, most of the cosmos cannot be seen, even with the most powerful telescopes. All but a tiny fraction of creation consists of two exotic, invisible ingredients called 'dark energy' and 'dark matter'." Astronomers admit they don't understand either of them.

Astronomers can never know the true size of the Universe; their powerful telescopes can view only the relatively small portion within their capabilities. Due to limitations of the speed of light, they can never know what lies far beyond their views. We cannot know whether the Universe had a beginning or will have an ending; we can only observe our neighborhood on which all the findings, data and speculations are based. However, the data we can gather tells us much about our tiny arena of the ever-expanding Universe; e.g., the ages and nature of its spheres. But the 13.4 billion years of age applies only to these spheres, not to those far beyond range.

Dark energy and dark matter, per se, are figments of the imagination; they are necessary only in vain efforts to support the mythical BB/Ac hypotheses. But black holes do exist as the densest form of energy; they are crucial to the origins and continuous support of galaxies. They form at the perimeter of the ever-expanding Universe in the absolute zero of space where energy particles, both old and newly formed, collapse into the densest form of (black) energy from which white-hot energy masses of various sizes later are spawned as stars, dust etc., in pinwheel galaxies. The observed acceleration of speed of universal galactic systems as astronomers look ever farther outward is an inherent phenomenon of the LB concept; lambda (cosmological constant), dark energy and dark matter are not needed to explain this natural phenomenon of the LB/FLINE model of origins and evolution.

Sean Carroll is correct: "We should be humble about dark energy. We haven't a clue as to what's going on." Conclusion stemming from false (BB) assumptions must always be wrong.

Superstrings: From Dark Void to Brilliant Light

Understanding our SS opens wide the door to relatively easy comprehension of origins of galaxies and their intimate association with black holes (actually spheres) from which they are birthed.

Such black spheres—the densest form of energy—are created at the perimeter of our ever-expanding-at-light-speed spherical Universe. Superstrings originate at the interface of our Universe and the absolute-zero coldness of the dark Space beyond, where they are activated as the dark energy comprising black, dense spheres.

These inactive black-energy spheres, growing ever larger, consist of superstrings tightly compacted, until spewed out as white-hot masses (stars of all sizes) embedded in hot gaseous dust-clouds, all in brilliant pinwheel fashion.

The Superstrings Theory, as explained by Brian Greene in *The Elegant Universe* (1999), may be the last frontier for understanding the origins and evolution of universal systems. The theory deals with the universally fundamental substance with which all energy and matter are structured: strings of electromagnetic energy in which gravity is inherent. How else could all matter remain attracted to all other matter at all stages of evolution throughout the Universe? They are Nature's perfect building blocks. While the Superstrings concept does not fit well in the BB hypothesis, it is a perfect link in the LB/FLINE model of universal origins.

The Beauty of String Theory

New Scientist, April 16, 2005

"What appeals... about string theory is that it is gorgeous. It is the only theory with a symmetry large enough to swallow both relativity and the standard model. It is the only game in town—all rivals have been shown to be inconsistent in string theory; we have a startling new picture: the subatomic particles of the universe—electrons, quarks, and neutrinos—are all different resonances of a string vibrating much like a rubber band."

"... Nature, at its most fundamental level, expresses itself in the beautiful, elegant and symmetrical way."

A Brief Summary of Beauty, Power and Precision

April 16, 2005

It may be puzzling to comprehend—but never the less true—that one of Nature's two most critical keys to understanding universal origins of beauty, power and precision lay dormant for 25 centuries. That key discovery, made around 2500 B.C., by Pythagoras, is the simple 3, 4, 5 triangle that made possible the discovery of the Fourth Law of Planetary Motion (1980-1999)—just 100 years after Einstein's revealing discovery, $E=mc^2$, and the Fifth Law (1980-2006).

But first, some background. Because the small band of Pythagoreans believed that Earth revolved around a central fire (Sun), they were killed by enraged neighbors, and much knowledge was lost. But Aristotle came to the rescue by recording the remaining knowledge.

In 150 A.D., Ptolemy XIII, brother of Cleopatra, issued an edict that Earth was the center of the Universe. This belief prevailed unscathed until Nicolas Copernicus, just before his death in 1543, bravely published a book placing the Sun at the center of our Solar System. Galileo knew the potential consequences of the Copernican truth, but forged ahead in full support of it, bravely suffering the consequences.

During the very early years of the 17th century, Johannes Kepler, the brilliant German astronomer, discovered the First Three Laws of Planetary Motion governing orbital movements of the six then-known planets around our Sun. This huge success followed his initial failure in 1595 to find the reasons the six planets are spaced in their geometric pattern around the Sun: a mystery destined to remain the greatest enigma of our Solar System for 400 years.

July, 2005 marked the 10th anniversary of discovery of the new Fourth Law of Planetary Motion that had remained so elusive throughout four centuries. The mathematical solution (1980-1995) to the Fourth Law, along with Kepler's First Three Laws, present a clear, detailed explanation of the manner in which the planetary orbits became geometrically spaced during the dynamic fiery birth of our SS some five billion years ago. Once in orbit, the fiery energy masses began their five-stage evolution through their easily observable stages: from energy to gaseous to transitional to rocky (e.g., Earth) to inactive (e.g., Mercury), all at rates in full accord with mass.

Other than a substantiated solution to the spacing of planets, the Five Laws (FL) put scientists on the right track for finding valid scientific solutions to all relevant solar system anomalies. After a decade of a relevant substantiated solution, scientists have yet to find an anomaly capable of escaping the realm of the Five Laws of Planetary Motion (FL), now a firmly embedded component of the LB/FLINE model of universal origins.

A good understanding of solar systems is crucial to the understanding of galaxies from which they were birthed in full accord with the new LB/FLINE model. It is essential that one fully understands the FLINE model before moving on to galaxies from which SSs are birthed in many fiery combinations of binary and multiple-planetary systems, all in the first stage of evolution: Energy. Viewing the beauty of pinwheel galaxies with their billions of stars embedded in gaseous dust-clouds spewed from a spinning black hole (sphere) is indeed an awe-inspiring sight. But what is the origin of multiple billions of black spheres that spawn these billions of majestic jewels of the night sky?

The simple and beautiful answer to this critical question—as revealed earlier in this book and in previous editions of the Energy Series (1973-2002)—resides at the expanding perimeter of our spherical Universe. At its interface with dark space at absolute-zero temperature where all atomic motion ceases, all dormant energy in the form of Superstrings coalesces as black, super dense spheres of many diverse sizes.

Thoughts on the Past, Present and Future of Scientific Beliefs

March 2003

Historical and modern-day myths include a flat Earth, fossil fuels, the Big Bang/Accretion hypotheses, planetary cores of rock, iron, etc., dirty-snowball comets, Earth's water coming from outer space via comets and asteroids, the asteroidal extinction of dinosaurs, and a flat Universe.

Of these myths, the ones posing the greatest obstacle to scientific progress are the antiquated Big Bang/Accretion hypotheses. Lacking any substantiated evidence, they inevitably lead scientists to ludicrous interpretations of, otherwise, excellent data in vain efforts to support these myths.

Ironically, that excellent data always fit precisely into the continuity of the imminent revolution in scientific thought: the Little Bangs/FLINE model of origins and evolution, now waiting in the wings to displace the mythical BB/Ac hypotheses.

The LB/FLINE model, underpinned with a plethora of overwhelmingly persuasive, well-substantiated evidence, leaves no room to doubt its validity as the imminent scientific revolution in beliefs about the ongoing origins and evolution of all things inanimate.

The enigmatic solution to the Fourth Law of Planetary Motion (1980-1995) brings closure to the Copernican Revolution initiated in 1543 when Copernicus placed the Sun at the center of the Solar System. In bringing the LB/FLINE model full circle, the new Law proved to be the crucial key to a definitive understanding of Solar System origins and the evolution of their planetary spheres. These findings predict a truly exciting future for science.

A Change in Direction of Scientific Thought

Two Journal-published papers and a book on origins and evolution of inanimate worlds by Alexander A. Scarborough may eventually change the direction of scientific thought on how everything came into being. After 30 years of in-depth research on the subject, the LaGrange author's two interrelated papers have been published in the *Journal of New Energy*, Vol. 7, No. 3, an International Journal of New Energy Systems, "devoted to publishing professional papers with experimental results that may not conform to the currently accepted scientific models." Both papers had been presented earlier at the International Conference of the Natural Philosophy Alliance, June 9-13, 2003.

The first paper, *Bringing the Copernican Revolution Full Circle via a New Fourth Law of Planetary Motion*, presents a definitive explanation of the manner in which the planets attained their geometric spacing around our Sun during the dynamic birth of our Solar System (SS) some five billion years ago. The paper unites the great discoveries of the past 2500 years in an overwhelmingly persuasive argument that brings the Copernican-Galilean Revolution full circle. While the revolutionary idea of a Sun-centered SS belongs to Copernicus, it was Galileo who suffered by keeping the idea alive after the death of Copernicus, in spite of knowing and undergoing the dire consequences.

The First Three Laws of Planetary Motion governing the functions of our SS were discovered by Johannes Kepler, a brilliant German astronomer, during the early years of the 17th century. But Kepler failed in attempts to discover the Fourth Law, and it remained unsolved until Scarborough's solution during the years of 1980 to 1995—nearly 16 years of persistent efforts that came to fruition 400 years after Kepler's initial attempt in 1595.

This discovery is important because it has proven to be the vital link in a definitive understanding of the origin and evolution of our SS. For the first time, it provides the necessary substantiated evidence and fundamentals to enable scientists to solve all anomalies of planetary systems, including the known extra-solar systems far beyond our own.

Scarborough's paper contains a list of 41 anomalies that are definitively explained only by his LB/FLINE model of origins and evolution. Three examples are: (1) why planetary orbits are inclined to the ecliptic; (2) why the Sun's equatorial mass has a 7° inclination to the ecliptic; (3) the 21 extremely powerful explosions of Comet SL9 segments at Jupiter's thin cloud-tops in 1994. His was the only accurate prediction in the write-in contest among scientists to predict the results of the fiery impacts; all other results, including those of the best supercomputers proved puny by comparison. His prediction had been the only one based on the fiery nature of comets; all others were based on "dirty snowballs" (from which Planet Earth allegedly formed, according to the Big Bang myth).

Discovery of the Fourth Law reveals the very rare combination of precise prerequisites for establishing a multiple-planet system like ours, thereby divulging the secret of why such systems are a rarity among the multitude of singular, binary and trinary solar systems, now being discovered on a regular basis by astronomers, just as predicted and explained by Scarborough's revolutionary LB/FLINE (1973-2002) model of origins and evolution of universal systems—which brings us to the second paper: *Origin and Evolution of Planetary Systems*.

Once the fiery energy masses were placed in geometrically-spaced orbits via the Four Laws of Planetary Motion, each began its five-stage evolution: from energy to gaseous to transitional to rocky to inactive, all at rates in proportion to size; e.g., the smallest planetary masses have reached their fifth (inactive) stage, while the largest, Jupiter and Saturn, are still in the second (gaseous) stage, and Earth is in its fourth (rocky) stage, with—after five billion years—about one-third of its energy core remaining to be transformed into planetary matter.

Just as an automobile is driven by an energy source, a planet must have an energy source to drive it through the five stages of evolution until the source is depleted. The continuous transformation of the core energy into virgin gaseous atmospheres, followed by virgin land and water to form a planet takes a very long time, measured in billions of years. The planets of our Solar System have required some five billion years to attain their current stages of evolution via this slow, but steady, transformation of energy-to-matter, known in scientific circles as Internal Nucleosynthesis. This fundamental knowledge is essential for understanding earthquakes and all other anomalies of planetary systems.

To explain how all planetary matter (atmosphere, land, water) was, and is, created, Scarborough's paper uses the example of the intimate relationship of the hydrocarbon fuels (gas, petroleum, coal), and details how they were (and still are) made via Internal Nucleosynthesis (IN) into atomic elements, followed by chemical reactions known as polymerization into ever larger molecules as they evolve from gaseous methane into liquid petroleum, often followed by solidification into coal whenever conditions dictate the changes. Planetary Evolution (E) would not be possible without Internal Nucleosynthesis; IN and E are inseparable.

Scarborough stated, "The great significance of this work is that it offers a long-sought, valid alternative to the Big Bang hypothesis that most scientists (79%) now recognize as myth. Like the Copernican idea of a heliocentric SS, it will force a change in direction of scientific thought, opening ever more pathways to understanding the fundamentals of origins and evolution of God's inanimate worlds in a forever-expanding Universe. In turn, this will enable scientists to make definitive interpretations of their spectacular discoveries."

Survey sheet to AAAS as a suggestion for improving the AAAS Meetings **_02-26-99_**

Rather than pursuing only the Big Bang and Accretion concepts of planetary origins and evolution, the AAAS Meetings should be more open to all factually-based concepts. Advocates of prevailing beliefs should not be permitted to block out newer concepts simply because of the challenge to their dogma. Such new ideas will open avenues of research leading to definitive answers to planetary anomalies. A prime example is the proposed Fourth Law of Planetary Motion (1995) that yet remains unknown to the greater scientific community.

Memo

Re: Mature Before Their Time: In the youthful universe, some galaxies were already old
(Ron Cowen, *Science News*, Mar 1, 2003)

Quoting Cowen, "A flurry of new reports suggests that a surprising number of galaxies grew up in a hurry, appearing old and massive even when the universe was still young. In mid-December, scientists announced in a press release that they had found a group of distant galaxies that were already senior citizens, chockablock with elderly, red stars a mere two billion years after the (alleged) Big Bang .. Some of those galaxies were nearly as large as the largest galaxies in the universe today ... astronomers may have to revise the accepted view of galaxy formation."

These findings should indeed force a change in the way astronomers view the origin and evolution of the Universe, along with the formation of galaxies. They confirm the fundamental principles of the Little Bangs (LB) theory of the ongoing creation of the ever-expanding Universe. This powerful evidence is exciting because such galaxies were both predicted and explained by the revolutionary LB concept over two decades ago. During the ensuing years, it has been presented to numerous science organizations, usually via ineffective poster sessions, but never permitted on the main program of our large organizations or to reach the public via the standard press release format.

Even with conjectural speculation, these findings do not fit into the Big Bang/Accretion (BB/Ac) hypotheses; rather, they contradict these hypotheses, while presenting anomalies whose solutions become definitive only via the LB concept. The impetus for these findings might well have stemmed from the principles and impeccable predictions of the LB/FLINE concept (1980-1996), available on the Internet at www.1stbooks.com. Certainly, they do confirm the predictions of the LB/FLINE, and thereby, render powerful support for this revolutionary concept that definitively explains why the youthful Universe contains old galaxies.

Why the Curvatures in Galactic Arms?

Even though dark matter has eluded all attempts at detection, most cosmologists of the BB persuasion are convinced it must be out there. They believe it is essential for understanding the galactic rotation curves, which describe the way stars rotate in curved arcs around the centers of galaxies. The problem: all stars in the middle and outer parts of galaxies orbit with the same speed, which begs the question of why don't the outer stars move more slowly than the inner masses—as planets do in our Solar System?

Rather than dark matter pushing these stars at equal speeds, physicist, Mordchai Milgram, could not accept the existence of a halo of dark matter controlling every galaxy. So he simply modified a Newton second law, F=ma, to arrive at MOND, a version of the law that increases the force acting on stars in the outer realms to move them faster along. His papers were rejected by the journals, one excuse being "the flat rotational curves will be resolved in other ways." The controversy between the two concepts continues growing.

This basic problem with both concepts of dark matter and MOND is the mythical BB on which they are based. Dark matter, other than black holes (spheres), will remain elusive; it is a convenient figment of imagination. In MOND's case, why try to change a fundamental law that will continue to survive?

But there is an answer, one that is inherent in the LB/FLINE model that needs no dark matter or modified principle to explain galactic arm curvatures. They occur naturally when the star-studded material is spewed from the two poles of a spinning black sphere. There is nothing to impede the original momentum (speed) of the stars in question; they simply have a longer distance to travel. Thus, the arms form into natural curvatures via original momentum of the stars. The older the galaxies, the greater the curvature of the arms.

Dark Matter and MOND

After many decades of research on galaxies, the dark matter and Milgram's MOND concept (*Discover*, Aug 2006) persist as the prevailing concepts for "understanding the so-called galactic rotation curves, which describe the way stars rotate around the centers of galaxies." The problem is that stars close to the galactic center should orbit faster than stars at the edge because of the differences in the pull of gravity on the inner and outer masses of stars—just as is true in our Solar System.

But there is a third alternative that circumvents the problems of the first two concepts: The Little Bangs (LB)/FLINE model of universal origins, a concept consistent with Einstein's general theory of relativity, and one that needs no dark substance to account for its inherent solution to the puzzling galactic rotation curves with all their stars equal in speed.

One cannot compare the principles of these curves with the orbital principles of our Solar System because—like comparing apples and oranges—they do not happen in the same way. While the galactic system stems from black-hole ejecta (LB), our Solar System planets are subsequent products of the interactions of two of the ejected masses of energy during the fiery birth of the Milky Way Galaxy, as definitively revealed by the Five Laws of Planetary Motion (FL). Scientists know that 0.5% of the ejected mass of the black hole remains at the center to hold the galaxy together.

In this model, the galactic rotation curves are an inherent factor of galaxy formation. The speed of ejected masses of energy (stars) embedded in gaseous dust-clouds must remain constant to produce the beautiful spiral arms of galaxies—and they do. Without such constancy, the curves could not possibly form in a BB as they naturally do; chaotic galaxies would rule the nights. Stars at the edge simply have greater distances to traverse in equal times; thus, they fall behind at rates proportional to distances from the galactic center in curvature fashion.

Analogy: Runners racing at equal speeds around a circular track would form the same pattern; runners in the outer lanes would gradually fall behind those in the inner lane, all at rates proportional to distances from the center of the circular track.

Please, do not change the physics of universes; they are constants. And why speculate with dark matter or dark energy, when definitive explanations are waiting in the wings?

Some Thoughts about this Book: Ten Points of Interest

July 25, 2005

1. This tenth edition in the Energy Series is the product of dedication to 36 years of a broad education in the sciences, followed by 32 years of researching the myths and facts of science.

2. During those 70 years, much has been learned about the great discoveries of the past 2500 years pertaining to beliefs about origins and evolution of universal systems.

3. I agree with the many scientists who openly admit that the Big Bang is pure myth, without an iota of substantiated evidence to support it.

4. The LB/FLINE model detailed in this book offers a valid scientific alternative to the BB myth—an alternative backed by voluminous substantiated evidence that leaves little, if any, room for doubt about origins and evolution of universal systems—all in tune with Einstein's relativity theories.

5. All the great discoveries of the past 2500 years form solid structural support for this alternative, and space discoveries continually confirm and add to its factual composition.

6. In bringing the Copernican Revolution full circle, this treatise may qualify as the most comprehensive book in the history of origins and evolution of universal systems.

7. Although currently suppressed by BB advocates, the well-substantiated LB/FLINE model appears destined, in the long run, to displace the BB myth in future textbooks.

8. The Universe functions via inviolate laws of chemistry, physics and mathematics. The fundamental keys to understanding our Universe must include the definitive Five Laws of Planetary Motion that detail how our geometrically-spaced Solar System came into being, and why and how the changes in orbital positions occurred.

9. But the biggest questions may always remain unanswered: Did God establish these laws, and then instill in mankind the brainpower to discover them one by one—a process we call science? And were mankind's souls instilled to discover and enact the purpose of our existence in the Universe—a process we call religion?

10. I believe this book to be one of the great gifts of a lifetime to those who have an interest in how the Universe began, and how it functions now and forever.

Science at the Crossroads: A Brief History of Dogma

Oct. 11, 2003

Abstract. This paper presents a brief history of controversial beliefs stemming from three great discoveries made by the Pythagoreans around 500 B.C. Their findings led to the Copernican Revolution that moved forward via the discoveries of Copernicus, Kepler, Newton, Einstein *et al.*, and now has been brought full circle by the discovery of the new Fourth Law of Planetary Motion, now firmly established via the crucial keys furnished by the Pythagoreans. Kepler's First Three Laws of Planetary Motion, along with the new Fourth Law, lead to a definitive model for understanding the origin of solar systems and the five-stage evolution of planetary spheres. The Fourth Law first eluded Kepler in 1595, solely because of his failure to recognize the discoveries of 500 B.C. as the crucial keys to comprehending the spacing of planets around our Sun. After 2500 years of dormancy, these findings pose a serious challenge to the Big Bang/Accretion hypotheses, and now seem destined to force a change in fundamental beliefs about the origin and evolution of solar systems and perhaps the Universe itself.

Text

This talk (paper) presents a brief history of the rise and decline of scientific truth. The rise began in 500 B.C. with three great discoveries that were destined to influence controversial beliefs for 2500 years, and even now they pose a serious challenge to beliefs of modern-day science.

A secret society of Pythagoreans in 500 B.C. put much faith in the Pythagorean Theorem and the Golden Ratio. They correctly rationalized that number is important for an understanding of nature; i.e., that number is the principle of all things. And they argued that Earth moves around a central fire, which may have influenced Copernicus some 2,000 years later. But because of their advanced beliefs, the Pythagorean community was destroyed by enraged neighbors, and much knowledge went to waste.

In 150 A.D., Ptolemy XIII, brother of Cleopatra, in order to settle the controversy, issued an edict that Earth was the center of the Universe. This decline in truth lasted 14 centuries before Copernicus, in 1543, placed the Sun at the center of our Solar System to re-ignite truth via the Copernican Revolution. Then in the early 17th century, Johannes Kepler discovered Three Laws of Planetary Motion governing the movements of planets around our Sun. However, Kepler failed in attempts to discover a Fourth Law explaining how the planets attained their geometric spacing around the Sun—a mystery that was to remain unsolved for 400 years. Kepler wrote, "Geometry has two great treasures: one is the theorem of Pythagoras; the other, the division of a line into extreme and mean ratio. The first we may compare to a measure of gold; the second (GR) we may name a precious jewel." For unknown reasons, Kepler failed to use this geometry in his search for the Fourth Law, which is the basic reason he failed to discover it.

In 1509, Luca Pacillo published a dissertation titled, *De Divina Proportione*, which was illustrated by Leonardo da Vinci. In his 20th century book, *The Divine Proportion*, H. E. Huntley wrote, "It was suggested in the early days of the present century that the Greek letter Phi ... should be adopted to designate the GR. It is a fascinating compendium of Phi's appearance in unexpected places. The pervasiveness of the GR was expressed best by C. Arthur Coan: "Nature uses this as one of her most indispensable measuring rods, absolutely reliable, yet never without variety, producing perfect stability of purpose without the slightest risk of monotony ... We shall find it flung broadcast throughout all Nature."

In 1644, René Descartes accurately wrote that Earth's interior is "Sun-like". Later in the same century, Isaac Newton discovered the gravity that holds our SS—and the Universe—together.

The next crucial discovery toward understanding planetary origins and evolution was Einstein's 1905 formula, $E=mc^2$, that reveals the intimate, interchangeable relationship between energy and matter; e.g., ongoing Internal Nucleosynthesis (IN), the engine that drives all Evolution (E) in universal spheres.

Finally, during the 15-year time frame of 1980-1995, my struggles with the Pythagorean and Golden Ratio geometry of the Fourth Law of Planetary Motion ended with a definitive solution to the 400-year mystery initiated by Kepler in 1595. These two great treasures of Pythagorean geometry made it possible to solve the mystery of how the planets attained their geometric spacing around our Sun, as was shown in the Seven Diagrams in previous talks, papers and books. Here you see their beauty again in our SS's geometry.

This geometric solution unites the great discoveries of 500 B.C.-1905 A.D. with all later discoveries of a geological or astronomical nature into a definitive LB/FLINE model that brings the Pythagorean/Copernican Revolution full circle. Backed by a plethora of substantiated evidence, the new model clearly reveals the dynamic origin of our SS, and the subsequent five-stage evolution of its planets and moons. In doing so, it opens pathways to definitive solutions to all planetary anomalies. This definitive model of planetary origins and evolution is simply the unification of all the brilliant historical works of the past, present, and predictably, the future.

The most serious decline in scientific truth began in 1796 when Pierre Laplace published his *Exposition of the System of the Universe* in which he hypothesized that a huge, lens-shaped cloud of gas rotated, cooled, contracted and accreted into planets, satellites and the Sun—a scene that used to be shown often on TV. Laplace's nebular hypothesis was accepted for a long time, but eventually was modified to the planetesimals hypothesis in which larger chunks of matter accreted into planets and moons. But the source of the accreting matter remained a mystery.

Then in 1848, Edgar A. Poe published his book, *Eureka,* in which he described how the Universe began with a "big bang" (BB)—anticipating, by 70 years, Willem de Sitter's theory of the expanding Universe, first published in 1917. So the BB became the source that furnished the matter necessary for accretion into planets. In 1950, scientists accepted Fred Whipple's hypothesis that comets are "dirty snowballs" left over from the BB, and that these represent the chunks that accreted eventually into planets, moons and our Sun. In reality, much evidence clearly reveals the fiery nature of comets, as was explained in a previous talk (and paper). But today, this knowledge remains suppressed.

Solely for lack of an alternative, the Big Bang/Accretion (BB/Ac) hypotheses gained a strong position in the scientific community during the 20th century. In a science meeting attended many years ago, I heard Michael Turner of the University of Chicago mathematically "prove" the BB. It sounded ludicrous then, and still does. But in spite of its illogical progression, it has become firmly entrenched in the scientific literature as the official concept in which interpretations of new findings are made—too often leading to ludicrous conclusions at which you won't know whether to laugh or cry.

A typical example will make *Alice in Wonderland* seem sensible by comparison. The article, *Mining for Cosmic Coal,* among too many to count, illustrates how this talk earned its title. Such foibles of the BB/Ac hypotheses are so numerous and so prolonged that it borders on the unbelievable. During the many decades of costly efforts to prove the BB/Ac hypotheses, not one iota of substantiated evidence supporting these beliefs has been found—this, in spite of billions of dollars poured into thousands of scientific projects during the past 40+ years. Quoting Carl Sagan, "Science is an ever-changing set of myths." Interpretations of excellent data, when based on mythical beliefs, must themselves remain mythical, which usually results in erroneous, often weird, interpretations. Contrariwise, all scientifically valid data usually fit precisely into the LB/FLINE model of origin and evolution of planetary systems.

Two revealing examples are: (1).The antiquated beliefs that gas, oil and coal came from outer space or from fossils; and (2). Our oceans of water came from outer space on comets. Valid scientific explanations of these examples are inherent in the ongoing Little Bangs (LB)/FLINE concept; i.e., the hydrocarbons (gas, oil, coal) and all water, are products of Internal Nucleosynthesis within Planet Earth, as was proven, beyond any doubt, by the author (1975-1982). This knowledge, too, remains suppressed, and the decline in truth continues unabated.

These two examples illustrate one of the primary predictions made in the book, *Structure of Scientific*

Revolutions, by Thomas Kuhn. Paraphrasing Kuhn: Advocates of prevailing dogma will go to any lengths, no matter how ludicrous, to protect against threats to their beliefs. And to paraphrase Max Planck, the imminent physicist: Science advances one funeral at a time; the old guard must die off before new ideas can be brought forward by younger generations. But hope is sustained in a valid 1999 statement by a famous old-guard witness concerning the BB/Ac hypotheses. Stephen Hawking, a world leader in physics, admitted, "Cosmology's hypotheses are conjectural when pressed to the limit."

In the author's humble opinion, the mythical BB/Ac hypotheses are the grandest and costliest myths in the history of science: an albatross around the neck of science.

Thomas Kuhn wrote that no scientific revolution has occurred, nor will occur, in the future unless there is a proposed alternative theory to replace the older one. The LB/FLINE model of origins and evolution, with its taproot reaching back to 500 B. C., is well qualified as the proposed scientifically valid alternative that stands firmly ready to displace today's weird beliefs. It clearly brings the Copernican Revolution (a definitive understanding of SS origins and the subsequent evolution of all planetary systems) full circle. Simultaneously, it fully substantiates the brilliant beliefs of the Pythagorean community so brutally destroyed by enraged neighbors in 500 B. C.

In defending Copernicus and the heliocentric Solar System, Galileo knew the potential consequences of the truth. Others who spoke it had been branded heretics, tortured and even burned at the stake. Today's methods are more subtle, but equally effective. BB/Ac advocates simply band together to control what gets into the news media and who is awarded grant money. Such protective measures are essential for survival of the status quo; they have proven extremely effective against the encroachment of new ideas during the past three decades of development of the many facets of a valid scientific alternative: the revolutionary LB/FLINE model. So, 2500 years after the Pythagoreans gave us the crucial keys to understanding planetary origins and evolution, Science stubbornly continues to struggle in vain to unlock Nature's mysteries without resorting to the use of these crucial keys—an utter impossibility.

Now at the crossroads of Conjectural Myths versus Substantiated Facts, Science will choose which road to take: either to continue dwelling in "Blunderland" by fulfilling Kuhn's prediction of shielding the status quo at any cost, or to heed the wise insight of the Pythagoreans, Sagan, Hawking, Planck *et al.*, and move boldly forward to the light at the end of the tunnel. While the first choice continues, there will be no chance that definitive solutions to the mysteries of origins and evolution can be found. If the second choice is made, definitive solutions soon will fall readily into place.

To shield or not to shield? That is the question. Meanwhile, the grandest and costliest hoax in the history of science remains today as an immovable obstacle to progress in understanding what the Pythagoreans initiated 2500 years ago; i.e., the critical significance of their geometry in unraveling the mysteries of Nature.

In closing, it is fortunate that the brilliant discoveries in 500 B.C. have survived the past 2500 years. Had this knowledge not been recognized, praised and preserved by a number of historical figures, especially Aristotle and his recordings, the Copernican Revolution might not have begun in 1543. And it might not have been brought full circle in 1995 by the discovery of the geometric solution to the Fourth Law of Planetary Motion that is destined to force a change in the fundamental beliefs of science.

I am honored to dedicate this paper (and talk) to the brilliant Pythagorean community that was so brutally destroyed by enraged neighbors around 500 B.C.

References and Notes:

1. Hetherington, *Encyclopedia of Cosmology*, (1993).

2. Colin Wilson, *Starseekers*, (1980).

3. Nigel Henbest, *The Planets: A Guided Tour of Our Solar System through the Eyes of America's Space Probes*, (1992).
4. R.A. Karam, J.M. Wampler, Georgia Institute of Technology, Neely Nuclear Research Center, *Characterization of Natural Gas, Oil and Coal Deposits With Regard to Distribution Depth and Trace Elements Content*, (1986).
5. Alexander Scarborough, *Undermining the Energy Crisis*, (1977, 1979).
6. Thomas Kuhn, *Structure of Scientific Revolutions*, (1956).
7. Alexander Scarborough, *The Spacing of Planets: The Solution to a 400-Year Mystery*, (1996).
8. R. G. Lerner, G. L. Trigg, *Encyclopedia of Physics*, Second Edition.
9. Alexander Scarborough, *New Principles of Origins and Evolution: Revolutionary Paradigms of Beauty, Power & Precision*, (2002).
10. William C. Mitchell, *The Cult of the Big Bang, Was There a Bang?* (1995).
11. Notes: Numbers 7 and 9 stem from the author's discovery of the geometric solution to the Fourth Law of Planetary Motion (1980-1995). Their new concepts are underpinned with well-substantiated evidence. #7 is available at Ander Publications, 202 View Pointe Lane, LaGrange, GA 30241. Phone 706-884-3239. #9 is available at www.Amazon.com or bookstores or from the author.

AAAS-SWARM-NPA Meeting, Tulsa, OK, April 3-7, 2006
Memo to: LDN

The Conference held at Tulsa University was a huge success, with a few participants presenting more than one paper. Dr. Robert Heaston presented five significant papers, with one being a major breakthrough that locked in well with the four presentations by Alexander Scarborough, assisted by his daughter, Kay Scarborough, a resident of Marietta.

Heaston showed the derivation of a critical mathematical formula that provides the missing link in Scarborough's Little Bangs concept of universal origins. The formula reveals that compression is Nature's initial key to nuclear fusion in spheres of all sizes: stars, suns, planets, comets. Such compression, according to the new concept, initiates in the super dense black spheres continuously forming at the spherical perimeter of a Universe expanding at the speed of light into the absolute-zero temperature of infinite Space where all atomic motion ceases at the interface with our Universe.

Eventually, the compression within the dense spheres becomes sufficient to trigger eruptions of billions of brilliant nuclear masses of all sizes, embedded in gaseous dust-clouds, from the two poles of each spinning sphere. The results are billions of beautiful pinwheel galaxies, each consisting of billions of stars; the smaller masses eventually evolve into planets, usually locked in orbit around larger masses—all in accord with the Four Laws of Planetary Motion. A small black fraction (0.5% of ejected mass) remains at the center to hold each galaxy together.

This Little Bangs (LB) concept explains the large varieties of strange solar systems scientists have discovered in galaxies—and why many erroneously believe that stars accrete from gaseous dust-clouds. It also explains the accelerating rate of expansion of the Universe, no matter in which direction astronomers look.

Scarborough's solution to the enigmatic Fourth Law of Planetary Motion proved to be the fundamental key to providing answers about origins of solar systems which, in turn, provided clues to the fiery origins of galaxies.

"This Little Bangs (LB/FLINE) concept is the product of a lifetime of study that began at age 12, and became more concentrated during the past 33 years. The first breakthrough occurred in 1973 when I realized the close relationship among the hydrocarbon fuels (gas, petroleum, coal), and how they were, and still are, made via universal laws. Our organization's ultimate goal is to ease science away from the Big Bang myth, and more toward substantial evidence," he added.

Educating via analyses of science in movies and TV

In *What's Wrong with This Picture?* (*Science News*, Oct 16, 2004) by Sid Perkins, scientists decry movies showing "inaccurate science on both big and small screens that misinforms viewers who may not distinguish what's fiction and what's fact. Scientists are using inaccurate scenes as ways to attract public attention to genuine scientific concepts."

"In an attempt to boost student interest in physics, Newport and his colleagues are developing an educational package, intended to reform the undergraduate physics curriculum that includes deconstructing scenes from films to teach scientific principles." As one scientist put it, "The facts of science are more interesting than anything you can make up."

In conjunction with these fine efforts, scientists should make every effort to exclude scientific myths and conjecture from textbooks and scientific literature. The Big Bang/Accretion hypotheses are a blatant example; they have become the grandest and costliest "fraud" in the history of science: an albatross around the neck of science.

Thomas Kuhn in his *Structure of Scientific Revolutions* makes it clear that no revolution can, or will, occur until there is a better alternative. The factual LB/FLINE model of origins and evolution meets and exceeds Kuhn's requirement. Its solution to the enigmatic Fourth Law of Planetary Motion and Kepler's First Three Laws are the keys to understanding the dynamic origin of our Solar System, the evolution of planets and moons, and subsequently, how everything came into being.

The interlocking facts of science are indeed more interesting—and far more exciting—than anything you can make up. Interlocking exciting facts are really the best way to boost student (and public) interest in the sciences.

New Type of Black Hole Holds Galactic Hint:
Validation of the Little Bangs (LB) Concept

Sept. 18, 2002

The article, *New type of black holes holds galactic hint*, is an excellent validation of two facets of the Little Bangs (LB) Concept (1941-1980) in which galaxies are created—usually in flat pinwheel fashion—via the continuous ejection of dust and brilliant stars from the two poles of a spinning black hole. This process continues until it reaches a balance between each black hole and its surrounding masses; as determined by Karl Gebhardt *et al.*, this balance occurs when each black hole reduces to a mass equal to 0.5% that of the stars surrounding it. The reason for such precision? Nature is governed by precise laws, which accounts for its beauty, power and perfection.

Black holes, the densest form of energy, are created continuously at the perimeter of the Universe expanding at the speed of light; this explains the fact that the farther out astronomers look, the faster the galaxies are moving away. Because of this fact of the LB, scientists will never be able to view the true edges of the Universe, or to estimate either its actual size or age.

Primordial Matter Possibly Re-Created

AP, *AJC*, June 19, 2003

In reality, due to limitations of the speed of light scientists have no way of knowing the actual size or age of the Universe, or whether it had a beginning or will have an ending. They do know that the Universe and all its component parts operate under specific laws, which are being discovered at a reasonable rate. However, the conjectural Big Bang/Accretion hypotheses always block the paths to proper interpretations of crucial data; e.g., the quark-gluon plasma (qgp) article on primordial matter.

According to the Little Bangs (LB)/FLINE model, nuclear-energy cores of all active spheres of the Universe consist of qgp; it is the energy source of their Internal Nucleosynthesis: the ongoing transformation of nuclear energy into all forms of radiation and atomic/molecular matter found in varying quantities in all stars, planets, moons, comets, etc. It is what makes the stars shine and spheres evolve. In all forms of energy, gravity is an inherent property that holds everything in orderly fashion.

Everything evolves; nothing is as it once was. But in the absence of qgp, evolution of inanimate and, subsequently, animate worlds would not be possible. Ongoing evolution and ongoing Internal Nucleosynthesis of the qgp core are inseparable: one cannot exist without the other. The ongoing source of the qgp via an ever-expanding Universe is explained by the LB/FLINE model.

Space Invaders

Space Invaders: The stuff of life has far-flung origins (*Science News*, May 1, 2004), by Alexandra Goho, conveyed a message "that the time between when Earth became habitable and when life emerged was too short to have generated the complexity required of self-replicating cells. However, if precursor molecules for making proteins, RNA and DNA, came from space, life-creating chemical reactions with Earth's primordial soup could have greatly accelerated." Further, "... there's growing evidence to suggest that dust particles spill onto Earth from comets. These icy bodies predate the solar system and bear the chemical signatures of their interstellar environment." Several other speculations abound in the Goho article predicated on the BB/Ac hypotheses.

The LB/FLINE model makes it crystal clear; all planets are self-sustaining entities that create their own atomic and molecular matter. They have no need to depend on substances from outer space: Earth's 4 1/2 billion years is sufficient time to provide suitable and ever changing environments that influence ever- changing species of life. Substantiated evidence clearly reveals that comets are relatively short-lived fiery energy ejecta from much larger energy masses. Over the centuries, they are born, run their fiery missions until either exhausted or consumed in spectacular, fiery crashes into larger masses. The best example of such crashes occurred in the 1994 nuclear explosions of Comet SL9 at Jupiter's thin cloud-tops. Only the LB/FLINE model accurately predicted, in writing, these powerful explosions. The world's best computer calculations, based on the "dirty snowball" belief, proved far too puny.

Back to the Beginning

Science News, May 12, 2001

Researchers have identified 26 of the youngest, most distant galaxies known, ablaze with the blue light of massive stars imaged soon after birth, some 13 billion light years away.

Initially, a pinwheel galaxy, as taught by the LB/FLINE model, contains young, brilliant blue masses of nuclear energy known as stars, which are partially buried in huge volumes of dust and gases, all ejected from the two poles of a spinning black hole, the densest form of energy, which had been created in the absolute-zero conditions at the perimeter of a Universe expanding at the speed of light. This explains why astronomers see the Universe expanding at ever-faster rates as they look ever farther into space: a version in which a lambda factor and dark energy—both essential speculations in the BB/Ac hypotheses—are not needed.

The 26 galaxies that may be the youngest and most-distant known, and containing brilliant blue stars—the hallmark of young, massive stars—fit precisely into this simple version of the LB/FLINE model. Over eons of time, such blue stars eventually fizzle into yellow, and finally, into red stars as their nuclear masses become more and more depleted.

Searching for signs of a force that may be everywhere...or nowhere

Science News - May 22, 2004, Vol. 165

"Ever since 1998, Robert Caldwell has been obsessed with something dark and repulsive. He spends nearly every waking moment trying to comprehend a mysterious entity that may be undermining gravity and pulling everything apart, making the universe expand at a faster and faster rate. This presumed force, sometimes called dark energy, might ultimately rip apart every object in the cosmos, from the tiniest of atoms to gargantuan clusters of galaxies." (*SN*, 2/28/04, p 132).

The Universe is indeed accelerating its rate of expansion—which is exactly the opposite of what is expected from a Big Bang (BB) origin. In trying to explain this mysterious discrepancy within the mythical BB/Ac version, astronomers must speculate profusely about an imaginary force that, in all probability, does not exist.

In reality, dark energy is not essential for explaining the accelerating expansion of the Universe via the ongoing Little Bangs (LB)/FLINE model of universal origins and evolution, with no need for speculation; the accelerated expansion is an intrinsic fact.

Cosmic Push

The article, *Cosmic Push*, reveals a number of fallacies, each stemming from belief in the mythical Big Bang/Accretion (BB/Ac) hypotheses. To quote, "Astronomers report new evidence that some mysterious force overcame gravity's tug about 6 billion years ago and ever since has been pushing galaxies apart at an accelerating rate." The evidence: "... scientists found that some distant exploding stars are dimmer, and therefore farther away, than expected. That was an indication that some cosmic push, referred to as dark energy, has been revving up the expansion of the Universe."

Further, "... both the supernova and the new galaxy-cluster studies suggest that the density of dark energy doesn't vary over time." But, "the findings don't entirely rule out a dark-energy density that increases with time."

These speculations are essential to survival of the BB concept, and they do improve its chances of being sustained for a few more years—in spite of the fact that Einstein recanted his lambda force (now called dark energy) as the greatest blunder of his life.

A more reasonable explanation of the accelerating expansion of the Universe resides in the LB/FLINE model, in which such expansion is an intrinsic factor that fulfills all the claims attributed to the speculative dark energy in the BB/Ac hypotheses.

The *Cosmic Push* article is a prime example of the many problems and mythical answers stemming from belief in the mythical BB/Ac hypotheses. The LB/FLINE model circumvents these problems by furnishing sound logic supported by a plethora of substantiated evidence.

Strange brew brings inorganic chemicals to life

According to the article, *Strange brew brings inorganic chemicals to life* (*Science News*, May 29, 2004), a team of chemists recently created a blob of cell-like structure from a mixture of inorganic chemicals. The blob had a semi-permeable membrane through which products of the reactions diffused out of the cell. "Mimicking division of living cells, daughter cells budded off from the original inorganic cells."

This experiment adds much weight to previous work done with the organic chemicals/life relationship. Like those experiments, it adds powerful evidence to the LB/FLINE model of origins and evolution in which Earth and all other planets are self-sustaining entities that create their own matter via Internal Nucleosynthesis (IN) that drives all Evolution (E) via the ongoing creation of all atomic and molecular substances, including those necessary for initiating and sustaining life.

Planets do not depend on debris from outer space to create their compositional atomic and molecular matter, including life itself.

Black holes made here

The article, *Black holes made here* (*Discover*, June 2004) by Karen Wright, raises a number of questions. If formed by the collapse of a massive star, how can the resulting black hole weigh more than millions of suns? But even if true, how could so many massive stars have been created via accretion? And why would each one collapse to form a black hole in the center of a galaxy? What is the significance of a black hole at the center of each galaxy, as observed by astronomers? Rather than depend on massive stars as a source of black holes, we might consider a more logical route for their creation—one that better fits astronomical observations and data: the Little Bangs/FLINE model of origins and evolution of universal systems.

In this revolutionary model, black holes are created as the densest form of energy via the collapse of exiting energies moving at the speed of light, and crashing into the absolute-zero temperature, in which all atomic motion ceases, at the spherical perimeter of the Universe. The tremendous density of the black hole exerts powerful gravity that rapidly attracts everything within its range, including new energies formed in space, as it grows ever larger and more powerful. At some point the gigantic mass is triggered into releasing its pent-up energy via its two poles, spewing out fiery energy masses called stars, embedded in huge volumes of gaseous dust-clouds, a mixture that can easily mislead observers into believing that stars are made via accretion of the gaseous dust. The galaxy's rapidly spinning motion accounts for its flat pinwheel pattern. A remnant of the black hole remains at the center of each galaxy to hold it all together.

This explanation of the origin of black holes and galaxies accounts for the increasing rate of expansion of the Universe—an intrinsic property of the LB/FLINE model—as astronomers look ever farther outward, no matter the direction, and obsoletes the need for the BB's lambda, cosmological constant, dark energy and gravitons.

In the author's opinion, gravity can indeed act at the quantum level and at all levels; it is intrinsic in all energy, no matter the form or mass. All forms of energy are intimately related and interchangeable; all function as a mass/gravity/distance relationship at all levels.

Presidential Address to the NPA Denver Meeting, April 7-10, 2004

I regret I cannot be with you in this important meeting. However, I want to thank all of you very much for your continued collaboration and interest in our NPA activities. This June, it will have been ten years since our first meeting in San Francisco. We can look back with satisfaction in many respects. But, we still have much to accomplish. Especially in the discernment and future focus in our areas of agreement as suggested in my editorial, *Our Challenge,* in a past Newsletter. We are aided in that now in our survey organized by Neil Munch.

Now is the time for change. A recent report from NASA is entitled: *Evicting Einstein.* Such words would have been heresy in the 20th century. But NASA is finally questioning Einstein's relativity. They plan an experiment to repeat with utmost precision the famous light bending effect of 1919 when Eddington supposedly confirmed Einstein's General Relativity Theory. Most of us at NPA know how biased that 1919 observation was. What is of higher interest now are the opening words of that NASA report.

"March 26, 2004: Sooner or later, the reign of Einstein, like the reign of Newton before him, will come to an end. An upheaval in the world of physics that will overthrow our notions of basic reality is inevitable, most scientists believe, and currently a horse race is underway between a handful of theories competing to be the successor to the throne."

This announces the next scientific "revolution", in the style of Thomas Kuhn's paradigm shifts. There has never been a better time for dissidents to speak up and make themselves heard. Such a declaration by NASA is long overdue and gives us joy, in fact, an immense joy.

However, a word of caution must be said. What comes next in Physics might not be necessarily better than the present situation. I am sure that many of us have some vague (or perhaps clear) idea of what that "Physics of the future" should be. But whatever it might be, I want to suggest our work at NPA should have fidelity to some schools of physical thought of the past. Many of us are "revisionists"; that is, scholars who look deeply not only into the flaws of Einstein's relativities but into the strong points of older theories that, for weak reasons, were abandoned or misunderstood. Building from this, many of us have also been able to find formulations that are superior to presently established ones, perhaps including improvements of our own.

I think we need to stress our spirit of "continuity" with the best of Physics in the past. And from there, to build the future. In this sense we are not "revolutionaries", or "outsiders", or "border line" physicists, much less "heretics". Rather, it was Einstein who fell into "heresy" respecting the rationality of Physics in the past. In the 1979 book, "General relativity—An Einstein Centenary Survey", one author tried to contrast Einstein with Lorentz, to show how Einstein accomplished what Lorentz could not. He said that Lorentz tried to "fix" the ship of Physics, by adding some pieces here and there. But Einstein did not repair the ship. He jumped out of it and created a new one altogether. So we see that Einstein was the "rebel", the side-liner, the true "heretic", not us. We, like Lorentz, are trying to really "repair the ship" which, as many of us think, still lies in the precarious conditions of 1900. So our work is not negative, is not mere criticism. It is very positive and not simply debunking a world personality and his science.

In a more profound sense, our critical yet constructive goals are even more positive. Because Einstein inspired a culture of relativism in the public at large and of mystified esoteric scientism among scientists we can say that his distrust in common sense has deep implications for the moral health of present day humanity. Nobody, I think, has expressed this danger so well as the French philosopher Jacques Maritain when in 1922 he wrote:

"It is pitiful for a civilization to have wise men lacking in their good common sense ... Einsteinianism, taken as a philosophical vision, given the fact that it is considered as Science with capital S in our century it is, by the same token, ***'a disorganizing agent of great power'*** *that tends to accustom peoples to the habit of accepting absurdity and to lose all trust in common sense, that is, in effect, to lose confidence in our intelligence and in our very human nature. This is truly an amputation of the intellective faculty..." (Reflexions sur l'intelligence,1922).*

We are dealing here with nothing less than our own human nature and our intellectual capacity that allows us to ponder such problems.

Finally, I want to encourage you always to be positive. Exactly what I reminded you about at a past Storr's meeting, in an address read by Domina Spencer. I meant the respect we all owe to the work and papers of others. Even in the case where we find superficiality or disagreement with our own views we should follow Aristotle's advice, namely: even then, we should admire the force of the mind being displayed in front of us. Each one says something valuable even if it is not enough for the final goal. Putting efforts together we can amass a greater amount of facts than we can individually accomplish. That is the purpose of the Survey being currently circulated among us. That is democracy at its best, without the temptation of mediocrity that usually afflicts democratic societies.

So, I leave you with three *positive* goals: **The three P's:**

1) Positively respect the valuable things in Physics of the past;
2) Positively respect the value of our human intelligence; and
3) Positively respect each others' work, even in the case of serious disagreements.

Thank you for paying attention to these words. Thank you Domina and Neil and Cynthia and Mahmoud. And thanks to all of you who make possible NPA's work. Have a very successful and enlightening meeting.

Francisco J. Muller
Miami, April 3rd, 2004

LB/FLINE Model: Bringing Unification Closer to Realization

September 14, 2004

Einstein wrote, "New theories are first of all necessary when we encounter new facts which cannot be 'explained' by existing theories. But this motivation for setting up new theories is, so to speak, trivial, imposed from without. There is another, more subtle motive of no less importance. This is the striving toward unification and simplification of the premises of the theory as a whole."

In a *Cosmic Conundrum* (*Scientific American*, Sep 2004), L. M. Kruss and M. S. Turner wrote, "The orange spheres represent the observable universe, which grows at the speed of light ... As expansion accelerates, fewer galaxy clusters are observable."

New facts which cannot be explained by the Big Bang/Accretion (BB/Ac) hypotheses always precisely fit into the simplified revolutionary LB/FLINE model of origins and evolution of universal systems (1973-1996). According to the new model, any Universe expands at the speed of light at its spherical perimeter. Gravity simply slows the expansion, but not enough to stop or reverse it. The cosmos will expand forever into Infinite Space saturated with frozen quantum energy particles (QEP). The QEP, along with the exiting light energies, form into cold, spherical, coal-black holes, the densest form of energy. Eventually, each spinning black hole is triggered into spewing from its two poles various sizes of fiery stars embedded in hot gaseous dust-clouds, all of which we see as pinwheel-shaped galaxies, each with a remnant of the black hole at its center (1979-1980). Thus, the Universe forever expands at the speed of light as it feeds and grows itself on QEP.

Unlike the Big Bang, the new model does not require lambda, dark energy or gravitons to explain the increasing rate of acceleration of the Universe: an inherent factor in the LB/FLINE model. As Einstein admitted, the lambda factor (a cosmological constant) was "the biggest blunder" he ever made in his life. Dark energy and gravitons should prove to be similar blunders in future astronomy; no mysterious form of energy is necessary to drive the accelerated cosmic expansion.

In the new model, gravity is inherent in all energy particles, atoms, molecules, and all matter therefrom. Apparently, gravity does arise from quantum mechanics rather than from relativity. Once past the BB/Ac roadblock, the simplified LB/FLINE model will change our fundamental understanding of the Universe, and bring unification closer to realization in "our quest for logical simplicity and uniformity in the foundation of physical theory."

"...impeccable logic, accurate facts, and sound evidence rewrite dogma ..." Dimitri Veras, Yale Scientific.
"...the most brilliant and fascinating work..." Brig. Gen. Gordon C. Carson III (Ret.).
"...a factual, easy-to-understand concept that makes a lot of sense. It provides answers..."
Morton Reed, Professor of Thermodynamics.

Letters and Memos

From: Ben Sheffield, Engineer. Jan. 14, 2001

"Established scientific theories die hard regardless of new facts. Whether you see changes, you may lay the groundwork for changes in the future. Hang in there!!!"

Philosophical Comments

*The greatest obstacle to advancement in the sciences of origins and evolution is the illusion of knowledge.
*Against speculative and unsubstantiated beliefs in origins and evolution, even the gods fight in vain.
*If you had to identify, in one word, the reason scientists have not achieved their goal of understanding universal origins and evolution, that word would be "consensus".
*In science, the greatest joy is discovery of a revolutionary truth; the saddest tragedy resides within the reasons for, and the consequences of, its suppression.

Alexander Scarborough, Jan. - Mar. 2001.

To: Editor, New Scientist, 84 Theobald's Road, London WCIX 8NS. September 24, 2005

The ten informative articles in *The World's Biggest Ideas* (*New Scientist*, 17 Sep 2005) presented very good explanations of each of the subjects. However, due to a number of recent events and more definitive interpretations, two articles, #1, The Big Bang (BB), and #9, Tectonics, warrant further considerations.

Many scientists openly recognize the BB as conjectural, with no substantiated evidence to support it. In reality, the CMB 2.7 K radiation pervading the Universe, like all other spaces on Planck's thermonuclear radiation curve, is short-lived, and therefore cannot be a leftover from the BB some 14 billion years ago. All such radiation must depend on its constant source of re-supply from ongoing thermonuclear reactions in all active universal spheres (e.g., stars, planets), as revealed by the newer LB/FLINE model of universal origins.

The same argument applies to the proportions of hydrogen, helium and deuterium in the Universe; they are the lightest and easiest elements produced quantitatively and proportionately in the same thermonuclear reactions in all active spheres of the Universe—all in compliance with Einstein's $E=mc^2$—a vital facet of the new model.

The article raises other questions: How did our Universe evolve into the immensely intricate cosmos we see around us? How and why was the Universe set up to expand the way it does? Why does the BB contradict one of the most powerful laws of physics, the second law of thermodynamics?

In reality, these questions and the second law pose no problems for the ongoing Little Bangs (LB/FLINE) model of universal origins and evolution. Researched in depth during the last 32 years, the answers are simply additional facets comprising the concept, now waiting in the wings for opportunities to become better known to inquisitive readers seeking more definitive science.

Concerning article #9, Tectonics, the new FLINE concept presents definitive explanation of the underlying basics of plate tectonics: the ongoing energy-matter relationship via Internal Nucleosynthesis that drives all universal evolution.

Tragically, the two biggest obstacles to comprehension of universal origins are belief in the BB myth and the suppression of newer, more substantiated ideas.

To: David W. Grogan, Articles Editor, Discover Magazine, New York, N.Y. 10011. October 28, 2005

In *Two against the Big Bang* by R. Panek, advocates of each of the two versions of universal origins—the Big Bang and a Static Universe—can honestly claim that the other's version has no substantiated evidence to support it. Excluding speculation, is there even one universal anomaly definitively solved by either version? Since myth begets myth, the answer is the inevitable resounding "NO", whether speaking of the past, present or future.

While the two prevailing concepts can never provide such evidence, waiting in the wings is the third version, the LB/FLINE model of universal origins, made viable via the new Fourth Law of Planetary Motion and Kepler's First Three Laws; together they clearly reveal the dynamic origin of our Solar System some five billion years ago. From this beginning, the evidence definitively reveals that every planet of every solar system is a self-sustaining entity that creates its own planetary matter via Internal Nucleosynthesis (IN) that drives all five-stage planetary Evolution (E). Now an established fact, IN and E are inseparable; one cannot exist without the other. Additionally, no definitive solution to planetary anomalies is capable of circumventing this well-substantiated evidence.

These little-known findings reveal definitive clues to the formation of galaxies, and to the processes whereby galaxies beget galaxies; processes that stem from the fundamental Superstring theory—but not in the manner suggested by the Steady State theory. The evidence also reveals "where those original hydrogen atoms came from", along with the nuclear-energy sources from which they continue to originate and pour forth.

We know from Galileo's ordeal that science must embrace new ideas, or it cannot advance up the slippery slopes of myths and conjecture embodied in prevailing beliefs. But when science inevitably does open up to the LB/FLINE model of universal origins, it will find no barriers to limitless advancement on the firm steps of substantiated evidence. To paraphrase Einstein, the right answers will be simple and beautiful—and they are. But more than this, serious readers find substantiated evidence far more exciting, informative and entertaining than myths or conjecture.

P.S. May I submit an article on the LB/FLINE model or one of its controversial ancillaries that remain ahead of their time?

To: Letters Editor, LaGrange Daily News, LaGrange, GA 30241. April 27, 2006

Your readers of the article, *Black holes called most energy efficient* (*LDN*, April 26, 2006), deserve to hear the rest of the story.

At the recent meeting of international scientists in Tulsa, OK, mathematical calculations and pictorial evidence clearly revealed the true nature of super-dense black holes (actually, black spheres). They are a vital phase in Nature's method of ongoing creation of the billions of pinwheel galaxies of our Universe. A black sphere accomplishes this by transforming its highly compressed form of energy into nuclear energy, spewing from its two poles as brilliant white masses of all sizes embedded in gaseous dust-clouds—fiery masses we see as stars. A calculated mass of the black sphere (0.5% of the ejected mass) remains at the center to hold the galaxy together.

Galaxies have no need for the alleged "zoning cops preventing too many stars from sprouting." The number of stars ejected is solely a function of the size of each black sphere.

Our Sun is an example of a moderate-sized star with ten smaller fiery masses—now planets and asteroids—locked in orbits around it. The smaller masses eventually evolved via Internal Nucleosynthesis (the transformation of nuclear energy into atomic matter) into their current observable stages of the five-stage evolution processes, all at rates in accord with size. Earth is now in the fourth (rocky) stage of its ongoing evolution; Mercury is in the fifth (inactive, barren) stage.

As the engines of conversion of super dense energy into active nuclear energy that lights our starry nights, and drives all universal origins and evolution, the black spheres are indeed the ultimate in energy efficiency.

To: Dr. Cynthia K. Whitney, Editor, Galilean Electrodynamics, Arlington, MA 02476. May 3, 2006

Thanks for sending the five papers from your earlier work on black holes during the 1980s and 1990s. I had not realized that you had made so much progress in this interesting field.

While I am not a mathematician or physicist, it seems to me that your formulae furnished a vital missing link in the manner in which nuclear energy is produced via the compression buildup in dense black spheres (a.k.a. black holes).

If so, this is powerful vindication of the probable manner in which nuclear energy masses of all sizes can be spewed from two poles of a spinning black sphere in pinwheel fashion as stars of all sizes embedded in spiral arms of hot gaseous dust-clouds—many locked in binary or trinary orbits, and in very rare cases, multiple orbits around a larger central mass; e.g., our Solar System in its original fiery stage of evolution into planets.

This vision in 1979-1980 is what inspired me to pursue the geometric solution to the Fourth Law of Planetary Motion for the next 15 years.

In my humble opinion, the Compression concept and the Fourth Law may be two of the most significant discoveries in astronomy during the past 100 years. Certainly, they do furnish sufficient substantiated evidence to warrant a change in direction of scientific thought from the Big Bang myth. They open many doors to understanding universal origins and evolution. And some day they might even prove to be Nobel Prize material.

I have been out of town much of the time since your letter of 19 April arrived, but will get Kay to send you the word-processor files of my three NPA papers for the Proceedings volume. If there are specifications she should know, please send them to her via e-mail. Thanks again.

To: Editor, Journal of the Royal Astronomical Society, Piccadilly, London, UK. August 1, 2006
Re: A Really Bright Future for Science!

This manuscript is a revised edition of the one sent earlier in July of this year.

The small changes and additions are significant enough to warrant this replacement, and its historical nature dictates that I follow through with the adjustments, no matter how small. They include one additional definitive factor, one additional reference, and a revised word count.

If the new fourth and fifth laws of planetary motion need further clarification, please let me know. They are clear to me only because of the 33 years spent in discovering their enigmatic solutions, but may prove more difficult to explain clearly for others to comprehend readily in much shorter time frames.

Thanks for your valuable time and efforts in this endeavor.

To: Glenn Strait, Science Editor, The World & I Magazine, Washington, D.C. 20002. August 17, 2004

I'm writing with the hope that this finds you after so long a time since our last correspondence in 2001. Since then, as you will see by the enclosed papers, a lot has transpired. Two papers from my latest book, *New Principles of Origins and Evolution* (2002) are scheduled for publication later this year.

By interlocking more and more substantiated evidence from space discoveries, the definitive LB/FLINE model of origins and evolution continues to be strengthened. Since 1995 when I first considered the evidence incontrovertible, I have become ever more convinced of its scientific validity as the concept destined to displace the speculative Big Bang/Accretion hypotheses in future textbooks. On this subject, Thomas Kuhn would be proud of his 20th-century prediction.

The new model incorporates key discoveries of the past 2500 years, each fitting precisely into the big picture without the need for speculation. The brilliant findings of Pythagoras, Copernicus, Kepler, Newton, Einstein, *et al.*, are the key ingredients melded by the new Fourth Law of Planetary Motion into the definitive LB/FLINE model (1973-1995) of origins and evolution.

According to Dr. Morton Reed, Professor of Thermodynamics, it is "a factual, easy-to-understand concept that makes a lot of sense. It provides answers to scientists who rethink current beliefs about planetary origins and evolution." In spite of its powerful evidence, the new model, unfortunately for science, students, and the public, remains ahead of its time.

The fundamental keys to comprehending universal systems are Einstein's 1905 formula and the new Fourth Law of Planetary Motion (along with Kepler's First Three Laws). For this reason, I now consider them the two most significant discoveries of the 20th century. Oddly, one was discovered five years into the century, and the other one five years before the century's end.

How can we work together on this exciting breakthrough?

Memo
On Black Holes, Brilliant Galaxies and Deadly Gamma Rays

On Tuesday evening, June 20, 2006 at 9:00 p.m., a video titled *Death Star* was featured on the *Horizon* program of GPB-PBS-TV in LaGrange, Georgia. Its primary message dealt with galaxies, black holes, and ongoing massive explosions that produce dangerous gamma rays. The evidence showed an evenly scattered distribution of ongoing powerful explosions throughout the far distant Universe, each emitting gamma rays lasting for a period of several days. Redshifted measurements revealed their distant locations at the perimeter of the expanding Universe.

In efforts to fit the findings into the prevailing BB concept, advocates attributed the cause of the ongoing powerful explosions to gigantic stars that had accreted from gaseous dust-clouds before exploding. But the projected immense size of such huge stars was far beyond any reasonable logic.

In reality, pictorial evidence clearly reveals these discoveries exactly as predicted and explained by the LB/FLINE model of universal origins. Each of the ongoing explosions heralds the dynamic birth of a new galaxy composed of white-hot stars of all sizes embedded in gaseous dust-clouds, all brilliantly visible (at times, bar-shaped) at the galactic center and throughout the full length of each of the spiral arms of the spinning pinwheel mass. A remnant—0.5% of the ejected mass—of the black-sphere source that births each galaxy remains at the center to hold everything together.

Obviously, these findings do not fit properly into the BB concept. But they do add powerful evidence in support of the scientific validity of the revolutionary LB/FLINE (1980-1996) model of universal origins and evolution, a model soundly based on the five laws of planetary motion that also do not fit properly into the BB concept.

If space discoveries are ever to be fully and accurately interpreted, discoverers must evaluate them in a concept other than the BB. The LB/FLINE model is a good starting point.

Thoughts for the Day

The BB cannot be tested in any meaningful way and so cannot qualify as time science. It is "truthiness", a new word which means the quality of stating concepts or facts one wishes or believes to be true, rather than concepts or facts known to be true.

The years may stretch into decades before scientific truth extricates itself from the quagmire of the BB myth.

But truth has a way of surviving; inevitably, it will lift this albatross from around the neck of science.

Conclusion

We live in a self-perpetrating universal ocean of electromagnetism: a most critical fundamental of everything—both inanimate and animate—in our Universe.

Electromagnetism is inherent in elementary particles of energy, whether in the nuclear stage or comprising the atoms of all matter. One might question whether electromagnetism and gravity are one and the same; logic and evidence indicate they are, indeed, the same. If so, this EM-G entity could well be the attractive force that brings and holds together all the components of universal systems throughout their complete cycles.

In the EM-G, the Universe operates entirely through electromagnetic force. Several cited factors explain the wide variations in electromagnetic field strengths of planetary spheres in accord with the LB/FLINE model of universal origins. This concept permits definitive solutions to universal anomalies; e.g., origins of solar systems, the Five Laws of Planetary Motion, the spacing of all planets, the 2.7 K radiation, lightning, etc.

References

(1) Charles W. Lucas, *A New Foundation for Modern Physics,* (1998). (*Common Sense Science*, Foundation of Science, Inc., Nov 2001).

(2) Alexander A. Scarborough, *The I-T-E-M Connection: How Planet Earth and Its Systems Were Made by Means of Natural Laws*, (1991), 7th Edition, Energy Series. (Ander Publications, 202 View Pointe Lane, LaGrange, GA 30241).

(3) Alexander A. Scarborough, *The Spacing of Planets: The Solution to a 400 Year Mystery*, (1996), 8th Edition, Energy Series.

(Ander Publications, 202 View Pointe Lane, LaGrange, GA 30241).
(4) Alexander A. Scarborough, *New Principles of Origins and Evolution: Revolutionary Paradigms of Beauty, Power and Precision,* (2002), 9th Edition, Energy Series. (Ander Publications, 202 View Pointe Lane, LaGrange, GA 30241).

CHAPTER VII

"THE RIGHT ANSWERS...SIMPLE AND BEAUTIFUL"

NPA 2007 Meeting
University of Connecticut
May 21-25, 2007

Abstract: The LB/FLINE Model of Universal Origins: From Superstrings to Planets (in reverse order)

The Little Bangs/FLINE concept (1973-2006) evolved in reverse order of the ongoing processes established by natural laws and sound logic. In 1973-1979, the definitive origin and intimate relationship of the hydrocarbon fuels (gas, oil, coal) (and all planetary matter) were established as products of Internal Nucleosynthesis (IN) that drives all Evolution (E). In 1980-1995, the Fourth Law of Planetary Motion and Kepler's Three Laws confirmed the dynamic origin of our geometrically-spaced planets as nuclear masses that are evolving through five stages of evolution. In 2006, the Fourth Law was modified and reworded; the Fifth Law was established as a separate entity from the original Fourth Law. All Five Laws (FL) are now included in the FLINE model.

Definitive evidence supplied by the FLINE revealed clues to the dynamic origin of galaxies that spew, from each pair of poles, hot gaseous dust-clouds embedded with nuclear-energy masses of all sizes; the smaller ones eventually evolve into planets, usually in some solar-system fashion. From this first stage, planets will evolve via Internal Nucleosynthesis through four more stages of evolution: gaseous, transitional, rocky, and inactive. Black spheres (a.k.a. black holes) form continuously at the perimeter of our spherical Universe at its interface with Dark Infinite Space at absolute-zero temperature. The dense, black spheres, the densest form of energy, and the starting point for nature's ongoing creations, consist of superstrings with which white-hot galaxies are spun: a black remnant remains at each galactic center to hold the system together.

A list of 23 selected validations from among many underpinning a definitive LB/FLINE model is included in this paper.

A Really Bright Future for Science

If the right answers are not to be found within the BB perspective, doesn't this tell us that the BB concept must be a major problem? Is science wrong in clinging to the BB in spite of its incapability of providing accurate interpretations to the many brilliant discoveries of space probes—interpretations likely to provide even more questions than answers? Is there an alternative in which new and old data can be interlocked precisely to provide answers to questions that may never be answered via prevailing beliefs?

Fortunately, there is an alternative that warrants full consideration: the LB/FLINE model of universal origins and evolution. The new model has been carefully constructed with factual evidence during the past 33 years (1973-2006). Findings of space probes continually fulfill its predictions, and verify its scientific validity. No anomaly appears capable of escaping its proclivity for solving the mysteries of universal origins and evolution.

To understand the revolutionary LB/FLINE model, one must first understand the Five Laws of Planetary Motion (FL) (described in previous papers).

Such knowledge and pictorial evidence provide powerful clues to the dynamic origins of solar systems consisting of fiery masses of energy (stars of all sizes) embedded in gaseous dust-clouds, all of which form the beautiful center and spiral arms of pinwheel galaxies: ejecta from the two poles of a spinning black hole (a.k.a. sphere). Scientists have discovered a black-sphere remnant (0.5% of the ejected mass) remaining at the galactic center to hold the system together. Such explosive actions from black spheres into brilliant white galaxies are known as Little Bangs (LB)—never ceasing, forever ongoing at the spherical perimeter of the Universe. They are first seen as quasars, later as galaxies.

As the newborn cloud-embedded energy masses (stars) are hurled outwards, many of them inevitably interlock in various types of solar systems, most commonly, in binary systems, and in lesser numbers of trinary systems. Multiple-planet systems like ours are very rare, but they can and do happen: geometrically spaced (Fourth Law of Planetary Motion), and functioning in accord with the five laws of planetary motion.

In turn, this knowledge and pictorial evidence lead to the realization that the smaller fiery masses soon begin their five-stage evolution into planets and moons as they revolve around larger masses of energy (stars, suns). The five stages are readily discernible; e.g., nuclear energy (Sun), gaseous (Jupiter), transitional (Uranus), rocky (Earth), and inactive (Mercury), all evolving at rates proportional to mass.

The knowledge of planetary Evolution (E) via Internal Nucleosynthesis (IN), in sync with the five motional laws (FL), will enable scientists to decipher the precise history of solar systems, and to predict future changes in planetary orbits and velocities as functions of time, and to solve definitively every anomaly of planets and solar systems; e.g., why Neptune and Uranus are drifting outward away from the Sun, while all other planets and asteroids are drifting inward toward the Sun (Fifth Law of Planetary Motion).

Understanding the formation of galaxies at the ever-expanding perimeter of the spherical Universe will enable scientists to determine how and why the dense black spheres form via buildup of attractive forces: electromagnetic, gravitational, and other energies (e.g., exiting light) at the absolute-zero interface with outer Space where all atomic motion ceases. Is gravity inherent in electromagnetism (EM)? Probably, else why are all universal bodies attracted to all other bodies of every size at all stages of their evolutionary existence, beginning at the superstrings level? Gravity and EM must be one and the same.

Definitive answers will bring closure to questions about an ongoing continuous source of energy that runs our Universe: strings of electromagnetic energy, perhaps the most fundamental substance.

In summary, the five laws (FL) of planetary motion provide sufficient evidence for a definitive understanding of the origin of solar systems and the evolution of their planets and moons via Internal Nucleosynthesis (IN)—the transformation of energy into atomic matter—that drives all planetary Evolution (E). This revolutionary FLINE model readily explains the dynamic origin of solar systems and the geometric spacing of planets (Fourth Law) before beginning their five-stage evolution into the currently observable stages. Every planet and moon is a self-sustaining entity that creates its own compositional matter via IN. This should bring closure to the prevailing question: What is a planet?

This knowledge provides clues to the formation of black holes, one of the densest forms of energy, and the source of galaxies, leading to an understanding of the origin, nature, and function of a black hole as it spews out a pinwheel galaxy from the two poles of its spinning mass. Scientists will learn definitively how these multitudes of black holes form at the perimeter of the ever-expanding Universe to supply our universal energy needs via galaxy formations. And that they are the beginning point for explaining the accelerating rate of expansion of the Universe without the need for dark energy or dark matter, and why galaxies are organized in vast sheets measuring in hundreds of millions of light years: inherent factors in the revolutionary LB/FLINE model.

Science needs definitive reasons the Universe expands, how the forces of Nature are unified, why and how everything in the Universe is connected to everything else, the true nature of planetary quakes, the abiogenic origin and source of the hydrocarbon fuels, and how quantum mechanics and gravity can be reconciled. If these needs are to be fulfilled, the BB must be forsaken and more feasible concepts of universal origins and evolution must be examined.

In sync with relativity theories, the LB/FLINE model of universal origins fulfills these needs. When examined closely, one finds two famous illusions to be inherent factors:

(1) With all other galaxies receding from it, our Milky Way galaxy appears to be at the center of the Universe. However, this illusion also holds true for all other visible galaxies when viewed from any one of them.

(2) The receding galaxies create the illusion of a static Universe.

Definitive solutions to universal mysteries via the LB/FLINE model predict a truly exciting future for the relevant sciences. The time is right for the BB to step aside and permit the vetting of younger ideas: a change in direction of scientific thought, one that bodes a really bright future for science for generations to come.

Origins of Universal Systems:
From Superstrings to Super Planets—The LB/FLINE Model

At the expanding interface of our spherical Universe and the absolute zero of Infinite Dark Space (IDS) enveloping it, countless numbers of black spheres (a.k.a. black holes) grow ever larger via accumulation of dormant superstrings energy from IDS, while maintaining their outward journey at speed-of-light velocity. This densest form of energy eventually is triggered into spewing nuclear-energy masses (stars of all sizes) embedded in hot gaseous dust-clouds from its two poles. The smaller masses eventually evolve into planets (usually in solar systems) that undergo four other stages of planetary evolution: gaseous, transitional, rocky, and inactive—all products of Internal Nucleosynthesis (IN), electromagnetism and gravity.

Structured with epoch-making discoveries, this revolutionary concept indicates the probability that Internal Nucleosynthesis, electromagnetism and gravity are one and the same.

The LB/FLINE Model of Universal Origins

The three assumptions of the LB/FLINE model of universal origins: S E T

1. Infinite Dark Space (I D S) at absolute-zero temperature.
2. Infinite Energy (I E), dormant in IDS until contacted by light.
3. Infinite Time (I T).

A few selected validations underpinning the LB/FLINE model of universal origins:

1. Kepler's three laws of planetary motion.
2. The new fourth and fifth laws of planetary motion.
3. The current spacing of planets in our solar system.
4. Why Uranus, Neptune and Pluto do not obey Bode's Law, but do obey the Fifth Law.
5. Why Pluto is the original tenth, and now the ninth planet in our solar system.
6. The five-stage evolution of planets at rates in accord with size.

7. The expanding Earth: why and how.
8. Oceans, continents, surface features: e.g., mountains, rivers, upheavals.
9. Earthquakes, planetary quakes.
10. Internal Nucleosynthesis (IN)—currently (or formerly) in all universal spheres.
11. The origin and fate of planetary rings.
12. The baseball-like seam of material outflows circling Earth.
13. The nature of Earth's core: nuclear energy (quark-gluon plasma).
14. The continuous increase in sea level and land.
15. The ongoing origin-extinction cycles of Earth's species via IN-E.
16. The multitudes of ejecta craters and sunken areas on moons, planets, asteroids, former comets. Proven by Allan Mills (Ref: *New Guide to the Moon*, by Patrick Moore, 1976).
17. The energy-matter relationship expressed by $E=mc^2$ (IN).
18. The fiery explosive power of comets; e.g., Comet SL9 on Jupiter in 1994, and the Tunguska event, June 30, 1908 (from a remnant of Comet Encke).
19. The disintegration of Planet(s) Asteroids (Olbers, 1802).
20. The accelerating expansion of our finite Universe (1998).
21. The universal 2.7 K radiation via the IN in Universal spheres (stars, etc.).
22. A black hole at the center of each galaxy.
23. The beautiful self-curvature arms of galaxies.

Note: While these validations can be explained definitively via the LB/FLINE model hypotheses, none appear to offer any definitive evidence in support of the Big Bang/Accretion hypotheses.

The Big Bang is to science what Alice-in-Wonderland is to politics: for sustainability, each must rely on impromptu impossibilities.

While discovery of the Unified Field Theory seems impossible via the Big Bang hypothesis, it does appear possible—even probable—via the LB/FLINE model of universal origins. Comprehension of the five laws of planetary motion is critical for understanding solar systems: their origins, functioning, history and future, along with the clues they present for continuity of relationship to other universal systems; e.g., galaxies, black holes, and superstrings—information that appears essential to a unified field theory or equation.

Hubble's Top 10 Discoveries

The top 10 discoveries of the Hubble Space Telescope are listed and briefly discussed in *Scientific American* (July 2006, pp 43-49).

Because each is viewed and interpreted in the BB perspective, no definitive interpretations are possible; they remain mysteries that demand speculation.

Of the ten discoveries, the seven listed below require no speculation when interpreted in the perspective of the LB/FLINE model of universal origins and evolution. Because each is an inherent factor in the new model, each has a definitive explanation, and each is intertwined intimately with the other six:

1. The Great Comet Crash (SL9 on Jupiter, July 1994).
2. Extra-solar Planets
4. Cosmic Birthing

5. Galactic Archaeology
6. Super-massive Black Holes Galore
8. The Edge of Space
10. The Accelerating Universe
The remaining three discoveries may not yet be definitive:
3. Death Throes
7. The Largest Explosions (GRB)
9. The Age of the Universe

What Lies Beneath

The article, *What Lies Beneath* (*Discover*, Aug 2002) by Brad Lemly, discusses Earth's nuclear-energy core, beginning with questions: "Why, precisely, is Earth's interior hot? What, exactly, is the source of the magnetic field, and why does it sometimes reverse? What, specifically, is the core made of?" "J. Marvin Herndon...whose research is featured in our cover story, 'Nuclear Planet', believes...that Earth is a natural fission reactor, with a massive ball of hot uranium at its center." [But what is the source of uranium?]

Herndon states that many other planets in our solar system have natural nuclear reactors, which are probably a general feature of planets.

Response: The idea of a "Sun-like" interior of Earth was first recorded by Descartes in 1644. But the idea was ignored while speculation abounded that Earth's central core is rock or iron, nickel or other metals. Uranium is the most recent one of such metals, none of which can furnish logical answers to questions of planetary origins, the five-stage evolution of planets, or the powerful electromagnetism of all active spheres of the Universe.

All metals, including uranium, are products of Earth's nuclear-energy core, and, as such, do not play an active role in the origin, evolution, and functioning of any sphere in the Universe. Only nuclear-energy cores (q-g plasma) that produce all planetary matter, including uranium, can provide the right answers, simple and beautiful—and well-substantiated.

The LB/FLINE model that first introduced Earth's nuclear core to science in 1973 may well be the only known concept capable of providing right answers. On the point at issue, one would believe Descartes and the new model with its quark-gluon (q-g) plasma core from which all the elements comprising Earth's hot mantle and beautiful surface features are made continually over the eons of time.

Every planet is a self-sustaining entity that created, and creates, its planetary matter (including uranium, water, land, etc.) via Internal Nucleosynthesis (IN) atom by atom within its quark-gluon nuclear core. Radioactivity is a by-product of Internal Nucleosynthesis, and contributes very little, if any, heat to the system.

A uranium core in the fission mode would fall far short of sufficient electromagnetic power or gravity to hold Earth and its systems in their normal functioning mode. Unfortunately, such a fission core sounds too much like the atomic bombs of World War II, but extremely far more powerful and threatening to our world. Fortunately, the evidence supporting a fiery nuclear core has grown indisputable since its introduction in 1644/1973.

Quoting *Dark Fingerprints* (*Science News*, Nov 18, 2006), "The mysterious cosmic push that's tearing apart the Universe began revving up about five billion years ago. But a new study reveals that several billion years earlier, the bizarre, elastic substance [dark energy] that fuels this push was lurking in the shadows and already beginning to fight gravity's tendency to pull things together."

The article further speculates that dark energy "might emanate from the cosmic vacuum and have a constant density," resembling the cosmological constant, and further complicating the situation. These "findings" are good confirmation of—and attempts to explain—the 1998 discovery of the accelerating expansion of the Universe; both are based on studies of exploding stars called supernovae.

According to the newer LB/FLINE model of universal origins, the only dark energy needed to drive the accelerating expansion of the Universe originates as black-energy spheres at its spherical perimeter. Here, the white energies moving outward at light-speed collide with dormant superstring energies residing in the absolute-zero coldness of the infinite black Space beyond.

As their energies coalesce, the black spheres grow ever larger and denser, while maintaining their original light-speed. Their increasing gravity and rapid momentum are the fundamental reasons behind the accelerating rate of expansion discovered in 1998. From the two poles of each spinning black sphere, a beautiful pinwheel galaxy is birthed: stars of all sizes buried in hot gaseous dust-clouds, a fact readily confirmed via observation.

Dark Fingerprints is another example of how the Big Bang forces scientists into speculations and misinterpretations of their spectacular discoveries. Rather than tearing apart the Universe, the dark energy spheres are the source of energy needed to build the Universe; they do so via gravity rather than push.

Answers: The Wrong and the Right

One of the miracles of modern-day science is the survival of the Big Bang for 159 years since its publication by Edgar Allan Poe in 1848. Belief in the concept, originally a laughing matter, grew stronger with time, due primarily to its acceptance by science in 1917 through the efforts of William de Sitter, and mainly because there was no viable alternative.

Speculations enhancing the concept increased slowly at first, but exponentially with time. Yet it survives without an iota of substantiated evidence; the miracle is that the Big Bang survives only on speculation and in spite of being called a myth by Carl Sagan, and conjectural by Stephen Hawking, two powerful leaders in science who recognized its vulnerable points.

These facts about the Big Bang appear set in stone, with no possibility of the concept rising above its level of Alice-in-Wonderland impossibilities. Now an albatross around the neck of science, it warrants removal as soon as possible. Even if that process started today, it would take years before science can extricate itself from the tangled web of speculation that had its beginning in 1848.

But when the process predicted by Thomas Kuhn in 1956 is completed, science will discover a viable alternative waiting in the wings: the LB/FLINE model of origins of universal systems that allows scientists to fulfill Einstein's prediction that "the right answers will be simple and beautiful".

Origins of Solar Systems and the Evolution of Planets

FL-IN-E Model	Accretion Model
1. Structured solely with facts; non-speculative	1. Remains highly speculative; not factual
2. Explains the spacing of planets in all solar systems	2. Cannot explain the spacing of planets in any SS
3. Explains planetary evolution via Internal Nucleosynthesis	3. Cannot explain planetary evolution via metal cores
4. Explains planetary energy sources that drive evolution	4. Does not explain these energy sources in planets
5. Explains compositional differences in all planets	5. Speculates on these compositional differences
6. Explains differences in sizes of all planets	6. Does not explain differences in sizes of planets
7. Explains the sources of moons around planets	7. Speculates on origins of planetary moons
8. Explains the origins of planetary rings	8. Speculates on origins of planetary rings
9. Capable of explaining all planetary anomalies of our SS	9. Incapable of explaining our planetary anomalies

10. Explains origins and differences of all moons	10. Speculates on origins and differences of all moons
11. Explains the sources, nature, anomalies of comets	11. Speculates on these sources, nature, anomalies
12. Explains sources, nature, anomalies of asteroids	12. Speculates on these sources, nature, anomalies
13. Incorporates three inseparable principles of Nature	13. Ignores these three inseparable principles of Nature
14. Explains electromagnetism via natural energy laws	14. Attributes electromagnetism to metallic cores
15. Based on works of Descartes, Buffon, Kepler, Olbers	15. Based on antiquated beliefs of Kant, Laplace, *et al.* Ignores works of Descartes, Buffon, Kepler, Olbers
16. Obeys the Second Law of Thermodynamics	16. Violates the Second Law of Thermodynamics
17. Fully explains all anomalies of extra-solar systems	17. Can't explain any of these anomalies; speculates
18. Explains why our SS layout is so rare among many SSs	18. Cannot explain the rarity of our SS's layout
19. Predicts and explains all anomalies of probes to planets	19. Cannot predict or accurately explain these results
20. Has an impeccable record of accurate predictions	20. Findings are usually surprising and puzzling
21. Supported by powerful corroborative evidence	21. Not supported by corroborative evidence

The abiogenic origin of the hydrocarbon fuels - not from fossils! - was proven beyond question during the ten years of 1973-1983. It is an excellent example *of* the ongoing transformation of all planetary elements and molecular matter via Internal Nucleosynthesis, polymerization, and subsequent interactions. Unfortunately, this definitive Energy Fuels concept has remained suppressed since its publication under the title of *Evolution of Gas, Oil and Coal* (1975-1983) in the Proceedings of the Sixth Miami International Conference on Alternative Energy Sources, Dec. 12-14, 1983, Miami Beach, Florida, pp 337-344.

The works of Copernicus, Kepler, Newton, Descartes, Einstein, *et al.*, and the discoveries and brilliant insight of the Pythagoreans around 500 B. C. finally mesh cohesively the new Fourth Law to bring the Pythagorean-Copernican Revolution—an understanding of the origin of solar systems and the evolution of planets and moons—full circle.

The Cosmic Microwave Background: A New Perspective on the 2.7 K Radiation NPA Meeting, May 23-27, 2005

The existence of a cosmic-radiation bath, discovered in 1965, is generally recognized to be the single most important observation in post-war cosmology. (1) Big Bang (BB) advocates interpret the CMB as a remnant of the de-coupling of energy and matter during the Big Event. The spectrum of this radiation would be "blackbody", or Planck-distributed into the full spectrum of Planck's Curve, with the maximum intensity at a wavelength of 0.18 cm and the temperature of 2.7K. (1) However, many scientists have believed there may well be other explanations, and in principle, the CMB can be explained in other ways.

While the CMB is identified as a relic of the BB, its true source remains debatable. The major problem with this pervasive radiation resides in its interpretation. In reality, it is blackbody radiation, which is short-lived when not emanating from a steady source of production. Such radiated production is characteristic of active spheres of the Universe. (4) (5)

"Consider the radiation in the space enclosed within a hollow steel ball, which is maintained at a constant white heat. The radiation within such a sphere is called 'blackbody' radiation because the radiation coming from each unit area of the walls of the enclosure is the same as that which comes from unit area of a perfect blackbody maintained at the same temperature. The radiation from each unit area of the walls of the enclosure is determined only by the temperature of the walls and not by the substance of which the walls are composed. In such an enclosure, the intensity of the radiation falling on a unit area of the walls equals the radiation coming from the unit area." (1)

In principle, the same thing applies to a planetary sphere in which the white heat equates to its nuclear-energy core and the mantle/crust equates to the walls. The radiated energy originates from the internal energy associated with atomic and molecular motion and the accompanying accelerations of electrical charges within the object (sphere).

Scientists realized that the results of the discovery of Penzias and Wilson were limited to only a few wavelengths clustered on the Planck curve. Other explanations of the background radiation, such as a combination of radio sources, could explain those data points. It was not until the mid 1970s that enough measurements at different frequencies had been made to prove that the background radiation actually follows Planck's distribution law for all thermonuclear reactions. (1)

Radiation is both absorbed and emitted by the walls of the enclosure, and by any gas, which might be in the enclosure. Planck suggested the idea that when radiation is emitted or absorbed, it is emitted or absorbed in multiples of a definite amount, identified as a quantum. The energy of a quantum of radiation of frequency v is proportional to v and so is equal to hv, where h is a proportionality constant known as Planck's constant. (1) (4)

The full curve of Planck's distribution law is the spectrum of the continuous radiation given out by a hot blackbody. The plotted curve shows the distribution function with respect to wavelength and/or frequency. It applies to any planetary mass containing a hot nuclear core, and is valid only when the blackbody radiates into a vacuum, as do all active universal spheres. (2) (3) (4)

The energy of the CMB radiation from energy cores corresponds to the excitation of CN; absorption spectra of CN and other molecules have provided much information about the CMB radiation (1)—which is indicative of an internal nuclear-energy source from which atoms and molecules are transformed from energy into matter.

No problem would exist here if scientists could accept the fact that the spheres of the SS are evolving from nuclear masses, and the CMB is in reality the radiation emanating from these internal nuclear reactions. (4) (5) Simply put, the relationship among the three characteristics of these internal nuclear reactions is best shown by the full curve of Planck's distribution: the spectrum of the continuous radiation given out by these hot black bodies.

Proponents of the BB theory claim evidence that the CMB originated at extremely high redshift. The evidence comes from the production of light elements. Knowing the present CMB temperature, and assuming entropy was nearly conserved back to high redshifts, they trace the thermal history of the Universe back in time to temperatures allegedly high enough to have driven thermonuclear reactions.

The LB/FLINE Concept of Dark Energy

May 29, 2007

"Will Dark Energy unlock the secrets of the Universe?" is a valid question posed and discussed in *The Secrets of Dark Energy* by Meg Urry (*Atlanta Journal-Constitution*, May 27, 2007). The simple answer is, "Yes."

While it cannot be explained in the perspective of a Big Bang, all data fit well in the revolutionary Little Bangs (LB) concept of a Universe expanding at an accelerated rate, all without the need for a lambda factor (cosmological constant that Einstein rejected).

Astronomers have verified the existence of a black hole at the center of galaxies, and calculated its mass at 0.5% of the brilliant mass (large or small) surrounding it—an inherent factor in the LB concept.

Black holes (a.k.a. spheres) continuously form at the interface of the speed-of-light universal space and

Infinite Space at absolute-zero temperature where all atomic motion ceases. In the absence of any heat, these are extremely dense spheres that retain their rapid momentum outward, spinning as they grow ever larger, until triggered into ejecting from their two poles billions of white-hot masses (stars) of all sizes embedded in hot gaseous dust-clouds, creating the illusion of stars being made via accretion. The smaller white-hot energy masses eventually evolve through four more stages of planetary evolution: gaseous (e.g., Jupiter), transitional (e.g., Uranus), rocky (e.g., Earth), and inactive (e.g., Mercury, Moon).

The masses of black spheres are sufficient to account for the calculated two-thirds Dark Energy of the Universe. Their massive momentum outward in all directions accounts for the accelerating rate of expansion of the Universe, and forever supplies the Universe with the unimaginable amounts of new energy demanded by a growing Universe.

Urry is right: "The answers are there... We are now at the beginning of a great new adventure to push forward the frontiers of understanding." We can start by recognizing the electromagnetic nature of our Universe.

Another Example of Big Bang Sophistry

Quoting from *New Look at Dinosaurs' Demise* (*Time*, Sep 24, 2007), "A recent study pinpointed the interstellar collision that eventually led to the dinosaurs' extinction. The event was so deep in space that the meteorite it created didn't slam into Earth and kill off the dinosaurs until nearly 100 million years later. While the findings, if true, would 'link the biological history of the earth to events far from earth,' according to William Bottke, who led the study, 'they would also alter what we know about the history of dinosaurs.'"

This article is an excellent example of Big Bang sophistry at its best (or worst). Its lack of substantiated evidence, the total disregard of factual scientific literature, and its willingness to utilize impossibilities fail to strengthen Big Bang beliefs that solutions to planetary anomalies come from outer space.

In reality, the right solutions will always come from within Earth (and within each planet in question) via Internal Nucleosynthesis (the transformation of energy into matter in accord with $E=mc^2$).

Now established beyond question, Earth's volcanism and resulting change in the composition of the atmosphere is the culprit responsible for the slow extinction of dinosaurs over a time frame of 500,000 years. The several layers of planet-produced iridium in Earth's crust play the key role in understanding the mysteries of all extinctions and other planetary anomalies.

Reasons for a Change in Direction of Scientific Thought

June 17, 2005

Before his untimely death from cancer, Carl Sagan called the Big Bang (BB) "a myth". In the summer of 1999, Stephen Hawking stated that cosmology's hypotheses are "conjectural". A recent poll revealed that 86% of a group of international scientists do not believe in the BB hypothesis.

Since its initial publication by Edgar A. Poe in 1848, the BB has failed to gain one iota of substantiated evidence in support of it. Yet, during the ensuing years (1848-2005), it has evolved via pure speculation into the grandest and costliest fraud in the history of science—an albatross around the neck of science—simply because, until now, there was no viable alternative.

Ten years have passed since the discovery of the geometric solution (1980-1995) to a 400-year mystery: the Fourth Law of Planetary Motion that first eluded Kepler in 1595. After this failure, Kepler went on to discover the First Three Laws of Planetary Motion. Together, the Four Laws reveal how the planets attained their geometric spacing around our Sun during the dynamic layout of the Solar System (SS) some five billion years ago. (l)(2)

(3)

This vital information will enable scientists to decipher the complete history of our SS, and to accurately predict future changes in orbital velocity and position of each of its planets as a function of time. As with the heliocentric SS of Copernicus (1543), the Four Laws are crucial for comprehending the anomalies of our SS and those of all other SSs. They present definitive evidence that planets do not form elsewhere, then migrate to their current orbital positions, as often speculated by BB advocates.

When the BB myth becomes displaced by the LB/FLINE model (1973-1996), scientists will be pleased by the rapidity and ease of solving the great number of anomalies that now plague current beliefs. In a recent paper on the new model, 41 examples of solved anomalies were listed. (2)(3)(4)

One example is the definitive explanation of the slight outward drifting of Uranus and Neptune, while the other planets are drifting inward toward the Sun: a mass/distance relationship. Another example is the definitive explanation of the geometric spacing of planets in our SS: the new Fourth Law of Planetary Motion (1980-1995). A third example: the accurate prediction (and explanation) of the powerful explosions of 21 remnants of Comet SL9 at Jupiter's cloud-tops in 1994—a prediction, in writing, that stood alone in its accuracy. Computer predictions based on BB beliefs were puny by comparison. (2)(3)(4)

The LB/FLINE model appears to be the only pathway to definitive solutions to these (and a great many other) anomalies. The first crucial step is an in-depth understanding of the Four Laws of Planetary Motion. Everything else then begins to make sense by simply interlocking into place in the scheme of things — much as it did when Copernicus placed the Sun at the center of our SS. The new model brings the Copernican Revolution full circle. (2)(3)(4)

For these reasons, I now consider the revolutionary Fourth Law (and its immense ramifications) incontrovertible, and one of the five most significant discoveries in the sciences of origins and evolution of universal systems. (3)

The new LB/FLINE model appears capable of withstanding the close scrutiny it warrants, and thereby changing the direction of scientific thought away from conjectural myth and more toward substantiated evidence. There is no better time than now to begin the change.

References

(1) Alexander A. Scarborough, **The I-T-E-M Connection: How Planet Earth and Its Systems Were Made by Means of Natural Laws,** (1991), 7th Edition, Energy Series. (Ander Publications, 202 View Pointe Lane, LaGrange, GA 30241).

(2) Alexander A. Scarborough, **The Spacing of Planets: The Solution to a 400-Year Mystery,** (1996), 8th Edition, Energy Series. (Ander Publications, LaGrange, GA 30241).

(3) Alexander A. Scarborough, **New Principles of Origins and Evolution: Revolutionary Paradigms of Beauty, Power and Precision,** (2002), 9th Edition, Energy Series. (Ander Publications, LaGrange, GA 30241).

(4) Alexander A. Scarborough, **Bringing the Copernican Revolution Full Circle via a New Fourth Law of Planetary Motion, Journal of New Energy**, Fall 2003. Proceedings, NPA, International Conference, June 9-13, 2003.

Thoughts on Electromagnetism and Gravity

April 7, 2005

(1) Energy strings are Nature's fundamental building blocks of all things universal.

(2) Gravity is inherent in the fundamental strings comprising all atomic and molecular matter throughout the Universe.

(3) The strength of electromagnetism is a function of mass and motion across lines of force.

(4) The strength of gravity is a function of mass and motion across lines of force.

(5) Therefore: Electromagnetism (EM) and Gravity (G) are one and the same.
(6) The combined electromagnetic-gravities of our Sun and its planets are sufficient for holding the SS together for billions of years without outside assistance.
(7) The combined EM-Gs of a central black hole and its spiraling galaxy of stars buried in dust-clouds are sufficient for holding the rotating system together without the need for dark energy or dark matter.
(8) These EM-G principles apply to the origins and functions of all universal systems.

Addendum:

Although lacking any substantiated evidence, the BB has thrived in the absence of a valid alternative—a situation that gives scientists no choice other than working within the limitations of the BB hypothesis. Today, the opportunities to work beyond these limitations are greatly expanded by the revolutionary LB/FLINE model of universal origins (1973-1996).

All universal systems continually evolve from energy into matter; nothing is as it once was, and nothing is where it was before—an impossibility within the rigid confines of the BB concept that places undue limits on the ongoing creation of atomic elements comprising universal systems. Evidence clearly shows that in the absence of Internal Nucleosynthesis (IN) via qgp in planets, Evolution (E) of planetary systems (comprised of virgin atomic matter) via IN would not be possible: qgp, IN, and E are inseparable; one cannot exist without the other. (3) (4) (5) (6)

Black Holes or Black Spheres: The Dark Energy of our Universe

October, 2007

In *The Discover Interview: Kip Thorne* (*Discover*, Oct 2007), the legendary astrophysicist explains his version of what black holes are made of: space-time. Thorne stated, "A big misconception is that a black hole is made of matter that has just been compacted to a very small size. That's not true. A black hole is made from warped space and time."

This new information is exciting; it is in precise agreement with the LB/FLINE model (1973-1996) in which space-time is the forever ongoing source of infinite energy that runs all functioning systems of our ever-expanding Universe. In reality, black holes—in accord with the revolutionary model—are black spheres of the densest form of energy, continually forming by the billions at the perimeter of a Universe expanding at the speed of light into the absolute-zero coldness (and curvature) of Infinite Dark Space where all atomic motion ceases.

In time, and sufficiently warmed by compression, the spinning black masses give births to billions of beautiful galaxies, usually in pinwheel fashion: billions of white-hot masses (stars) of all sizes embedded in hot gaseous dust-clouds. The smaller white-hot masses of energy eventually evolve into self-sustaining entities that undergo the five stages of planetary evolution via Internal Nucleosynthesis, usually as planets and moons in a variety of solar systems.

Thus, our self-sustaining Universe is designed to run forever, continuously creating billions of dense, black spheres that spawn billions of galaxies that, in turn, birth all universal systems; all are products of curved space and time: space-time.

Thorne's belief in space-time validates both this crucial heart of the revolutionary model and its mathematical solution (2005-2006) by Dr. Robert Heaston. In turn, the LB/FLINE model vindicates Einstein's prediction that "the right answers will be simple and beautiful".

A Brief Historical Sequence of the Evolving LB/FLINE model:

1. Evolution of Gas, Oil and Coal (1935-1980) (**I**nternal **N**ucleosynthesis)
2. Evolution of Planets (1973-1983) (**E**volution)
3. Birth of the Solar System (1980-1995) (**L**ittle **B**angs)
4. The Fourth and Fifth Laws of Planetary Motion (1980-1995) (**F**ive **L**aws)
5. From Void to Energy to Universe (1980) (LB) (Space-time = Black Energy Spheres.)
6. From space-time (dense black energy) to white-hot galaxies (1980-2006)
7. Universal Law of Creation of Matter (1980-2000) (**L**ittle **B**angs)

Numbers 1, 2, 3: Presented initially at the AAAS Meeting, January 3-8, 1982, via a four-page bulletin, Washington, D.C.; and at Miami Energy Conference (1983).
Number 4: Presented numerous times at meetings of AAAS, AGU, GSA in Poster Sessions (1982-1999), and NPA meetings (1999-2007) in regular programs.
Number 5: Written in manuscript form, unpublished (1979-1980).
Number 6: Presented at NPA meetings (1999-2007).
Number 7: Presented at AAAS and NPA meetings (1995-2005).

Letters and Memos

To: Dr. Carlyle B. Storm, Director, Gordon Research Conferences, University of R.I. 04-15-97
Re: Gordon Research Conference: Origins of Solar Systems, June 15-20, 1997

Thank you for the information in your letter (undated) received April 14 stating that the subject Conference has received more applications than can be accommodated, and that my name has been placed on a waiting list at this time. With a sense of great urgency, I respond with some pertinent information.

The primary reason I have pushed for such a conference is to introduce the new Fourth Law of Planetary Motion (FL) to the scientific community for a critical in-depth examination. You will recall that the First Three Laws were discovered by Kepler after his initial attempt in 1595 to solve the enigma of the geometric spacing of the six then-known planets. The elusive enigmatic FL finally was discovered in 1994-1995. *Intimately entwined with Kepler's First Three Laws, together the Four Laws* clearly reveal the only manner in which our Solar System could have formed—while conclusively showing how each and every planet attained its present orbital position around the Sun.

My previous writings since 1975 had clearly revealed how planets evolve through five common stages of evolution from energy to gaseous to transitional to rocky to inactive (dead) spheres via Internal Nucleosynthesis (IN). Begun in 1973-1975 with the initial pamphlet on abiogenic fuels, *this non-speculative FL/IN concept has been carefully and factually pieced together* as a *new paradigm of the origin of our Solar System and the evolution of its planets and moons. The new FL that ties it all together was the final piece of the puzzle in which every facet intermeshes with all others.*

This exciting concept appears incontrovertible; if any flaw does exist, it remains elusive to everyone familiar with the new standard. Definitive and testable, the new paradigm is capable of standing on its own merits until and unless it fails a test. Meanwhile, it should be given a thorough in-depth examination by the scientific community. The ideal place to begin is the GRC in June.

If the GRC evaluation finds no fatal flaw, the FL (and its inherent IN partner) could be a candidate for *one of the most significant discoveries of all time.* As such, it will open the floodgates to understanding how solar systems (including exo-planets, binary systems, etc.) came into being. If it is not discussed at the GRC, the floodgates will remain closed and this understanding might remain dormant a long time.

If the purpose of the Meeting is to find valid scientific answers to origins of solar systems, the new FL/IN paradigm warrants a high priority at the GRC in June. Won't you use your influence to permit a thorough examination there?

To: Dr. Carlyle B. Storm, Director, Gordon Research Conferences, University of R.I. 05-16-97

I write again with a great sense of urgency concerning the Gordon Research Conference on *The Origins of Solar Systems,* June 15-20, 1997.

For the past 24 years, I have researched the subject diligently and faithfully. These efforts came to full fruition with the discovery in 1995 of the Fourth Law of Planetary Motion that explains precisely (and beyond doubt how each and every planet attained its current orbital position around the Sun). In explaining the spacing of planets, the new Law will enable scientists to decipher the history of the Solar System and to predict the future displacement and fate of each and every planet.

To reach this stage of understanding our planetary origins, it was necessary to research all ancillary subjects pertaining to origins and evolution of planets. The results can be summarized in thirteen inter-locking facets of a new FL/IN paradigm that appears destined to displace prevailing beliefs about such origins early in the 21st century. The concept was first presented in totality at the annual meeting of the Georgia Academy of Science (of necessity in Poster session format rather than in a 15-minute paper) on May *26,* 1997. It is scheduled for another presentation at the University of Connecticut on June 11. Additionally, the National Science Foundation now is giving my grant application to expand its scope their "fullest consideration", according to their recent letter.

This new perspective opens the floodgates to a provable understanding of the origins of solar systems. *Truly, it is the key to understanding our planetary origins and evolution. Please be assured that without this crucial key, the search for answers to such origins and evolution will continually lead to speculation and futility.*

I cannot expect anyone to believe or accept this monumental discovery until he/she has a chance to learn its multitude of intermeshing facts. I am confident that it will withstand all in-depth examinations. I feel equally certain that the attendees at the meeting in June would benefit immensely from such a discussion. Won't you permit them to learn of this new knowledge—and benefit therefrom? Certainly, its factual nature warrants full consideration for a high priority on any relevant program.

To: Glenn Strait, Science Editor, The World & I Magazine, Washington, D.C. 20002. December 29, 1997

Thanks for the two press releases on *Asteroid Discoveries from Near-Mathilde Flyby* dated 97-12-19.

Herewith is a copy of my letter of July 19, 1997 in response to the initial articles on the subject in *Science* and *Science News*. It is self-explanatory, and seems to provide answers to the anomalies, or unexpected and surprising characteristics, of the asteroid. Both publications responded in the same manner as has been done over the past two decades: they regret that "we are not able to publish it."

My records indicate that blind copies were sent to the following:

Dr. Neal Lane, NSF

Dr. Pankotin, NSF

Lou Harvath, Editor, LaGrange Daily News
UGA
Dr. John Chappell, Jr., San Luis Obispo, CA
Editor Glenn Strait (It's possible that the copy was not mailed to you.)

The anomalies of 253 Mathilde cannot be explained logically in the perspective of current beliefs. However, the answers do fall readily into place, interlocking precisely, like pieces in a jigsaw puzzle, in the perspective of the FL/IN paradigm.

Reading the explanations given in the two press releases gave my associates and me (and other readers, I'm sure) a few hearty laughs; they are truly comical responses that appear to be born out of desperation to force-fit them into current beliefs about our planetary and universal origins. But seriously, it is more of a tragedy than a comedy.

But some day, recognition of the true natures of asteroids and of comets will be the major factor in undermining prevailing beliefs about our origins. When that time finally arrives, your files should contain sufficient information on the new concept for many good articles on both sides of the subject.

Glenn, I've finally gotten a fax machine: note the number in the address heading. So you now have another choice of method for sending messages. But please feel free to use the one that is most convenient to you in any situation.

Thanks for the information, past, present and future.

To: Glenn Strait, Science Editor, The World & I Magazine, Washington, D.C. 20002. May 8, 1998

Thanks again for the e-mail regarding *Oil to Gas? Or Gas to Oil? Asteroid Belt Collision...,* and the two moons of Uranus, along with some suggestions and comments.

I appreciate your kind offer to help in my communication with other scientists and I see no problem there. However, I must tell you that my efforts to correspond with many of them during the past 25 years have failed, while others have had some small success. Some years ago I wrote to Mango twice in response to his news releases about the oil and gas, but received no reply. But on very rare occasions, some three or four scientists did correspond briefly on various subjects, but to no avail.

Yes, I had read and noted specifically the *Science* article, *Catalytic Explanation for Natural Gas.* Schoell is right in stating that Mango's "mechanism does not explain what we observe in nature." Nor can it account for the formation of the unimaginable volumes of natural gas found deep within Earth's crust; nor for the three-layered system of deep methane, medium-depth petroleum and surface coal. Volumes of accumulated evidence in the literature now clearly reveal that methane forms first from carbon and hydrogen atoms, then polymerizes to ethane, propane and butane, followed by pentane, the initial liquid component of petroleum. Polymerization and cross-linking continues on through the lightweight, crude oil and coal stages of evolution of hydrocarbon fuels as functions of prevailing conditions of pressure, temperature, length of entrapment (or exposure), and the presence of catalytic metals that assist in the polymerization and cross-linking functions. Polymerization of hydrocarbons and other compounds via metal catalysts are tried and true processes covered by a number of patents. I do not understand Mango's contrary persistence.

The original atoms of carbon and hydrogen that form the methane gas (as well as all atoms that form our atmosphere, crust, mantle, etc.) are created from Earth's "Sun-like" core (Descartes)—which accounts for the vast stores of deep methane. *Without such Internal Nucleosynthesis (IN) to supply atoms, there could be no evolution (E) of planets or moons.* <u>IN and E are inseparable in all active spheres: one cannot exist without the other</u>. *The proposed Fourth Law of Planetary Motion, along with Kepler's First Three Laws, clearly reveal how the original Sun-like masses attained their orbital spacing around the Sun before beginning their evolution through the five stages common to all planets.* These five stages of evolution can be identified simply by close observations of the spheres of our Solar System; each stage complies with all natural laws.

Frankly, the Kortenkamp *et al.*, cosmic-dust connection to mass extinctions appears based on speculation and force-fitted data. Published photographs of asteroids clearly reveal their solid, dustless and thermally-formed surfaces, indicating that Olbers was correct in his interpretation in 1802 of the source of the orbiting fragments: the explosion of a [actually three] former planet[s]. They do not congregate as "rubble-pile" asteroids that "might throw out enormous quantities of dust into interplanetary space". Comets would be a far better source for the alleged cosmic dust. However, such dust falling on Earth would amount to less than a small fraction of 1% of the massive outpourings from some 600 active volcanoes, along with huge underwater vents throughout Earth's history. I cannot visualize any scenario in which such dust would affect Earth's climate sufficiently to affect mass extinctions. *In reality, IN, <u>the</u> crucial factor in evolution, can and does affect both climate and mass extinctions to a degree not yet accepted.*

I was somewhat surprised to read that "the AAAS has posted a list of experts on asteroids, and their effects, on the journalist's password-protected section of 'Eurekalert!'—to access this list for contact data on qualified commentators, go to the 'Resources' section and click on 'Experts by Topic'". This is a very powerful and assiduous form of censorship that should be highly successful in preserving the status quo while keeping out oppositional ideas presented by ignoramuses like myself who cannot possibly know as much about the subject as the listed experts.

My communication problem seems to reside in knowing that everything is connected to everything else while not knowing where to end the conversation about these endless connections. No matter where the conversation ends, the reader or listener is left with only a partial view of the complete picture, and thus cannot be expected to see the big picture clearly enough to understand it to the point of acceptance. Thus, whenever attempts to explain any single facet of the FLINE paradigm are made, frustration is the usual result. After 25 years of separating the myths and facts of science while enduring the stream of rejections, and at age 75, perhaps I have reached the "burned-out" stage for now. There are simply too many relevant articles in the news now that warrant responses in which a new perspective would help the authors via more accurate interpretations and better understanding of their findings.

During the next five weeks, I have three oral presentations scheduled in meetings, and am still awaiting word from the referees of my paper on the proposed *Fourth Law of Planetary Motion.* If you will bear with me until some of the smoke clears, perhaps I can regain a little of the fire to carry on. Thanks again.

Memo to File **June 10, 1998**

The short article, *Franklin Institute Bestows Medals (Physics Today*, June 1998, p 87), states that "Martin Rees was also the first to propose the fantastic—and now widely accepted—theory that the engines driving the high-energy, deep-space quasars seen by the Hubble Space Telescope are actually enormous black holes." The article contained no date of Rees' initial proposal: the date may be crucial.

In February 1980, my 1979 manuscript entitled *CREATION: From Void to Energy to Universe* was copyrighted, but remained unpublished. In its basic concept, both the creation and coalescing of energies occurs at the perimeter of the Universe in the form of black holes, which continue moving at nearly the light-speed of the collapsing electromagnetic energies it attracts. The in-falling electromagnetic energies are accelerated onto the black holes at the speed-of-light squared, thereby accounting for the stunning brilliance arid eerie sounds of the quasars created in this manner. Eventually, their speed carries the black-hole-quasar object beyond the control of the magnetic forces holding them together, and they simply explode into a bubble-shaped formation of a beginning series of galaxies. A remnant of the black-hole-quasar remains at the center of each galaxy as a source from which the fiery masses are continually ejected to form the billions of fiery stars and dust comprising each galaxy.

Herewith is a more detailed copy of the original manuscript (pp 34-36). Since it consists of a continual series of explosions at the perimeter of an ever-expanding Universe, this concept was given the nomenclature of *The Little Bangs Theory* (LBT) of 1980. Since then, my discovery of the proposed Fourth Law of Planetary Motion (FL) and the previous recognition of Internal Nucleosynthesis (IN) as the driving force behind the five-stage evolution (E) of planets from fiery energy masses to gaseous to transitional to rocky and finally to cold, inactive spheres, added powerful corroborative chapters to this FLINE paradigm. John Muir best expressed such connections: "When we try to pick out anything by itself, we find it is hitched to everything else in the Universe."

Thus, it is important to know the date of Rees' proposed theory of quasars driven by black holes in order to know which concept first established the theory "that the engines driving the quasars are actually enormous black holes". However, I do not wish to collide with Rees over the issue at this time, since he, as UK's astronomer royal, might be one of the referees of my paper, *On the Spacing of Planets: The Proposed Fourth Law of Planetary Motion,* currently under review at the Royal Astronomical Society, London. Suffice it to say that the black-holes-quasars concept is a crucial key to understanding the creation of energy and matter at the perimeter of a Universe expanding at the speed of light: i.e., the Little Bangs Theory (LBT). In its perspective, all the evidence interpreted as supportive of the Big Bang (BB) can be readily interpreted as even more powerful evidence for the LBT. Through the years, this revolutionary concept has been offered as a viable alternative to the more speculative BB—with little, if any, significant success. The pending Fourth Law of Planetary Motion (FL) of 1995, in conjunction with Kepler's First Three Laws, supplied the last remaining pieces of the puzzle detailing how our planets attained their current spacing around the Sun in order to begin evolving via Internal Nucleosynthesis (IN) through the five stages of Evolution (E) common to all planets: the FLINE paradigm—a product of 25 years of painstaking research into the origins of solar systems and the evolution of planets (1973-1998).

At this writing, the accumulated evidence supporting both the FLINE paradigm and the LBT appears incontrovertible. However, problems remain, problems that are best described by Dara Horn in her article: *The Shoulders of Giants* (*Science*, 29 May 1998).

To: Dr. John Chappell, Jr., San Luis Obispo, CA. July 30, 1998

Here is the information you requested in our telecom of July 29, along with some additional information on the status of my research and writings on the origins of solar systems and the evolution of planets.

Since I first met you at the AAAS Annual Meeting in February 1995, I have succeeded in putting the finishing touches on the solution to the proposed Fourth Law of Planetary Motion. There were more missing pieces that revealed more critical information than I had realized. Even after publishing my latest book, *The Spacing of Planets: The Solution to a 400-Year Mystery* (March-April 1996), more and more pieces of the puzzle of planetary origins and evolution continue to be uncovered by relevant discoveries reported in science magazines and newspapers. Time is the limiting factor now in putting all the new pieces in place. The new evidence always seems to fit precisely into the FL/IN/E paradigm in which the Four Laws of Planetary Motion (FL), Internal Nucleosynthesis (IN) and Evolution (E) are strongly entwined as the backbone, heart and soul of this revolutionary concept.

It is frustrating that these exciting developments are not permitted (even in Poster Sessions) a voice in science meetings simply because they oppose prevailing, but readily refutable, dogma. Mark Twain said it best: "What people don't know isn't nearly so aggravating as what they do know that isn't true." Since that 1995 Meeting, all of my abstracts on the subject have been rejected by peer reviewers in several science organizations. In each case, no effort was made to give a valid reason for the rejection—perhaps because peer reviewers could find no reason other than the papers' posing a serious challenge to the beliefs in which they were schooled.

However, there is one bright side: these persistent rejections have given me more time to enhance the FLINE paradigm with a great amount of additional supportive evidence. This revolutionary concept is now a no-nonsense, factual, non-speculative, testable, definitive and viable alternative to prevailing, but misguided, beliefs about planetary origins. If a significant flaw exists therein, it remains elusive at this writing. If mankind is to truly comprehend our planetary origins, evolution and all other enigmatic anomalies of planets and planetary matter, scientists must be willing to examine this viable alternative to current beliefs. At this point in time, that possibility appears highly unlikely.

To the best of my memory, following is a summary of abstract rejections since 1995:

Organization	Years
AAAS	1996, 1997, 1998 AGU 1996, 1997 (No submission in 1998)
Gordon Research Conf.	1996, 1997, 1998
Ga. Acad. of Science.	1998

Since the AAAS Meeting in 1995, the three NPA Meetings I have attended and two GAS Meetings were the only times I was permitted to present this critical information to scientists. In earlier years, the NSF refused my grant application on two separate occasions, and the Chapman Conference leaders refused open discussions on these topics on two or three occasions. My paper on the Fourth Law was submitted to the Royal Astronomical Society (RAS) (Jan 12) and to the American Astronomical Society (Jul 7). The only response to date is the letter of receipt from the RAS. Additionally, the AAAS has twice ignored my request for a membership application. To summarize, since the completion of the Fourth Law of Planetary Motion in 1995-1996, only the NPA and GAS have permitted its presentation to a scientific group. To preserve the scope and history of these findings for posterity, self-publications in 1975, 1977, 1979, 1986, 1991 and 1996 were essential. As Michener wrote, "If it's not written, it doesn't exist."

Herewith is the latest three-page essay on the subject.

To: Mr. K.C. Cole, Journalist, Los Angeles Times, Los Angeles, CA 90053. **August 14, 1998**

Your excellent article headlined *33-year-old changing way astronomers see universe (Atlanta Journal-Constitution,* Sep 13, 1998) featured Andrea Ghez's presentation in August at Rutgers University in New Jersey. Her paper adds corroborative evidence to a revolutionary concept that has evolved during the past 25 years: the little-known LB/FLINE paradigm of origins of the Universe and its solar systems and the orbital spacing and subsequent evolution of planets. The lack of familiarity with this valid scientific concept stems directly from its serious threat to the present domineering beliefs about such origins.

The FLINE paradigm explains why "a massive black hole sits at the center of the Milky Way" and at the center of all galaxies—and why a black hole is essential to the formation of each and every galaxy. In the perspective of the Little Bangs Theory (LBT) of 1980—a vital part of the FLINE paradigm—the explosive black hole spews out in pinwheel fashion the dust, gases and fiery masses of energy we call stars. Obviously, such close initial proximity accounts for Ghez's "discovery that most newborn stars appear to be twins". Seldom can they remain single stars. Clearly, they usually were born double; however, some can drift apart later as single stars—which accounts for their older ages.

In similar manner since 1979, the evolving FLINE paradigm has contradicted the prevalent theories of how stars form from dust and gases—which, in reality, are by-products created simultaneously with the fiery stars comprising the galaxy ever since their explosive ejections from the twin outlets of a central black hole. Obviously, this concept adds dramatically to the large amount of evidence already contradicting the Big Bang hypothesis of the origin of our Universe.

The new paradigm provides the perfect (and perhaps the only logical) mechanism to explain the formation of double stars, including the recently discovered exo-planets. Ghez's belief that the current theory "very nicely produces our sun and planets" continues to ignore the Four Laws of Planetary Motion (FL) detailing how the planets attained their orbital spacing around the Sun. The Four Laws clearly reveal how our dynamic Solar System formed as a special and rare multi-star system via the same explosive forces that formed all binary-star systems. Planetary Evolution (E) is a fact of Nature, but such ongoing changes are not possible without a source of energy to drive them forward through the five stages of evolution common to all planets: from energy to gaseous to transitional to rocky to inactive (dead) spheres. In scientific terms these processes are known as Internal Nucleosynthesis (IN) (the creation of atoms from energy, per $E=mc^2$) and polymerization (the combining of atoms to form all planetary matter). The complete details are found in my writings of the past 25 years.

In further confirming the FLINE paradigm, Ghez clearly reinforces this revolutionary concept to the extent that it now seems destined to displace prevailing beliefs about planetary and universal origins early in the 21st century. Perhaps her paper and the many corroborative findings of space probes will encourage scientists finally to give more serious consideration to the FLINE paradigm of 1973-1998. If you care to delve deeper into this new concept for more stories, I would be happy to work with you.

To: Dr. John E. Chappell, Jr., Director, NPA, San Luis Obispo, CA 93406. August 28, 1998

Thanks for the NPA Newsletter of August, 1998. I appreciate all that you are trying to accomplish for the good of science, and would like to add a few comments.

As you know, I have experienced the same frustrations through our science leaders concerning my manuscript, letters and abstracts on *The Spacing of Planets: The Proposed Fourth Law of Planetary Motion* during the three years since its completion in 1995. I must agree that their usual recourse of complete silence is indeed the most demeaning answer possible. However, there is some consolation in knowing that they are unable to pinpoint a single fatal flaw in the FLINE paradigm—which is one of the reasons for the silence. But it is discouraging (and somewhat encouraging) to watch the documentary video films on TV getting ever closer to the principles comprising the concept; inevitably, they will be forced into the same discoveries and conclusions underpinning it. Meanwhile, I'm consoled somewhat by the belief that a closed mind is unwittingly the cradle of bias, the taproot of blissful ignorance—an immovable object.

I, too, find that many laymen can't accept the ridiculous [but lucrative] contentions of the Big Bang theory, as publicized in recent years. To complicate matters, a full page ad in last Sunday's *Parade* magazine was selling videos on *The Young Age of the Earth* based on equally false claims and misinterpretations. Ironically, these false statements are no more misleading than the claims supporting prevailing scientific beliefs about the big bang, the origins of solar systems and the evolution of planets. When analyzed in the perspective of the FLINE paradigm, both views appear equally ludicrous: both misinterpret and force-fit data into preconceived notions that stem from antiquated beliefs. And in both cases, the alleged supportive evidence blithely ignores three fundamental realities of Nature: (1) the Four Laws of Planetary Motion (FL); and (2) the Internal

Nucleosynthesis (IN) that is absolutely essential to the third reality, planetary and universal (3) Evolution (E). In the circumstance of creation of our Solar System, the three realities are interdependent: any one of these ongoing and inseparable events is not possible in the absence of either of the other two. For all active spheres in solar systems, FL + IN = E (an absolute).

If science ultimately is to win out over the powerful forces behind the beliefs expressed in the *Parade* ad, we must first put aside preconceived notions, and then establish our beliefs on a more solid foundation of provable facts, while excluding speculation. A viable alternative, the FLINE paradigm offers science this opportunity. But I suspect that it will lay dormant until billions of dollars of grant money are used in arriving finally at its absolute truths.

Herewith is a copy of my abstract, *The FLINE Paradigm: Critical Insights into the Origins of Solar Systems and the Orbital Spacing and Subsequent Evolution of Planets,* which is being submitted to the next AGU meeting in December. I'm not optimistic that it will be any more welcome than its relevant abstracts rejected in 1996 and 1997. These same submissions/rejections apply to the AAAS Meetings. However, I hope to use it in the next NPA Meeting. Also, I'm enclosing my dues of $30; later I plan to send a ¼-page ad to go with the next newsletter, to be billed at $25. Regarding the D-B, please change my zip code to 30241 and add Phone or Fax 706-884-3239. Replace the *1986, New Concepts of Origins,* with *1998, The FLINE Paradigm: Critical Insights into the Origins of Solar Systems and the Orbital Spacing and Subsequent Evolution of Planets* (Paper $6).

To: David L. Bergman, Foundations of Science, Common Sense Science, Inc., Kennesaw, GA 09-04-98

Thanks for the premier issue of *Foundations of Science* (July, 1998, Vol. 1, No. 1). It's exciting to see the voice of reason and common sense circumventing the censorship of peer reviewers trained in the absurd, yet prevalent, beliefs about the Big Bang (BB), the nature of the Universe, the origins of solar systems and the evolution of planets.

Kepler's laws did indeed save celestial mechanics from even more absurdities about planetary orbits. However, the BB has put orthodox physics right back into a situation not unlike the one that existed before Kepler's Three Laws of Planetary Motion in the early 17th century: the necessity of having to be explained by complicated assumptions that lead to ever more critical misinterpretations of observations and data. The consensus method in which force-fitting is essential has gotten completely out of hand, thereby giving meaning to the expression, "Oh what a tangled web we weave..."

Another opportunity for scientists to understand the workings of our Solar System is available now to scientists via the solution to the mystery that first eluded Kepler in 1595: the Fourth Law of Planetary Motion (1995) detailing how the planets attained their orbital spacing around the Sun. Because its factual nature and fundamentally sound structure poses a very serious challenge to current dogma, the new Law has been unable to get past peer reviewers schooled in the strongly entrenched accretion beliefs about the origin of our Solar System and the evolution of planets and moons. As with Kepler's Three Laws, the new Fourth Law can save celestial mechanics once again from ever more absurdities about planetary origins, orbital spacing and evolution—but it will continue to be a difficult struggle.

To: Neil E. Munch, Munch Engineering Corporation, Montgomery Village, MD 20886. 09-29-98
Re: Assumption Shifts in Special Relativity and Their Consequences

Thanks for the copy of your paper pertaining to Special Relativity.

While its contents are out of my field and beyond my capabilities to render any significant comments, I can empathize with your statement that "it's hard to believe that something as cosmic in nature as special relativity could be flawed by something as mundane as inadequate control of assumptions."

Since getting more deeply involved in the planetary and astronomical sciences during the past 25 years, I have become disillusioned by the total lack of tolerance of peer reviewers for any new perspective outside their realm of understanding—especially if it challenges establishment-dogma based on such assumptions.

The same problem you discuss prevails in my fields of science, and I suppose, in most any arena of science. Usually, it's easy to pass the situation off as human nature. However, as John says, intolerance and censorship should have no place in science. But I'm convinced they are here to stay.

Herewith is a copy of my latest abstract, *The FLINE Paradigm: Definitive Insights into the Origins of Solar Systems and the Orbital Spacing and Subsequent Evolution of Planets.* If accepted by peer reviewers schooled in prevailing dogma, I would be genuinely surprised.

Thanks for your efforts in furthering the valiant cause of the NPA.

Distribution: EOS, AGU, AAAS, Strait, Astronomical Journal, Science News, UGA, Herndon
Subject: Re: Examining the Overlooked Implications of Natural Nuclear Reactors (Herndon, AGU, Eos, Sep. 22, 1998) vs. The Revolutionary FLINE Model **Oct. 5, 1998**

During the past two decades beliefs about the composition of Earth's core have undergone significant changes. In the early 1980s scientific papers concentrated on how and why a rocky center formed in our planet. Later in the decade the emphasis was placed on an iron core as the more logical composition to generate a magnetic field. The latest concept (Herndon, Eos, Sep 22, 1998) promotes the idea of uranium from outer-space missiles as the core ingredient that furnishes the internal heat via fission alone.

However, all of these theories have problems; all are speculative and contain fatal flaws. For example, a uranium core decaying via fission alone as the source of the heat within would fall far short of the energy needed to drive planetary evolution, and would provide no valid source for Earth's electromagnetism. We have only to observe our Sun to understand that the lighter elements, primarily hydrogen and helium, created via Internal Nucleosynthesis (IN) in nuclear masses (the true source of heat and electromagnetism in all active spheres) are the initial building blocks for fusion reactions into all of the atomic elements of which planetary spheres are composed—often including the last and heaviest one, number 92, uranium. Theories based on this specific IN perspective offer more logical solutions to all planetary anomalies.

The most logical concept of Earth's core was first defined by Descartes as being "Sun-like". When one ponders all possible solutions to the mysteries of planetary evolution, this old idea takes on the reality of truth. Some historical information on the subject will fill in the missing pieces of Herndon's article, and perhaps help maintain an accurate record. My initial publications in 1975 on the powerful supportive evidence for nuclear cores in active planets include three titles: *Birth of the Solar System, Evolution of Planets,* and *Evolution of Gas, Oil and Coal.* These subjects were first presented in poster sessions at the AAAS Meeting in 1982 in Washington, D.C. and again in the 1983 Meeting in Detroit, Michigan.

The subject was expanded to booklet form in 1977 under the title, *Undermining the Energy Crisis,* then to book form in 1979 under the same title—always emphasizing the critical role of nuclear-energy cores in the evolution of planets via Internal Nucleosynthesis of the atomic matter comprising their atmospheres, crust, fuels, etc. During the ensuing years, these revolutionary concepts were presented at various meetings, including the AAAS, the AGU, the AGS, other professional groups and civic organizations. Many abstracts appear in the records of these meetings. However, my manuscripts and news releases continued to be rejected, and on occasions were handled unfairly by officials in charge of the programs.

So I was forced to continue writing and publishing in ever-expanding book form as fast as new discoveries were made (and finances permitted it). In all situations, the new knowledge fitted precisely into the ever-evolving concept. These books include: *From Void to Energy to Universe* (1980) (manuscript copyrighted, not published), *New Concepts of Origins: With White Fire Laden* (1986), *The I-T-E-M Connection: How Planet Earth and Its Systems Were Made by Means of Natural Laws* (1991), and *The Spacing of Planets: The Solution to a 400-Year Mystery* (1996). The additional confirmed evidence was continually submitted to various organizations and editors, but to no avail except in one case: *Evolution of Gas, Oil and Coal* was published in *Alternative Energy Sources VI* (Vol. 3, pp 337-344) by the Clean Energy Research Institute, University of Miami (1983). Unfortunately, a partial version of my *Energy Fuels Theory* was published in the *New York Times* the same day, and thereby became erroneously known as Gold's theory. However, Gold's incomplete concept did not include the nuclear-energy source of the elements comprising hydrocarbon fuels; his source was missiles from outer space. Fourteen specific facts discussed in my writings validate this internal energy source of the hydrocarbon atoms.

Herndon's article cites the Oklo uranium deposit, as predicted by Kurado (1956), as having achieved a self-sustaining neutron chain reaction. To quote further, "The discovery of the intact remains of natural nuclear reactors has profound scientific implications, especially in proving the existence in nature of an energy source capable of producing more than an order of magnitude more energy than radioactive decay alone [radioactivity is a relatively cool by-product of extremely hot nuclear reactions]. Curiously, the subject appears never to have been addressed in the pages of *Science* or the *Journal of Geophysical Research,"* thus, the erroneous title of Herndon's article. However, the implications of natural nuclear reactors have not been overlooked; during the past 25 years they have been fully utilized in the evolving IN/E paradigm in which no planetary Evolution (E) is possible without Internal Nucleosynthesis (IN) in planetary cores. We have only to observe the elements created within each and every star and planet to recognize the full power and extent of thermonuclear fission/ fusion reactions. The types and amounts of atomic elements produced in each of our active spheres are determined by the sphere's size, distance from the Sun, and stage of evolution—factors that determine the pressures and temperatures within. For example, Earth, in the fourth stage (rocky) of evolution, can produce all 92 elements, including uranium.

The IN/E relationship led inevitably to the solution (1980-1995) to the Fourth Law of Planetary Motion that first eluded Kepler in 1595. Together the Four Laws (FL) and the IN/E relationship form the FLINE paradigm that clearly reveals how planets were spaced in a geometric progression and now are evolving through five common stages of evolution at rates in full accord with size. The vast amount of supportive evidence assures that the new concept will remain irrefutable under the closest scrutiny.

To quote Eos again: "Herndon [1992]...proposed the idea of planetary-scale nuclear fission reactors as energy sources for the giant planets. ...Herndon [1993, 1994] extended the concept to non-hydrogenous planets, especially the Earth." However, Herndon strays far off course when he hypothesizes that Earth's core is a mass of uranium that came from outer space in the form of meteorites and condensed at the planet's center, an hypothesis that totally ignores the basic principles of the FLINE paradigm—especially the Four Laws of Planetary Motion—and their application to all planets of all solar systems. However, he is correct—and in full agreement with the FLINE paradigm in suggesting "that the variable and intermittent changes in the intensity and direction of the [Earth's] geomagnetic field have their origin in nuclear reactor variability," [*but in a nuclear-energy core rather than from uranium, one of the 92 elemental products--via IN--of the nuclear-energy core*]. *All active planetary cores are nuclear fission/ fusion reactors capable of producing elements in full accord with the conditions prevailing within and without the sphere at any given time.* To emphasize: *planetary cores are not only fission reactors; fusion is more dominant than fission in such cores, or else there could be no planetary evolution.*

The Eos article is somewhat akin to the FLINE paradigm, but is modified by tying everything in with the prevailing accretion hypothesis, much as did Gold's erroneous version of creation of the hydrocarbon fuels. However, by attempting to tie their beliefs in with prevailing beliefs rather than sticking to more factual concepts, both authors ended up with theories that contain too many fatal flaws.

To understand all the mysteries of planetary origins and evolution, scientists must think in terms of the Four Laws of Planetary Motion that put the fiery masses of energy (from a potential binary system) into properly spaced orbits, followed by the inseparable interactions of Internal Nucleosynthesis and Evolution of each sphere via the "Sun-like" energy core of each and every planet. A quote from my writings of 1980 seems appropriate: "We must forsake such beliefs that Earth is flat, that its core is iron or rock, that fuels were made from fossils, that energy is finite, and that planets are created from dust and gases. Such concepts have too long misled scientists down blind alleys." (To this we can add "that comets are dirty snowballs or ice-balls.") Yet in 1998 scientists are still being misled by prevailing beliefs based on the accretion hypothesis that all planetary matter came ready-made from outer space. But if we can recognize that each planet is—and always has been since the geometric layout of the Solar System—a self-sustaining entity that evolves through five common stages of evolution via Internal Nucleosynthesis, the solutions to all planetary anomalies will fall readily into place. The FLINE paradigm offers a valid scientific understanding of planetary cores and their crucial roles in the origin and evolution of all planets and moons.

Memo: 1998 Fall Meeting, American Geophysical Union, Washington, D.C. 20009. Oct. 26, 1998
Re: Abstract for Fall Meeting, 1998. Critical Insights into the Origins of Solar Systems and the Orbital Spacing and Subsequent Evolution of Planets

No response to my submission of the subject paper has been received, so I must assume that the abstract has been rejected once again without notification and without a valid reason. If so, this will be the third consecutive year of rejections since my mathematical solution to the proposed Fourth Law of Planetary Motion was finalized in 1995. Meanwhile, corroborative evidence supporting this revolutionary discovery continues to mount with the advent of each relevant finding in the fields of astronomy and geology. I'm left with the feeling that reviewers and editors are not willing to take the time to learn of its substantial evidence, and are perhaps dubious of its genuine challenge to prevailing beliefs—and therefore reject it out of hand without a valid reason. There is no room in science for hasty pre-judgments without the facts.

The strictly factual FLINE paradigm is proving to be clearly superior to all other concepts at explaining solar-system origins, orbital spacing and the evolution of planets and moons, etc. There is no authentic reason to doubt its scientific validity or to deny it a fair hearing and a thorough evaluation against prevailing beliefs. If any major progress in the planetary sciences is to be made any time soon, scientists must be allowed to learn of this viable alternative to current dogma—an alternative that permitted me to accurately predict today's fuel prices and the reasons therefore during the darkest days of the energy crisis of the 1970s—a feat that could have been accomplished only in the perspective of this revolutionary concept. The same holds true for all other anomalies of solar systems and planets. Further, every bold prediction made via the IN/E paradigm during the energy crisis and the reasons behind it have proven deadly accurate.

We must put aside preconceived notions based on ever-modifying assumptions, and recognize that planetary Evolution (E) is not possible without Internal Nucleosynthesis (IN)—the engine that drives each of the spheres through their five common stages of evolution. Recognition of the true nature of planetary cores would eliminate the erroneous beliefs that Earth's core consists of rock (early 1980s), iron (early 1990s) and uranium

(1998). In any active sphere, IN and E are inseparable. Both IN and E interlock precisely with the proposed Fourth Law of Planetary Motion (1995), whose solution, along with Kepler's First Three Laws (FL), clearly detail how our planets (and the exo-planets) attained their spacing around a sun. Only in this perspective can all planetary anomalies be understood and proven beyond doubt.

I now consider as incontrovertible each of the three facets of the FLINE paradigm and the precise manner in which they interlock. But your help is needed in finding someone who is willing to examine the evidence and challenge any aspect of the paradigm on an impartial basis. I am confident that it will withstand the closest scrutiny and the longer test of time, just as Kepler's First Three Laws have done. Your membership would benefit immensely if given opportunities to understand and challenge the concept. At this crucial stage, we need to know whether or not it contains any fatal flaw that has managed to escape detection during the 23 years of developing this revolutionary FLINE paradigm.

If my paper and abstract are not permitted beyond peer review, and since this was my primary reason for attending, please cancel my registration and refund the fee. However, if I'm wrong about the status of my paper, I do apologize.

c: Conveners: X. Song & L. Stixrude (U17).

To: Ms. Sandra Blakeslee, Science Reporter, The New York Times. November 30, 1998

Scientific Meetings Produce Clash of Agendas (*Science,* 30 Oct 1998, p.867) was read with a great deal of interest, especially the part about your experiences to get the story to the public. While an embargo on scientific news may be only a temporary ban, it sometimes drifts into the realm of censorship, especially when the findings pose a serious challenge to prevailing beliefs. Good examples can be found in the histories of Copernicus, Galileo, Semmelweiss, etc. A more recent example is the censorship for the past three years of my discovery of the proposed Fourth Law of Planetary Motion (1980-1995) explicitly revealing how the planets attained their orbital spacing around our Sun. This solution had proven elusive to astronomers since 1595, the year Kepler first recorded his attempts to solve the mystery of the geometric spacing of the six then-known planets.

As explained in the flyer herewith, this Fourth Law was the final link that bonded together all previous substantiated findings into a factual and irrefutable concept of the origin of our Solar System and the evolution of planets via the processes of Internal Nucleosynthesis. The revolutionary paradigm consists of three chronological, inseparable and ongoing realities:

1. The Four Laws of Planetary Motion (FL).
2. Internal Nucleosynthesis (IN).
3. Evolution (E).

No solar system is possible without undergoing these three realities during the prime of its existence, until eventually each of the cycles reaches its natural ending.

The main problem with the FLINE paradigm? It "slam-dunks" prevailing beliefs about planetary origins and evolution, and poses a serious challenge to the Planetary Accretion hypothesis (and other aspects) of the speculative Big Bang concept of universal origins. Because it has not been permitted a presentation at any national science meeting since its completion in 1995, most scientists remain unaware of this paradigm, but the few who have learned of it can find no valid rebuttal against any of its facets -- perhaps because none are possible. From a strictly factual scientific viewpoint, this three-prong theory appears rock-solid. It seems destined to displace prevailing beliefs early in the next century.

Its abstract finally has been accepted by the AAAS for a Poster Session on January 23, 1999 at their annual meeting in Anaheim. Herewith is a copy. I sincerely believe that history will rank the Fourth Law of Planetary Motion, along with the FLINE paradigm of planetary origins and evolution it finalized, among the greatest discoveries of all time. In my biased opinion, it should be the most monumentally significant paper to come out of the Anaheim meeting—but only if it can reach enough of the right people to become known. The reason for this letter to you as a reputable reporter: it is a genuine opportunity for the big scoop of a lifetime. I will work with you toward this end, if you so desire.

In this situation, no embargo exists, although manuscripts have been submitted to editors. And through the 25 years of putting the facts together, I have been forced to self-publish each facet of the evolving concept in order to save it for posterity. They are summarized in *The Spacing of Planets* (1996).

To: Dr. Hadiro Takahashi, 2-13-18 Takamatsu, Morioka 020-0114, Japan. December 31, 1998

It was good to hear from you again, and especially to learn about your progress with the T-B Law at science meetings and in a publication. Thank you for updating me with the beautiful card.

Your suggestion of sending a copy of your two papers is appreciated, and I look forward to receiving them. Meanwhile, I'm sending herewith a confidential preview copy of my news release pertaining to my work since the last time I saw you at the AAAS Meeting. It is self-explanatory, and is scheduled for release the day of my Poster presentation on January 23, 1999. After that time please feel free to take any action you deem desirable with it.

The finishing touches to my solution to the proposed Fourth Law of Planetary Motion were completed in 1995, but my abstracts for the 1996, 1997 and 1998 Meetings understandably were rejected by advocates of prevailing beliefs. So this will be the first time I've been able to present the completed concept to the larger scientific community. Meanwhile, I felt forced to self-publish my latest book, *The Spacing of Planets: The Solution to a 400-Year Mystery* (1996) for posterity. Herewith is my latest flyer.

My paper on the solution to the proposed Fourth Law was submitted to the Royal Astronomical Society in London (January, 1998), and to the American Astronomical Society (later in 1998). The RAS notified me of its receipt and assigned ID number; however, since then I have heard nothing from them. The AAS rejected it on the grounds of insufficient data—a poor excuse, since more than 80% of it is data and graphs. The real problem, I suspect, is that it upsets current textbook beliefs about the origins of solar systems and the evolution of planets, as well as offering a serious challenge to the Big Bang—and they are not yet ready to take that giant leap, regardless of its voluminous substantiated (and apparently irrefutable) evidence.

At this time, I plan to be at the meeting only for that one day. I hope to see you there.

To: Dr. Michael S. Strauss, AAAS Meetings Office, Washington, D.C. 20005. January 29, 1999

Thank you again for permitting my presentation on planetary origins at the AAAS Meeting in Anaheim on January 23. About 15 scientists expressed a genuine interest in my Poster Session showing the definitive aspects of the FLINE paradigm that features the proposed Fourth Law of Planetary Motion explaining the mathematical spacing of planets and the subsequent Internal Nucleosynthesis that drives planetary Evolution. But to overcome the hurdles encountered by any monumental change, I still need your help.

On Friday, January 22, I left approximately 150 News Releases and 150 single-page briefs in the AAAS newsroom clearly explaining the new concept—a factual, non-speculative, easy-to-understand explanation of the origins and evolution of solar systems based on the proposed Fourth Law of Planetary Motion. Yet there was a total lack of response from reporters. On January 25, Ellen Cooper and I searched both the AAAS newsroom and the media newsroom without finding a trace of leftovers of the two FLINE papers. I'm still puzzled as to how that many releases could get by so many good reporters without any response to its historical revelations on the origins of solar systems and the evolution of planets via Internal Nucleosynthesis. Can you help me understand why this monumentally significant and fully corroborated breakthrough is experiencing such difficulties in the science and news media arenas?

Each of the many precisely intertwined facets of this 1995 paradigm was carefully and factually developed during the previous 23 years to the point of being indisputable. The potential benefits to science are immeasurable. The most recent example is the article by Richard A. Kerr entitled *Earth Seems to Hum Along With the Wind (Science,* 15 Jan 1999, p.321). To quote: "The whole planet vibrates with a deep soft hum...imperceptible to all but the most sensitive seismographs. ...it was clear that seismologists now accept the reality of the hum, and one group presented data suggesting that the winds of the atmosphere, rather than something within Earth, excite the hum." However, the two causes, in reality, are inseparable.

In my writings of the early 1990s, this hum was identified as emanating from within Earth's nuclear energy core. The same hum will be found in every planet with a still-active core, in our Sun and in all stars throughout the Universe. This ubiquitous sound of nucleosynthesis was aptly named the *Hum of Creation* (1995). To understand its full significance, scientists need only to understand the FLINE paradigm. There now seems little, if any, doubt that it is destined to become the prevailing concept of the 21st century—unless someone is able to find a fatal flaw in it (which now seems highly improbable).

Understanding a partial facet of the FLINE paradigm might be difficult without having the full picture—especially of its main feature: the proposed Fourth Law of Planetary Motion explaining the mathematical spacing of planets. Since my writings cannot get past peer reviewers, perhaps a meeting among a select few of the AAAS scientists could be arranged to evaluate this crucial, but easy-to-understand concept at your headquarters. In the interest of science, can you arrange such a meeting in the near future?

c: Ellen Cooper, AAAS

To: Dr. Michael Strauss, AAAS, Washington, D.C. 20005. March 8, 1999

The FLINE paradigm is a factual non-speculative, easy-to-understand concept of planetary origins and evolution. Soundly based on Kepler's First Three Laws of Planetary Motion and the proposed Fourth Law (1995) that first eluded the great astronomer in 1595, the FLINE paradigm presents definitive mathematical explanations of how the nebulous planetary masses attained their orbital spacing around our Sun and evolved into their present stages of evolution. This new perspective, derived through careful research during the past 25 years, consists of three chronological, inseparable and ongoing realties:

1 .The Four Laws of Planetary Motion (FL)
2. Internal Nucleosynthesis (IN)
3. Evolution (E).

Without these inviolate principles, no solar system is possible. They reveal that planetary evolution is not possible with any type of core other than that hypothesized by Descartes as "Sun-like"—a concept now corroborated by abundant substantiated evidence. Such evidence mounts with each relevant discovery of planetary anomalies, both of Earth and of space probes. From "Sun-like" mass to inactive sphere, every planet is a self-sustaining entity creating its own atomic compositional matter via Internal Nucleosynthesis throughout its first four stages of evolution (from plasma to gaseous to transitional to rocky to inactive)—at a pace in full accord with size. To see examples of each stage, we have only to observe with insight the Sun, planets, moons, comets and asteroids.

Since the completed solution to the Fourth Law in 1995 gave undeniable validity to the FLINE paradigm, I am disturbed that it has been unable to get past peer reviewers or editors and into the greater scientific community for in-depth examinations. Its potential for a revolutionary change in scientific thought appears equal to that of the Copernican idea that placed our Sun at the center of the Solar System. And like that idea, its advantages to scientists trying to understand planetary anomalies are too great to be ignored at this time. As a matter of principle, they need in-depth examinations that will either confirm or refute its scientific validity.

I would like to propose the subject matter as a seminar or symposium at the 2000 AAAS Meeting in Washington, but am unfamiliar with the procedure. Will you be so kind as to send the necessary information, forms, etc., for submitting the proposal on the origins and evolution of planetary systems? Or perhaps use this letter as the submittal?

c: Stephen J. Gould

To: Editor, The Washington Post, 1150 15th Street, NW, Washington D.C. 20071. April 26, 1999
Re: The FLINE Paradigm: On the Origin, Orbital Spacing and Evolution of Planets

The annual meeting of the NPA-SWARM section of the AAAS was held April 11-15 in Santa Fe. On April 14, my paper on the FLINE paradigm featuring the proposed Fourth Law of Planetary Motion was allowed forty minutes for presentation, plus 15 minutes for questions and answers. The audience numbered about 35 international scientists who saw 23 overhead projections of the crucial points throughout the talk. Upon returning to my seat, I was paid a high compliment by Dr. Roland Kron from Minnesota who stated that it was "spellbinding". As has been the case throughout the past 26 years of the evolving concept, there were no valid rebuttals or unanswerable questions about this revolutionary concept of the origin arid evolution of planets. This was earlier summarized by the *Yale Scientific:* "Scarborough leaves no holes in his persuasive argument. His impeccable logic, accurate facts, and sound evidence rewrite dogma in a manner that would stifle any attempt at rebuttal."

On April 15, the news media brought forth the exciting announcement of the discovery of a multiple-planet solar system. This was followed on April 16 with a news article in the *Wall Street Journal* (*WSJ*) about an oil field that "grows even as it's tapped". Both discoveries were predicted and explained clearly years ago by the FLINE paradigm; both are corroborative evidence confirming facets of the concept that were developed during the years of 1973-1995. Had they been in the news a week earlier, I would have added these crucial confirmatory discoveries to my long list of corroborative evidence of the FLINE paradigm (on April 14) that goes much deeper into the explanations of these anomalies.

The reasons for writing this letter are twofold: First, it should help to keep the historical record straight when the concept inevitably displaces prevailing dogma in the next century, and second, it makes a terrific (scoop) article for your paper. Herewith is the rest of the story: a copy of the *WSJ* article, my letter of response, a summary sheet of the FLINE paradigm, and a flyer advertising the book (now at Bookland in LaGrange).

I realize that while the concept is easy for scientists to understand, it may seem somewhat confusing to others. If you so desire, I would be happy to work with you or your best reporter on the project, either long- term or short-term to make it even clearer to your readers. Or you may prefer to edit and use this letter in your letter-to-the-editor section.

Thanks for your interest, consideration and valuable time.

To: Gordon I. Rehberg, 1727 Kimberly Park Drive, Dalton, Georgia 30720. Apr. 29, 1999

Thanks for the suggestions in your letter of April 19 concerning going on the Internet. I have put book ads with content description, etc., on-line on three different occasions via vendors, but just those minor excursions cost more than they returned. I imagine the cost of a scanned book would wreck my budget with little if any compensation. But I can't rule it out yet.

I do have some good news, enclosed herewith. It's self-explanatory and exciting because it is the first time an outside source has described the manner in which an oil well renews itself from a hot, high-pressure source beneath it—just as the Energy Fuels Theory (EFT) of 1973 predicted and explained. This is a giant leap forward for the new paradigm, and a powerful bit of evidence against the fossil fuels hypothesis and, unfortunately, the Big Bang theory. Even more than ever, I consider the evidence incontrovertible and the concept irrefutable.

April 30, 9:25 a.m. Gordon: I just got off the phone with some really great news. My daughter, Kay, in Shreveport called and told me that she had received another e-mail from Glenn Strait, Science Editor of *The World & I* magazine, Washington, DC, requesting a copy of my paper, *The FLINE Paradigm: Definitive Insights into the Origins of Solar Systems and the Orbital Spacing and Evolution of Planets.* I will send a full copy with the 23 illustrations as presented on April 14 at the NPA-SWARM-AAAS Meeting in Santa Fe. It includes the fuels concept as absolute proof of the Internal Nucleosynthesis that produces all planetary matter—which, in turn, strengthens the proposed Fourth Law of Planetary Motion while identifying the fundamentals of Evolution. As you know, the FLINE paradigm consists of *three chronological, inseparable and ongoing realities: The Four Laws of Planetary Motion (FL), Internal Nucleosynthesis (IN) and Evolution (E).*

I believe I told you that Glenn and I have been corresponding for several years during which we exchanged information on current beliefs vs. facets of the paradigm. I was able to explain to him each of the relevant space discoveries within the scope of this new perspective. Apparently, he has enough information now to convince him of the validity of the concept. If he does print the full version, he will have the scoop of the century—maybe of several centuries—in the planetary sciences; perhaps the biggest breakthrough since the Copernicus-Kepler-Newton era of enlightenment. Certainly, it will open up whole new avenues of research in the scientific community. Things are finally reaching an exciting level—and just when discouragement was beginning to displace my enthusiasm for fitting all the new pieces into the concept as fast as they were discovered. It has become overwhelming.

Thanks for your interest and faith in the concept. It has helped a great deal.

To: Science Editor, Wall Street Journal, New York, N.Y. May 7, 1999

On the consecutive days of April 15 and 16, two exciting and highly significant news items were little noted, but will be long remembered. The initial news release via the AAS revealed the discovery of the first multiple-planet solar system other than our own. The second news item was headlined *It's No Crude Joke: This Oil Field Grows Even as it's Tapped* (*Wall Street Journal*, April 16, 1999). Strange as it seems, the two discoveries are intimately linked and that linkage, coincidentally, had been presented in a science meeting in Santa Fe on April 14 in my paper entitled *The FLINE Paradigm: Definitive Insights into the Origins of Solar Systems and the Orbital Spacing and Evolution of Planets*.

The news release about the distant solar system tells of three giant planets orbiting a star. The innermost planet contains at least three-quarters of the mass of Jupiter and orbits only 0.06 AU from the star. The middle planet contains at least twice the mass of Jupiter and takes 242 days to orbit the star once. It resides approximately 0.83 AU from the star, similar to the orbital distance of Venus. The outermost planet has a mass of at least four Jupiters and completes one orbit every 3.5 to 4 years, placing it 2.5 AU from the star. This information is powerful evidence against current beliefs about how solar systems form, and it will force scientists to alter their beliefs about such formations occurring from dust, gas, planetesimals, etc.

Conversely, the findings fit precisely into the FLINE model of planetary origins. The sizes, velocities, distances and gaseous nature of the three huge Jupiter-like planets account for their strange orbital positions and their being simultaneously in the second stage of the five common stages of planetary evolution. Although single and multiple-planet systems are abundant throughout the Universe, solar systems with nine planets that include an Earth-like planet and three distinct Asteroids belts like ours, should be very rare. This is assured by the strict prerequisites imposed by mathematics and natural laws on the manner in which these systems are created.

The exciting article about the oil field that grows was sent to me by Grady Traylor, a local geologist who is familiar with my struggles against the ludicrous fossil fuels hypothesis during the past 26 years. It describes how very hot oil is being forced under very high pressure up through the bottom of the oil field it is refilling—precisely as predicted and explained by the Energy Fuels Theory (EFT) (1973-1980). The hot petroleum, a polymerization product from virgin methane made deep within Earth, continually seeks the upward paths of least resistance in its journey towards the surface. This scenario, backed by a large amount of incontrovertible evidence, further exposes the obsoleteness of the FFH that originated in the 1830s, while adding more corroborating facts to those already in powerful support of the EFT.

So the three days in mid April can be remembered as moments in truth about the origins of planets --not from dust, gases, planetesimals, etc. —and about the origin of hydrocarbon fuels—not from fossils! The major significance? The FLINE model clears the path to understanding planetary origins and evolution, and it guarantees us a vast renewable supply of fuels (gas, petroleum, coal) for a very long time.

To: Dr. Morton Reed, 49 Post Oak Drive, LaGrange, Georgia 30240. May 11, 1999

Your letter steers me onto the fringes of the FLINE paradigm, so some answers to your questions must be mixed with speculation—which is not my forté. But I will try, even while knowing that speculation through the years has proven to be the most effective roadblock to progress in the planetary sciences.

I've always believed that Nature is, by far, the most dominant cause behind environmental changes, and that Earth's inhabitants can and do play minor roles in those ongoing changes. All species must either adapt to Nature's changes or die out, as so many of them (e.g., dinosaurs) have done in the eons past. With these ever-changing conditions—driven by the planet's energy core—millions of species come and go, and have done so since long before the arrival of humans. While mankind perhaps can alter the time frame for a given species, we can never make more than a dent in Nature's set-in-concrete plans for our evolving planet. I believe that carbon dioxide emissions fall into this category, with perhaps a 90-10 ratio (a pure guess) of Nature's production to that of its inhabitants (which per se is a large amount). The same things apply to global warming, which began long before humans arrived on the scene, although the ratio might change to perhaps 95-5 (approximately) in this case. But speculation usually is misleading.

The FLINE model tells us that gas, being the easiest hydrocarbon for Nature to make internally, always will be our most plentiful fuel, while petroleum will always be a distant second, with coal in a very distant third place.

Since coal is more limited and its composition is very valuable as a source for chemicals, medicines, dyes, etc., its use as a fuel is not a wise long-term choice.

Thus, Earth appears to have more than enough hydrocarbon fuels to get us well into the nuclear-fusion era of the future. I must agree that the future appears to reside in a hydrogen economy in conjunction with fusion reactors. The thermal pollution inherent with reactor usage could grow into a definite threat to aquatic life, and proper disposal of the radioactive waste, as you know, is a major problem. I do not have the answers, but I hope that someone does soon. In my opinion, electric cars, or a combination thereof, must and will arrive on the scene in the distant future. Like the FLINE model, their time has not yet arrived—but future acceptance of each of the two does appear inevitable.

Thanks for your interest and concern. Perhaps together we can make a significant dent in the armor of speculation, and add to the factual knowledge of science.

To: Dr. John Chappell, Jr., P.O. Box 14014, San Luis Obispo, CA 93406. May 19, 1999

Thanks for the NPA Newsletter of May 14, 1999. Throughout history, traditionalists have stood in powerful opposition to factual challenges against their speculative beliefs. I fear that as long as they control the news media and the budget money, no serious challenge to the status quo will be granted a fair hearing or permitted any degree of publicity. Whether we need to keep fighting or to face the inevitable does appear to be in question.

The only recourse remaining for me is to continue self-publications of the FLINE model as fast as the corroborating evidence pours in. This keeps me extremely busy, since every discovery pertaining to planets falls in this category, and my pleas for help continue to fall on deaf ears. So my next updated book is scheduled in January 2000.

Through the years, you have done a highly commendable job with the NPA and, more specifically with the Special Relativity Theory. If I were now in a position to do so, I would not hesitate to sponsor the NPA financially to keep it going. Perhaps someday soon, we might get a break that would make this possible. I hope so. I do not understand how Foundations, especially the NSF, can overlook factual concepts while pouring billions into the support of speculative beliefs that inevitably will be displaced.

By now you have probably realized that my half-day Symposium on the FLINE model of origins of solar systems and the orbital spacing and evolution of planets has been rejected by the AAAS. Herewith is a copy of the letter of rejection and my response to the flimsy (even pitiful) excuses of the peer reviewers. You can judge for yourself the quality, or lack thereof, of peer reviewers used to evaluate these pioneering efforts. Obviously, they were unable to look beyond their own bias toward prevailing beliefs in which they were schooled.

Thanks again for what you have done for all of us and for science in general. Some day it will become more recognized.

P.S. Since the Santa Fe Meeting, I have gained three influential allies here in Georgia. So there is hope.

To: Dr. Carlyle B. Storm, Director, Gordon Research Conferences, University of R.I. June 2, 1999

I am deeply concerned that for the past number of years my annual application to the Gordon Research Conference on Origins of Solar Systems has been rejected. The latest rejection was received on June 1, 1999.

Since my last letter of appeal of March 24 to you, some highly significant events crucial to your next meeting have occurred. Herewith is a news release concerning an oil well that grows even as it's tapped (*WSJ*, April 16) and the discovery of the first multiple-planet solar system other than our own (AAS, April 15). In all probability, you are aware of both events, but perhaps unaware that both were predicted and are explained factually by the revolutionary FLINE model of solar system origins.

The crucial link between these predicted findings was presented in a science meeting in Santa Fe on April 14 in my paper, *The FLINE Paradigm: Definitive Insights into the Origins of Solar Systems and the Orbital Spacing and Evolution of Planets* (1973-1999), featuring the solution to the enigmatic Fourth Law of Planetary Motion (1980-1995) explaining how the planets attained their spacing around the Sun.

The historical and scientific significance of the new model of origins and evolution would be difficult to overstate. Each and every relevant discovery about Earth and of the space probes continually serves as additional corroborating evidence. There now is little, if any, room to doubt that this most revolutionary concept of our Solar System since Copernicus will displace the current model early in the 21st century when the true natures of comets and asteroids are confirmed by the scheduled space probes.

Two of the most crucial subjects for critical examination at your meeting should be: (1) the proposed Fourth Law of Planetary Motion, a definitive statement on the enigmatic spacing of planets; and (2) the manner in which planets evolve through five common stages of evolution. Understanding these two anomalies will lead to understanding much about the recently discovered multiple-planet system and why the oil well grows even as it's tapped.

The FLINE model inevitably will prove of tremendous value to scientists in understanding every anomaly of solar systems and the evolution of their planets, moons, comets, etc. Why not let it all begin at your next meeting on Origins of Solar Systems? In a full and open discussion someone might finally be able to detect a fatal flaw in this revolutionary concept—which, of course, does not appear likely. Am I mistaken in believing that research conferences exist primarily for the purpose of exploring new ideas, especially those based soundly on facts?

To: Gordon Rehberg, Dalton, Georgia 30720. **June 3, 1999**

Thanks for the two interesting articles from the *News Weekly*; I enjoyed reading them. I'm not much into the financial, political and business arenas for fear of overloading my mind at age 76. But that may have already happened.

The feedback of questions was even more interesting. I will try to answer them as briefly as feasible.

(1) You are correct in your statements about water being evidence as strong as that of gases for the IN concept; both are among the simplest and easiest compounds made via IN. We can go a step further by saying that all planetary matter is equally powerful evidence of IN, the transformation of energy into atomic elements comprising all such material—the smaller the element, the easier to make. All planetary matter is continually being produced via IN and polymerization, which accounts for all volcanic actions through the eons and for the constant cracking of crust as Earth very gradually grows in size via the accumulation of the greatly expanded volume of the virgin material (relative to the volume of its original compacted energy). The book's section on planetary quakes gives more details on how all of this ties in with earthquakes and volcanoes.

(2) The principles of the IN concept dictate that petroleum will be distributed throughout Earth's crust in both the northern and southern hemispheres. Economics has dictated that the northern latitudes—in which the vast majority of oil has been consumed—will be the leading source for this essential product. There is no reason I know of to suspect that essentially the same amount of gas and oil found above the equator does not exist in the southern hemisphere. "Why explore so far away, when so much supply is so close to the consumers?" may be a logical question.

(3) The energy core within each active planet is one source of the sea of electrons in which we live. During a thunderstorm, the atmospheric electrons are attracted to the positive charges in clouds at the moment that a small jet of electrons emanate from the ground and escape upward through either the highest point [or the route(s) of least resistance]. In these cases, the two bolts meet a relatively short distance above the ground to become a single or branched bolt of lightening. This happens instantaneously, thus giving the impression of a bolt from the sky. The branched flashes confirm that atmospheric electrons can be attracted into the instantaneous discharge of the primary flash; the branches usually do not touch ground as they coagulate in the atmosphere and discharge into clouds. The fact that the two bolts meet above ground has been confirmed by scientists working on grants (as seen in TV documentaries). The presence of buried fulgurites (paths fused by lightning) is confirmation that electrons do exit from Earth's crust, while the more recent discovery of energy sprites high above the cloud-tops confirms the upward momentum of the energy that left the ground and the lower atmosphere in such a hurry.

As to tornadoes, the same basic principles seem to apply. The positive charges in low-lying clouds sometimes are low enough to the ground to attract a swarm of spinning electrons emanating from the ground. (Small examples of the latter often can be observed as small whirlwinds of dust moving across an open field.) Again, the two meet at a distance above the ground, at times forming a spinning mass with a small or large diameter: a tornado. Lightning discharges (usually or always?) can be seen within the swirling mass, attesting to the similarity of the two subjects in question. I strongly suspect that the driving force behind the swirling mass is electrons-driven, although I have no proof positive. But close observation of tornadoes shows them spinning rapidly only at the bottom while the upper part remains dormant (at least that's the way they look on TV, the only place I've seen them). I also suspect that the final answers to the causes reside within the parameters of the conditions of the clouds at the time they initiate the contact with the spinning mass of electrons emanating from the crust.

(4) It is my understanding that drilling into the immense storehouse of oil and natural gas beneath Los Angeles is prohibited because of safety and environmental reasons, along with the fact that the world's supply is sufficient without taking those risks.

(5-8) While I am unaware of the Bilderberger Society, I do know that some groups would be adversely affected by this knowledge of planetary origins and the Internal Nucleosynthesis that drives Evolution. Among those groups, I've been told and also suspected, are scientists and oil companies who already benefit under the current, but erroneous, beliefs. It seems natural that those persons who would be adversely affected by any change would attempt to prevent that change. So I must agree with you that "as long as these organizations you are appealing to for recognition and review have strong financial ties to government money for their existence, the truth will be hidden from view." An example of this can be seen in my annual letter mailed yesterday to the Gordon Research Conference on *Origins of Solar Systems.* Can you imagine my work being rejected year after year by a Research Conference on the subject it covers so factually? Rocking the boat is strictly forbidden. Do not confuse us with facts. We prefer the status quo and further speculation and the supporting grant money; the taxpayers don't really mind. Why would anyone, if seriously interested in solar system origins, continually refuse to examine the proposed Fourth Law of Planetary Motion and the substantiated FLINE model it supports, if only to try to find a fatal flaw in them?

While writing this, I received a call from a friend, William Mitchell, Carson City, NV, who is busy writing another book against the Big Bang model. He was excited about an article in the June issue of *Discover* magazine that tells of the vast supplies of oil in Earth's crust. While the article does corroborate my writings of the past 26 years, it stops short of the Internal Nucleosynthesis concept of petroleum origins. Bill is aware of the powerful evidence against the Big Bang that resides within the FLINE model, and probably will utilize it in his new book, but I did not question him on this point. I'm sending him a copy of the article, *It's No Crude Joke: This Oil Field Grows Even as it's Tapped (WSJ,* April 16, 1999). As you know, it is powerful verification of my concept as first described in *Fuels: A New Theory* (1975). Whether this article finally will change the years of laughter, doubt and apathy into a serious approach to in-depth examinations of the many facets of the FLINE model remains to be seen.

Thanks again for all of your help. Have a good trip to NJ.

To: Gordon Rehberg, Dalton, Georgia 30720. **June 29, 1999**

Thanks for the exciting article, *Using the Phi Relationship for a New Propeller Design (Professional Pilot,* May 1999, pp 33-36), by John L. Petersen, President, The Arlington Institute. Jayden Harman, the inventor of the recessive impellers that produce no cavitation, and can spin 6000 rpm and produce five times more thrust, should be well rewarded for this major breakthrough that puts Nature's Golden Ratio (GR) to great practical use. His thought processes must have closely followed mine when solving the Fourth Law of Planetary Motion (FL) via use of the same Phi mathematics.

Of course, his concept will be readily accepted because it is easy to demonstrate and to understand. However, if scientists do become willing to take the time to learn the FL solution, it will become as easy to understand as Harman's propeller—and relatively easy to prove. But first, they must have the will to examine it with open minds—and that is proving to be far more difficult than with Harman's invention.

One of the key points in these breakthroughs via Nature's Phi geometry is the fact that they are only two among thousands of potentially practical applications waiting for someone to discover. Maybe these two solutions will be an inspiration for other problem solvers to utilize the geometry that pervades throughout the Universe. One example is the *Image of a Planet: Too Hot to be True? (Science News,* June 26, 1999) (Copy enclosed). Believe it or not, on the assumption that the photo was correctly aligned, and suspecting that Phi was involved, I measured the angle from the center of the large, bright mass to the center of the smaller mass; it measures precisely 26 ½°, one of the first critical angles in Phi geometry that I learned about during my initial trials in 1980. Along with other data from the past, I feel safe in concluding that the smaller mass is indeed a planet in the first (energy) stage of evolution. As was revealed in Comet Halley photos (1986-87), such fiery masses do indeed have the spectrum and pattern identical to that of the background stars.

While I finally and correctly did guess that the GR had to be the key to the origin of our Solar System, the solution proved elusive for some 15 years of off-and-on trying. But now everything checks, double-checks and cross-checks, just as is taught in the Georgia Tech Engineering classes. While I can find no flaw in it, there is some comfort in realizing that as yet, neither can (or has) anyone else.

I must agree with your statement: "I am constantly amazed that in this world there are countless people professing objectivity and a willingness to consider new ideas, but will do anything and everything to suppress information counter to their own preconceived notions." It complements Max Planck's belief that science advances one funeral at a time.

Thanks for all the interesting information and for your great interest in new ideas.

Enclosures: (1) History and predictions of the FLINE model of planetary origins/evolution.
(2) *Science News* article.

To: Gordon Rehberg, Dalton, Georgia 30720. **July 2, 1999**

Thanks for the five science articles sent with your letter of June 28th. All were informative and interesting to read. However, since they all were written in the perspectives of the Big Bang (BB) and the Accretion hypotheses, most of the interpretations of data were erroneous, and at times amusing. Their findings make a lot more sense when interpreted correctly in the perspectives of the Little Bangs Theory (LBT) and the FLINE model of planetary origins and evolution. I will comment briefly on each article.

In *Tidal Forces May Cause Europa Cycloids,* Hoppa and Tufts did an excellent job of explaining how the cycloidal tracks on the surface of Europa formed and are forming. I agree with their findings, which are based on the fact of a subsurface layer of water encircling the globe. As you know, the FLINE model explains how the vast amount of water was (and is) created and forced to the surface to form these ever-changing cracks; only in this respect do the two beliefs differ: water does not come from outer space. And unfortunately, the author still clings to the belief that Europa's core is metallic and rock, rather than an energy core that produces the water and all the other planetary matter of Europa.

The photos in *More Than Pretty Pictures* speak more in favor of the LBT than the BB; however, since no attempt to interpret them was made, no other comment is in order. Goldin's quote in *Astrobiology Team Taking Shape at Ames* is a classic: [The objective is] to "bring a new understanding of fundamental life processes on Earth and throughout our Universe, *if it exists.* [It does exist.] It is a revolution that *will require its own revolution...in scientific thinking"*. Comment: To accomplish this, scientists must first recognize that Internal Nucleosynthesis (IN) is the essential driving force behind evolution (E): No IN, no E; the two are inseparable in all active spheres. When and if this is accomplished, the BB and Accretion concepts will be automatically and hopelessly undermined. In any case, let's hope, even if in vain, that the $10MM to $100 MM annual budgets will be spent wisely.

The last two articles, *Satellite Seeks Clues to Birth of Universe* and *What is FUSE?,* discuss facets of the same subject. Here again the FUSE program will attempt to understand its findings strictly in the BB perspective; e.g., "One of the key objectives of FUSE's three-year, $190-million mission will be to better understand the conditions that prevailed during the universe's formative stages" of the BB. This will be done by measuring the amount of deuterium in existence, thereby enabling astronomers to work backward to calculate how much deuterium was present during the BB! The absurdity is obvious to those who realize that the BB never happened and that all elements have been (and still are being) produced via IN in all active spheres. Opinion: the BB concept, structured entirely with speculations based on false assumptions, is the grandest scientific hoax ever perpetrated on an unsuspecting public.

I ran across the amusing article, *Life's Far Flung Raw Materials (Scientific American,* July 1999, pp.42-49), describing how the raw materials for Earth's life came from outer space. Throughout the past few decades, too many of Earth's anomalies have been "solved" by simply saying that "it came from outer space". Life on any planet where it exists is there because the conditions were and are suitable for it to evolve. To accept the belief that gas, petroleum, water, life, etc., came from outer space is a cop-out. The Accretion hypothesis is absurd. Yet the three future comet missions are designed to collect data revealing the connection between comets and how planets were made therefrom. However, upon completion, the results, if correctly interpreted, will show no connection to our origins—other than the damage comets can inflict on planets. The 21 powerful nuclear explosions of Comet SL9 on Jupiter in 1994 are examples. Perhaps the time and money spent on the three missions will not be a total waste.

To: John E. Chappell, Jr., Director, NPA, San Luis Obispo, CA 93406. **December 14, 1999**

Thanks for the November Newsletter and the NPA Directory-Bibliography dated October 1999. I was pleased to see that the membership continues to grow at a faster pace. Objectivity and tolerance in science are excellent and attractive goals that are attainable.

I will try to attend the next Meeting in June, but cannot be certain at this time. In any case, I will submit a paper on the new FLINE model of planetary origins and evolution, which continually builds a solid structure with each discovery about Earth, moons and other planets, including the 28 known exo-planets.

Factually based and without the need for speculation, the revolutionary FLINE paradigm of planetary origins and evolution is structured on three inseparable and indisputable principles of Nature: the Four Laws of Planetary Motion (FL) and the Internal Nucleosynthesis (IN) that drives Evolution (E) in all active spheres of the Universe. Imbued with these qualifications, the FLINE model appears destined to become one of history's most significant and crucial scientific breakthroughs in the planetary sciences.

By firmly interlocking the great discoveries of Copernicus, Galileo, Kepler, Newton, Descartes, Dutton, Einstein and many others, this new model stands alone in bringing their significance full circle. Other known models neither establish these critical connections, nor are capable of explaining every planetary anomaly, including those of the exo-planets. Scientists will find that every discovery—past, present, and future—has its taproot deeply embedded in the principles of the FLINE model. It is the only model which every facet can, and will, be proven beyond doubt, and the only model in which no fundamental flaw has been, or can, be found. In its capability for explaining every anomaly of solar systems, the new model stands alone and shines a powerful spotlight on the absurdities of the Accretion concept of planetary origins—an expose of immense implications. In the opinion of a number of advocates and on the basis of its substantial and substantive evidence, *The Spacing of Planets: The Solution to a 400-Year Mystery,* featuring the Fourth Law of Planetary Motion, will rank high among the most significant books in the history of discoveries about planetary systems. As such, may I suggest that it be added to our list of especially important books? Also, I believe Mitchell's book on the Big Bang, although still ahead of its time should be on the list. Both books expose the absurdities of prevailing beliefs. Both seem ideal for work-type discussion sessions. If possible, I would like to speak and answer questions in one session.

Why scientists cling to antiquated ideas instead of exploring more promising concepts remains the greatest mystery of all. Why spend billions of dollars in vain efforts to prove and promote absurd concepts, and not one penny to explore these definitive ideas? What has happened to true science? Has it evolved into the pseudoscience to which true scientists are adamantly opposed? Why are speculative writings revered over more factual articles? What is wrong with our peer review system that permits this farce to continue year after year? Will the situation ever improve other than "by one funeral at a time"?

Herewith is some additional information on the subject. We appreciate your tireless efforts in the NPA.

AAAS-SWARM-NPA 2008 MEETING

University of New Mexico

Alexander Alan Scarborough

April 7-11, 2008

Abstract: 20 questions about our universe are posed, and 20 definitive answers are presented in accordance with the new LB/FLINE model. Some items discussed include: what we know and don't know, and may never know; the intimate connection between $E=mc^2$ and all universal spheres; and the role of black holes. The conclusion is drawn that the answers can replace many things that are utilized in the Big Bang concept that do not seem to be definitive, provable or believable.

AAAS-SWARM-NPA—Paper #2
20 Questions with Short, Definitive Answers to
The Mysteries of Origins and Evolution of Universal Systems
April 7-11, 2008

1. Will we ever know the true size of our Universe?
   No; we may know only the true size of the visible Universe. Due to limitations of the speed of light, we may never know the true size of the invisible Universe that lies, perhaps, very far beyond the visible part.

2. Are assumptions of finite universal space enveloped in infinite outer space (at absolute zero) reasonable?
   Yes, according to the LB/FLINE model of universal origins that visualizes a finite Universe enveloped in infinite outer space, dark and at absolute-zero temperature where all atomic motion ceases. The expanding interface of the two spaces is the source of space-time and the inexhaustible source of universal energy in the form of massive black holes (spheres), each of which later spews out a galaxy of brilliant stars from its two poles.

3. What don't we know, and will never know about the Universe?
   We don't know whether the Universe had a beginning or whether it will ever end. We do know that our Milky Way galaxy appears to be at the center of our Universe, an optical illusion. But the illusion of being at the center will hold true for any other galaxy when viewed from that galaxy.

4. Why is the Universe expanding at an accelerating rate?
   The massive momentum at around 99% of c speed and the massive gravities of black holes (spheres, the densest form of energy) that form continuously at the perimeter of our spherical Universe account for its increases in the rapid expansion rate observed as astronomers look ever farther outward in all directions.

5. What is the relationship among universal space, the Outer Space that envelops our Universe, and space-time?
   The curvature of space (and space-time) is simply the spherical shape of our Universe as it expands at c speed like a bubble into infinite space. Results of these interfacial collisions are black holes (spheres), the densest form of energy: the source that supplies all energy necessary to run universal systems, perhaps forever, according to the LB/FLINE model.

6. What does the accelerating rate of expansion tell us about universal origins?
   The accelerating rate of expansion of the Universe is powerful evidence of the ongoing manner in which our Universe functions as a self-sustaining entity now, and perhaps forever, with its endless supply of energy: a powerful argument against the Big Bang.

7. Why does it appear to be a static Universe?
   The optical illusion of being at the center of the Universe (see question 4), and all galaxies moving almost

directly away from the viewer gives the illusion of a static Universe.

8. Why does Earth appear to be at the center of the Universe?
   Earth appears to be at the center of the Universe because of the optical illusions in questions 4 and 7; i.e., no matter which direction astronomers look, the view is the same: objects appear to be all moving away in like manner.

9. Is the Universe spherical, and if so, why?
   If indeed the Universe is expanding in all directions at c speed away from Earth, and into the absolute-zero temperature and vacuum of Outer Space, sound logic dictates its spherical shape. Universal systems would be unable to function, or even exist, without the constant supply of energy at its spherical interface with the dark outer space that envelops it.

10. What are black holes, and what key role do they play in origins of universal systems?
    Black holes (a.k.a. spheres) are the densest form of energy, forming at the perimeter of the Universe expanding at c speed. Initially at c speed, their massive momentum and massive gravities account for the accelerating rate of expansion of the Universe. Eventually the spinning spheres are triggered into spewing from their two poles brilliant white masses of energy (a.k.a. stars) embedded in hot gaseous dust-clouds: beautiful pinwheel galaxies.

11. What processes are continuously producing billions of galaxies, each with billions of stars of a variety of sizes?
    These galaxy-forming processes are continuously producing billions of galaxies, each with billions of stars of a variety of sizes embedded in hot gaseous dust-clouds.

12. How are solar systems created in a wide variety of mysterious formations?
    Many of the smaller masses of energy in galaxies are captured in orbits around a larger mass in a wide variety of formations known as solar systems. Eventually, they evolve into planets through four additional stages of evolution (gaseous, transitional, rocky (e.g., Earth), and inactive, all at rates proportional to size.

13. What is the connection between $E=mc^2$ and all universal spheres?
    The four stages of planetary evolution (E) are driven by Internal Nucleosynthesis (IN, the transformation from energy into atomic matter) in accord with $E=mc^2$. Every sphere is a self-sustaining entity that creates all its compositional matter.

14. Why do galaxies generally form with spiral arms in curvature patterns?
    The beautiful curvature of galactic spiral arms is attributed to the discovery that all ejected masses (stars) continue to orbit the central mass (black hole remnant) at the same speed. The progressively greater distance traveled by the progressively farthest stars accounts for their falling behind in curvature fashion.

15. Why and how do planets evolve through five stages of evolution?
    The five laws of planetary motion, along with definitive evidence from other solar systems, clearly reveal

that all planets began as fiery masses of energy, and now are undergoing four more stages of evolution at rates in accord with mass (size). The second stage is gaseous; Jupiter is a prime example. The third stage is transitional, as best illustrated by Uranus. Earth is the best example of a planet in the fourth (rocky) stage of evolution. Mercury and our Moon are good examples of the fifth (inactive, dead) stage of spherical evolution. All planetary evolution is driven by Internal Nucleosynthesis: the transformation from energy into atomic matter that later polymerizes to the molecular matter comprising all universal systems.

16. What are the fundamental reasons for origin and extinction of species?
    During the stages of their evolution, planets undergo a large variety of environmental changes, particularly during the fourth (rocky) stage Earth is experiencing. If favorably located in the right orbit, a planet can, and will, produce millions of species that adapt to prevailing conditions. When conditions change over time, species that cannot adapt soon become extinct; e.g., dinosaurs are the best-known example.

17. What is the most impossible aspect of the Big Bang?
    Why and how is it remotely possible to pack all universal matter into a tiny mass before a Big Bang?

18. Could a Big Bang begin in the same way the Little Bangs began?
    A Big Bang may be quite possible as the first phase of the LB/FLINE model, perhaps as simple as triggering it with a small amount of light in the dark, absolute-zero coldness of the vacuum of Outer Space.

19. How may the Big Bang and Little Bangs be merged into a definitive concept?
    After a Big Bang, the principles of the new model would take over from a Big Bang to keep the Universe growing and functioning in accord with its basic principles of natural laws.

20. Is the LB/FLINE model a viable alternative to the Big Bang/Accretion hypotheses?
    Yes! Based on sound logic and overwhelmingly persuasive evidence, the LB/FLINE model is a definitive alternative to the Big Bang/Accretion hypotheses, and does appear to be irrefutable.

The LB/FLINE model is in tune with laws involving complicated transformations, called the Lorentz symmetries, in which space and time are inseparable. The main action occurs at the perimeter of a finite Universe expanding symmetrically at c rate while retaining its perfect spherical shape. Here the exiting energy/ matter collide with the dormant energies at absolute-zero temperature in the cold black darkness of Outer Space, seeding billions of small black spheres that continue their growth into huge black masses of the densest form of energy. It is here that the Lorentz transformation equations come into play, giving the relations between space and time coordinates of a single event as measured by observers.

AAAS-SWARM-NPA 2008 MEETING

University of New Mexico

Alexander Alan Scarborough
April 7-11, 2008

Abstract: Unification of the Big Bang and the LB/FLINE Model of Origins of Universal Systems.

Due to lack of a viable alternative, the unsubstantiated BB concept has become well-entrenched in scientific literature. While most scientists recognize its mythical (Sagan) and conjectural (Hawking) status, they are forced to interpret their amazing discoveries in the BB perspective. Since myth begets myths, futility is too often the reason that science remains stymied in spite of such brilliance. However, most, if not all, discoveries do fit precisely into the revolutionary LB/FLINE model of origins of universal systems. This paper presents a viable alternative that offers a way out of the maze. Altering the initial stage of the BB via elimination of the impossibility of all universal energy/matter being contained in a very small mass will enable science to get back on the right train of thought. Once past the initial stage of the BB inflation, the LB/FLINE will take control via its processes of creating black spheres of space-time, the densest form of energy, the source of all galaxies, each containing billions of stars embedded in hot gaseous dust-clouds. The smaller stars eventually evolve into planets of various sizes through four readily observable stages of evolution via Internal Nucleosynthesis. The LB/FLINE model is soundly based on the five laws of planetary motion that offer substantial evidence for definitive solutions to planetary and, eventually, all universal anomalies.

AAAS-SWARM-NPA Paper #1
The Proposed BB/LB/FLINE Concept:

Unifying a Modified BB with the LB/FLINE Model of Universal Origins and Evolution

April 7-11, 2008

Most scientists are aware of the shaky foundation of the Big Bang that begins with an impossibility, and is subsequently sustained step-by-step by other impossibilities, each with a WOW factor—a beautiful fairy tale that captures imaginations at each impasse in a fashion similar to the magic of *Alice in Wonderland*. The BB is indeed a captivating myth—the grandest and costliest in the history of science, and one that can, and must, be modified into a valid scientific concept without a mythical aura. But how can this be accomplished?

To begin the unification, we must recognize the impossibility of packing all universal matter, or even a minute fraction of it, into a tiny mass (according to Marilyn vos Savant and common sense) that explodes with a big bang, releasing all present-day universal energy and matter. This concept provides no logical way of explaining definitively the accelerating rate of expansion of the Universe, or consequently the ongoing generation of universal space-time. Thus, it does not provide a logical source for the unimaginable and forever-ongoing supply of new energy necessary for universal expansion, momentum and functioning of its many systems.

As it stands today, the BB is incapable of explaining even the five laws of planetary motion that govern the origins, planetary spacing and functioning of solar systems. Neither does it offer any knowledge that can lead to definitive solutions to the anomalies of all solar systems. In the BB perspective, scientists will remain unable to decipher definitively the dynamic formation and history of our Solar System, or to predict its future.

In the LB/FLINE model, none of the many BB problems present an obstacle to understanding the origins and functions of universal systems. When we retrace our steps, and modify the BB, we get some definitive answers and a viable connection to the LB/FLINE model.

In the beginning, in the black void of Infinite Space at absolute-zero temperature, a significant spark of light is all that's needed to set off an instantaneous expansion at the speed of light in all directions, forming a spherically-shaped miniature Universe already expanding at the same rate (c speed) it does today, but containing only the white energy of light—no other matter or energy is present at this point.

At the interface of the perimeter of the expanding sphere of light and the absolute-zero coldness of the dark Infinite Space beyond—and where all atomic motion ceases—strange things begin to happen. Many tiny black spheres of extremely dense energy begin forming while continuing their rapid momentum into Infinite Space, increasing in mass as functions of time by generating the space-time energy that later serves as seeds for galaxies.

These processes of generating universal space-time in the form of dense masses of superstrings energy (black spheres) are the initial functions of all new-born universes, according to the LB/FLINE model of ongoing origins of universal systems.

Eventually, each massive, spinning black sphere of dense energy is triggered into spewing from its two poles billions of white-hot masses (stars) of all sizes embedded in hot gaseous dust-clouds. The results are billions of beautiful spiral galaxies, each containing a multitude of solar systems, usually with one or more smaller masses orbiting a larger central mass, all in tune with the laws of planetary motion.

The smaller, orbiting masses eventually evolve into planets going through five easily detectable stages of evolution: from energy to gaseous, to transitional, to rocky, to inactive, as functions of mass and time, all via Internal Nucleosynthesis.

Earth is about midway in the fourth (rocky) stage, while Mars is progressing (or has already moved) into the fifth (inactive) stage. Pluto, the original tenth planet, is now the ninth planet, but still in the tenth orbit in our Solar System—an event that happened whenever the Asteroids planet(s) disintegrated (H. Olbers terminology, 1802). Because of its tiny size, comparatively speaking, it has evolved quickly through the five stages of evolution; it, too, is in the fifth (inactive) stage.

The plethora of substantiated evidence (1973-2007) heavily favoring the LB/FLINE model leaves little, if any, doubt about its scientific validity. However, while this modified spark-of-light version of the BB origin might never be substantiated, it is a good starting point for the revolutionary model of universal origins. It supplies the continuous source of unimaginable quantities of energy essential for all functions of the Universe—perhaps forever.

With this unified model, no anomaly—planetary and universal—appears capable of eluding a valid scientific solution. Further, it gives us solid assurance that planets and solar systems are ubiquitous in a large variety of combinations and sizes in each and every one of the billions of brilliant galaxies that beautify our night skies.

Addendum: Black Holes or Black Spheres? (Nov 1, 2007)

In *The Discover Interview* with Kip Thorne *(Discover*, Oct 2007, pp 51-53), the legendary astrophysicist explains what black holes are made of: space-time.

Thorne states: "A big misconception is that a black hole is made of matter that has just been compacted to a very small size. That's not true. A black hole is made from warped space and time."

This information is exciting, because it is in precise agreement with the LB/FLINE model (1973-1980) in which space-time is the forever-ongoing source of infinite energy that continuously runs the ever-expanding spherical Universe. In reality, black holes must be black spheres of the densest form of energy, continually forming by the billions at the perimeter of a Universe expanding at the speed of light.

In time, when triggered by warmed compression, the spinning masses give births to beautiful galaxies, usually in pin-wheel fashion: billions of white-hot masses (stars of all sizes) embedded in hot gaseous dust-clouds.

The smaller white-hot masses eventually evolve into self-sustaining entities that undergo the five stages of planetary evolution via Internal Nucleosynthesis.

Thus, our self-sustaining Universe is designed to run forever, continuously spawning black holes (spheres):

space-time energy that drives all universal systems. As expansion continues creating distant subdivisions, Earth's subdivision (Milky Way) will eventually die of old age as it consumes all its energy via nucleosynthesis into matter, and turns cold in death.

A Time for Change

A quarter-century of painstaking research into the origins of solar systems and the evolution of planets has failed to uncover sufficient substantiated evidence in support of the Big Bang (BB). However, the research did uncover sufficient substantiated evidence against the BB and in powerful support of an ongoing Little Bangs (LB) concept of creation. In these contrasting concepts, the LB remains steadfast in its capability to explain the stream of enigmas encountered in the BB concept.

The LB, supported by the new Fourth and Fifth Laws of Planetary Motion and Kepler's First Three Laws, presents powerful arguments against the concepts of planetary accretion, cores of iron, rock or silicate in active planets, primordial asteroids, "snowball" comets, BB nucleosynthesis, a cosmological constant, finite energy and "fossil" fuels. Each of these antiquated theories is structured with speculation and misinterpretations, with no substantiated facts. Each is fundamentally flawed, a house of cards destined to crumble in light of the LB and its FLINE model's three intimate and inviolate principles: the Five Laws of Planetary Motion (FL) and the ongoing Internal Nucleosynthesis (IN) that continually drives all Evolution (E).

In contrast to the BB, the LB and the revolutionary FLINE model offer viable alternatives that factually circumvent fundamental flaws and give assurance that current antiquated concepts cannot long endure. For example, the recent discovery that our Universe is expanding at an ever faster rate as astronomers look ever farther into space is very powerful evidence against the BB, which predicted just the opposite: the ever slowing rate of expansion at ever greater distances. Simultaneously, this crucial discovery—an inherent principle of the LB concept—is powerful substantiated evidence for the ongoing LB theory that accurately predicted and explains why the expansion rate increases with distance. No cosmological constant is necessary.

The recent exciting discoveries of two groups of young planet-sized objects unattached to any mother stars, along with the 50 known extra-solar systems, were predicted and explained by the LB as early as 1980. Perhaps recognition of the LB/FLINE concept will come when scientists learn the true nature and sources of asteroids and comets in the first decade of this century.

But the path will not be easy. Obviously, the strong entrenchment of the prevailing Accretion Disk Theory (the dust aggregation/planetesimals/accretion hypothesis) in the scientific literature presents one of history's most formidable barriers to change in direction of scientific thought. However, history does teach that unsatisfactory concepts are replaced eventually by more satisfactory ones.

CONCLUSION

The LB-FLINE-BEC Model of Universal Origins

June, 2008

Abstract: The LB-FLINE-BEC (Little Bangs/Five Laws/Internal Nucleosynthesis/Evolution/Bose-Einstein-Condensation) model of universal origins had its beginning in 1973 via *Fuels: A New Theory*. Realization and substantiation of the intimate relationship of the hydrocarbon fuels (gas, oil, coal) led to the ongoing source of Earth's atomic and molecular matter: a nuclear energy core, later found to be in agreement with Descartes' belief (1644) in Earth's "Sun-like" core. This led to a definitive understanding of the five stages of planetary

Evolution (E) driven by Internal Nucleosynthesis (IN) at rates in accordance with mass. Recognition of the fiery origin of our Solar System, and subsequent discovery of the Fourth and Fifth Laws of Planetary Motion (FL), provided clues that led to definitive evidence of galactic formations spewed (Little Bangs, LB) from Black Holes (a.k.a. Black Energy Spheres, BES), the densest form of energy, previously formed at the perimeter of our spherical Universe expanding at c speed into the black, absolute-zero-temperature Void enveloping it. The conditions here are perfect for creating space-time energy via the Bose-Einstein-Condensation (BEC). This final link in the LB-FLINE-BEC model of ongoing universal origins provides, and explains, the steady source of energy essential for driving forever the ongoing functions of all universal systems: from *Void to Energy to Universe* – coincidentally, the title of the unpublished 1980 manuscript in the author's ten-edition Energy Series (1975-2008). The updated model offers unlimited opportunities for scientists to find, as Einstein predicted, "... the right answers...simple and beautiful" for the Theory of Everything (TOE).

Bibliography

* Albert Einstein, **The Meaning of Relativity**, Fifth Edition (1953). Princeton University Press.
* Thomas Kuhn, **The Structure of Scientific Revolutions**, (1956).
* Alexander Scarborough, **Fuels: A New Theory** (First Edition, 1975), Ander Publications, LaGrange, Georgia 30241. Presented at annual meetings of AAAS, AGU, GSA, Georgia Academy of Sciences, (1982-1999).
* Alexander Scarborough. **Undermining the Energy Crisis,** (1977, 1979), Ander Publications, LaGrange, Georgia 30241.
* Colin Wilson, **Starseekers**, (1980). Doubleday & Co., Bellew & Higton Publishers, London, W1P6JD.
* Alexander Scarborough, **Evolution of Gas, Oil and Coal** (1975-1983), **Alternative Energy Sources VI, Vol. 3**, University of Miami, Edited by T. Nejat Veziroglui, 1983. Hemisphere Publishing Corp., Washington, New York, London.
* R. A. Karam, J. M. Wampler, Neely Nuclear Research Center, Georgia Institute of Technology, Atlanta, Georgia. **Characterization of Natural Gas, Oil, and Coal Deposits with Regard to Distribution Depth and Trace Elements Content**, (1986).
* Alexander Scarborough, **New Concepts of Origins: With White Fire Laden** (1996), 6th Ed, Energy Series, Ander Publications, LaGrange, GA 30241, USA.
* Cynthia Whitney, **Distribution of Stars as Test Particles in a Two-Body Background Field** (1988), **Hadronic Journal**, Nonantum, MA 02195, USA.
* Donald K. Yeomans, **COMETS: A Chronological History of Observations, Science, Myth, and Folklore**, (1991). John Wiley, NY.
* R. G. Lemer & G. L. Trigg, **Encyclopedia of Physics** (1991), Second Edition, VCH Publishers, NY.
* Alexander A. Scarborough, **The I-T-E-M Connection: How Planet Earth and its Systems Were Made by Means of Natural Laws,** (1991), 7th Edition, Energy Series, Ander Publications, 202 View Pointe Lane, LaGrange, GA 30241.
*Nigel Henbest, **The Planets** (1992). Published by the Penguin Group, Viking.
*N.S. Hetherington, **Encyclopedia of Cosmology** (1993), Garland Publishing, NY & London.
* William C. Mitchell, **The Cult of the Big Bang: Was There a Bang?** (1995), **Common Sense Books**, Carson City, Nevada.
*Cynthia Whitney, ***How Can Spirals Persist?*** **Astrophysics and Space Science** 227:175-186, 1995.
* Alexander A. Scarborough, **The Spacing of Planets: The Solution to a 400-Year Mystery**, (1996), 8th Edition, Energy Series Ander Publications, LaGrange, GA 30241.
* Charles W. Lucas, **A New Foundation for Modem Physics**, (1998). **Common Sense Science**, Foundations of Science, Nov 2001.
* Alexander A. Scarborough, **New Principles of Origins and Evolution: Revolutionary Paradigms of Beauty, Power and Precision,** (2002), 9th Edition, Energy Series. Ander Publications, LaGrange, GA 30241.
* Alexander A. Scarborough, **Bringing the Copernican Revolution Full Circle via a New Fourth Law of Planetary Motion**, **Journal of New Energy**, Fall 2003. Proceedings of the NPA Conference, June 9-13, 2003.

* Robert J. Heaston, **The Characterization of Gravitational Collapse as a Mass-Energy Phase Change**. Proceedings of the Natural Philosophy Alliance, ISSN 1555-4775, Vol. 1, No.1, pp. 33-42 (Spring 2004).

* Alexander A. Scarborough, **A New Fourth Law of Planetary Motion**, Proceedings of the NPA, Vol. 1, No. 1, Spring 2004.

* Robert J. Heaston and Peter Marquardt, **The Constant Gravitational Potential of Light - Part 1: Theory**. Presented at joint AAAS-SWARM-NPA Conference, University of Tulsa, Tulsa, OK (3-7 April 2006).

* Alexander Scarborough, **From Myth to Reality in 12 Giant Steps and 2500 Years,** John E. Chappell Memorial Lecture, Presented at joint AAAS-SWARM-NPA Conference, University of Tulsa, Tulsa, OK (3-7 April 2006).

* Wm. L. Hughes, **The Electromagnetic Nature of Things,** (Not Dated). Dakota Alpha Press, Rapid City, ND 57702.

The 5-5-5-0 Stages of the SO-FLINE-BEC Model of Origins and Evolution of Universal Systems

Alexander Alan Scarborough

Recipient of the Sagnac Award by International Peers (2009)

From the Pythagoreans around 500 B.C. to modern-day space-probe discoveries, the revolutionary SO-FLINE-BEC model of universal origins and evolution is an ongoing illustrative integration of the greatest discoveries (old and new) in the history of science. Guiding readers from Void to Energy to Universe, each definitive step along the way provides beautiful continuity—a joy to relish for your lifetime—all stemming from the author's fourth and fifth laws of planetary motion that first eluded Kepler in the early 17th century.

Definitions

SO: The five-stage **S**equence of **O**rigins: from (**1**) Black Void at Absolute Zero to (**2**) Black Energy Spheres via BEC to (**3**) brilliant, noisy Quasars, the explosive mothers of Galaxies, to (**4**) expanding Galaxies creating (**5**) billions of fiery Solar Systems buried in gaseous dust-clouds.

FL: The **F**ive **L**aws of Planetary Motion (three by Kepler, two by Scarborough) governing solar systems; e.g., the geometrical spacing of orbits and their subsequent moderate changes that apply to all solar systems, no matter how weird, via their mass-velocity-distance-time relationships.

IN-E: The observable five-stage evolution of planets: from (**1**) Energy to (**2**) Gaseous to (**3**) Transitional to (**4**) Rocky (e.g., Earth) to (**5**) Inactive—all via Internal Nucleosynthesis (**IN**) that drives all Evolution (**E**) of universal spheres; e.g., Earth creates its hydrocarbon fuels (gas, oil, coal)—not from fossils! All planets are self-sustaining entities creating their compositional matter in accord with $E=mc^2$.

BEC: The **B**ose-**E**instein-**C**ondensate process at the perimeter of our expanding Universe supplies the Black Energy Spheres at Absolute Zero (**0**) (a.k.a. Black Holes) that are essential for universal origins and evolution, and for the accelerating rate of expansion of the Universe via momentum-gravity principles.

This revolutionary model is the product of 37 years of research in origins and evolution of universal systems. Its principles assure the capability of readily discovering definitive solutions to universal anomalies—solutions that remain unattainable via current BB beliefs.

The SO-FLINE-BEC Model of Universal Origins and Evolution: Ten unprecedented revelations about planets and solar systems

Alexander Alan Scarborough
Recipient of the Sagnac Award by international peers

1. All planets undergo a five-stage evolution from Energy to Inactive via Internal Nucleosynthesis at rates in accord with mass.
2. Earth is approximately 65% through its fourth stage (rocky) of evolution, with about 35% remaining before entering its fifth and final stage (inactive).
3. All planets and spherical moons are self-sustaining entities that create their own compositional matter during the first four stages of planetary evolution; e.g., Earth's hydrocarbon fuels (gas, oil, coal)—not from fossils (1975-1979)—are created throughout its fourth (rocky) stage of evolution.
4. The Five Laws of Planetary Motion reveal the dynamic fiery birth of our Solar System with 10+ original planets some five billion years ago.
5. The new Fourth Law (Scarborough 1980-2006) reveals precisely how the 10+ planets attained their geometric spacing around our Sun.
6. Scarborough's new fifth law explains why Uranus and Neptune are the only two planets drifting slowly away from the Sun, while the remaining planets, including the remnants of Planet Asteroids, are drifting slowly towards the Sun.
7. Planet Earth has drifted 38% closer to the Sun during its five billion years of existence, while Uranus has drifted 7% farther away.
8. All planets go through a cycle of planetary rings consisting of ejecta during the latter part of the second stage of evolution; e.g., Saturn's rings are currently at their peak.
9. Space probes have revealed the initial slight rings of Jupiter that will increase as time goes by: a mass-time relationship.
10. Photos from space probes reveal Io as the youngest moon of Jupiter, leaving a dark trail leading to and over the Red Spot, perhaps indicative of Io's ejection from Jupiter in recent geological times.